Surgical Atlas of the Musculoskeletal System

Surgical Atlas of the Musculoskeletal System

Authors
M. Llusá, MD, PhD
À. Merí
D. Ruano, MD, PhD

Translation Editors
Miguel E. Cabanela, MD, MS
Sergio Mendoza-Lattes, MD
Joaquin Sanchez-Sotelo, MD, PhD

Board of Directors, 2008-2009

E. Anthony Rankin, MD
President

Joseph D. Zuckerman, MD
First Vice President

John J. Callaghan, MD
Second Vice President

William L. Healy, MD
Treasurer

Frederick M. Azar, MD
Treasurer-Elect (Ex-Officio)

Richard J. Barry, MD

Thomas C. Barber, MD

James H. Beaty, MD

Kevin J. Bozic, MD, MBA

Leesa M. Galatz, MD

John T. Gill, MD

Christopher D. Harner, MD

M. Bradford Henley, MD, MBA

William J. Robb III, MD

Michael F. Schafer, MD

James P. Tasto, MD

G. Zachary Wilhoit, MS, MBA

Karen L. Hackett, FACHE, CAE
(Ex-Officio)|

Staff

Mark W. Wieting, *Chief Education Officer*

Marilyn L. Fox, PhD, *Director, Department of Publications*

Laurie Braun, *Managing Editor*

Mary Steermann Bishop, *Manager, Production and Archives*

Courtney Astle, *Assistant Production Manager*

Susan Morritz Baim, *Production Coordinator*

Suzanne O'Reilly, *Graphics Coordinator*

Anne Raci, *Production Database Associate*

Abram Fassler, *Page Production Assistant*

Charlie Baldwin, *Page Production Assistant*

Laura Khoshaba, *Publications Assistant*

Juliet Orellana, *Publications Assistant*

The *Surgical Atlas of the Musculoskeletal System* is a translation from the original Spanish title *Manual y Atlas Fotográfico de Anatomía del Aparato Locomotor* by M. Llusá, À. Merí, and D. Ruano, published by Editorial Médica Panamericana.

Editorial Médica Panamericana played no role in the translation of *Manual y Atlas Fotográfico de Anatomía del Aparato Locomotor* from Spanish into the English language, and disclaims any responsibility for any errors, omissions and/or faults and/or possible faults in translation.

The publication of this work has been supported by a grant from the Spanish State Office for Books, Archives and Libraries, a division of the Ministry of Culture.

The material presented in the *Surgical Atlas of the Musculoskeletal System* has been made available by the American Academy of Orthopaedic Surgeons for educational purposes only.

All rights reserved. No part of this publication may be reproduced, stored in a retrieval system, or transmitted, in any form, or by any means, electronic, mechanical, photocopying, recording, or otherwise, without prior written permission from the publisher.

Published 2008 by the
American Academy
of Orthopaedic Surgeons
6300 N. River Road
Rosemont, IL 60018

Copyright © 2008
by the American Academy of
Orthopaedic Surgeons

ISBN 10: 0-89203-394-0
ISBN 13: 978-0-89203-394-2

Authors

Manuel Llusá Perez, MD, PhD
Profesor titular del Departamento de Anatomía y Embriología Humana
Facultad de Medicina, Universidad de Barcelona
Especialista en Cirugía Ortopédica y Traumatología
Hospital de Traumatología del Vall d'Hebron

Àlex Merí Vived
Fisioterapeuta
Diploma de Estudios Avanzados en Organogénesis y Anatomía Clínica Aplicada
Profesor de la EUIFN Blanquerna, Universitat Ramon Llull

Domingo Ruano Gil, MD, PhD
Catedrático del Departamento de Anatomía y Embriología Humana
Facultad de Medicina, Universidad de Barcelona
Académico de Número de la Real Academia de Medicina de Catalunya

Translation Editors

Miguel E. Cabanela, MD, MS
Professor of Orthopaedic Surgery
Department of Orthopaedic Surgery
College of Medicine
Mayo Clinic
Rochester, MN

Sergio Mendoza-Lattes, MD
Associate Professor
Orthopaedic Surgery and Rehabilitation
The University of Iowa
Iowa City, IA

Joaquin Sanchez-Sotelo, MD, PhD
Consultant and Associate Professor
Orthopaedic Surgery
Mayo Clinic
Rochester, MN

Reviewers

John G. Anderson, MD
Associate Clinical Professor
Michigan State University
College of Human Medicine
Codirector – Grand Rapids Orthopaedic
Foot and Ankle Fellowship
Assistant Program Director – Grand Rapids
Orthopaedic Residency
Orthopaedic Associates of Grand Rapids
Grand Rapids, MI

Philip E. Blazar, MD
Assistant Professor of Orthopaedic Surgery
Harvard Medical School
Brigham and Women's Hospital
Boston, MA

Donald R. Bohay, MD, FACS
Associate Professor
Department of Orthopaedic Surgery
College of Medicine
Michigan State University
East Lansing, MI

Charles D. Bukrey, MD
Associate Professor
Orthopaedic Surgery
Michigan State University
College of Human Medicine
Grand Rapids, MI

Terrence J. Endres, MD
Associate Clinical Professor
Michigan State University
Orthopaedic Trauma Surgery
Orthopaedic Associates of Grand Rapids
Grand Rapids, MI

Gregory J. Golladay, MD
Clinical Assistant Professor
Michigan State University
Orthopaedic Assoicates of Grand Rapids, P.C.
Spectrum Health
Grand Rapids, MI

Michael R.F. Jabara, MD
Orthopaedic Associates of Grand Rapids
Grand Rapids, MI

Clifford B. Jones, MD
Associate Clinical Professor
Michigan State University
College of Human Medicine
Grand Rapids, MI

Thomas A. Malvitz, MD
Chairman
Department of Orthopaedic Surgery
Spectrum Health
Grand Rapids, MI

John Maskill, MD
Orthopaedic Associates of Grand Rapids
Grand Rapids, MI

James R. Ringler, MD
Orthopaedic Associates of Grand Rapids
Grand Rapids, MI

Scott S. Russo, MD
Orthopaedic Surgery
Orthopaedic Associates of Grand Rapids
Spectrum Health Hospitals
Grand Rapids, MI

James R. Stubbart, MD
Assistant Professor
Michigan State University
Department of Orthopaedic Surgery
Orthopaedic Association of Michigan
Grand Rapids, MI

Preface

The original edition of this book, *Manual y Atlas Fotográfico de Anatomía del Aparato Locomotor* by Llusá, Merí, and Ruano, was published in Spanish by Editorial Medica Panamericana. A colleague and I were asked to review this book to assess whether its translation into English might be worthwhile. Upon review, we found the content succinct but valuable and up to date, and, importantly, we found the photographic images abundant and superb. I then recommended to the Publications Committee of the American Academy of Orthopaedic Surgeons (AAOS) that the book be translated into English for the US market.

After much hard work by many people, we are pleased to offer you the English edition, the *Surgical Atlas of the Musculoskeletal System*. I think all involved in orthopaedic surgery (surgeons, residents, and medical students) will find this book to be a very valuable source to review the anatomy of the musculoskeletal system. The amazing multiplanar photographic images will aid in the study and review of each anatomic area. Also, it can be used as a quick resource in the daily practice of orthopaedic surgery.

Many people contributed to the translation and publication of this book. Although I cannot name them all, I want to express my gratitude to Dr. Carlos Garcia-Moral of the Oklahoma Hand Surgery Center, who joined me in the initial review of the Spanish text. I especially want to thank Dr. Joaquin Sanchez-Sotelo (Mayo Clinic) and Dr. Sergio Mendoza (University of Iowa) for their translation of the text and figures. We are very appreciative to both of them for adding this additional task to their already full plates. Also, I would like to acknowledge the contributions of Dr. Philip Blazar and Dr. John Anderson and colleagues, who reviewed the final manuscript. Finally, my sincere thanks go to the staff of the Publications and International Departments at AAOS for their dedication to this project.

Miguel E. Cabanela, MD
Chairman
International Committee
AAOS

Table of Contents

Section 1: GENERAL INFORMATION

Chapter 1: Terminology 3
Chapter 2: Osteology 7
Chapter 3: Arthrology 15
Chapter 4: Myology 25
Chapter 5: Neurology 31
Chapter 6: Angiology 37

Section II: SCAPULAR GIRDLE AND UPPER LIMB

Chapter 7: Anatomic Regions 45
Chapter 8: Osteology 49
Chapter 9: Arthrology 79
Chapter 10: Myology 117
Chapter 11: Neurology 171
Chapter 12: Angiology 195

Section III: HEAD AND TRUNK

Chapter 13: Topographic Regions and Surface Anatomy 213
Chapter 14: Osteology 217
Chapter 15: Arthrology 239
Chapter 16: Myology 257
Chapter 17: Neurology 279
Chapter 18: Angiology 283

Section IV: PELVIC GIRDLE AND LOWER LIMB

Chapter 19: Topographic Regions and Surface Anatomy 295
Chapter 20: Osteology 299
Chapter 21: Arthrology 315
Chapter 22: Myology 341
Chapter 23: Neurology 377
Chapter 24: Angiology 391

Section 1

General Information

Chapter 1

Terminology, Planes, and Axes

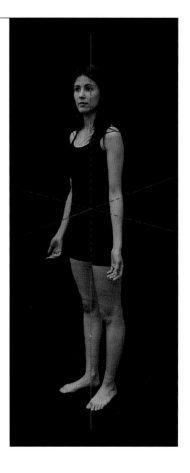

Terminology and Body Planes

Terms Used for Direction
When body parts are described in terms of location within the body, the three-dimensional space needs to be taken into account. It is important to refer to the anatomic axes and planes in which space is organized. Thus, it is critical to learn the terms used to express directions to facilitate these references and locate an anatomic structure in space.

Anatomic Position
A subject is said to be in *anatomic position* when standing with the lower limbs parallel and the toes pointing forward, the upper limbs at the side of the trunk with the palms of both hands facing forward, the shoulders in slight abduction, and the head facing forward, as shown in Figure 1-1.

An imaginary line that is drawn perpendicular to the floor and through the center of the body will define two halves that are approximately symmetric. This imaginary line is called the midline. All structures along the midline or relatively close to it are called *medial* or *median*; examples are the median sacral artery and the medial collateral ligament.

Anatomic Directions
The most common terms used for general anatomic directions are as follows (Figure 1-1):
- *Superior* or *cranial*—to the head.
- *Inferior* or *caudal*—to the feet.
- *Anterior* or *ventral*—to the front. The term *ventral* is used mainly in the abdomen.
- *Posterior* or *dorsal*—to the back.
- *Medial*—to the midline.
- *Middle*—at the midline.
- *Lateral*—away from the midline.
- *Central*—to the center of the body.
- *Peripheral*—to the body surface.
- *Deep*—in the inside.
- *Superficial*—at or near the surface.
- *Apical*—the superior part, vertex, or tip (apex) of an anatomic structure.

The following terms are used when describing the limbs:
- *Proximal*—close to the root (origin) of the limb, ie, the trunk.
- *Distal*—away from the trunk.
- *Medial*—to the midline; synonym for *ulnar* in the upper limb and *tibial* in the lower limb.

American Academy of Orthopaedic Surgeons

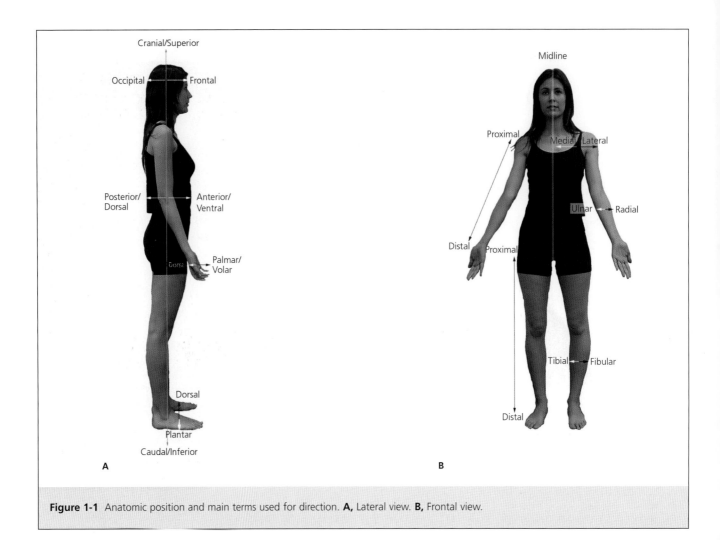

Figure 1-1 Anatomic position and main terms used for direction. **A,** Lateral view. **B,** Frontal view.

- *Lateral*—away from the midline; synonym for *radial* in the upper limb and *fibular* in the lower limb.
- *Palmar* or *volar*—relating to the palm of the hand; synonym for *plantar* in the lower limb.
- *Dorsal*—refers to the dorsum of the hand or foot; opposite of *palmar*.

Example: The elbow is medial and proximal to the hand and lateral to the spine.

The following terms are used when describing the face:
- *Frontal*—relating to the face, anterior.
- *Occipital*—relating to the posterior region of the head.

Axes, Planes, and Motions

The three axes of the body are (Figure 1-2):
- The longitudinal or vertical axis, which corresponds to the y axis, is perpendicular to the floor. The longest longitudinal axis, which runs from the top of the head through the trunk, is called the main axis. Rotational motions are performed around the longitudinal axis.
- The transverse or horizontal axis, which corresponds to the x axis, is parallel to the floor. Flexion-extension motions are performed around this axis.
- The sagittal or anteroposterior axis, which corresponds to the z axis, is perpendicular to the other two and oriented from back to front. Abduction-adduction motions are performed around this axis.

The planes of the body are as follows (Figures 1-3 and 1-4):
- Frontal or coronal plane—this plane divides the body into anterior and posterior parts. It contains the transverse and longitudinal axes.
- Transverse or horizontal plane—this plane divides the body into superior and inferior parts. It contains the sagittal and transverse axes.
- Sagittal plane—this plane is perpendicular to the floor. It divides the body into right and left sides. It contains the sagittal and longitudinal axes.
- Axial plane—this plane is defined as the plane

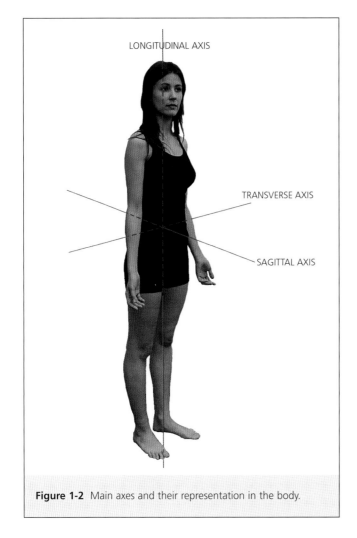

Figure 1-2 Main axes and their representation in the body.

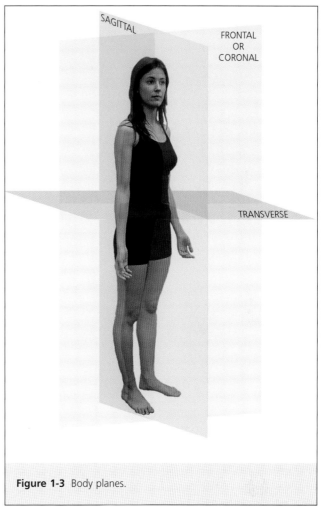

Figure 1-3 Body planes.

perpendicular to the long axis of any structure. In the anatomic position, the long axis of the body is the vertical axis, so the axial plane corresponds to the transverse plane for the whole body except for the feet. In the anatomic position, the long axis of the feet is the sagittal axis, so the axial plane is the frontal plane.

The most common body positions are standing; sitting; and decubitus, or lying horizontally.

Other positions include:
- *Prone*, which means with the front surface facing down. For example, prone decubitus describes the position where the person is lying on his or her belly; hand pronation means palm facing downward, in medial rotation.
- *Supine*, which means with the front surface facing up. For example, supine decubitus means that the person is lying on his or her back; supination with regard to the hand means with the palm facing upward, in lateral rotation.

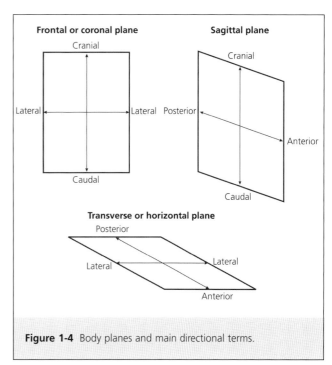

Figure 1-4 Body planes and main directional terms.

Chapter 2

Osteology

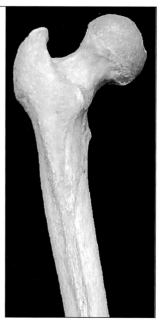

Introduction

Osteology (from *osteon*, "bone," and *logos*, "study") is the branch of anatomy that deals with the study of bones. Bones comprise most of the skeleton, and their main functions are to provide structural support and to protect the body. Bones are resistant yet lightweight, providing strength to withstand forces without representing a heavy load. Despite their rigidity, bones are living tissue that remodel throughout life.

The basic functions of bones are as follows:
- Support—Bones provide body structure and muscle attachments that allow motion through the joints.
- Protection—Bones protect internal organs such as the brain inside the cranium and the heart and lungs inside the thorax.
- Mineral homeostasis—Bone tissue accumulates minerals and releases them into the bloodstream if their levels are low.
- Hematopoiesis—The production of blood cells, or hematopoiesis, occurs inside the bones, in the so-called red bone marrow.
- Energy accumulation—The yellow bone marrow is formed primarily by adipose cells, which may be used as a source of energy.

Structure of Bones

The skeleton has two major components: the osseous component, formed by the bones, and the cartilaginous and membranous component, a softer structure. The skeleton is also divided into two parts: the axial skeleton, which is derived from the sclerotome induced by the notochord, and the appendicular skeleton. The axial skeleton includes the head, neck, and trunk and forms the body axis. The appendicular skeleton includes the extremities (the upper and lower limbs), which articulate with the body through the scapular and pelvic rings (Figure 2-1).

Features of Bones

The features and components of bones are described here. Several of these features, including the diaphysis, epiphysis, and metaphysis, are unique to the long bones (Figure 2-2).

Diaphysis—The diaphysis is the body, stem, or central part of a long bone, which usually has a triangular cross section. In the hand and foot, the diaphysis of the long bones (metacarpals, metatarsals, and phalanges) is called the body.

Epiphysis—The epiphysis is the end of a long bone. A proximal and a distal epiphysis are present in most bones. They are usually covered by hyaline cartilage forming articular facets. The epiphyses may also contain eminences and roughened areas for insertion of the ligaments that reinforce articular capsules, as well as muscles and tendons that produce motion. In the hand and foot, the proximal epiphysis of the long bones (metacarpals, metatarsals, and phalanges) is called the base and the distal epiphysis is called the head.

Metaphysis—During skeletal growth, areas of cartilaginous tissue called physes, which are responsible for growth in length, are present near the end of the long bones (Figures 2-3 and 2-4). The ossification process takes place in the metaphysis, which is located

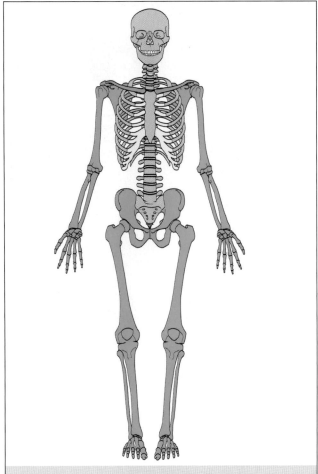

Figure 2-1 The human skeleton, with the appendicular portions in blue and the axial portions in brown.

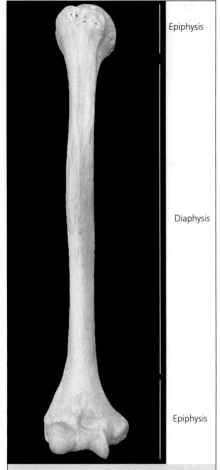

Figure 2-2 Long bone (humerus), with the epiphyses and diaphysis labeled.

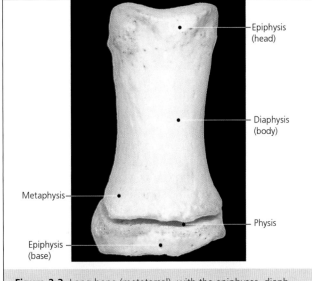

Figure 2-3 Long bone (metatarsal), with the epiphyses, diaphysis, metaphysis, and physis labeled.

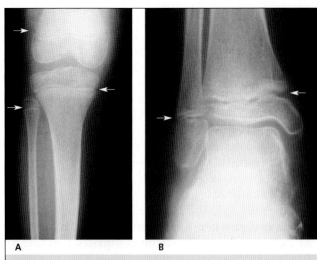

Figure 2-4 Radiographs of the knee **(A)** and ankle **(B)** of a skeletally immature individual show the growth physes (arrows).

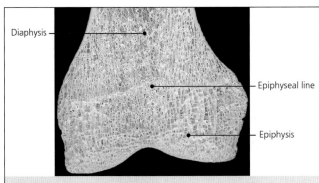

Figure 2-5 Frontal section of the distal aspect of the femur showing the epiphyseal line.

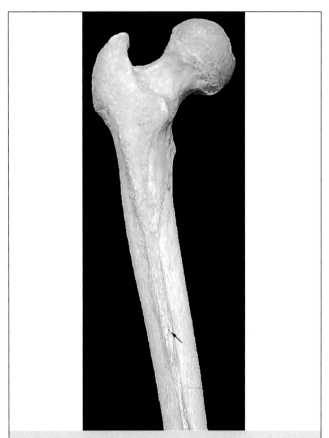

Figure 2-6 Dorsal view of the proximal half of the femur. The arrow points to the diaphyseal nutrient foramen.

between the physis and the diaphysis. Once ossification is complete, the physis is replaced by a region of sclerotic bone called the epiphyseal line (Figure 2-5).
Nutrient foramen—This is a small orifice (Figure 2-6) that provides access to blood vessels after they enter the nutrient canal.
Articular cartilage—The articular cartilage is the tissue that covers the epiphyses where the bones articulate, reducing friction and absorbing impact during motion (Figure 2-7). Cartilage is a type of connective tissue that lacks nerve endings and blood vessels. There are three types of cartilage. Hyaline cartilage, the most common, is composed of a homogeneous matrix and type II collagen fibers oriented along lines of force. It is present not only in articular facets but also in the cartilage of the nose, larynx, sternum and ribs, and tracheal rings. Fibrocartilage is composed mainly of type I collagen with a small amount of extracellular matrix. It is present in several articular structures, including menisci, labra, and intervertebral disks. Finally, elastic cartilage contains elastic fibers and is present in regions that allow deformation, including the ears and the epiglottis.
Medullary canal—This space inside the long bones (Figure 2-8) is filled with bone marrow.
Bone marrow—Before birth, the medullary canal of the long bones and those areas formed by trabecular and woven bone are filled with red bone marrow, the main function of which is blood cell production (hematopoiesis). Over the years, red bone marrow is replaced by yellow bone marrow. In the adult, red marrow is found mainly in the sternum, ribs, vertebrae, and pelvis.
Layers of bone coverage—The periosteum is a layer of connective tissue that covers the external surfaces of bone except where the bone is covered with articular cartilage. The periosteum comprises two layers: an external (fibrous) layer that provides support to the blood vessels, lymphatic vessels, and nerves entering the bone; and an internal (osteogenic) layer that

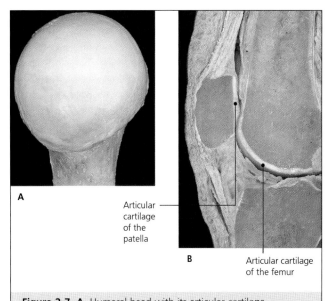

Figure 2-7 A, Humeral head with its articular cartilage. **B,** Sagittal section of the knee joint showing the articular cartilage of the patella and femur.

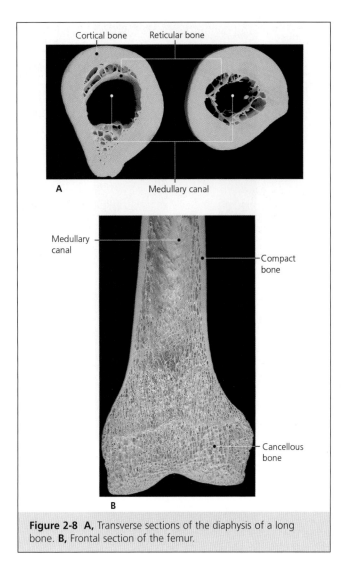

Figure 2-8 A, Transverse sections of the diaphysis of a long bone. **B,** Frontal section of the femur.

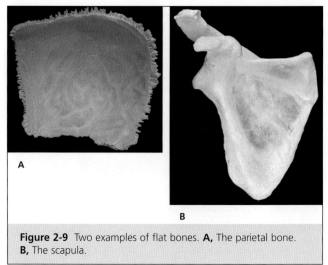

Figure 2-9 Two examples of flat bones. **A,** The parietal bone. **B,** The scapula.

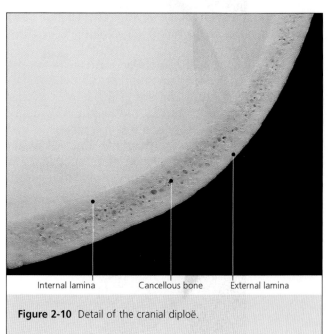

Figure 2-10 Detail of the cranial diploë.

contains vessels and bone cells essential for bone nutrition, repair, and growth. Sharpey's fibers provide a strong bond between the periosteum and cortical bone, forming a structure that permits the attachment of tendons, muscles, and ligaments at regions called entheses. The endosteum covers the medullary canal and contains osteoprogenitor cells.

Classification of Bones

Bones are classified according to their morphology into long, flat, and short bones.

Long bones (Figure 2-2) are characterized by the predominance of one dimension over the other two. Long bones are located in the appendicular skeleton; examples include the femur, humerus, radius, tibia, and phalanges.

Flat bones (Figure 2-9) are those in which two dimensions predominate over the third dimension. Often they have one concave face and one convex face. They are usually covered by a thin layer of cortical substance denser on the edges, separated by cancellous substance inside. The scapula, sternum, and several of the cranial bones are examples of flat bones. In the cranium, the cancellous substance that separates the inner and outer cortical layers is called the diploë (Figure 2-10).

Short bones are those in which all dimensions are approximately equal (Figure 2-11). They are covered by a thin layer of cortical bone and contain cancellous bone and marrow inside. Carpal and tarsal bones are short bones.

Many bones do not fit accurately into this morphologic classification and are classified as irregular bones (vertebrae, for example) (Figure 2-12). Some of these,

Osteology

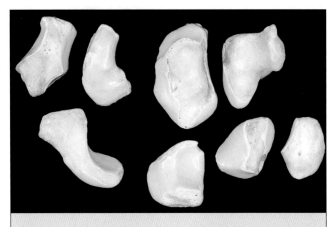

Figure 2-11 Carpal bones are examples of short bones.

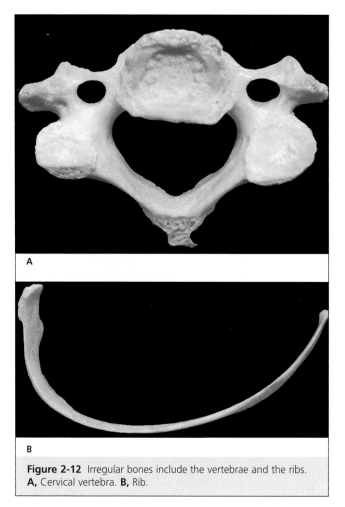

A

B

Figure 2-12 Irregular bones include the vertebrae and the ribs. **A,** Cervical vertebra. **B,** Rib.

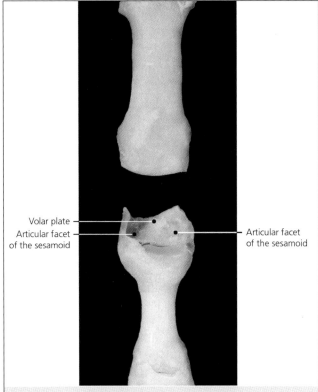

Figure 2-13 The metacarpophalangeal joint of the thumb, opened to show the articular facets of the sesamoids.

External Features of Bones

Most bones have external features such as rough areas, eminences, tuberosities, tubercles, lines, or crests that are variably prominent. These areas usually provide attachment sites for ligaments, tendons, and muscles.

Although most eminences are genetically programmed, they continue to develop according to tension and compression stresses applied to bone, which, being a dynamic tissue, has the ability to respond and adapt. In some cases, when bones are not subjected to muscle forces, they remain smooth, without major areas of roughness.

Likewise, bones have spaces or excavations such as channels, grooves (open), conduits (closed), fossae, or holes. These features are usually the site of the origin of muscles (fossae) or allow the passage of different anatomic structures. In addition, various faces, edges, and angles define and characterize the different bones.

Internal Features of Bones

Bone matrix is composed of approximately 25% water, 25% proteins, and 50% minerals. It contains four

such as the ribs, are described as flat and long. Other names are used as well.

There are also the so-called sesamoid bones (Figure 2-13), which are short bones located inside some tendons. Most authors consider the patella a sesamoid bone (Figure 2-7).

Some cranial bones have cavities inside them, and they are called pneumatic bones (Figure 2-14). These cavities are created by resorption of previously deposited mineral.

American Academy of Orthopaedic Surgeons

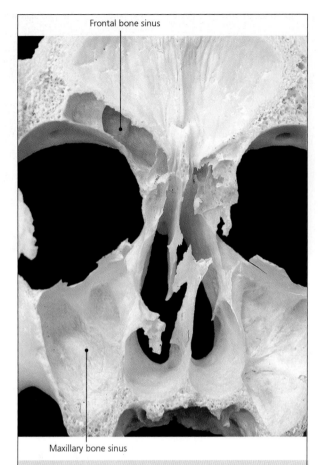

Figure 2-14 The cranium contains several sinuses formed by pneumatic bones. This illustration shows a frontal section of the cranium.

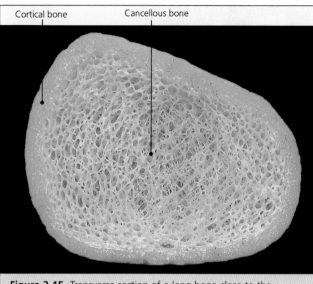

Figure 2-15 Transverse section of a long bone close to the epiphysis.

major cell types: osteoprogenitor cells, which may be transformed into bone cells; osteoblasts, responsible for the synthesis of the organic components of bone matrix; osteocytes, mature bone cells derived from osteoblasts; and osteoclasts, which participate in bone degradation and resorption. All these cells provide bone the dynamic features of living tissue that undergoes degradation and renewal.

Cortical bone (Figures 2-8 and 2-15) represents the external layer of bones, which lies underneath the periosteum. It is composed of close layers without intervening spaces (except for haversian canals, Volkmann's canals, and some additional small canals). The term compact bone is applied to regions of substantial thickness, mostly in the diaphysis of long bones.

Cancellous bone (Figures 2-8 and 2-15) is characterized by a tissue type formed by trabecular bone that contains spaces filled with marrow. It constitutes the majority of flat and short bones and only the epiphysis of long bones. The architecture and organization of trabecluae correspond to lines of force. In 1892, Wolff analyzed the organization of bony trabeculae in the femoral neck and pronounced Wolff's law. According to this law, bone density and growth are greater in areas subjected to load, responsible for the areas of roughness and eminences described above.

Reticular bone (Figure 2-8) is similar to cancellous bone, but the spaces are larger, and it is located mainly in the medullary canal of long bones, and tends to disappear at mid-diaphysis.

Cortical substance porosity rarely exceeds 30%, whereas cancellous substance porosity ranges from 50% to 90%.

Bone Vascularity

Blood vessels (Figure 2-16) enter bones in different areas. The following arteries are found entering long bones:

- Nutrient artery—Each bone may have one or more nutrient arteries. They enter compact bone at the diaphyses through nutrient foramina (Figure 2-6) and are divided inside the bone into longitudinal branches that irrigate the bone and bone marrow to the metaphyseal level. A nutrient artery is usually a branch of a nearby major systemic artery.
- Periosteal branches that irrigate the outer third of cortical bone at the diaphyses.
- Metaphyseal and epiphyseal vessels (Figure 2-17) from rich periarticular vascular networks. They perforate cortical bone and irrigate the ends of long bones as well as the bone marrow.

Veins and lymphatic vessels exit bones. Some veins exit as emissary vessels, which are typically found in some cranial bones. Many vessels are joined by sensory and vasomotor nerves, especially at the periosteum.

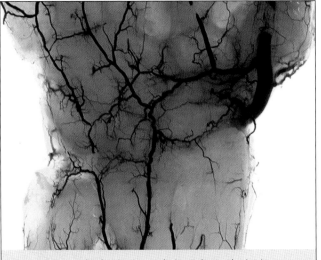

Figure 2-16 A diaphanization technique shows the intricate network of blood vessels supplying the bones of the wrist.

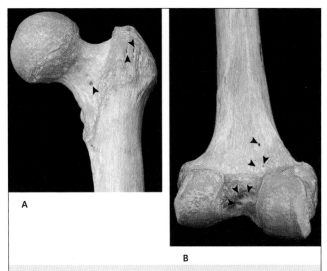

Figure 2-17 Proximal **(A)** and distal **(B)** ends of the femur. The arrowheads indicate the location of small epiphyseal and metaphyseal nutrient foramina.

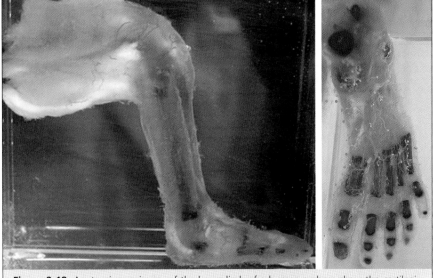

Figure 2-18 Anatomic specimens of the lower limb of a human embryo show the cartilaginous anlages for the future bones (by endochondral ossification) with their main blood vessels.

Ossification

Bones are formed from primitive mesenchymal tissue. There are two kinds of ossification: intramembranous and endochondral.

Intramembranous ossification is the process of direct ossification of dense mesenchymal tissue, which serves as a model. Ossification starts at the center of the model and continues to the periphery of the bone.

Endochondral ossification is an indirect process (Figure 2-18). Chondroblasts differentiated inside primitive mesenchymal tissue create a new model of hyaline cartilage and perichondrium. Osteoprogenitor cells differentiated in the center of the diaphysis are transformed into osteoblasts, which produce bone tissue where minerals are deposited. This primary center of ossification is called the bone collar, which, along with the periosteum and blood vessels, is critical for later development. The secondary centers of ossification are located in the epiphyses. The growth plate, or epiphyseal plate, separates a secondary center from the primary center of ossification until they coalesce. The epiphyseal line remains in adult bone.

Chapter 3

Arthrology

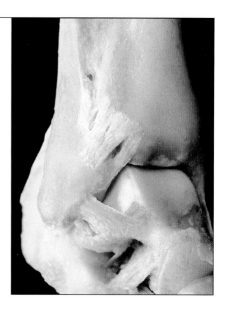

Introduction
A joint is defined as the collection of elements that join bones to each other. Different types of motion occur at the joints, making the skeleton a mobile structure. Joints play a fundamental role in proprioception.

Classification of Joints According to Structure
Joints can be classified according to structure. The three types are fibrous, cartilaginous, and synovial.

Fibrous
In fibrous joints, or synfibroses, the junction is made of fibrous tissue (connective tissue). There are three types of fibrous joints: (1) sutures, of which there are four types: dentate (Figures 3-1 and 3-2), squamous (Figure 3-3), flat (Figure 3-4), and schindylesis (Figure 3-5); (2) gomphoses, the alveolar cavities into which teeth insert (Figure 3-6); and (3) syndesmoses (Figure 3-7).

Cartilaginous
In cartilaginous joints, the junction is made of cartilage. There are two types of cartilaginous joints: (1) synchondroses (Figure 3-8), such as the costochondral joints, where the bones are joined through a fibrocartilage; and (2) symphyses (Figure 3-9), where the joint is formed by a cartilaginous disk.

Synovial
Synovial joints, or diarthroses (Figure 3-10), are the classic joints. They comprise two distinct structures: the articulation, formed by the articular facets and articular cartilage, and the surrounding tissues, including the joint capsule, joint cavity, synovium, and ligaments.

The articulation comprises the articular facets and articular cartilage. Cartilage prevents direct bone-to-bone contact and modifies the articular shape. Cartilage is devoid of nerves and blood vessels, and its mechanical properties are between those of a solid and those of a liquid.

The intra-articular pressure within a synovial joint is slightly negative (suction), inhibiting dislocation by traction. The need for the joint capsule to accommodate to joint movements explains the existence of articular recesses or pouches in some locations.

Ligaments
The purpose of these resistant, almost nonelastic structures is to keep the bones together and to prevent dislocation. They reinforce the joint, limiting excessive excursion, which might damage the articulation. Three types of ligaments are found in synovial joints: capsular, extracapsular, and intracapsular (Figure 3-11). Capsular ligaments are reinforcements of the joint capsule. Most often they are thickenings of the fibrous capsule. Extracapsular ligaments are outside and at a distance from the articular capsule. Intracapsular ligaments are inside the joint but outside the synovial cavity (limited by the synovial membrane). Some are called interosseous ligaments, as they originate and insert on closely adjacent bony structures.

General Information

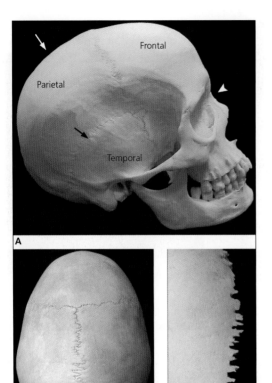

Figure 3-1 A, The skull, showing the location of the various types of sutures. The white arrow indicates dentate sutures, the black arrow indicates squamous sutures, and the arrowhead indicates flat sutures. **B,** Dentate sutures at the top of the skull. **C,** Teeth of a dentate suture.

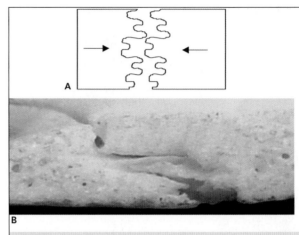

Figure 3-2 Drawing **(A)** and detail **(B)** of a dentate suture. The teeth fit together and are reinforced by fibrous tissue joining the dentate suture.

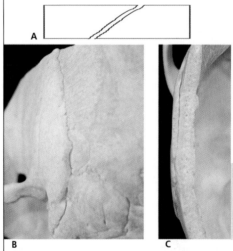

Figure 3-3 Drawing **(A)** and details **(B** and **C)** of a squamous suture, in which one lamina overlaps another lamina. In this example, the temporal bone overlaps the parietal bone.

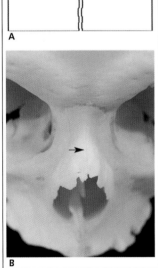

Figure 3-4 Drawing **(A)** and detail **(B)** of a flat suture. The nasal bones articulate in the internasal suture through relatively flat and regular surfaces.

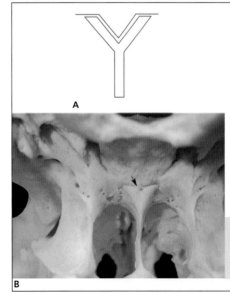

Figure 3-5 Drawing **(A)** and detail **(B)** of a schindylesis. In this location, the base of the vomer has a small slot for the sphenoid crest.

ARTHROLOGY

Figure 3-6 Articulation of a tooth with its alveolar cavity, called a gomphosis.

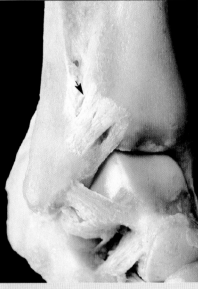

Figure 3-7 The distal tibioperoneal joint is an example of a syndesmosis.

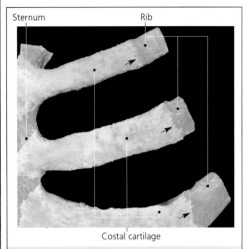

Figure 3-8 Costochondral joints (arrows) and sternocostal, or sternochondral, joints.

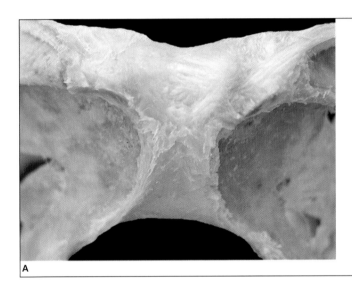

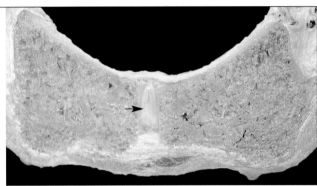

Figure 3-9 Anterior view **(A)** and transverse section **(B)** of the symphysis pubis. The arrow points to the interpubic disk.

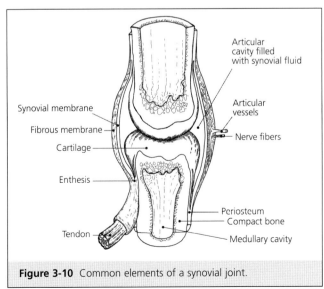

Figure 3-10 Common elements of a synovial joint.

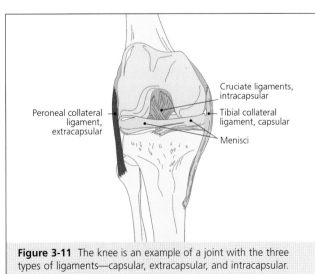

Figure 3-11 The knee is an example of a joint with the three types of ligaments—capsular, extracapsular, and intracapsular.

American Academy of Orthopaedic Surgeons

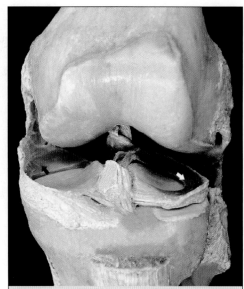

Figure 3-12 Anterior view of the knee joint. The anterior cruciate and collateral ligaments have been sectioned to distract the joint. Arrows point to the menisci.

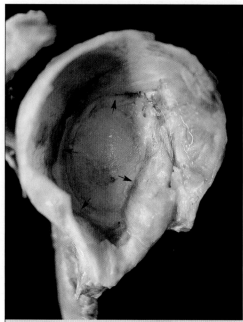

Figure 3-13 Scapular glenoid cavity (arrows), demonstrating the large glenoid labrum for the humeral head.

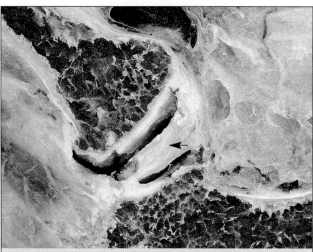

Figure 3-14 The sternoclavicular joint disk facilitates a perfect fit for the articular surfaces.

Articular Fibrocartilage

Elements made of articular fibrocartilage increase joint congruity. They include the articular meniscus, the labrum, and the articular disk. The articular meniscus (Figure 3-12) is usually concave, has a triangular cross section, and does not divide the entire articular cavity. The labrum (Figure 3-13) expands the articular cavity when one of the articular surfaces cannot be contained in another. The articular disk (Figure 3-14) is a kind of fibrocartilage that completely separates both articular surfaces and may divide the synovial cavity into two chambers. Disks and menisci both represent mobile articular facets that increase the surface area of the joint and facilitate a better distribution of articular pressure. Both tend to degenerate because of limited vascularization.

Classification of Joints According to Function

Joints can also be classified according to the type of motion they allow. The three types are synarthroses, amphiarthroses, and diarthroses (synovial joints).

A synarthrotic joint is formed by mesenchymal tissue and allows no motion. These joints have no cavity or joint capsule. There are different kinds of synarthroses depending on how the mesenchymal tissue develops. They include fibrous and bony joints.

An amphiarthrotic joint is intermediate between a diarthrosis and a synarthrosis. There is a small cavity in this type of joint, which allows minimal motion. These joints have no joint capsule and are subjected to impact, which they buffer. Cartilaginous joints are amphiarthroses.

Diarthrotic joints are synovial joints, as described above. They allow motion and include different types, which can be categorized according to whether they have zero, one, two, or three axes of motion.

Types of Synovial Joints
Zero Axes of Motion
A flat joint, or arthrodia (Figure 3-15), does not have an axis of motion. It allows sliding motion in all directions.

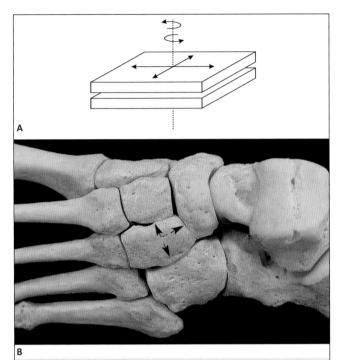

Figure 3-15 A, Diagram of a flat joint. B, Intertarsal joints (arrows) are examples of flat joints.

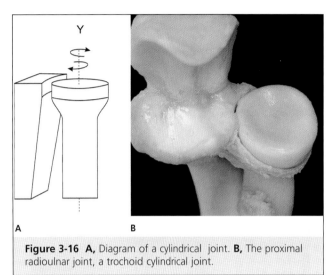

Figure 3-16 A, Diagram of a cylindrical joint. B, The proximal radioulnar joint, a trochoid cylindrical joint.

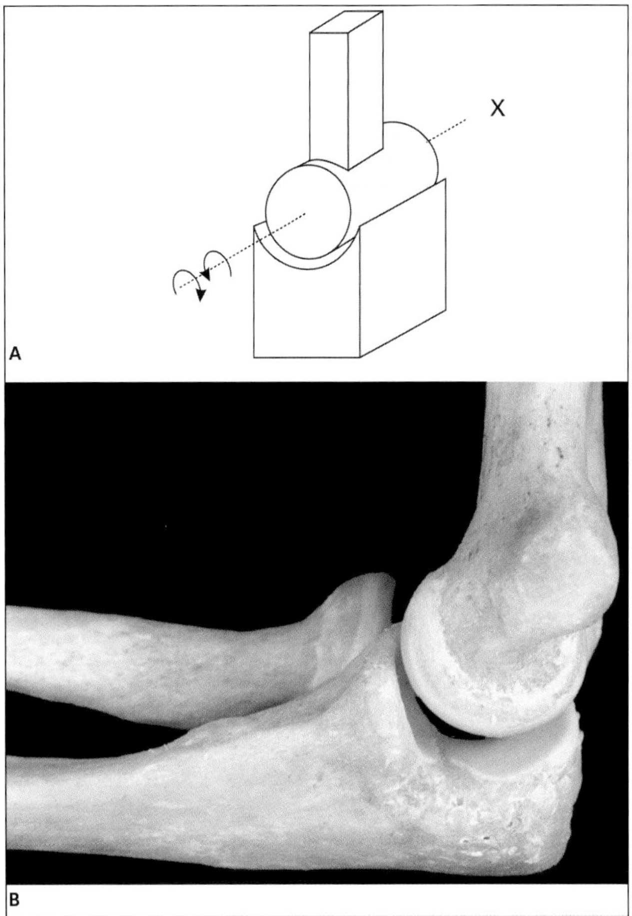

Monoaxial
Cylindrical joints are monoaxial; that is, they allow motion along one axis. They include trochoid joints (Figure 3-16), which allow rotation, and trochlear, or ginglymus, joints (Figure 3-17), which allow flexion and extension.

Biaxial
Biaxial joints include condylar, or ellipsoidal, joints (Figure 3-18), which allow flexion, extension, abduction, and adduction; and saddle joints (Figure 3-19).

Triaxial
Triaxial joints allow motion in three planes. Called spherical joints or enarthroses (Figure 3-20), these spherical, triaxial joints allow flexion, extension, rotation, abduction, and adduction.

BLOOD AND NERVE SUPPLY
Blood is supplied to joints through networks of arteries, veins, and lymphatic vessels around the joint. Usually they branch out from nearby major arteries, the vessels of which reach the periphery of bones (epiphyseal vessels).

Articular nerves contain sensory fibers rich in joint mechanoreceptors and free nerve endings present in both the joint capsule and the ligaments. According to Hilton's law, a joint is supplied by the same nerves that supply the muscles acting on the joint.

Figure 3-17 A, Diagram of a trochlear joint. B, The ulnohumeral joint, a trochlear joint.

MOVEMENTS AT SYNOVIAL JOINTS
Synovial joints allow various types of movements, including sliding, angular motion, rotation, combined movements, and special movements performed by specific joints.

American Academy of Orthopaedic Surgeons

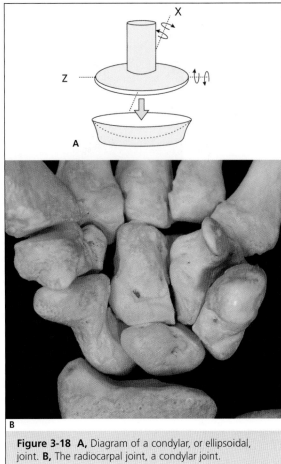

Figure 3-18 **A,** Diagram of a condylar, or ellipsoidal, joint. **B,** The radiocarpal joint, a condylar joint.

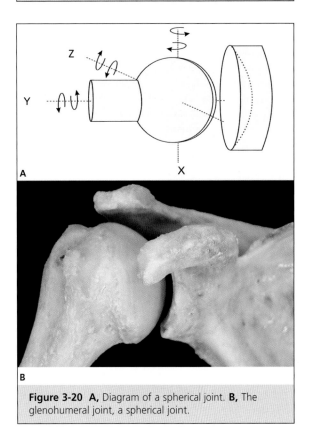

Figure 3-20 **A,** Diagram of a spherical joint. **B,** The glenohumeral joint, a spherical joint.

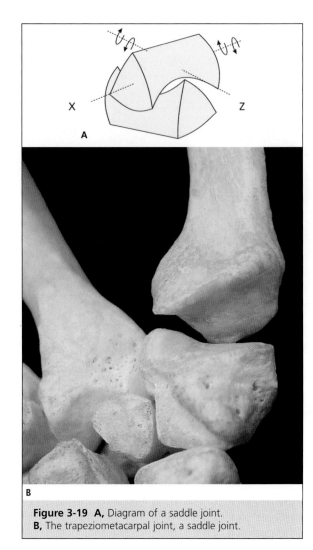

Figure 3-19 **A,** Diagram of a saddle joint. **B,** The trapeziometacarpal joint, a saddle joint.

Sliding

Sliding occurs when an articular facet slides on another articular facet. In most cases, the articular facets are flat.

Angular Movements

Transverse axis Three types of joint motion occur through a transverse axis (Figure 3-21). The first type is flexion, or anteversion. Flexion implies a decreased angle between the bones. In the foot, the term plantar flexion is used to describe the movement required to stand on the tiptoes. A second type of joint motion that occurs through a transverse axis is extension, which implies an increased angle between the bones. The term dorsiflexion is used to describe the movement that takes the dorsum of the foot close to the leg, which corresponds to extension. It should be noted that some authors relate this movement to foot flexion because it decreases the angle between bones, but that would be a physiologic or kinesiologic definition of motion. In fact, the plantar aspect of the foot is embryologically

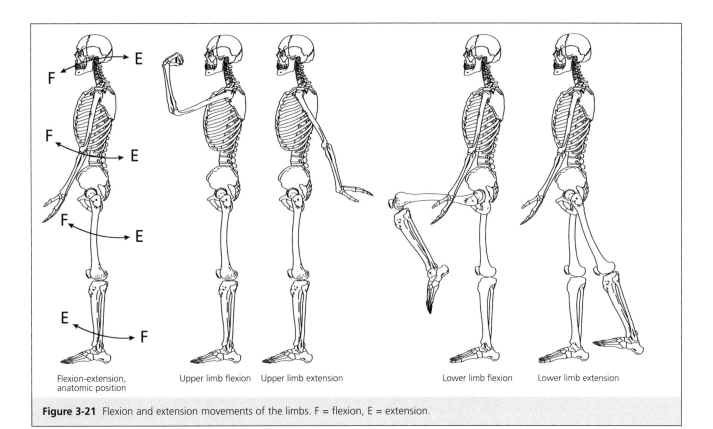

Figure 3-21 Flexion and extension movements of the limbs. F = flexion, E = extension.

equivalent to the palm of the hand, but substantial rotation occurs during fetal development, leading to the familiar position of the foot seen in the adult. This rotation explains the location of the extensor muscles, which produce dorsiflexion, in the anterior region of the leg, and the flexor muscles, which perform plantar flexion, in the posterior region. This accounts for the differences between the anatomic and the physiologic definitions. Finally, hyperextension, which is an increase in the joint angle beyond the anatomic position, also occurs in the transverse plane.

Sagittal axis Three types of joint movement occur through the sagittal axis (Figures 3-22 through 3-24): abduction, adduction, and inclination. Abduction is movement of a body part away from the midline. In the hand, the midline is defined as a line drawn though the column of the long finger. Based on this line, wrist abduction corresponds to radial deviation; finger abduction is achieved by separating the fingers. Conversely, adduction is approximation of a body part to the midline. Wrist adduction from the hand midline is called ulnar deviation; finger adduction is achieved by moving the fingers together. The third type of sagittal joint movement is inclination, which describes lateral movements of the axial skeleton.

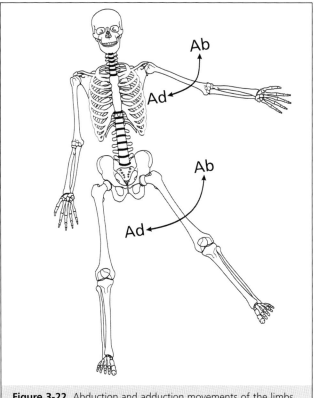

Figure 3-22 Abduction and adduction movements of the limbs. Ab = abduction, Ad = adduction.

General Information

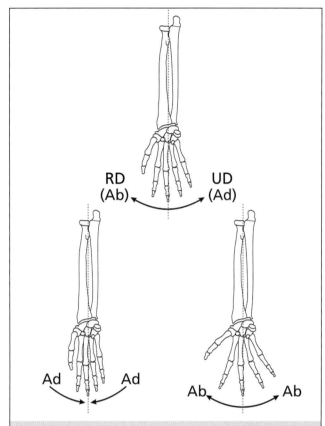

Figure 3-23 Radial/ulnar deviation and abduction/adduction movements of the hand. RD = radial deviation, UD = ulnar deviation, Ab = abduction, Ad = adduction.

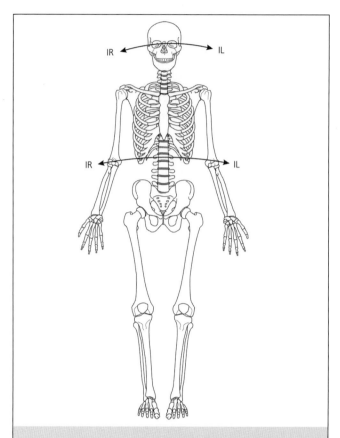

Figure 3-24 Inclination movements of the axial skeleton. IR = inclination to the right, IL = inclination to the left.

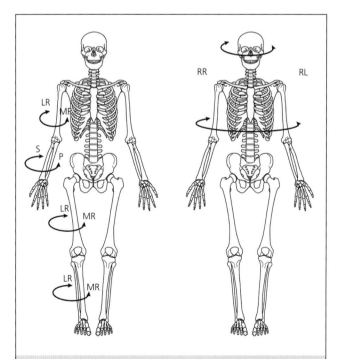

Figure 3-25 Rotational movements of the axial and appendicular skeletons. LR = lateral rotation, MR = medial rotation, S = supination, P = pronation, RR = rotation to the right, RL = rotation to the left.

Rotation
Movement of a bone around its longitudinal axis either medially or laterally is called rotation (Figure 3-25). When the muscles on one side rotate to the same side, the rotation is called ipsilateral; when they rotate to the opposite side, the rotation is called contralateral.

Combined Movement
Combined movement that generates an ellipsoidal movement is called circumduction. Circumduction occurs only in biaxial or triaxial joints, where it combines all the possible joint motions.

Special Movements at Specific Joints or Joint Complexes
- Inversion—motion that positions the sole of the feet opposite to each other. It combines plantar flexion, adduction, and supination.
- Eversion—movement that takes the sole of the foot outwards. It combines dorsiflexion, abduction, and pronation.
- Antepulsion (Figure 3-26)—movement of a body part ventrally.
- Retropulsion—movement of a body part dorsally.
- Supination (Figure 3-27)—movement of the fore-

Arthrology

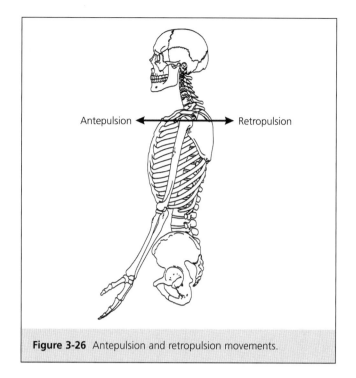

Figure 3-26 Antepulsion and retropulsion movements.

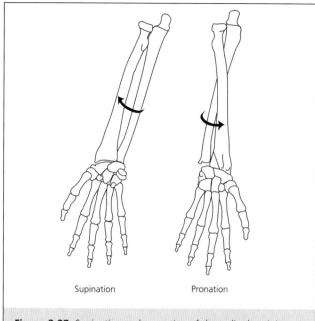

Figure 3-27 Supination and pronation of the radioulnar joint.

arm that positions the palm of the hand facing the front; equivalent to lateral rotation.
- Pronation—movement of the forearm that positions the palm of the hand facing the back; equivalent to medial rotation.
- Elevation (Figure 3-28)—movement of a body part cranially.
- Depression (descent)—movement of a body part caudally.
- Opposition—to place one body part facing another.
- Gliding—rotational motion in the sagittal arcs of the scapula against the thoracic wall.

Range of Articular Motion

Articular range of motion is determined by several of the anatomic factors mentioned above, including the shape of the articular facets and the elasticity of the capsule and ligaments of the joint complex. It also depends on physiologic factors, including the tone of the muscles opposed to the motion, as well as the presence of muscles able to generate motion. Thus, the physiologic limit of articular range of motion is observed with active motion, whereas the anatomic limit is observed with passive motion (with an external force).

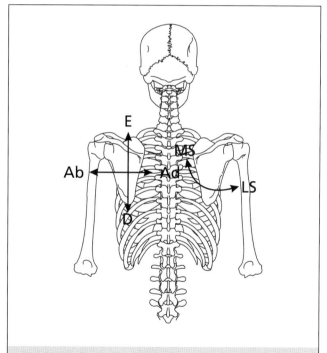

Figure 3-28 Scapular movements. Ab = abduction, Ad = adduction, E = elevation, D = depression, MS = medial shifting, LS = lateral shifting.

American Academy of Orthopaedic Surgeons

Chapter 4

Myology

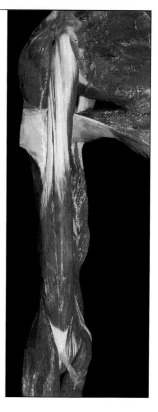

Introduction

Myology (from *myo*, "muscle," and *logos*, "study") is the branch of anatomy dedicated to the study of muscles and associated tissues (tendons, fascia, etc).

The three types of muscle are smooth, cardiac, and skeletal. Smooth muscle is involuntary; an example is the smooth muscle in the digestive system, which allows peristaltic movements in the gastrointestinal tract. The other two types of muscle, cardiac and skeletal, are striate; that is, the fibers appear striated when viewed under the microscope. Cardiac muscle is found in the heart. It undergoes involuntary contraction and is characterized by autoexcitement, which allows the heart to beat. This book concentrates on skeletal muscle, which is striated and voluntary, although it may also have reflex contractions, and has muscle tone.

The main functions of skeletal muscle are as follows:
- Maintenance of position—muscle tone and contraction allow maintenance of a given position, such as the standing position.
- Motion—muscular contractions shorten the muscles, producing motion through the joints by pulling on bone attachments.
- Thermogenesis—muscular contractions also generate heat. For example, shivering increases body temperature.

Muscles are organized into muscular compartments (Figure 4-1, *A*) encased by bones and fasciae (Figure 4-1, *B*). From a structural point of view, muscle fibers or cells are organized into fascicles, which form muscles. The fascial layers that cover the muscles are called the endomysium (around muscle fibers), perimysium (around fascicles), and epimysium (around muscles). Gross anatomy, which is the subject of this book, is concerned with the epimysium and the compartment fasciae.

Skeletal muscle fiber has the following properties (Figure 4-2):
- Excitability—the ability to respond to some stimuli, such as the neurotransmitter acetylcholine, by the production of action potentials.
- Contractility—the ability to shorten and generate tension in response to an excitatory stimulus.
- Extensibility—the ability to lengthen without sustaining damage.
- Elasticity—the ability to return to its original length after contraction or lengthening.

Anatomic Classification Systems

Muscles can be classified according to various characteristics, including fiber arrangement or number of muscle bellies. Several classification systems are described briefly below.

Muscle Bellies and Tendons

Muscles may be classified according to the number of heads that form the muscle as a whole. Most often,

General Information

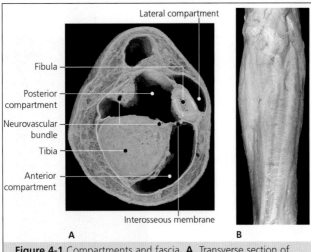

Figure 4-1 Compartments and fascia. **A,** Transverse section of the leg after removal of the muscles, leaving empty fascial compartments. Note the individual tracts for vessels and nerves through the fascial tissue. **B,** Muscular fascia of the forearm.

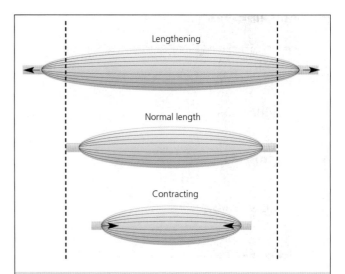

Figure 4-2 Properties of skeletal muscle. Viscoelasticity is the ability of the muscle to return to its normal length after lengthening or contraction. With muscle contraction, the muscle belly enlarges.

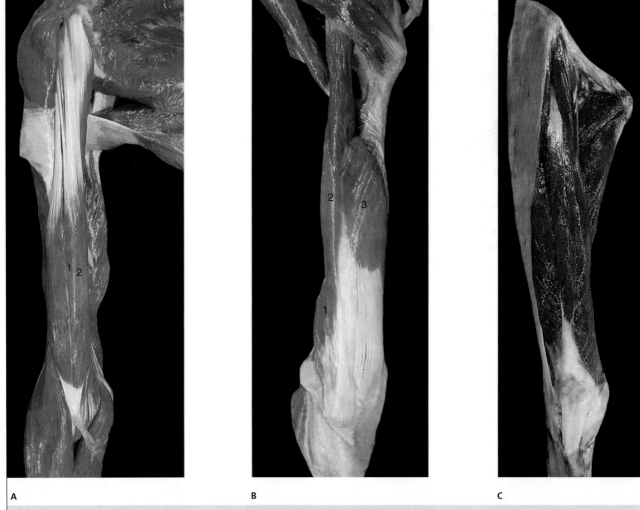

Figure 4-3 Examples of muscles with several heads, which are numbered. **A,** Biceps brachii (two heads). **B,** Triceps brachii (three heads). **C,** Quadriceps (four heads). The fourth head of the quadriceps is deep underneath the rectus femoris (1).

Myology

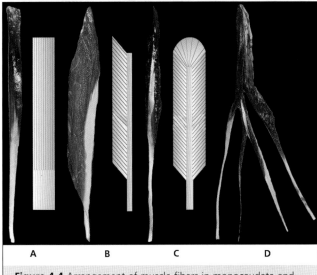

Figure 4-4 Arrangement of muscle fibers in monocaudate and polycaudate muscles. **A,** Fusiform arrangement. **B,** Semipennate arrangement. **C,** Pennate arrangement. The muscles shown in **(A)**, **(B)**, and **C** are monocaudate; the muscle shown in **D** is polycaudate.

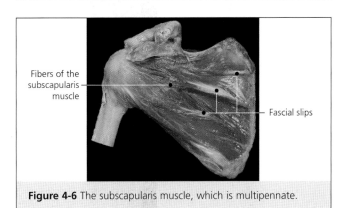

Figure 4-6 The subscapularis muscle, which is multipennate.

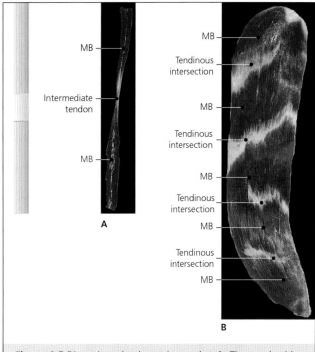

Figure 4-5 Digastric and polygastric muscles. **A,** The omohyoid, a digastric muscle. **B,** The rectus abdominis, a polygastric muscle. MB = muscle belly.

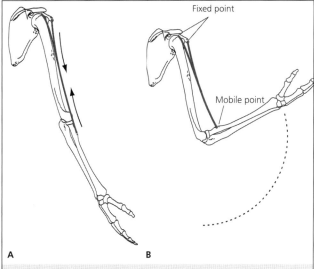

Figure 4-7 A, Biceps contraction (arrows) results in elbow flexion. The proximal insertions usually act as the fixed point and the distal insertion as the mobile point **(B)**.

the number of bellies is referred to when there are two or more, as in biceps (from *bi-*, "two," and *-ceps*, "heads"), triceps (three heads), or quadriceps (four heads) (Figure 4-3). Each of the heads has an independent origin; they coalesce to form a common insertion tendon. Depending on the number of terminal tendons for attachment, muscles are classified as monocaudate (one terminal tendon) or polycaudate (more than one terminal tendon) (Figure 4-4).

The muscle belly is the portion of the muscle made of contractile tissue; that is, the nontendinous portion. Most muscles have a single belly, which usually is located in the central region. Muscles with two bellies separated by an intermediate tendon (Figure 4-5, *A* and *B*) are called digastric muscles (two bellies); examples are the digastric and omohyoid muscles of the mandible. Muscles with several bellies separated by tendinous intersections, such as the rectus abdominis, are called polygastric muscles (Figure 4-5, *C*).

Arrangement of Fibers

Biomechanically, muscle force depends on the number of fibers. Muscles continue as tendons, so the number of muscle fibers that can be inserted depends on tendon width. Muscles that are elongated in shape are called fusiform muscles (spindle form) (Figure 4-4, *A*). Semipennate (semi-feather-shaped) muscles have a long tendon with muscle fibers on one side, allowing the insertion of a larger number of fibers (Figure 4-4, *B*). Pennate mus-

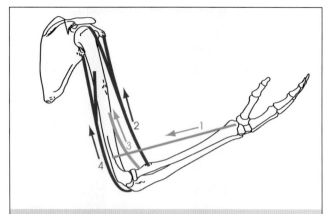

Figure 4-8 Muscle functions. The biceps (2) and brachialis (3) perform a similar function and are considered agonists in elbow flexion. The brachioradialis (1) provides only moderate help in elbow flexion, but it helps stabilize the joint. The triceps (4) performs the opposite function, elbow extension, and is considered an antagonist muscle.

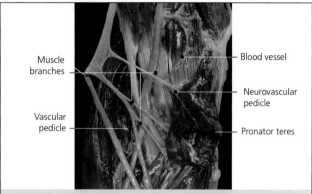

Figure 4-9 Detailed elbow dissection showing the vascular and neurovascular pedicles.

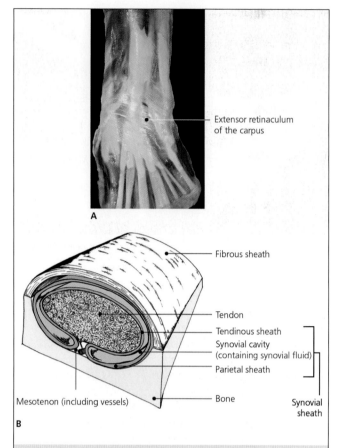

Figure 4-10 Tendon vascularity in the wrist. **A,** Dorsal view of a wrist with latex injected (green areas) to demonstrate the synovial sheaths for the extensor tendons. **B,** Cross section of a tendon and the structures forming the tunnel within which it glides.

cles have muscle fibers attached on both sides of the tendon, which provides great strength in a limited space (Figure 4-4, *C*). Multipennate muscles are a subtype of pennate muscles (Figure 4-6).

Muscle Function

Muscle contraction is responsible for the body motions described in chapter 3. Functionally, a muscle approximates its two attachment points, acting on each joint crossed. Usually there is a fixed point and a mobile point (Figure 4-7).

In addition, each movement is the result not of a single muscle action but of a coordinated motor program. Learning a movement requires coordination of muscles located both nearby and at a distance (muscle loops), which are responsible for the final smooth motion.

Muscles that contribute to the same motion are called agonists (from the Greek word meaning "fighter") (Figure 4-8). Usually one of the agonists functions as the primary motor muscle, which is mainly responsible for the movement. Antagonist muscles (from the Greek word meaning "rival") are opposed to agonist muscles and perform the opposite action. In addition, other muscles act as joint stabilizers, functioning to immobilize one area so that movement can occur in another area, or synergistically (from the Greek term *synergos*, meaning "working together"), to help produce motion.

During the physical examination, neuromuscular function is evaluated by asking the patient to perform the motion for which the muscle being analyzed is the primary motor, while isolating it from other agonist muscles as much as possible. In addition, muscle strength may be evaluated by asking the patient to contract the muscle against resistance.

Vascularity and Innervation

Muscle vascularity and innervation are provided by pedicles spread regularly along the muscles (Figure 4-9). These pedicles may include blood vessels, nerves, or both (neurovascular pedicles).

Blood vessels and nerves are located inside fascial neurovascular compartments and perforate the different layers covering the muscles until they reach the muscle

Myology

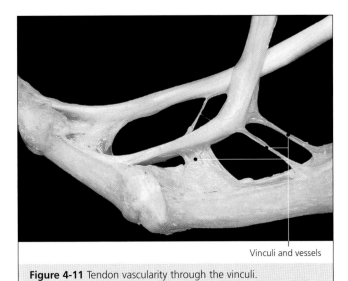

Figure 4-11 Tendon vascularity through the vinculi.

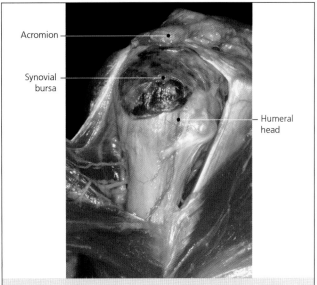

Figure 4-12 Subacromial bursa, which prevents supraspinatus tendon friction at the subacromial space.

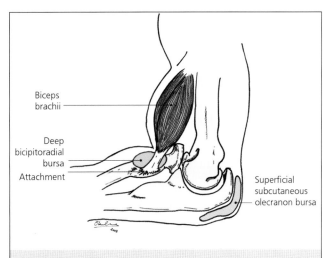

Figure 4-13 Deep and superficial bursae in the elbow. Superficial bursae decrease friction on subcutaneous bony structures.

fibers. Veins drain muscles and usually are given the name of the associated artery.

Motor innervation (efferent innervation) originates at the anterior horn of the spinal cord and penetrates the muscles through muscular rami. Sensory information (afferent innervation) is conducted through peripheral nerves to the posterior horn of the spinal cord.

Muscles located in the dorsum of the trunk are innervated by the posterior branches of the spinal nerves, whereas the remaining muscles are innervated by the anterior branches as well as some of the cranial nerves. These topics are covered in chapter 5.

Tendons, Aponeuroses, and Synovial Bursae

Tendons and aponeuroses are formed by regular dense connective tissue with numerous collagen fibers arranged along the lines of tension generated by muscles.

Muscles have three ways of attaching to bone: through muscle tissue, tendons, or aponeuroses. With the first type, the muscle fibers attach to the bone periosteum. With tendon attachments, tension generated by the muscle is transmitted to the bone through a tendon. This is the type of insertion seen with long muscles. Finally, aponeurotic attachments are flat; these insertions are typical of flat muscles.

Tendons are composed of fiber fascicles that are covered by the paratenon. Some tendons slide over bony eminences or inside bone grooves and may be prone to wear. For this reason, a tendon sheath commonly covers the tendon in these areas.

Tendon gliding is accommodated by grooved areas in the bone. A fibrous sheath transforms the channel into a tunnel or osteofibrous conduit. The tendon glides inside, protected by a synovial sheath (Figure 4-10). This sheath has one parietal layer and one tendinous layer as well as a virtual cavity filled with synovial fluid. The layers coalesce in synovial recesses. The entire complex may be covered by a retinaculum, a thickening of the fascial covering that protects the tendons.

Tendon vascularity is provided through the mesotenon, which is usually located in the deep tendon surface. Vascularity may also be provided through vinculi (Figure 4-11) or through blood vessels at sites of bony attachments.

Synovial bursae are located in regions where tendons may be subjected to friction with other muscles, bones, or ligaments (Figure 4-12). Synovial bursae contain synovial fluid, which dissipates loads and facilitates sliding. Occasionally, these bursae communicate with a nearby joint (Figure 4-13).

Muscles are separated by fat accumulations that facilitate sliding. These may coalesce into fat pads.

Chapter 5

Neurology

Introduction

Neurology (from *neuro*, "nerve"; and *logos*, "study") is the branch of anatomy dedicated to the study of the nervous system. Because this book focuses on the musculoskeletal system, the main features of the peripheral nerves are discussed, but the cranial nerves are not.

The nervous system is broadly divided into the central nervous system and the peripheral nervous system. The central nervous system comprises the brain and the spinal cord. The peripheral nervous system comprises the nerves that connect the central nervous system with other body systems.

The spinal cord (Figure 5-1) is organized into segments, or functional metameres, each of which controls a given body region. The gray matter, located in the center of the spinal cord, contains the neural cell bodies. It is surrounded by the white matter, which contains the ascending (to upper segments or the brain) and descending (to lower segments) axons, which coordinate different actions. The gray matter is organized into an anterior horn, which contains the motor neurons, and a posterior horn, which receives axons of the sensory neurons. The axons of motor neurons innervate muscles and glands; these axons are called efferent nerve fibers. Afferent nerve fibers are the axons of sensory motor neurons located in the dorsal spinal ganglia; they are responsible for the transmission of nerve impulses from peripheral receptors to the central nervous system (Figure 5-2).

Spinal (Rachidial) Nerves

From the spinal cord, two dorsal roots (one for the right side of the body and one for the left) and two ventral roots (also one for the right side and one for the left) emerge. The dorsal root emerges from the dorsal horn and contains motor neurons; the ventral root transmits impulses from peripheral organs to the ventral horn of the spinal cord. The dorsal and ventral roots meet, forming a mixed spinal nerve; thus, most spinal nerves are both sensory and motor.

Nerves are a collection of thousands of nerve fibers (axons). Each nerve fiber is surrounded by a covering layer called the endoneurium. Nerve fascicles are surrounded by the perineurium, and each nerve is surrounded by the epineurium, which is similar in structure to muscles (Figure 5-3).

Each spinal nerve corresponds to one spinal functional segment and controls a particular body region (Figure 5-4). A dermatome is defined as the skin area innervated by fibers from a single nerve from a single spinal cord segment; a myotome is a group of muscles controlled and innervated mainly by one spinal nerve. Similar terms apply to regions of the skeleton (sclerotome) and viscera (viscerotome).

The spinal nerves correspond to the spinal cord organization and are intimately related to it (Figure 5-5). Thus, there are eight cervical nerves (C1 through C8), 12 thoracic nerves (T1 through T12), five lumbar nerves (L1 through L5), five sacral nerves (S1 through S5), and one or two coccygeal nerves

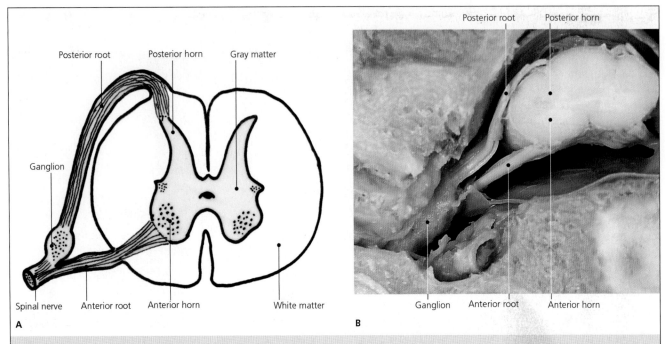

Figure 5-1 A, Drawing of a cross section of the spinal cord. **B,** Transverse section of the spinal cord and the roots that form the spinal nerve.

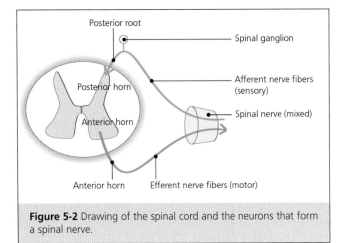

Figure 5-2 Drawing of the spinal cord and the neurons that form a spinal nerve.

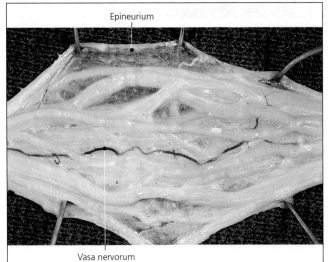

Figure 5-3 Opened nerve showing multiple nerve fibers and some interconnections. The external layer covering the nerve is the epineurium. The central structure, injected with black latex, is a blood vessel (vasa nervorum) that irrigates the nerve fibers.

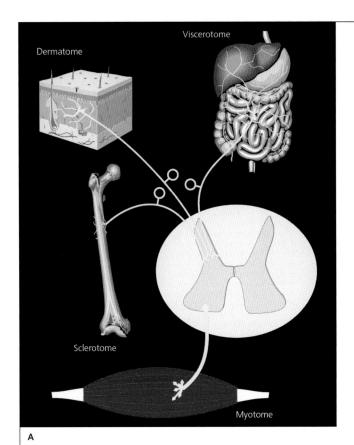
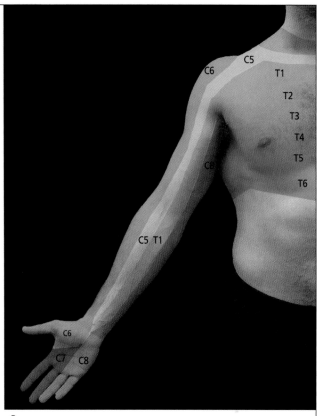

Figure 5-4 A, Each medullary segment controls a different region. **B,** Dermatomes of the upper limb and cranial aspect of the thorax. This is the classic representation, although in reality the dermatomes overlap such that if one level is injured, the adjacent dermatomes continue to innervate that area.

Figure 5-5 Schematic representation of the spinal cord segments and the spinal nerves. The spinal cord ends at the L1-L2 level.

(Co1 and Co2). There are eight cervical spinal nerves and only seven cervical vertebrae because nerves exit between vertebrae: nerve C1 exits between the occipital bone and the atlas (C1), and so on down the vertebrae, such that nerve C8 exits between the seventh cervical and the first thoracic vertebra. From this point down, each spinal nerve exits through the intervertebral foramen (below) of the corresponding vertebra (Figure 5-6).

The spinal cord has several intumescentiae, or swellings, at the cervical and lumbar levels due to the formation of neural plexuses for the upper and the lower limb. The brachial plexus is formed by nerves C5 to T1; the lumbosacral plexus is formed by nerves L1 to S4. The spinal cord ends at level L1 or L2; the multiple nerve fibers arising at the end of the spinal cord are named the cauda equina ("horse tail") and correspond to the spinal nerves below L2.

Once formed, the spinal nerve exits through the intervertebral foramen and is divided into two rami (Figures 5-7 and 5-8). The dorsal ramus is directed posteriorly, to the vertebral channels, where it divides into a medial branch and a lateral branch. The medial branch is directed to the region around the spinal

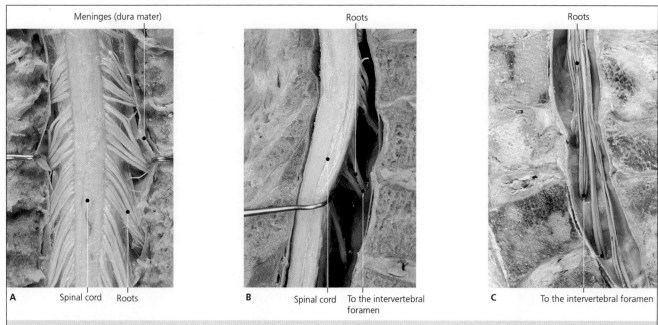

Figure 5-6 A and **B,** The spinal cord and its roots exiting to the intervertebral foramen. **C,** At the lumbar spine level the spinal cord is replaced by the nerve fibers exiting through each intervertebral foramen.

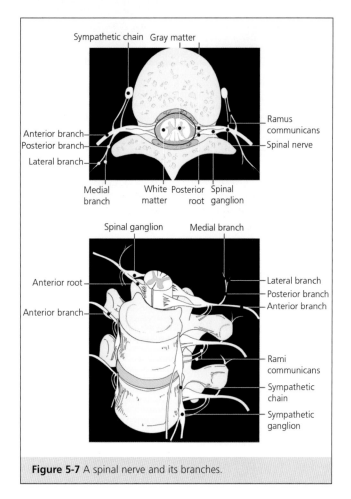

Figure 5-7 A spinal nerve and its branches.

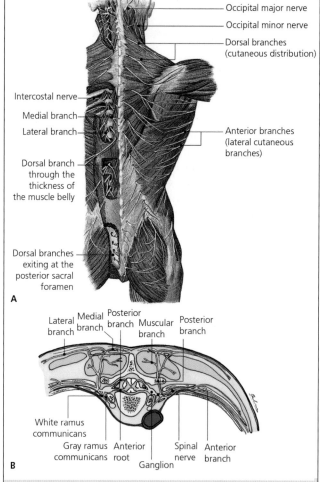

Figure 5-8 A, Portion of an old lithograph showing the posterior branches of the nerves of the trunk. **B,** Transverse section of a typical intercostal spinal nerve and its branches.

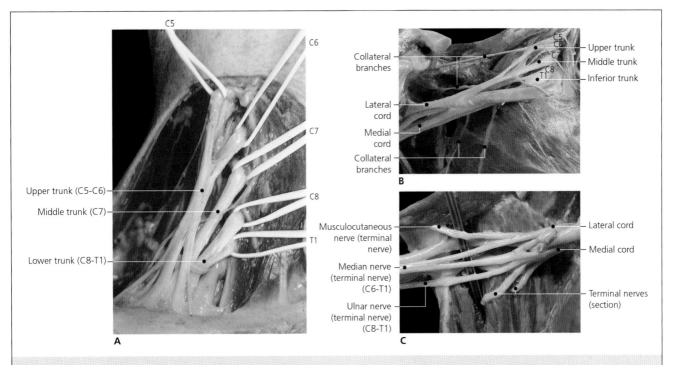

Figure 5-9 The brachial plexus, showing its root contributions, connections into trunks (**A** and **B**), split into divisions, and cord reorganization into cords and terminal nerves (**C**).

apophysis and supplies the muscles and skin in the medial tract.

At the sacral level, the dorsal rami exit through the posterior sacral foramens. The ventral ramus supplies the limbs and the anterior and lateral regions of the trunk. At the sacrum, these branches exit through the anterior sacral foramens. Meningeal branches return to the spinal canal to supply the meninges, blood vessels, ligaments, and vertebrae. Finally, communicating branches communicate with the autonomic nervous system. The vascular supply for the nerves is provided by the vasa nervorum ("vessels for the nerves").

Plexuses

The anterior branches of the spinal nerves form the cervical, brachial, lumbar, and sacral plexuses. The intercostal nerves, which supply the spaces between the ribs, are the only nerves from T2, in the upper thorax, to T12, in the abdominal region, that are organized in a differentiated manner.

Plexuses form a complicated network for spinal nerve interconnection. They are organized into trunks, fascicles, collateral branches, and terminal nerves that contain the nerve fibers originating in several spinal cord segments (Figure 5-9). For example, one of the terminal nerves, the obturator nerve, contains fibers from L2 to L4.

Chapter 6

ANGIOLOGY

INTRODUCTION

Angiology (from *angio*, "vessel," and *logos*, "study") is the branch of anatomy dedicated to the study of the cardiovascular system, which is formed by the heart, blood vessels, and lymphatic system.

The heart is an organ located in the mediastinum. Its main function is pumping blood to the pulmonary and systemic circulatory systems (Figure 6-1). The pulmonary circulatory system carries blood between the heart and the lungs; its main goal is blood oxygenation. The systemic circulatory system carries blood between the heart and the rest of the body; it delivers oxygenated blood and nutrients to the tissues and retrieves metabolic debris and carbon dioxide.

Blood vessels are the conduits that allow the passage of blood pumped by the heart. Blood vessels that exit the heart and distribute blood from the ventricles to the tissues are called arteries. Blood vessels that carry blood from the tissues to the ventricles of the heart and send it back are called veins. It is sometimes mistakenly assumed that if a vessel carries blood with a high carbon dioxide content, it must be a vein; however, in the pulmonary circulatory system, the vessel that carries blood with a high carbon dioxide concentration from the heart to the lungs is the pulmonary artery. Thus, in general terms, veins and arteries are distinguished by whether they carry blood to the heart (veins) or from the heart (arteries).

Blood vessels are usually named according to anatomic location or relationship with a nearby bone or organ. Blood vessels are formed by various tissues that themselves require an appropriate blood supply. The blood vessels that supply other vessels are called vasa vasorum (vessels of the vessels). Similarly, blood vessels that supply nerves are called vasa nervorum.

The circulatory system contains blood, which is composed of plasma with carrier proteins and other solutes, as well as cells such as leukocytes, thrombocytes, and erythrocytes. The main functions of blood are the transport of oxygen and nutrients to the tissues and the retrieval of metabolic products and carbon dioxide to be eliminated; the regulation and maintenance of internal pH and osmotic pressure; and the protection of the body through mechanisms such as blood clotting in wounds and defensive cells such as antibodies and leukocytes.

The transportation of blood to and from the tissues, called circulation, is necessary for cell survival. Vessels course to their destinations inside differentiated fascial compartments.

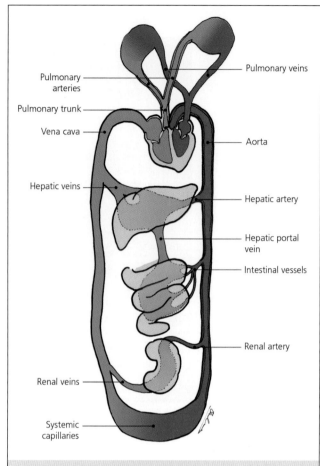

Figure 6-1 Drawing illustrating the pulmonary and systemic circulatory systems.

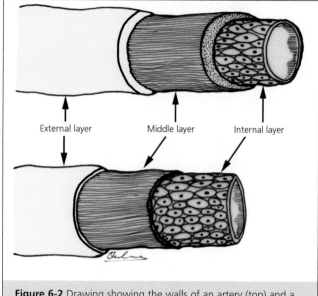

Figure 6-2 Drawing showing the walls of an artery (top) and a vein (bottom).

ARTERIES

Arteries are blood vessels that carry blood from the heart to the tissues. Arteries (from *aer*, "air," and *tereo*, "carrier") were so named because the arteries of a cadaver contain no blood, leading early anatomists to the incorrect conclusion that arteries transported air.

Arteries are conduits consisting of a wall and a central lumen that blood passes through. The walls of arteries are composed of three layers—the external, middle, and internal layers (Figure 6-2):

- The intima, or internal layer, formed by the endothelium (simple flat epithelium), is in contact with the blood. On the external aspect of this layer is the basal membrane of the endothelium, which is covered by an internal elastic lamina.
- The media, or middle layer, is a thick layer formed by smooth muscle tissue. This layer provides the vasomotor ability of arteries; along with the elastic layers, the middle layer provides the ability of arteries to recover their original shape when they are deformed. The external elastic lamina is superficial to the middle layer and separates it from the adventitial layer.
- The adventitia, or external layer, is formed by elastic fibers and collagen.

The large arteries that branch off the aorta are called elastic arteries, or conducting arteries. They are characterized by their large size and high elastic fiber content, needed to accommodate the high blood flow and pressure changes that accompany heartbeats. Examples of elastic arteries include the brachiocephalic trunk, the common carotid arteries, the subclavian arteries, and the common iliac arteries (Figure 6-3).

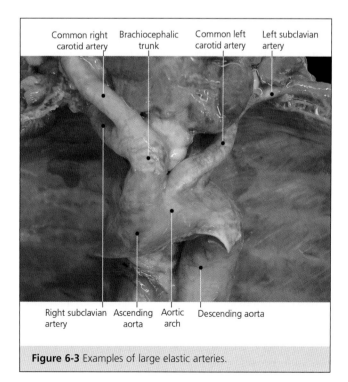

Figure 6-3 Examples of large elastic arteries.

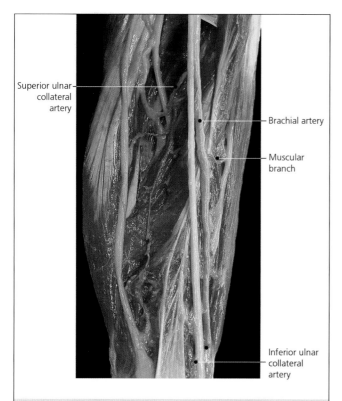

Figure 6-4 The brachial artery is a muscular artery that distributes blood to the upper limb. Note the collateral and muscular branches.

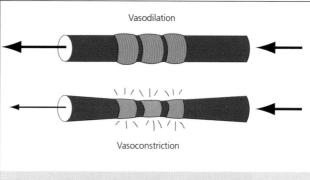

Figure 6-5 The precapillary sphincters control blood flow through the arterioles by means of vasodilation (top) and vasoconstriction (bottom).

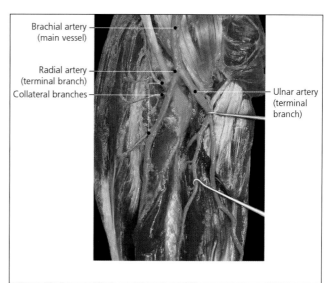

Figure 6-6 This illustration shows the dissection shown in Figure 6-4 continued into the elbow, revealing the distal branches of the brachial artery.

The elastic arteries transport blood to the muscular arteries (distributing arteries) (Figure 6-4). Compared with elastic arteries, distributing arteries have a more developed internal layer with more smooth muscle and fewer elastic fibers, which allows the regulation of blood flow through vasodilation and vasoconstriction. Examples of distributing arteries include the axillary, brachial, and femoral arteries.

Arteries get smaller as they near the target tissue. Arterioles are small arteries that carry blood to the capillaries. As arterioles near the capillaries, they lose most of the outer layers and only the endothelial layer remains. In addition, at the end of an arteriole is a collection of smooth muscle fibers that form the precapillary sphincter (Figure 6-5), which controls the flow of blood to the tissues through the mechanisms of vasodilation and vasoconstriction. Finally, capillaries are microscopic vessels formed by a thin endothelial layer that allows the exchange of substances and gases between the cardiovascular system and the tissues.

Several terms relating to arteries require definition. Blood flows through arteries known as main vessels, which give off branches called collateral vessels (Figures 6-4 and 6-6). In some locations, the main artery ends with a division into several branches that are called terminal vessels (Figure 6-6).

An anastomosis is the communication between blood vessels. Anastomoses can be arterioarterial (Figure 6-7) or arteriovenous.

Arteries supply blood to bones through epiphyseal, metaphyseal, and diaphyseal branches that penetrate the bones through nutrient orifices, as described in chapter 2. In the muscles, the muscle fibers are penetrated by neurovascular pedicles (Figure 6-4), which branch out inside the muscle belly. Finally, the vasa vasorum and vasa nervorum supply blood to the nerves.

Veins

In contrast to arteries, veins direct blood from the tissues to the heart. Thus, veins originate in the venous capillaries, which anastomose with the arterial capillaries. Veins increase in size as they near the heart.

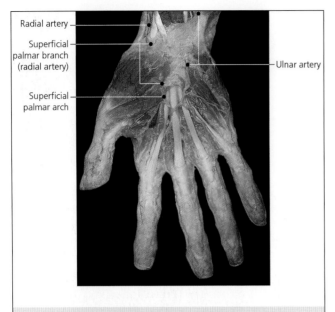

Figure 6-7 The superficial palmar arch is an arterioarterial anastomosis between the radial and ulnar arteries.

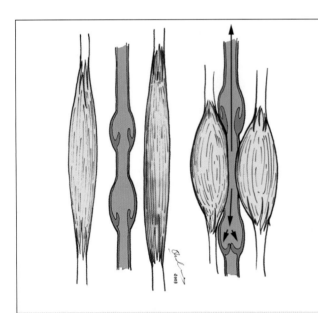

Figure 6-9 Muscle contraction compresses the veins, facilitating the return of blood to the heart (arrow pointing up), as valves prevent blood return (arrows pointing down).

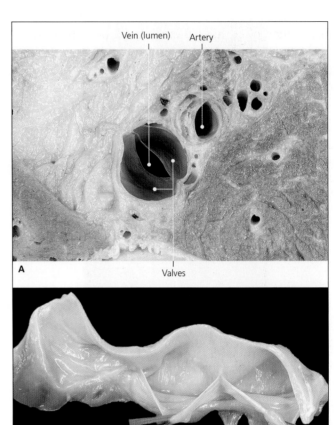

Figure 6-8 A, Transverse section of a limb showing an artery and a vein with a two-leaved valve. **B,** Pulmonary valve with its three leaves open. This valve prevents pulmonary blood from returning to the right ventricle.

The capillary veins drain to the venules, which drain to the veins, which have a thinner intermediate layer than in arteries. This is due to the lower blood pressure in the venous system, which does not require a thick muscle layer.

In the arterial system, the heart acts as a pump, forcing blood to the tissues. In contrast, the venous system does not have a pump to help return blood to the heart. Instead, four factors cause the flow of blood to the heart: the lower pressure of the right atrium, respiratory motion, venous valves, and muscle contraction.

Blood flows from high-pressure zones to low-pressure zones. After it pumps blood to the right ventricle, the right atrium has a very low pressure, which facilitates the return of blood to fill the atrium. In addition, during inspiration the diaphragm descends, increasing the abdominal pressure and impulsing blood through the inferior vena cava and the azygos system to the thorax (heart). Finally, veins, especially those in the lower limbs, contain two- or three-leaved valves that allow blood to flow in only one direction (Figures 6-8 and 6-9).

There are two main groups of veins, superficial and deep veins (Figure 6-10). Superficial veins are located superficial to muscular fasciae, at the subcutaneous level; they often have their own names. In contrast, deep veins most often accompany arteries (two or three veins per artery), are located underneath superficial fasciae, and have the same names as the arteries they join; these are known as venae comitantes. In addition, there are veins that communicate between the superficial and deep systems. When a vein is occluded, blood is diverted through a collateral venous network, which may increase in size to compensate for the abnormal circulation.

Veins may also anastomose with other vessels. There are venovenous anastomoses, arteriovenous anastomoses, and venous-lymphatic anastomoses.

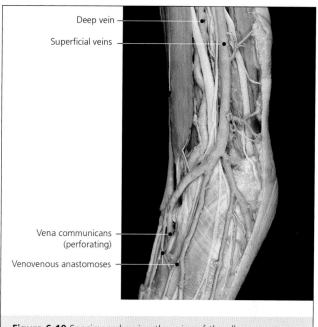

Figure 6-10 Specimen showing the veins of the elbow.

Section 2

Scapular Girdle and Upper Limb

Chapter 7

Anatomic Regions

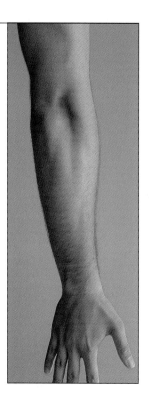

Above the clavicle are two fossae. The minor supraclavicular fossa (Figure 7-1) is a small triangular area bordered at its base by the clavicle and at its sides by the clavicular and sternal origins of the sternocleidomastoid muscle. Lateral to this triangle is the lateral cervical region, which is also a triangular area; its borders are formed by the clavicle at the base, the sternocleidomastoid muscle medially, and the trapezius laterally. This region is very important because it contains the brachial plexus. The lateral cervical region is divided into the omoclavicular triangle and the major supraclavicular fossa, separated by the muscle belly of the omohyoid muscle.

Anteriorly, the presternal region is anterior to the sternum; this forms the medial border of the pectoral region. This region covers the whole extent of the pectoralis major muscle. Inside this region, especially in women, is the mammary region (Figure 7-1), which includes the breasts, and the inframammary region, inferior to the breasts.

Below the clavicle is the infraclavicular fossa, formed at the clavipectoral (or deltopectoral) triangle, bordered by the clavicle and the space between the clavicular portions of the deltoid and pectoralis muscles. The deltoid region is lateral to this region and covers the deltoid muscle.

Medially, the axillary region (Figure 7-2) contains the axillary fossa, which is shaped like a quadrangular pyramid with the apex located between the clavicle and the first rib. The medial wall is formed by the thorax, covered by the serratus anterior muscle. The posterior wall is formed by the subscapularis, teres major, and latissimus dorsi. The anterior wall is bordered by the pectoralis major and minor muscles. The lateral wall, in the anatomic position, is bordered by the arm muscles, the coracobrachialis and short head of the biceps. Finally, the base is formed by the axillary fascia. The axillary fossa contains the brachial plexus, blood vessels including the axillary artery and vein as well as its branches, and a large number of lymphatic vessels and nodes.

The arm (Figure 7-3) is defined as the region of the upper limb located between the shoulder and elbow joints. The anterior brachial region, as indicated by its name, is located anteriorly and above the biceps muscle. This muscle, especially when it is contracted, delineates two grooves. The lateral bicipital groove corresponds to the lateral edge of this muscle, and the medial bicipital groove corresponds to the medial edge. Posteriorly, the posterior brachial region covers the triceps brachii.

The elbow (cubital region) is divided into a posterior cubital region and an anterior cubital region (Figure 7-4). The latter region contains the antecubital fossa. Topographically, the anterior fossa resembles a triangle with its base virtually limited by the elbow flexion crease, its lateral border formed by the brachioradialis muscle, and its medial border formed by the pronator teres. The distal biceps tendon divides the antecubital

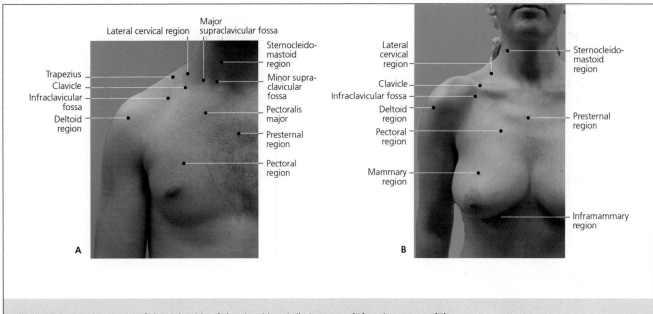

Figure 7-1 Anterior aspect of the right side of the shoulder girdle in a man (**A**) and a woman (**B**).

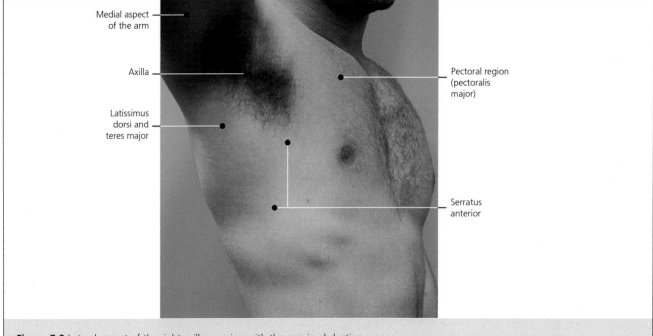

Figure 7-2 Lateral aspect of the right axillary region with the arm in abduction.

fossa into two halves, the medial half being covered by its aponeurotic expansion. The floor of the antecubital fossa is formed by the brachialis muscle.

The forearm (antebrachial region) is defined as the region between the elbow and wrist. Anatomic terms used for the forearm are similar to those used for the arm, with the forearm divided into an anterior antebrachial region (Figure 7-5), which covers the anterior muscles and the lateral aspect of the forearm, and a posterior antebrachial region, which covers the dorsal muscles and part of the lateral muscles. The medial region, between the anterior and posterior regions, is the ulnar (medial) border; the lateral zone, between the anterior and posterior regions, is the lateral (radial) border.

The wrist (carpal region), a small region formed by the carpal bones, is divided into an anterior carpal region and a posterior carpal region. The anterior carpal region includes the carpal tunnel and the flexor tendons, and the posterior carpal region includes the dorsal carpal grooves, which contain the extensor tendons.

The hand (Figure 7-6) is clearly divided into the palm (anteriorly) and the dorsum (dorsal region). The palm of the hand includes two important muscular regions: the thenar region, including the thenar eminence, which is lateral and contains the thumb intrinsic muscles; and the hypothenar region, including the hypothenar eminence, which is medial and contains the intrinsic muscles of the little finger. The middle palmar region, covered by the palmar fascia, is located between these two eminences; it contains the flexor tendons for the fingers and other intrinsic hand muscles, which are described in chapter 10.

In the hand, the metacarpal region contains the metacarpal bones, which can be felt on the dorsum; the

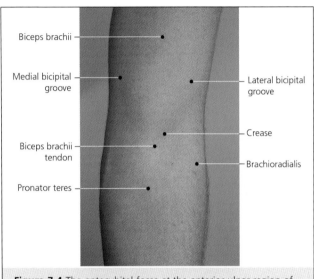

Figure 7-3 Medial aspect of the left arm in abduction.

Figure 7-4 The antecubital fossa at the anterior ulnar region of the left upper extremity.

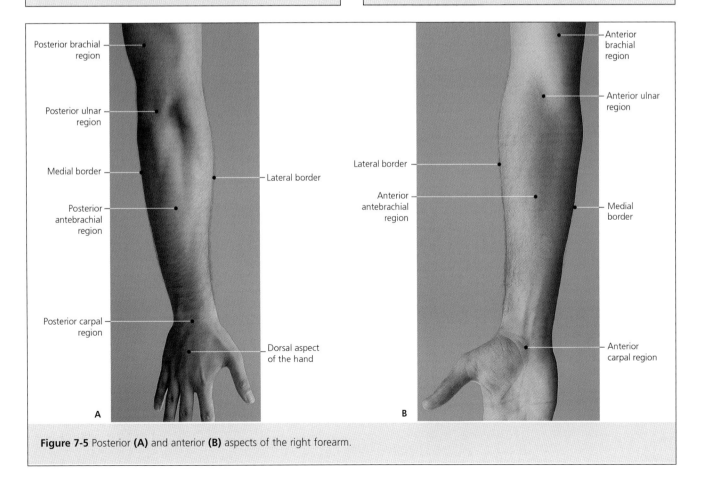

Figure 7-5 Posterior **(A)** and anterior **(B)** aspects of the right forearm.

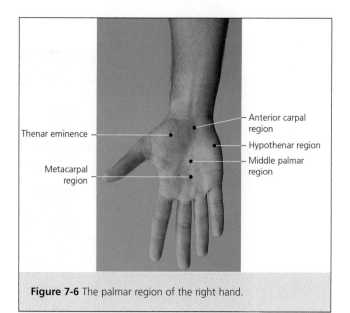

Figure 7-6 The palmar region of the right hand.

heads of the metacarpal bones can be seen forming the knuckles. Finally, the fingers of the hand are formed by the phalanges and are named as follows: thumb; second (II), or index, finger; third (III), or long, finger; fourth (IV), or ring, finger; and fifth (V), or little, finger. The fingers are divided into palmar aspects, which contain the flexor tendons, and dorsal aspects, which contain the extensor mechanism of the fingers. The ungual region, located in the fingertip, contains the nail.

Chapter 8

OSTEOLOGY

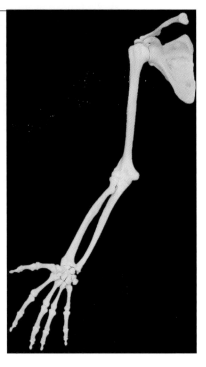

INTRODUCTION
The shoulder girdle is responsible for the articulation of the upper limb with the thorax. Classically, it is considered to be formed by the clavicle and the scapula, but functionally, it is closely related to glenohumeral motion. The bones of the free portion of the upper limb include the humerus, radius, ulna, carpal bones, metacarpals, phalanges, and sesamoids.

CLAVICLE
The clavicle (Figures 8-1 and 8-2) is a flat, long bone that lies transversely between the sternal manubrium and the scapular acromion. Viewed from above, it is S-shaped, with a medial convexity and a lateral concavity. It has a superior and an inferior face, the latter containing a number of anatomic details that are described below. The superior face contains no important anatomic features and is related only to the skin and the supraclavicular sensory nerves.

The medial, or sternal, end is either triangular or round and has a sternal articular facet that articulates with the sternal manubrium through articular fibrocartilage (Figure 8-3). This articular surface extends inferiorly to articulate with the cartilage of the first rib. On the inferior side, close to the joint, is a prominence for the costoclavicular ligament. Muscle attachments to the medial side of the clavicle include the pectoralis major anteriorly and the sternocleidomastoid and sternohyoid muscles posteriorly.

On the lateral side, the clavicle flattens, forming an anterior and a posterior border that provide origins for the pectoralis major and deltoid anteriorly and the trapezius posteriorly.

The central portion of the clavicle is called the body. Its most relevant detail corresponds to the sulcus for the subclavius muscle, located in the inferior face where this muscle attaches. The small edges of this sulcus provide insertion for the clavipectoral fascia.

The lateral end of the clavicle is called the acromial end; it articulates with the acromion through the acromial articular facet. Two eminences at the inferior face provide insertion for the fascicles of the coracoclavicular ligaments—the conoid tubercle for the conoid ligament and the trapezoid tubercle for the trapezoid ligament.

SCAPULA
The scapula, or shoulder blade (Figures 8-4 through 8-7), is a flat, triangular bone that lies at a 60° angle to the clavicle to accommodate the thorax outline. The clavicle extends from the level of the first or second rib (superior angle) to the seventh or eighth rib (inferior angle).

The scapula has a superior, a lateral, and a medial (also called vertebral) border. These borders meet to form the inferior, superior, and lateral (also called glenoid) angles.

Two distinguishing features are found in the superior border. One is the coracoid (from Latin *corvus*, "raven") process, so named because it resembles a raven's beak. The coracoid process provides insertion for three muscles. These muscles are, from lateral to medial, the short head of the biceps, the coracobrachialis, and the pectoralis minor. Nearby is the scapular or coracoid notch, the shape and depth of which vary widely. The notch is

Scapular Girdle and Upper Limb

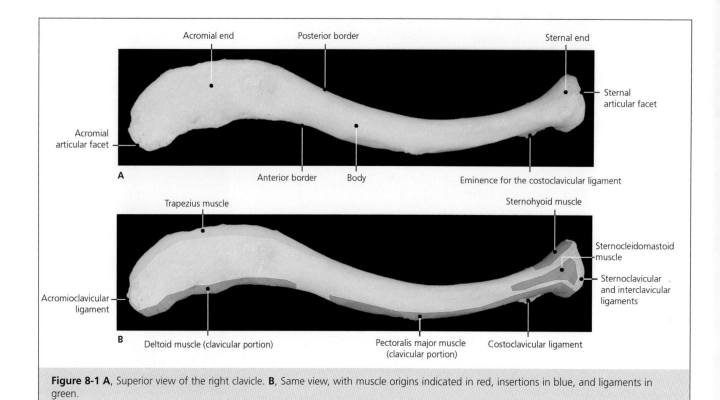

Figure 8-1 A, Superior view of the right clavicle. **B,** Same view, with muscle origins indicated in red, insertions in blue, and ligaments in green.

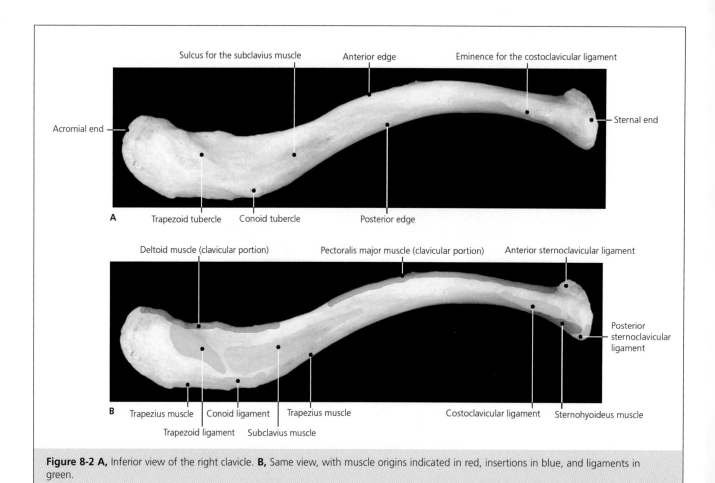

Figure 8-2 A, Inferior view of the right clavicle. **B,** Same view, with muscle origins indicated in red, insertions in blue, and ligaments in green.

OSTEOLOGY

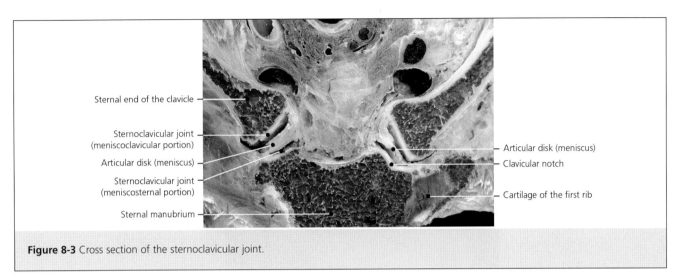

Figure 8-3 Cross section of the sternoclavicular joint.

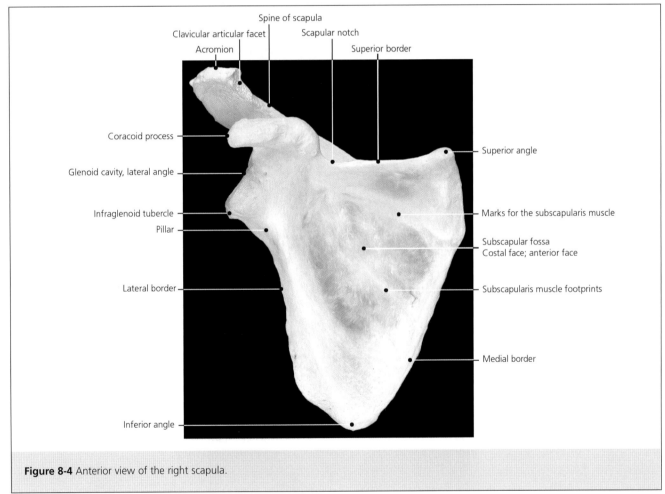

Figure 8-4 Anterior view of the right scapula.

closed superiorly by the superior transverse ligament. The suprascapular nerve passes through this opening, and the suprascapular artery lies over the ligament. The omohyoideus muscle attaches medial to this notch. This muscle is described in chapter 16, where the trunk is discussed, because it has only minimal actions on the scapula.

The medial border provides attachments for the rhomboid muscles and, anteriorly, for the serratus anterior muscle. The lateral border provides insertion for the teres major and minor on two small lateral facets, one for each muscle.

The inferior angle is the inferior vertex of the scapula, which provides insertion for some fibers of

Scapular Girdle and Upper Limb

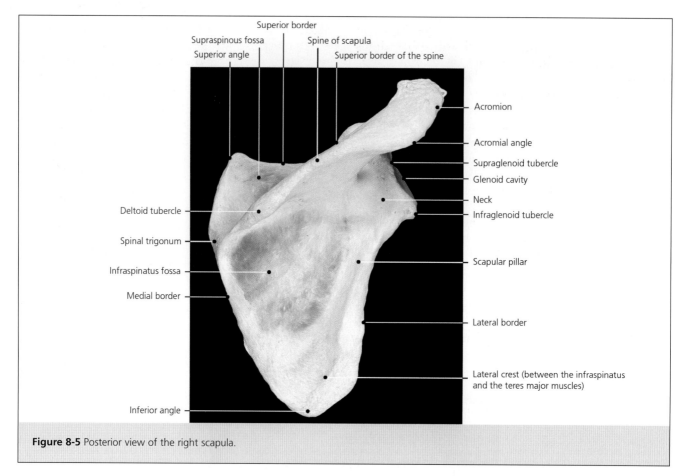

Figure 8-5 Posterior view of the right scapula.

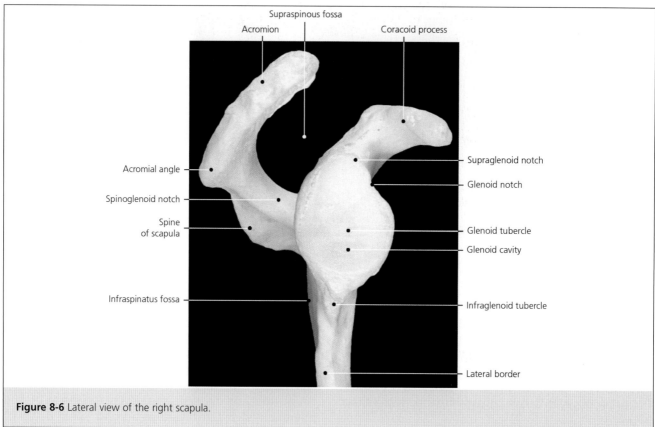

Figure 8-6 Lateral view of the right scapula.

the latissimus dorsi. The superior angle varies in shape, sometimes being rounded and other times pointed. It provides attachment for the levator scapulae muscle.

The lateral angle (Figure 8-6) contains most of the important structures. First, it contains the articular facet for the humerus, the glenoid cavity. The glenoid is pear shaped, narrowing in the middle at the glenoid notch. The glenoid tubercle is located in the center of the glenoid. There are two additional tubercles: the supraglenoid tubercle, for insertion of the long head of the biceps, and the infraglenoid tubercle, for insertion of the triceps. The lateral angle narrows close to the glenoid cavity, forming the scapular neck. On its anterior face is an elongated and denser bone eminence called the scapular pillar.

The scapula has an anterior costal face and a posterior face. The most important anatomic feature of the anterior face is the subscapular fossa (Figure 8-4), a slightly concave area with small oblique ascending crests that provide origin to the subscapularis muscle, covering the whole fossa. The serratus anterior muscle attaches to the medial region and border.

The posterior face (Figure 8-5), in contrast to the anterior face, is convex and is divided into two regions by a large structure, the spine of the scapula. It is shaped like a triangle with a dorsal subcutaneous base and a vertex that fuses with the scapula but floats freely on its lateral side. The superior region is the supraspinatus fossa, which includes the superior aspect

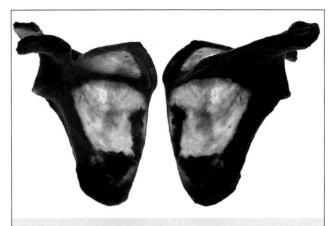

Figure 8-7 The scapulae shown under bright light demonstrate the thin central cortical bone.

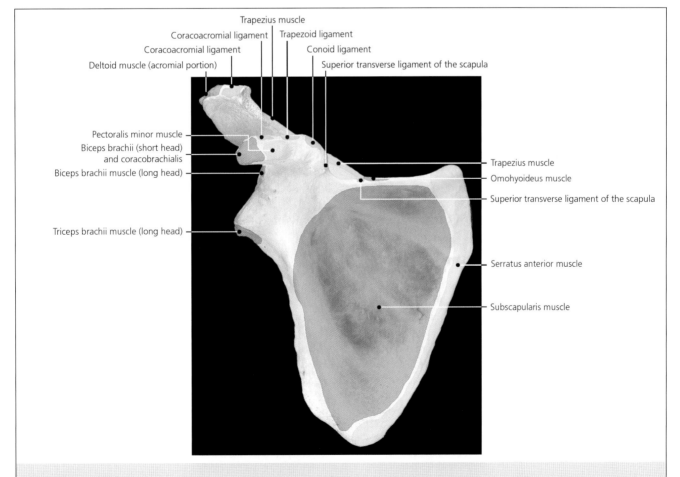

Figure 8-8 Anterior view of the right scapula. Muscle origins are indicated in red, insertions in blue, and ligaments in green.

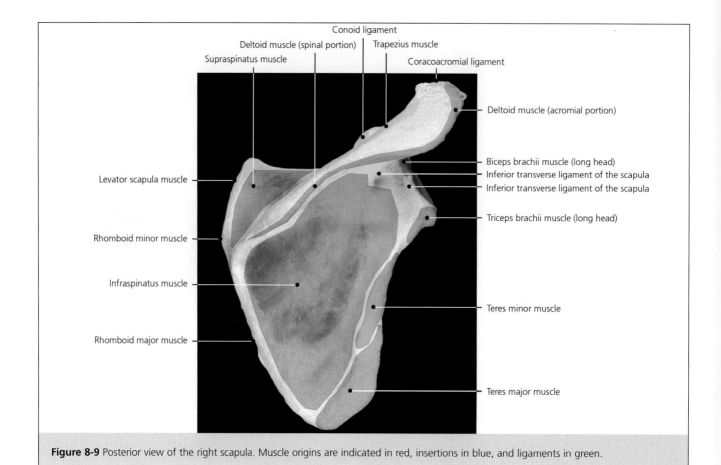

Figure 8-9 Posterior view of the right scapula. Muscle origins are indicated in red, insertions in blue, and ligaments in green.

of the scapula and of the spine of the scapula; it provides origin for the supraspinatus muscle. The inferior region, or infraspinatus fossa, includes the inferior two thirds of the posterior face of the scapula and provides origin to the infraspinatus muscle. In some individuals a lateral crest is present, which separates the origins of the infraspinatus and the teres minor and major muscles.

The spine of the scapula has a superior border, onto which the trapezius muscle inserts, and an inferior border, on which the deltoid muscle originates. Close to the medial border, the spine fuses with the rest of the scapula. In this region is a small triangle, the spinal trigone, which allows sliding of the ascending fibers of the trapezius muscle and also delineates the zone of insertion of the rhomboid muscles, with the rhomboid major inserting above the spinal trigone and the rhomboid minor inserting below it. Laterally, the spine of the scapula projects away from the body of the scapula and continues as a flat free end called the acromion. The area of reflection between the spine of the scapula and the acromion is called the acromial angle. The region at the scapular neck where the two posterior fossae communicate is called the spinoglenoid notch. The acromion provides insertion for the trapezius internally and the deltoid externally. The anterior region, also known as the acromial beak, contains the clavicular articular facet of the acromion, part of the acromioclavicular joint.

The subcutaneous acromial bursa is located between the superior aspect of the acromion and the skin.

Figures 8-8 and 8-9 show the origins and insertions of muscles and ligaments on the scapula. Figures 8-10 and 8-11 are radiographic views of the shoulder that demonstrate the relationship of the scapula, clavicle, and acromion to the glenohumeral articulation with the arm at the side (AP view) and in abduction (axillary view).

Humerus

The humerus (Figure 8-12) is the long bone in the free portion of the upper limb that articulates with the shoulder girdle through the scapula. External and cross-sectional features of the humerus are shown in Figures 8-12 through 8-21. Muscle origins and insertions are shown in Figure 8-22. Like all long bones, the humerus has a diaphysis and two epiphyses. The proximal epiphysis, called the humeral head (Figures 8-13 and 8-14), is hemispheric. The medial side is covered with articular cartilage to form the glenohumeral joint. The area covered by cartilage is separated from the rest of the epiphysis by a groove called the anatomic neck.

OSTEOLOGY

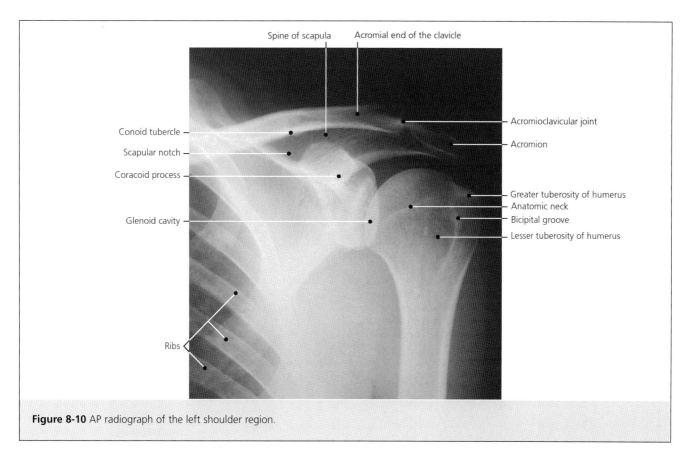

Figure 8-10 AP radiograph of the left shoulder region.

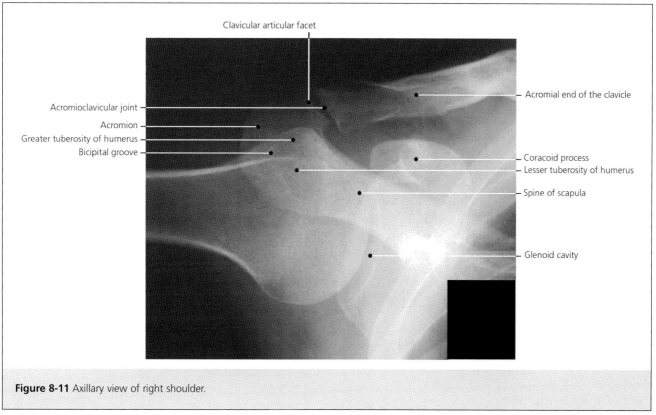

Figure 8-11 Axillary view of right shoulder.

Scapular Girdle and Upper Limb

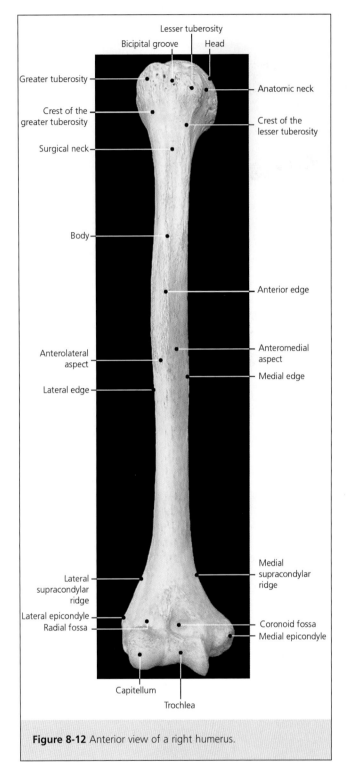

Figure 8-12 Anterior view of a right humerus.

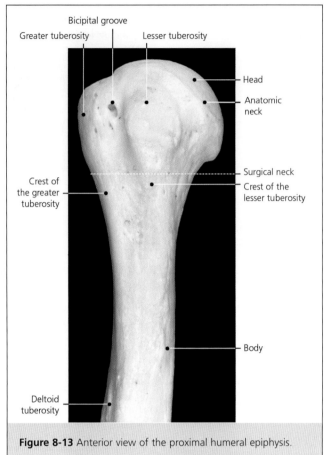

Figure 8-13 Anterior view of the proximal humeral epiphysis.

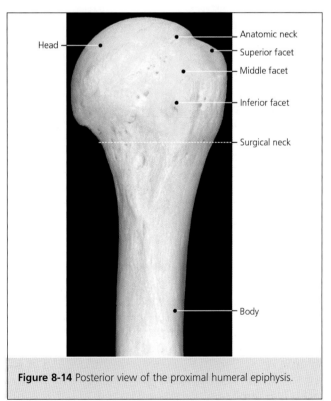

Figure 8-14 Posterior view of the proximal humeral epiphysis.

Osteology

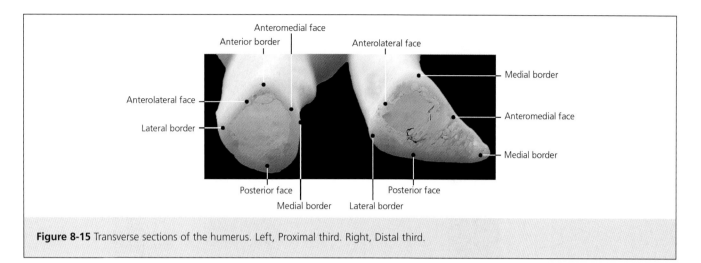

Figure 8-15 Transverse sections of the humerus. Left, Proximal third. Right, Distal third.

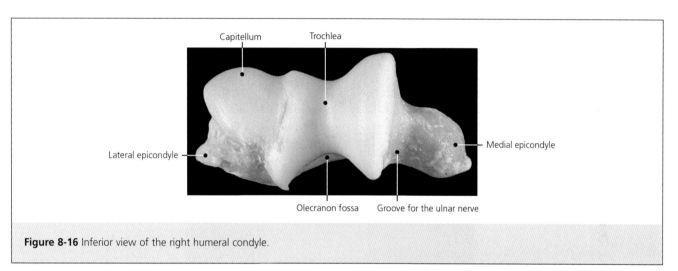

Figure 8-16 Inferior view of the right humeral condyle.

Anteriorly, the humerus has two eminences, one medial and one lateral. The first eminence, called the lesser tuberosity, provides insertion for the subscapularis muscle. Distal to the lesser tuberosity, the crest of the lesser tuberosity provides insertion for the teres major and latissimus dorsi. On the lateral side, the greater tuberosity, or trochiter, extends to the posterolateral aspect of the proximal humerus. It has three facets, for insertion of the supraspinatus, infraspinatus, and teres minor muscles. There is also a crest on the anterior aspect of the greater tuberosity, which provides insertion for the pectoralis major muscle. Between the tuberosities and crests is a bone groove closed by the transverse humeral ligament; the long head of the biceps is located in this groove. This channel is known as the intertubercular or bicipital groove. Beneath the lesser and greater tuberosities, the proximal epiphysis is separated from the diaphysis at the surgical neck. The humeral body has a triangular cross section at its midportion (Figure 8-15), with anteromedial, anterolateral, and posterior faces and medial, lateral, and anterior borders; the borders are poorly defined, however, and can be distinguished only on cross sections.

In the middle portion of the humeral body, the course of the radial nerve defines the radial, or spiral, groove. The rotation of the humerus during embryonic development explains the course of the radial nerve and the course of the radial or spiral groove. The origin of the lateral head of the triceps lies proximal to the radial groove; the origin of the medial head of the triceps lies distal to the groove. The deltoid tuberosity, located proximal to the groove on the anterolateral face, provides a V-shaped insertion for the deltoid muscle. The insertion of the coracobrachialis muscle is located approximately at the same height on the anteromedial face. Finally, the anterolateral and anteromedial faces of the distal half of the humerus are covered by the brachialis muscle origin.

The distal humeral epiphysis (Figures 8-16 through 8-18) is flat and is angled about 45° with respect to the humeral body (Figure 8-19). This epiphysis has two subcutaneous bony eminences, one medial and one lateral. The medial eminence, called the medial

Scapular Girdle and Upper Limb

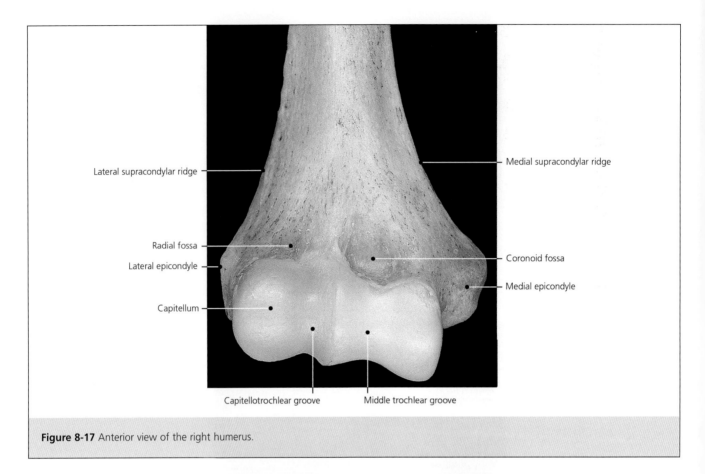

Figure 8-17 Anterior view of the right humerus.

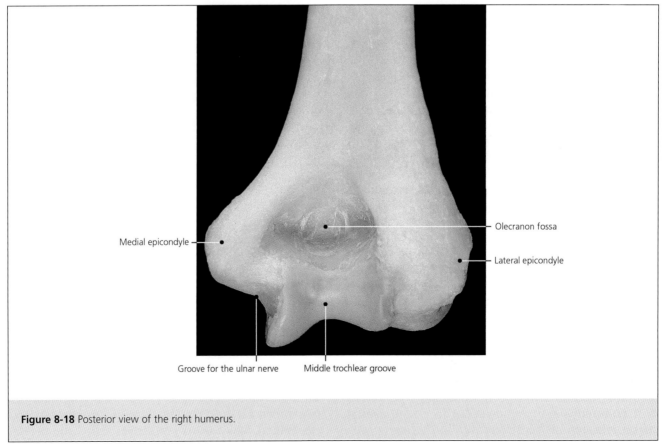

Figure 8-18 Posterior view of the right humerus.

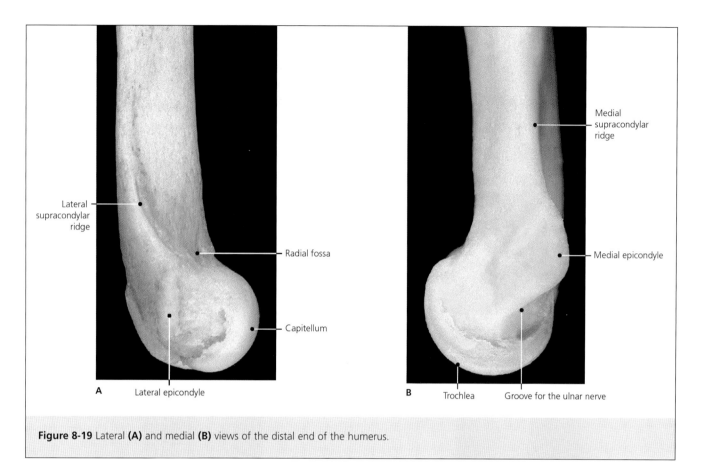

Figure 8-19 Lateral (A) and medial (B) views of the distal end of the humerus.

epicondyle or epitrochlea, is more prominent; the lateral eminence is called the lateral epicondyle. These two eminences are located outside the articular capsule and provide origin for the collateral ligaments and several muscles: the flexor-pronator group arises from the medial epicondyle, and the extensor-supinator group arises from the lateral epicondyle. The lateral supracondylar ridge (extension of the lateral humeral border) and the medial supracondylar ridge (extension of the medial humeral border) are located above these eminences. Rarely, a medial supracondylar process is present (Figure 8-20), located at the medial side of the distal third of the humerus. In some individuals, a ligament of Struthers extends from this apophysis to the medial epicondyle.

On the ventral side of the distal humeral epiphysis, between the medial and lateral epicondyles, are articular surfaces for the forearm bones. The humeral trochlea, located medially, articulates with the ulna; it has two borders and a central sulcus. On the lateral side, a small capitellotrochlear sulcus separates the trochlea from the capitellum; this sulcus extends to the edge of the radial articular circumference. The capitellum, or humeral condyle, which is lateral to this sulcus, is hemispheric and articulates with the radial head concavity. The trochlea contains a coronoid fossa, which accommodates the coronoid apophysis during

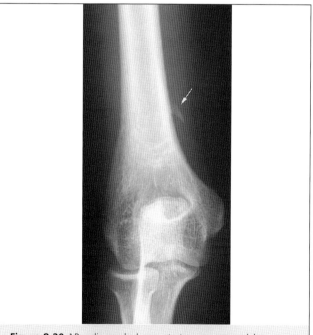

Figure 8-20 AP radiograph demonstrates a supracondylar apophysis (arrow).

elbow flexion. The radial fossa, located above the humeral condyle, accommodates the radial head.

The olecranon fossa, which accommodates the olecranon during elbow extension, is on the posterior aspect

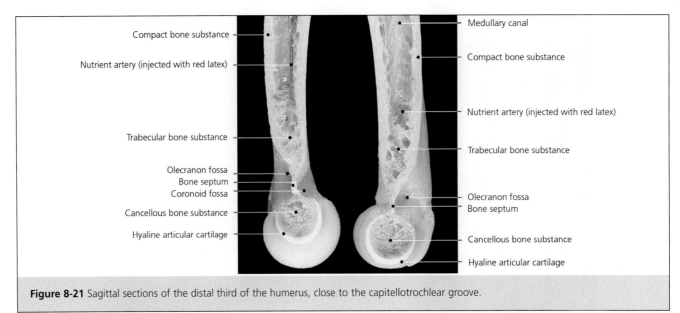

Figure 8-21 Sagittal sections of the distal third of the humerus, close to the capitellotrochlear groove.

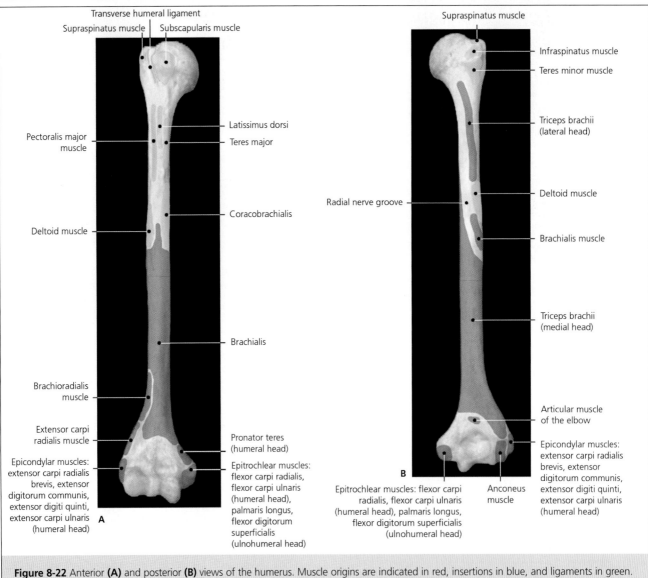

Figure 8-22 Anterior **(A)** and posterior **(B)** views of the humerus. Muscle origins are indicated in red, insertions in blue, and ligaments in green.

Osteology

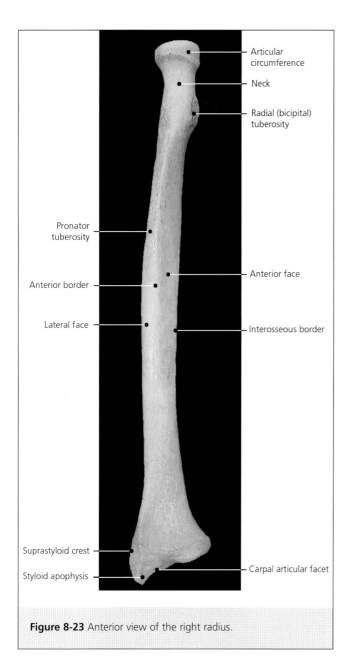

Figure 8-23 Anterior view of the right radius.

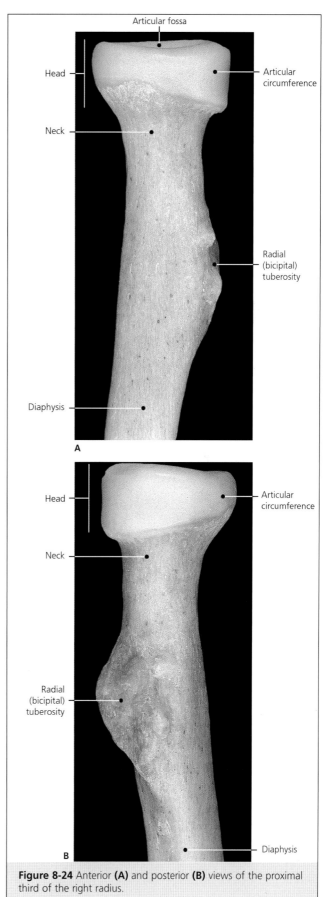

Figure 8-24 Anterior (A) and posterior (B) views of the proximal third of the right radius.

of the distal humerus. A sulcus for the ulnar nerve lies on the posterior aspect of the medial epicondyle.

Figure 8-22 shows the muscular origins and insertions on the humerus.

RADIUS

The radius (Figure 8-23), like the humerus, is a long bone. Like all long bones, it has a diaphysis and two epiphyses. The proximal epiphysis (Figures 8-24 and 8-25) has an almost circular head that has a proximal articular facet for the capitellum and an articular circumference that articulates with the ulna. The radial head narrows into the radial neck before it continues into the diaphysis.

The diaphysis of the radius is bowed, allowing forearm pronation. Although the radius appears to be cir-

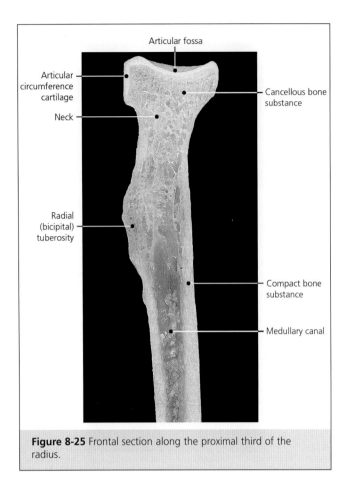

Figure 8-25 Frontal section along the proximal third of the radius.

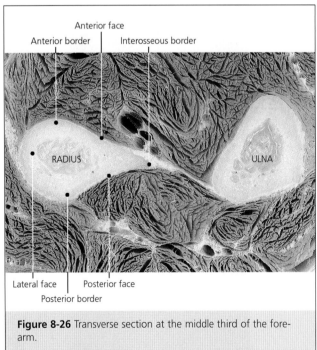

Figure 8-26 Transverse section at the middle third of the forearm.

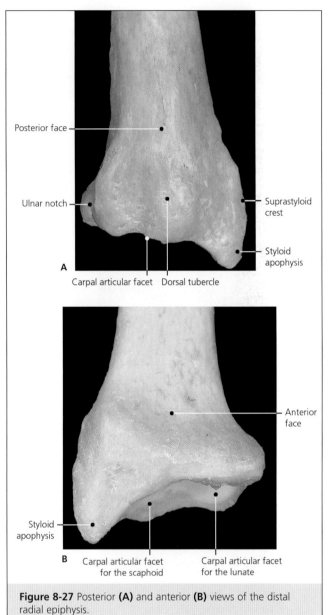

Figure 8-27 Posterior (**A**) and anterior (**B**) views of the distal radial epiphysis.

cular, its cross section at the midportion is almost triangular (Figure 8-26), with the following borders: an interosseous border that faces the ulna, which provides insertion for the interosseous membrane; and a posterior and an anterior border. These borders delineate the three faces of the radius: lateral, posterior, and anterior. At the proximal end are two tuberosities: a medial tuberosity for the distal biceps tendon, called the radial or bicipital tuberosity, and a lateral tuberosity called the pronator tuberosity, for insertion of the pronator teres.

The radius widens at the distal epiphysis (Figures 8-27 and 8-28), where its cross section is a polygon with five faces (Figure 8-29): an anterior face; a medial face; a lateral face; and two dorsal faces, one dorsolateral and one dorsal. The lateral face has a short extension called the suprastyloid crest, which ends in the styloid apophysis, for insertion of the brachioradialis muscle. On the medial face is a concave surface

Osteology

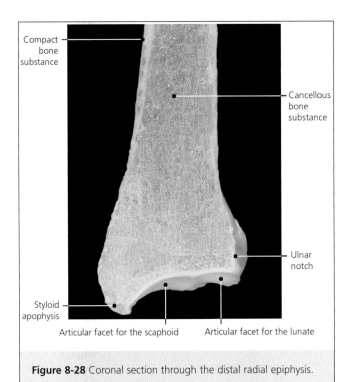

Figure 8-28 Coronal section through the distal radial epiphysis.

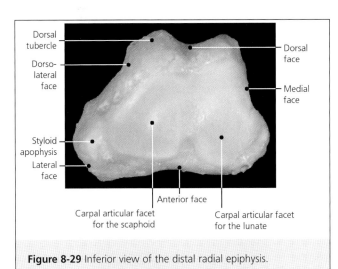

Figure 8-29 Inferior view of the distal radial epiphysis.

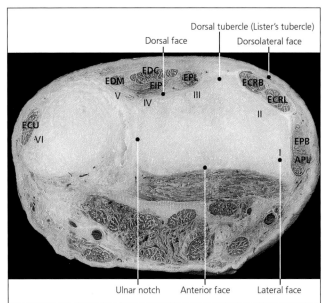

Figure 8-30 Transverse section of the distal third of the forearm. The grooves for the extensor muscles are labeled. APL = abductor pollicis longus, EPB = extensor pollicis brevis, ECRL = extensor carpi radialis longus, ECRB = extensor carpi radialis brevis, EPL = extensor pollicis longus, EDC = extensor digitorum communis, EIP = extensor indicis proprius, EDM = extensor digiti minimi, and ECU = extensor carpi ulnaris.

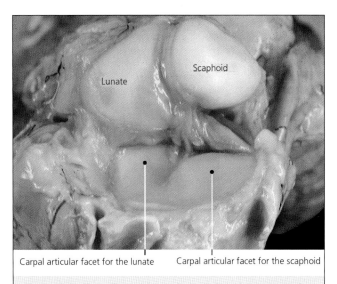

Figure 8-31 The radiocarpal articular capsule is shown opened with the wrist flexed, revealing the articular surfaces.

for the ulna, called the ulnar notch, which allows pronation and supination through the distal radioulnar joint. Dorsally, there are grooves for the extensor muscle tendons (Figure 8-30). The third carpal groove is separated from the second by the dorsal, or Lister's, tubercle, which can be felt subcutaneously. The anterior face is concave to accommodate the flexor tendons, and it forms the floor of the carpal tunnel. This floor is partly covered by the pronator quadratus. The distal carpal articular surface has two articular facets for the scaphoid and lunate carpal bones, respectively (Figure 8-31), separated by a bony crest, the intrafossal ridge.

Ulna

The ulna (Figure 8-32) is a long bone. Its proximal epiphysis (Figure 8-33) is characterized by a dorsal prominence called the olecranon that provides insertion for the triceps. It has a small anterosuperior tip that enters the olecranon fossa when the elbow is in extension. Anteriorly, the concavity of this eminence, called the trochlear notch or greater sigmoid cavity, articulates with the humeral trochlea to form the ulnohumeral

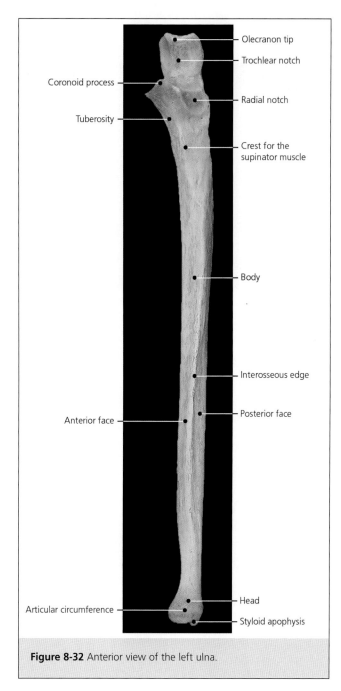

Figure 8-32 Anterior view of the left ulna.

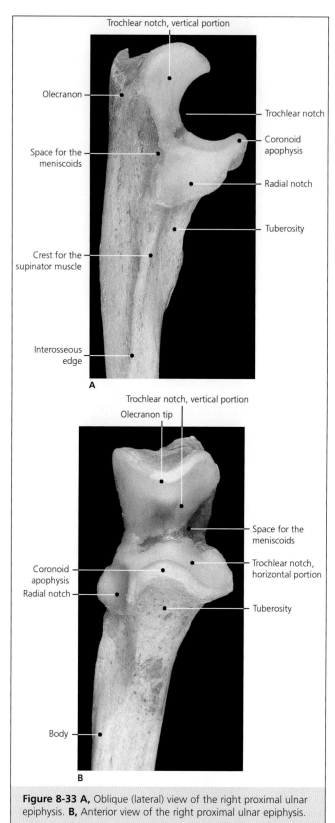

Figure 8-33 A, Oblique (lateral) view of the right proximal ulnar epiphysis. **B,** Anterior view of the right proximal ulnar epiphysis.

joint. This greater sigmoid cavity may be thought of as being divided into a vertical and a horizontal portion, separated by a transverse line devoid of cartilage that extends medially to accommodate a synovial meniscoid. A sagittal middle crest extends from the coronoid apophysis to the tip of the olecranon, dividing the cavity into a medial and a lateral side. Laterally, the radial notch, or lesser sigmoid notch, articulates with the radial head. Distal to this articular facet is a crest for the origin of the supinator muscle. On the anterior aspect, the coronoid process enters the coronoid humeral fossa. Distal to the coronoid is a small, rough tuberosity for insertion of the brachialis muscle.

The ulnar body has a triangular shape with three borders (Figure 8-34): interosseous (facing the radius), anterior, and posterior. These borders delineate the medial, anterior, and posterior faces.

OSTEOLOGY

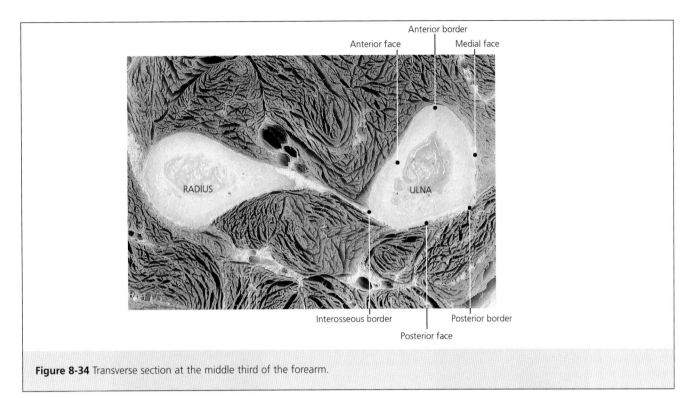

Figure 8-34 Transverse section at the middle third of the forearm.

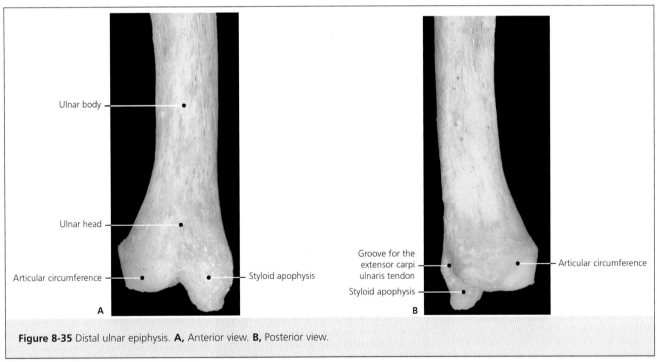

Figure 8-35 Distal ulnar epiphysis. **A,** Anterior view. **B,** Posterior view.

The distal epiphysis, or ulnar head (Figure 8-35), has an extension called the ulnar styloid. On the lateral side is an articular circumference for the radius. A small posterior groove, next to the ulnar styloid, facilitates the passage of the extensor carpi ulnaris tendon through the sixth dorsal groove (Figure 8-30).

Figure 8-36 shows the origins and insertions of muscles and ligaments on the bones of the forearm. Figures 8-37 and 8-38 are radiographic views of the elbow demonstrating the relationship of the bones to each other with the elbow in extension and in flexion.

CARPUS

The carpus (Figures 8-39 through 8-42) is formed by eight short bones arranged in two transverse rows: one proximal and one distal. The proximal carpal row is formed by the following bones, from radial to ulnar: scaphoid, lunate, triquetrum, and

American Academy of Orthopaedic Surgeons

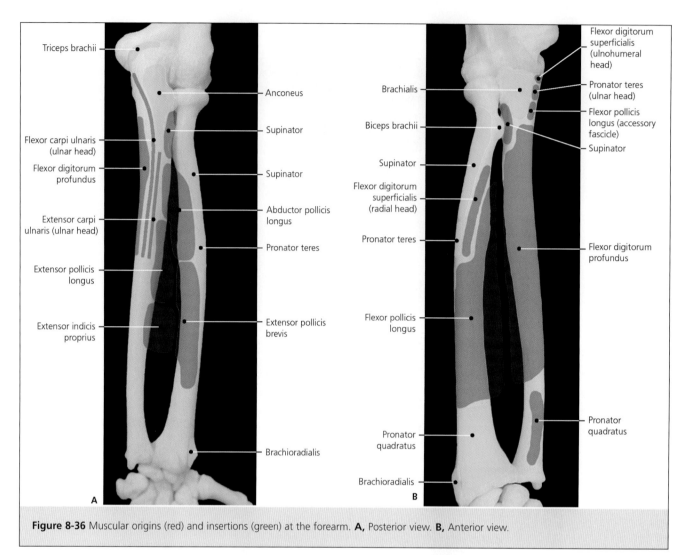

Figure 8-36 Muscular origins (red) and insertions (green) at the forearm. **A,** Posterior view. **B,** Anterior view.

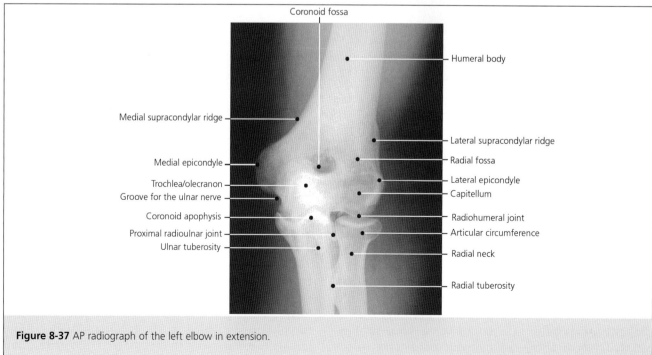

Figure 8-37 AP radiograph of the left elbow in extension.

Osteology

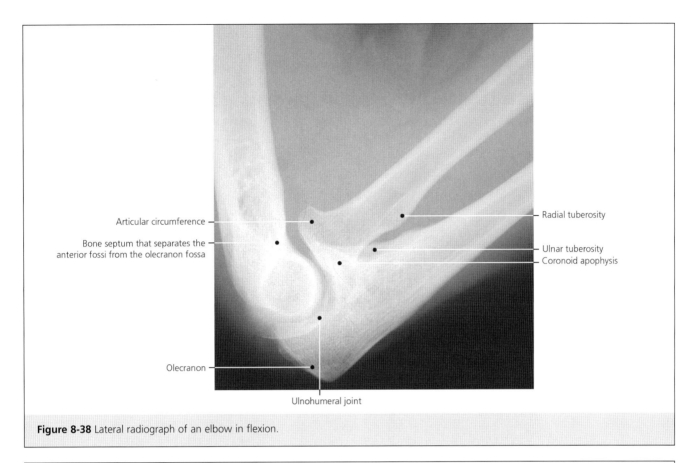

Figure 8-38 Lateral radiograph of an elbow in flexion.

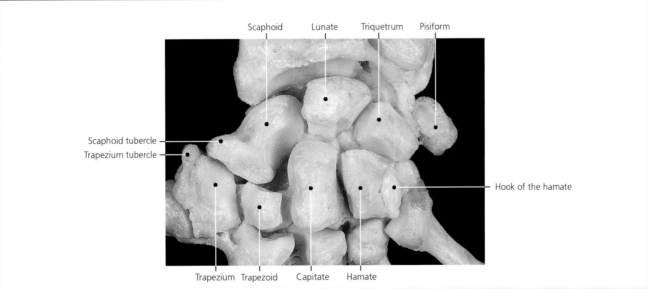

Figure 8-39 Anterior view of the carpal bones. The anterior and interosseous ligaments have been resected, but the posterior ligaments have been preserved to maintain the structure of the region. The pisiform has been displaced (its dorsal face is shown) to allow visualization of the underlying triquetrum.

pisiform (anterior to the triquetrum). The distal row includes the trapezium, trapezoid, capitate, and hamate. The carpus as a whole presents an anterior (palmar) concavity, forming the bony part of the carpal tunnel, which allows the passage of multiple structures. The posterior (dorsal) aspect of the carpus is slightly convex and is covered by the extensor tendons.

Scaphoid

The scaphoid (Figure 8-43) (from Greek *skaphos*, "boat," because of its boat shape) is the largest bone of the prox-

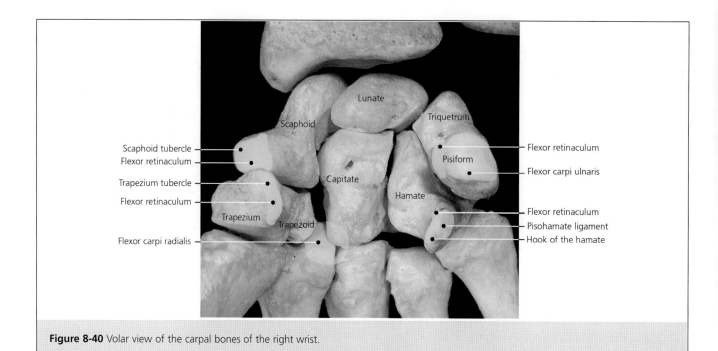

Figure 8-40 Volar view of the carpal bones of the right wrist.

Figure 8-41 Dorsal view of the carpal bones of the left wrist.

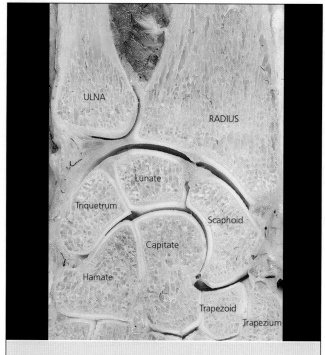

Figure 8-42 Frontal section of the wrist region.

imal row. The scaphoid has a waist, or neck; a proximal pole; and a distal pole. It articulates proximally with the scaphoid facet of the radius, radially with the lunate and the capitate, and distally with the trapezium and the trapezoid. The scaphoid tubercle provides insertion for the radial collateral ligament and the flexor retinaculum. It also serves as a turning point for the flexor pollicis longus tendon through the carpal tunnel. This bone may be felt at the base of the anatomic "snuff box."

Lunate

The lunate (Figure 8-44) (from Latin *luna*, "moon," because of its crescent shape) is the second bone of the proximal row. It articulates proximally with the lunate facet of the radius, distally with the capitate and hamate, radially with the triquetrum, and ulnarly with the scaphoid. It is important to note that this bone is located in the concavity of the midcarpal joint, so it may not be seen in axial cross sections including the scaphoid and triquetrum. The lunate and scaphoid form part of the bony floor of the carpal tunnel.

Osteology

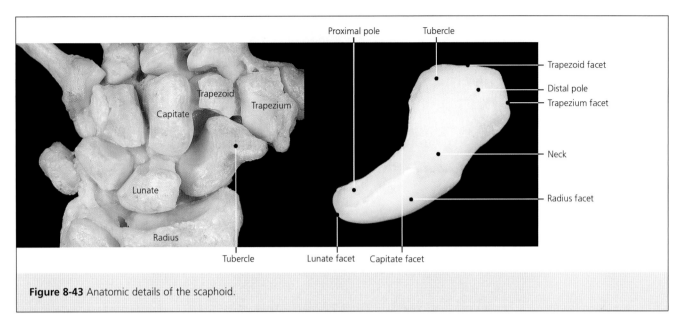

Figure 8-43 Anatomic details of the scaphoid.

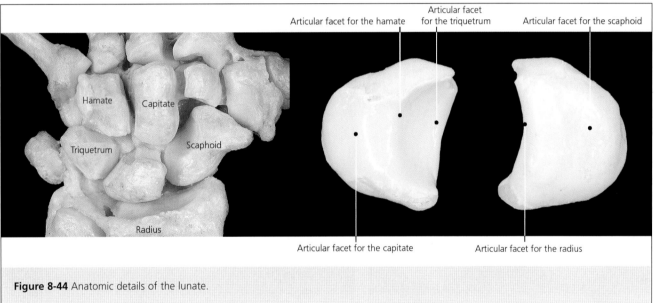

Figure 8-44 Anatomic details of the lunate.

Triquetrum

The triquetrum (Figure 8-45) (from Latin *triquetrus*, "three-cornered") is shaped like a pyramid with a superolateral base. This bone articulates with the ulna through a fibrocartilage rather than directly. The triquetrum has a small dorsal tubercle for insertion of the posterior fascicle of the lateral collateral ligament. It articulates ulnarly with the lunate, distally with the hamate, and palmarly with the pisiform.

Pisiform

The pisiform (Figure 8-46) (from Latin *pisum*, "pea," because of its pea shape) is a sesamoid bone found within the flexor carpi ulnaris tendon. It articulates with the anterior face of the triquetrum. The pisiform provides insertion for the flexor retinaculum, the ulnar collateral ligament, and the aponeurosis that covers Guyon's tunnel. It also provides origin for some of the muscle fibers of the hypothenar eminence. The pisiform can easily be felt subcutaneously.

Trapezium

The trapezium (Figure 8-47) is the most radial bone of the distal carpal row. It articulates proximally with the scaphoid and distally with the first and second metacarpals, through two small facets separated by a bony crest. The tubercle of the trapezium delineates an anterolateral channel for the flexor carpi radialis tendon. This tubercle also provides insertion for the flexor retinaculum.

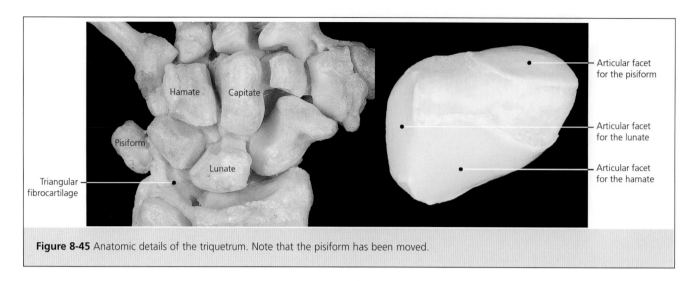

Figure 8-45 Anatomic details of the triquetrum. Note that the pisiform has been moved.

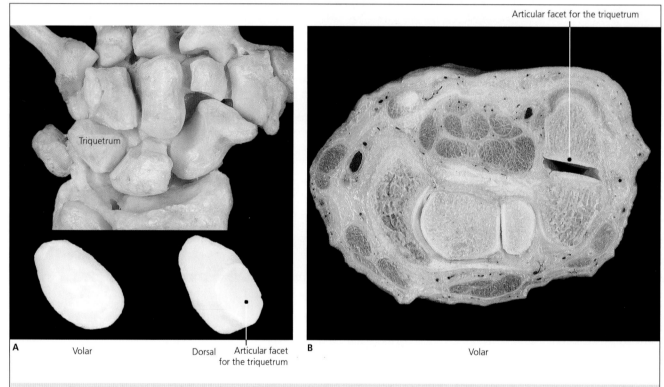

Figure 8-46 A, Dorsal view of the left wrist with the pisiform moved to show the triquetrum and the inferior articular facet of the pisiform (its dorsal face). **B,** Transverse carpal section.

Trapezoid
The trapezoid (Figure 8-48) articulates with four bones: the scaphoid proximally, the second metacarpal distally, the trapezium ulnarly, and the capitate radially.

Capitate
The capitate (Figure 8-49), located in the center, is the largest carpal bone. It is divided into a head, a neck, and a body. Proximally it articulates with the scaphoid and lunate, distally with the three central metacarpals, ulnarly with the trapezoid, and radially with the hamate.

Hamate
The hamate (Figure 8-50) is the most radial bone of the distal carpal row. It articulates proximally with the lunate and triquetrum, distally with the two ulnar metacarpals, and ulnarly with the capitate. Its palmar surface is characterized by a hooklike process, the hamulus, or hook of hamate, that provides insertion for the flexor retinaculum as well as ligaments and

Osteology

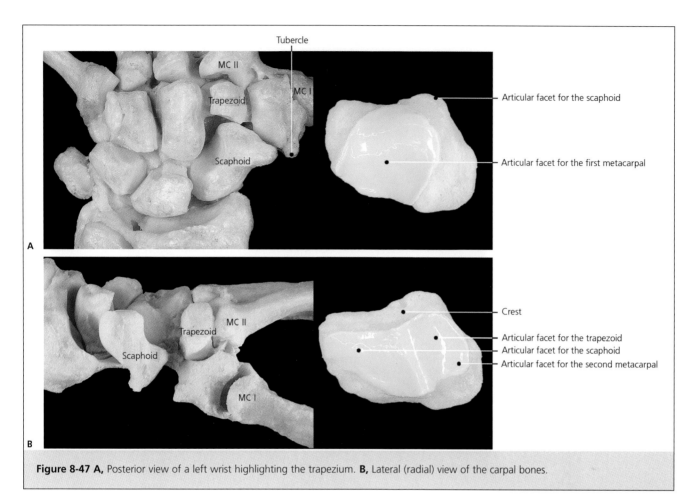

Figure 8-47 A, Posterior view of a left wrist highlighting the trapezium. **B,** Lateral (radial) view of the carpal bones.

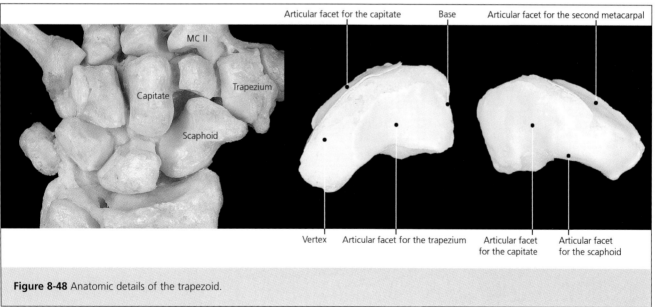

Figure 8-48 Anatomic details of the trapezoid.

muscles of the hypothenar eminence. This process can be felt subcutaneously.

Metacarpals

The metacarpals (Figure 8-51) form the true skeleton of the palm. These five long bones connect the distal carpal row with the base of each phalanx. The metacarpals are numbered from I to V, with I being the thumb metacarpal and V being the metacarpal of the little finger. The metacarpals follow the concavity present at the carpus.

Like other long bones, each metacarpal has a body

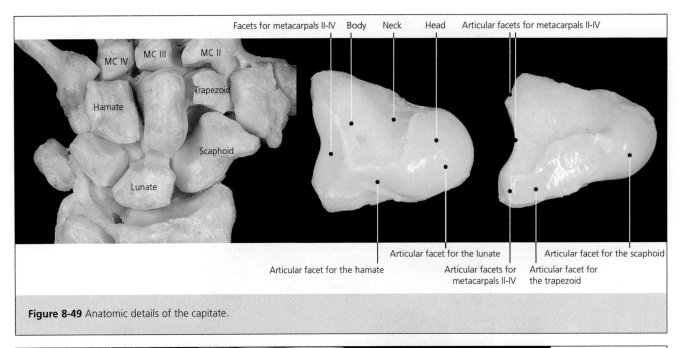

Figure 8-49 Anatomic details of the capitate.

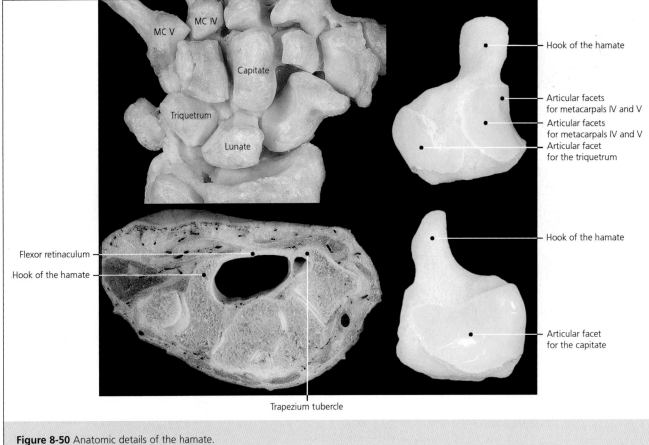

Figure 8-50 Anatomic details of the hamate.

and two epiphyses (Figure 8-52). The proximal epiphysis is called the base, and the distal metacarpal is called the head; between the base and the head is the body. Proximally, the head narrows slightly into the neck. The base articulates with the distal carpal row, and the head articulates with the base of the phalanx. When the hand is closed into a fist, the metacarpal heads are visible as the knuckles. On both sides of the head are small tubercles that provide attachment for the metacarpophalangeal collateral ligaments. Several

Osteology

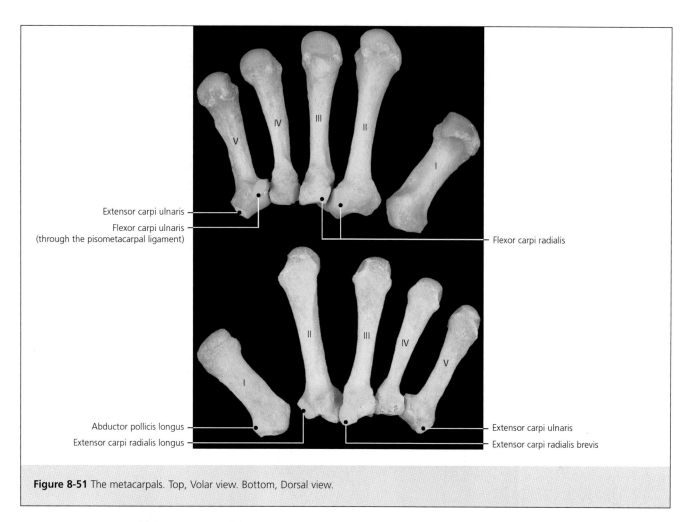

Figure 8-51 The metacarpals. Top, Volar view. Bottom, Dorsal view.

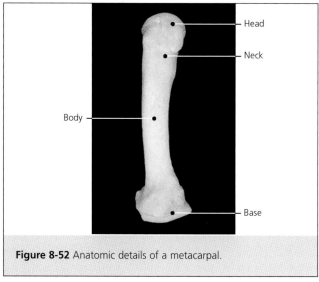

Figure 8-52 Anatomic details of a metacarpal.

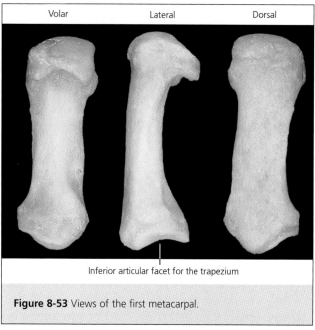

Figure 8-53 Views of the first metacarpal.

ligaments and muscles of the hand are located between the metacarpals.

The length of the metacarpals, in decreasing order, is as follows: second, third, fourth, fifth, first. Clinically, the third metacarpal is the most prominent. This is because of its relationship with the distal carpal row, as the second metacarpal-trapezoid articulation is more proximal than the third metacarpal-hamate articulation.

The first metacarpal (Figure 8-53) articulates with

American Academy of Orthopaedic Surgeons

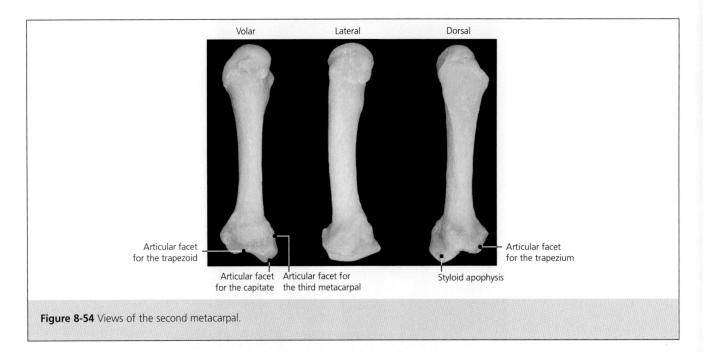

Figure 8-54 Views of the second metacarpal.

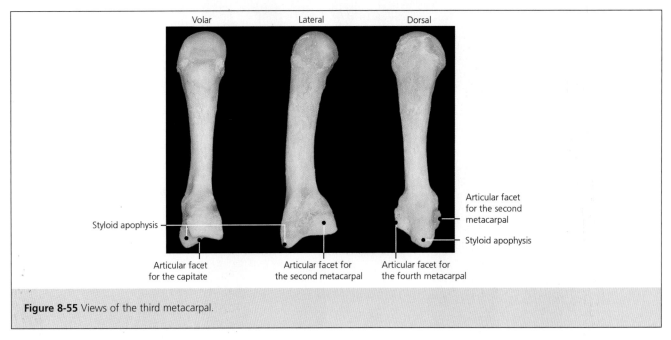

Figure 8-55 Views of the third metacarpal.

the trapezium in a saddle-type joint; its inferior articular surface is concave, and it has no ulnar articular facets. Several tendons from the forearm muscles and the thenar eminence attach to this bone. The first metacarpal is stronger than the others and has a round cross section, whereas the other metacarpals have a triangular cross section, with a dorsal face and two volar faces (anterolateral and anteromedial).

The second metacarpal (Figure 8-54) has three inferior facets, which articulate with the first three bones of the distal carpal row—trapezium, trapezoid, and capitate—and an ulnar facet that articulates with the third metacarpal. The second metacarapal also has a small styloid apophysis for the palmar insertion of the flexor carpi radialis.

The third metacarpal (Figure 8-55) is located at the middle axis of the hand. A styloid apophysis is present on the dorsal aspect for insertion of the extensor carpi radialis brevis. The third metacarpal has two ulnar facets, which articulate with the second and fourth metacarpals, and one proximal articular facet, which articulates with the capitate.

The fourth metacarpal (Figure 8-56) has no styloid process. It has two ulnar facets that articulate with the third and fifth metacarpals and two proximal articular facets for the capitate and the hamate.

Osteology

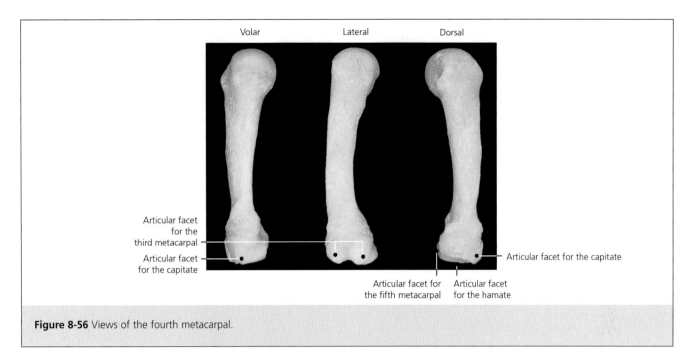

Figure 8-56 Views of the fourth metacarpal.

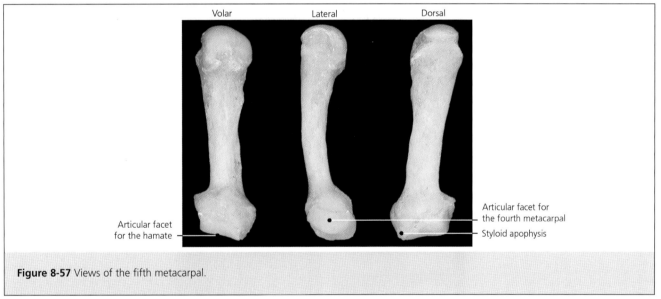

Figure 8-57 Views of the fifth metacarpal.

The fifth metacarpal (Figure 8-57) has only one ulnar articular facet, for the fourth metacarpal, and one proximal facet, for the hamate. It has a small posteromedial styloid apophysis for insertion of the extensor carpi ulnaris.

Phalanges

The digits are named and numbered from radial to ulnar (Figure 8-58): thumb (I), index (II), long (III), ring (IV), and little, or small (V). The phalanges are long bones (Figure 8-59), with a proximal end, or base, and a distal end, or head, which has a trochlea. The base and head are connected by the phalangeal body.

Each finger has three phalanges (Figures 8-60 and 8-61); the thumb has only two. The phalanges, which resemble columns of decreasing height, are called the proximal phalanx, the middle phalanx, and the distal phalanx. Like the metacarpals, the phalanges are numbered from radial to ulnar, starting with the thumb phalanx.

The base of the proximal phalanx articulates with the corresponding metacarpal head, and its head articulates with the base of the middle phalanx. On each side of the head are small tubercles for insertion of the ulnar collateral ligaments of the proximal interphalangeal joint. Palmarly, on the ulnar aspects of the diaphyses of the first and the second phalanges (P1 and P2), are prominent crests at the phalangeal borders. These crests provide insertion for the fibrous digital sheaths.

Scapular Girdle and Upper Limb

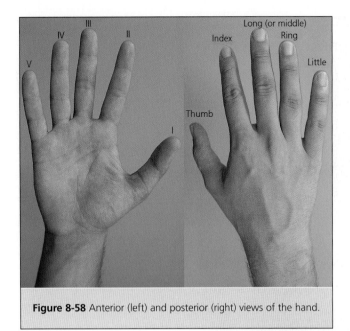

Figure 8-58 Anterior (left) and posterior (right) views of the hand.

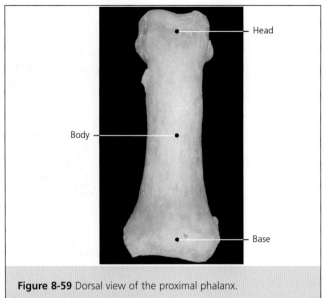

Figure 8-59 Dorsal view of the proximal phalanx.

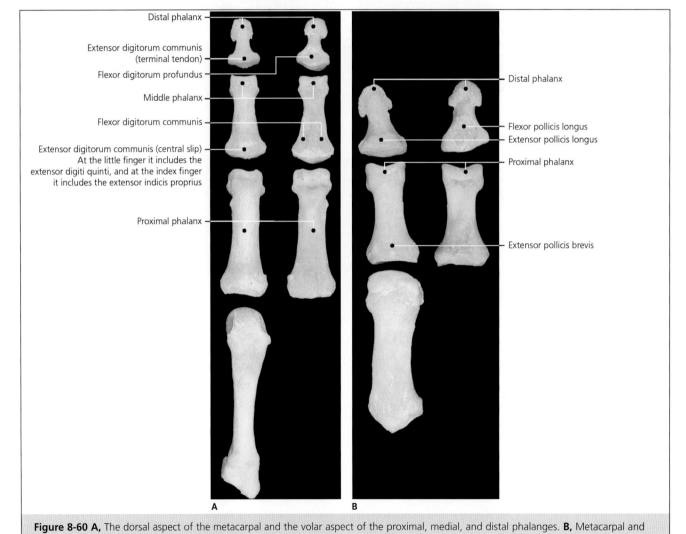

Figure 8-60 A, The dorsal aspect of the metacarpal and the volar aspect of the proximal, medial, and distal phalanges. **B,** Metacarpal and phalanges of the thumb (size not proportional).

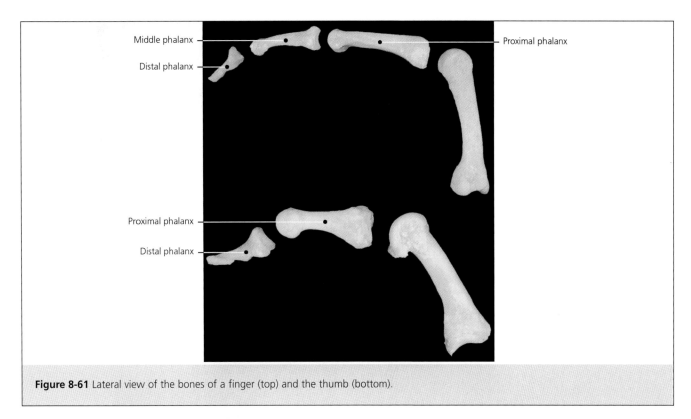

Figure 8-61 Lateral view of the bones of a finger (top) and the thumb (bottom).

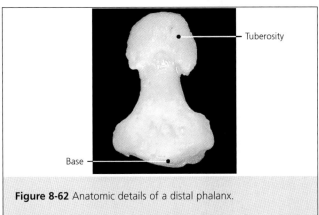

Figure 8-62 Anatomic details of a distal phalanx.

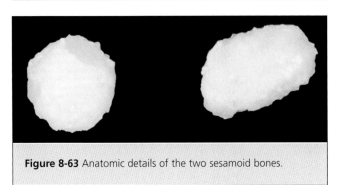

Figure 8-63 Anatomic details of the two sesamoid bones.

The base of the middle phalanx articulates with the head of the proximal phalanx and provides insertion for the tendon of the flexor digitorum communis (on fingers II to V). The head of the middle phalanx articulates with the base of the distal phalanx. This head also has two small tubercles for insertion of the ulnar collateral ligaments for the distal interphalangeal joint.

The tendon of the flexor digitorum profundus attaches to the base of the distal phalanx (Figure 8-62). The head of the distal phalanx is characterized by the distal phalangeal tuberosity. The distal phalanx of the thumb provides insertion for the flexor pollicis longus.

Finally, several sesamoid bones are found in the hand (Figure 8-63). A pair of sesamoids lies close to the first metacarpal head, and in about 80% of individuals another sesamoid is present at the level of the head of the fifth metacarpal. Occasionally, sesamoids are found at the level of other metacarpophalangeal or even interphalangeal joints.

The ulnar sesamoid of the first metacarpophalangeal joint provides insertion for the abductor pollicis and the flexor pollicis brevis. The radial sesamoid provides insertion for the adductor pollicis brevis.

Figures 8-64 and 8-65 are AP radiographs of the carpal region and hand, respectively.

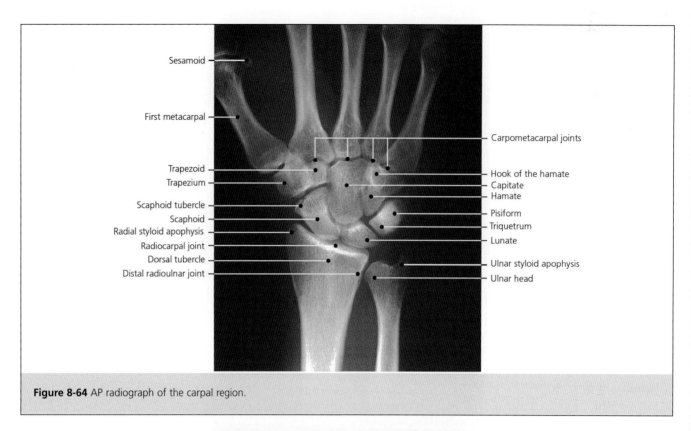

Figure 8-64 AP radiograph of the carpal region.

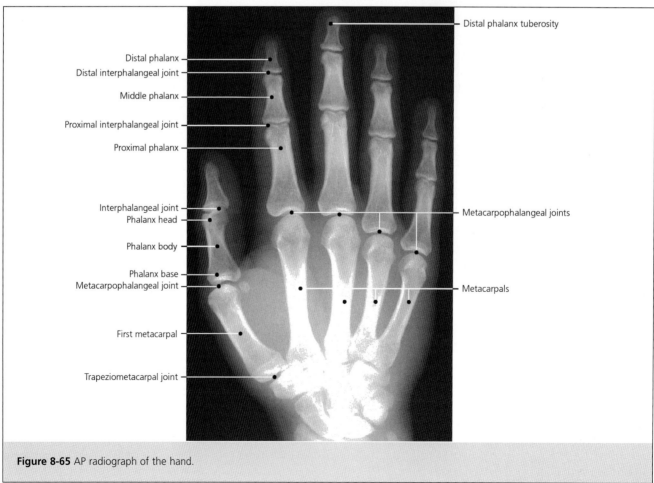

Figure 8-65 AP radiograph of the hand.

Chapter 9

Arthrology

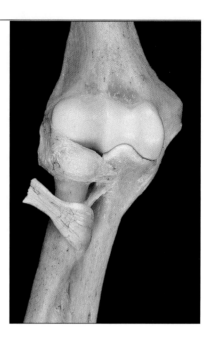

Shoulder

This section describes the sternoclavicular, acromioclavicular, glenohumeral, and scapulothoracic joints.

A few of the scapular ligaments deserve special mention. First, the superior transverse scapular ligament (Figure 9-1) closes the scapular notch, creating a hole that provides access for the suprascapular nerve; the suprascapular artery and vein cross over the top of the ligament. This neurovascular bundle courses from the supraspinatus to the infraspinatus fossa through the spinoglenoid notch. This notch is limited dorsally by an arch formed by the inferior transverse scapular ligament (Figure 9-2), located between the spine of the scapula and the glenoid cavity.

The coracoacromial ligament (Figures 9-3 and 9-4), as indicated by its name, is attached to the lateral aspect of the coracoid process and the acromion. The coracoacromial ligament commonly bifurcates, with the medial portion extending into the acromioclavicular joint. In addition, this ligament forms the roof of the subacromial space, where the subacromial bursa and the supraspinatus tendon are located.

Sternoclavicular Joint

The sternoclavicular joint includes the entire sternocostoclavicular complex (Figure 9-5). The sternal end of the clavicle has a sternal clavicular facet that extends inferiorly to articulate with the first rib. On the sternum is a clavicular notch for the clavicle and, nearby, a costal notch for the costal cartilage of the first rib. The sternoclavicular joint is a saddle joint. The articular surfaces are separated by an articular disk fixed superiorly to the clavicle and inferiorly to the sternum and first rib. This disk allows increased conformity between the convex clavicular facet and the slightly concave sternal facet. Because of this disk, some authors distinguish two synovial cavities, the meniscosternal cavity and the meniscoclavicular cavity; the latter is much more mobile. These two cavities sometimes communicate through a hole in the center of the disk.

The articular capsule is reinforced with the anterior and posterior sternoclavicular ligaments (Figure 9-6). An additional ligament, the so-called interclavicular ligament (Figure 9-7), reinforces the superior aspect of the joint, extending from one clavicle to the opposite over the jugular notch of the sternum. This ligament has short fibers, from the clavicle to the upper portion of the sternal manubrium, and long fibers, which join the clavicles.

The costoclavicular ligament is attached to the inferior face of the sternal end of the clavicle (Figure 9-8) and connects the clavicle to the chondral region of the first rib. It has no direct relationship with the articular capsule.

The movements of the shoulder joint are transmitted through this sternocostoclavicular complex, which has three degrees of freedom due to the articular disk. Although these movements are complex, they can be summarized as elevation-descent, forward flexion, extension, and internal/external rotation.

Scapular Girdle and Upper Limb

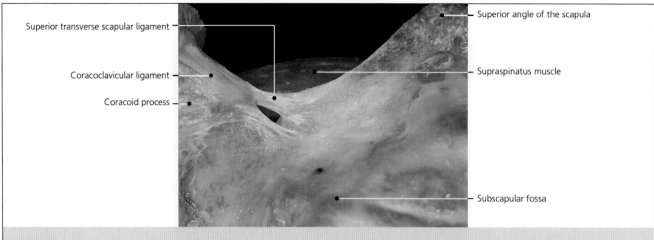

Figure 9-1 Detail of an anterior view of the superior transverse scapular ligament.

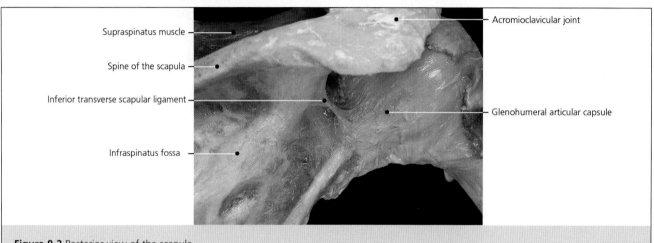

Figure 9-2 Posterior view of the scapula.

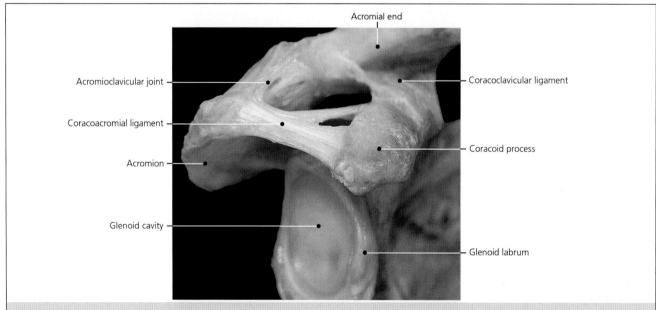

Figure 9-3 Oblique view of the coracoacromial ligament, showing an oblique fascicle from the coracoid tip and a second one from the neck. Some fibers reach the acromioclavicular joint.

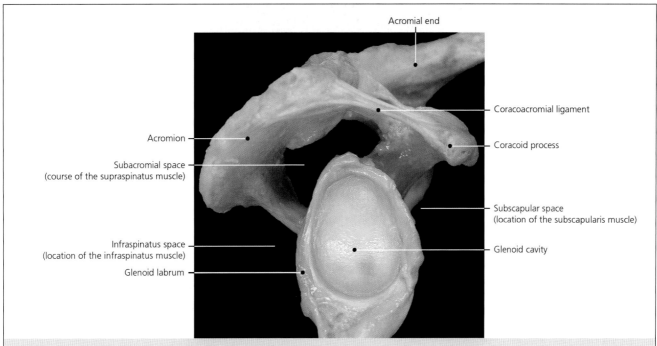

Figure 9-4 Lateral view of the scapula showing the roof formed by the acromion and the coracoacromial ligament. This roof, along with the glenoid cavity, delineates the subacromial space.

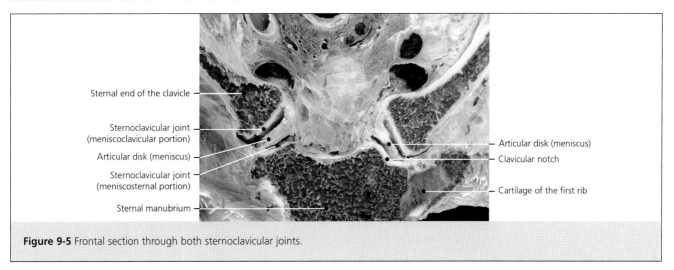

Figure 9-5 Frontal section through both sternoclavicular joints.

Acromioclavicular Joint

The acromioclavicular joint is a flat joint formed by the clavicular articular facet for the acromion and the acromial articular facet for the clavicle (Figure 9-9). An articular meniscus lies between the two articular surfaces (Figure 9-10).

The articular capsule is reinforced by two acromioclavicular ligaments (Figure 9-11), the superior acromioclavicular ligament and the less dense inferior acromioclavicular ligament, which lies under the superior ligament.

The coracoclavicular ligament also reinforces this joint at a distance (Figure 9-12). It connects the superior aspect of the coracoid process to the inferior face of the acromial end of the clavicle. Two fascicles are present: the conoid ligament, which is located posteromedially, is triangular, and extends from the coracoid process to the conoid tubercle of the clavicle; and the trapezoid ligament, located anterolaterally, which extends from the inner edge of the coracoid process angle to the trapezoid line in the clavicle. A serous bursa lies between the two ligaments. These two ligaments transmit weight and forces from the free end of the upper limb to the clavicle and the thorax.

Glenohumeral Joint

The glenohumeral joint is a spherical articulation that joins the humerus with the scapula. The articulation is formed by the hemispherical portion of the humeral head covered by articular cartilage and by

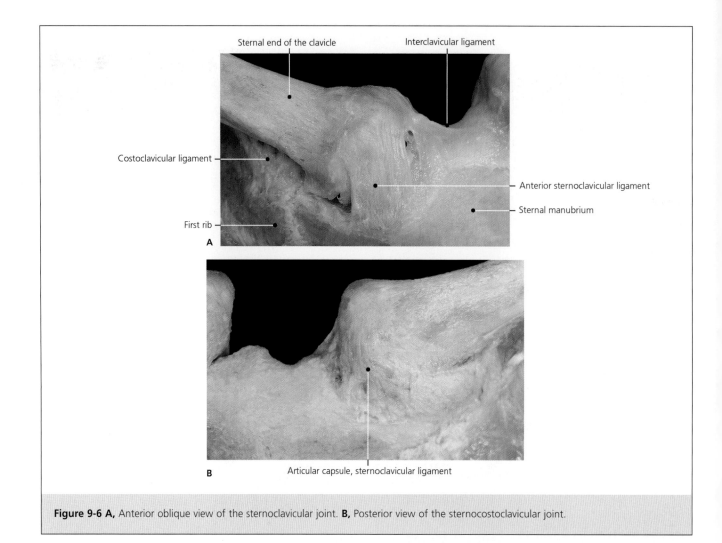

Figure 9-6 A, Anterior oblique view of the sternoclavicular joint. **B,** Posterior view of the sternocostoclavicular joint.

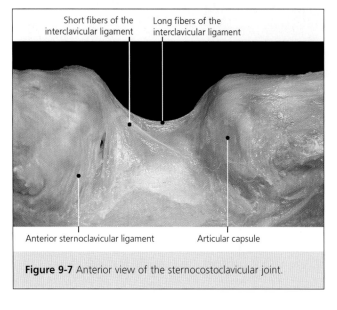

Figure 9-7 Anterior view of the sternocostoclavicular joint.

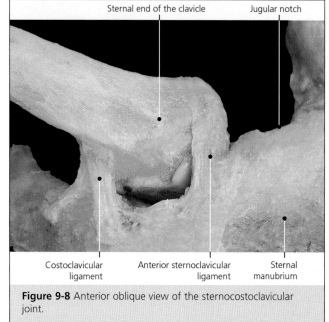

Figure 9-8 Anterior oblique view of the sternocostoclavicular joint.

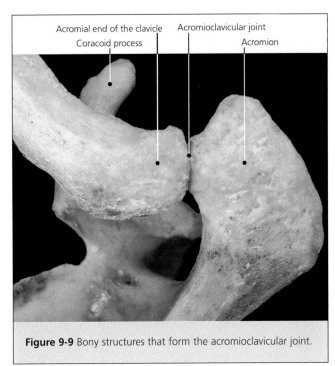

Figure 9-9 Bony structures that form the acromioclavicular joint.

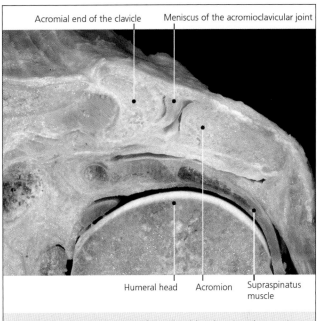

Figure 9-10 Frontal section of the shoulder showing the meniscus of the acromioclavicular joint.

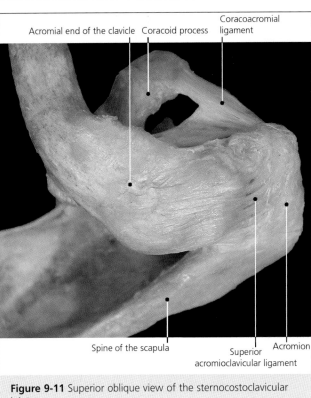

Figure 9-11 Superior oblique view of the sternocostoclavicular joint.

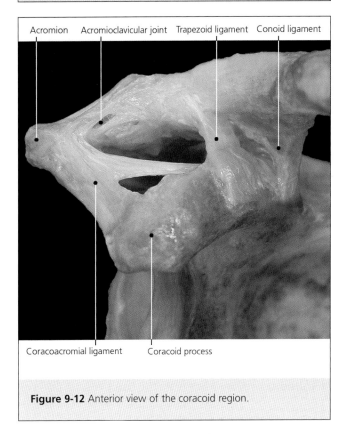

Figure 9-12 Anterior view of the coracoid region.

the glenoid cavity of the scapula (Figure 9-13). The humeral head is clearly larger than the glenoid, which results in some degree of incongruity and instability. A funnel-shaped fibrocartilage called the labrum enlarges the articular cavity, reducing the incongruity (Figure 9-14). In addition, the synovial fluid has a "suction effect," which adheres the humeral head to the glenoid cavity.

The articular capsule attaches to the periphery of the glenoid labrum, the scapular neck, and the anatomic neck sulcus (Figure 9-15). The tendon of the long head of the biceps perforates the gleno-

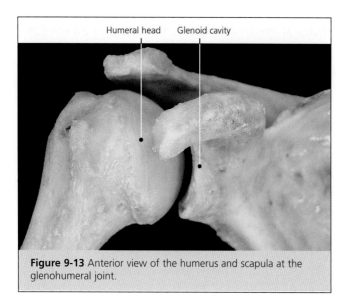

Figure 9-13 Anterior view of the humerus and scapula at the glenohumeral joint.

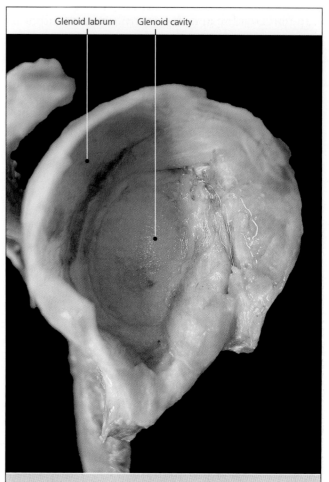

Figure 9-14 Lateral view of the glenoid cavity and its labrum.

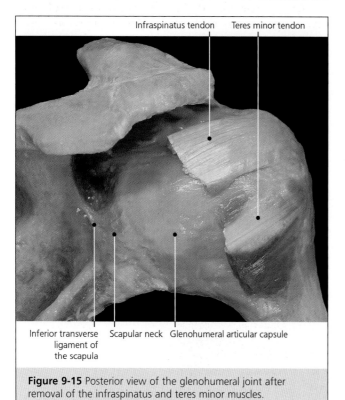

Figure 9-15 Posterior view of the glenohumeral joint after removal of the infraspinatus and teres minor muscles.

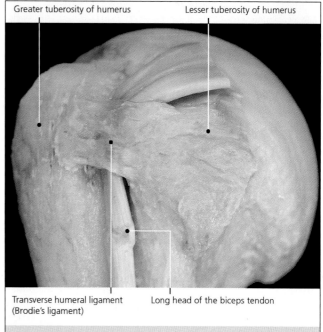

Figure 9-16 Anatomic details of the proximal humeral epiphysis.

humeral capsule but remains outside the synovial membrane. On the humerus, the transverse humeral ligament (Brodie's ligament) (Figure 9-16) connects the two tuberosities as an extension of the fibrous capsule and the superior glenohumeral ligament. The transverse humeral ligament closes the intertubercular sulcus to form a conduit for the passage of the long head of the biceps and its surrounding synovial sheath. The articular capsule is connected to several periarticular synovial sheaths and bursae (Figure 9-17), including the subtendinous bursa for the subscapularis tendon and the intertubercular tendinous sheath.

The capsule is reinforced anteriorly by the glenohumeral ligaments, which can be observed only from the inside of the capsule (Figure 9-18). The superior glenohumeral ligament originates between the external aspect of the glenoid labrum and the part of the glenoid extending above the notch and attaches into the superior region of the lesser tuberosity. It is separated from the coracohumeral ligament through a small space. The middle glenohumeral ligament originates at the same location but attaches on the inferior aspect of the lesser tuberosity. There is a capsular triangle between these two ligaments where the capsule is not reinforced, with a small oval opening—Weitbrecht's foramen—which communicates with the subtendinous bursa of the subscapularis tendon. The inferior glenohumeral ligament is the longest and strongest of the three. It originates at the anteroinferior region of the glenoid cavity, below the notch, as well as at the scapular neck. It extends obliquely to the anteroinferior aspect of the surgical neck of the humerus.

The glenohumeral joint is reinforced superiorly by the coracohumeral ligament (Figure 9-19), which extends from the base of the coracoid to the upper portion of the humeral greater tuberosity. There is a deeper fascicle known as the coracoglenoid ligament. Llorca has speculated that the coracohumeral ligament might be a remnant of the pectoralis minor muscle, which may have been connected to the humerus in the past. The inferior region is devoid of reinforcements and forms the axillary recess of the articular capsule, which facilitates the range of shoulder motion.

The shoulder is the most mobile joint of the human body and also is stable because the ligaments described are somewhat lax and do not limit motion excessively. Thus, the deep scapular muscles both move the shoulder and provide stability due to resting muscle tone. This is why the rotator cuff muscle group has strong attachments in the capsule, providing both mobility and stability (see chapter 10).

Scapulothoracic Joint

The scapulothoracic joint is not a joint in the classic sense because it does not link two bones. The costal

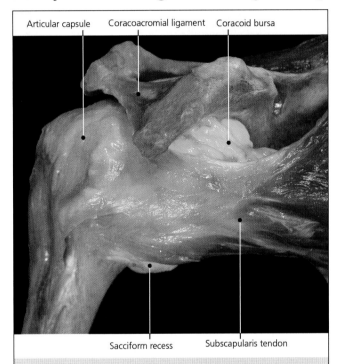

Figure 9-17 Articular capsule of the shoulder joint and anterior bursae. The bursae are injected with green latex.

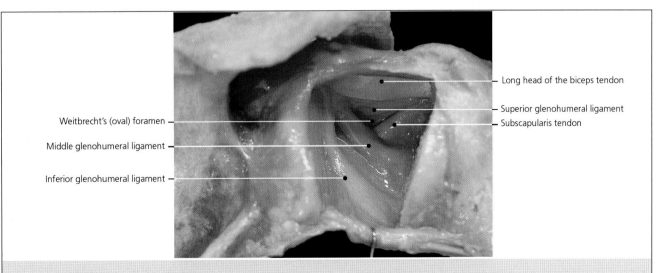

Figure 9-18 Posterior view of the glenohumeral (anterior) articular capsule after opening the joint and removing the humeral head.

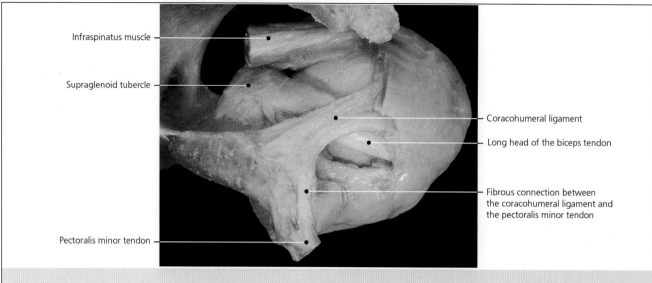

Figure 9-19 Superior view of the glenohumeral joint with the articular capsule open to show the long head of the biceps tendon.

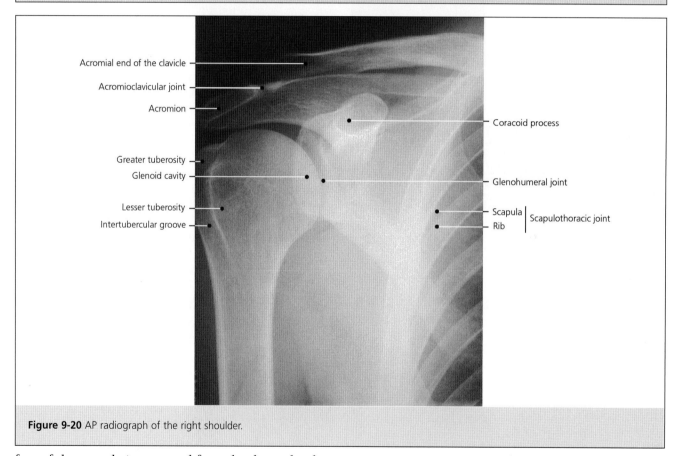

Figure 9-20 AP radiograph of the right shoulder.

face of the scapula is separated from the thorax by the serratus anterior and the subscapularis, and it glides on the thoracic wall similar to a diarthrodial joint. Some authors use the term syndesmopexy to refer to this joint.

Scapular movements are integral to shoulder motion. For example, 60° of the full 180° of abduction in the scapular plane is due to scapular rotation.

Figure 9-20 is an AP radiographic view of a right shoulder, showing the acromioclavicular, glenohumeral, and scapulothoracic joints.

Elbow

The elbow includes three joints inside the same capsule (Figure 9-21)—the ulnohumeral joint, the radiohumeral joint, and the proximal radioulnar joint.

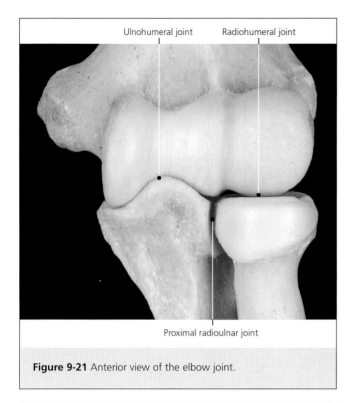

Figure 9-21 Anterior view of the elbow joint.

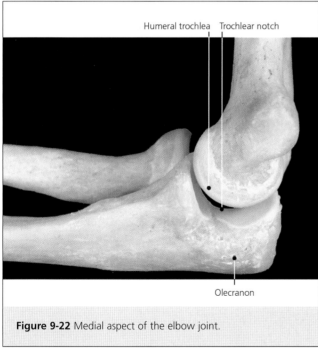

Figure 9-22 Medial aspect of the elbow joint.

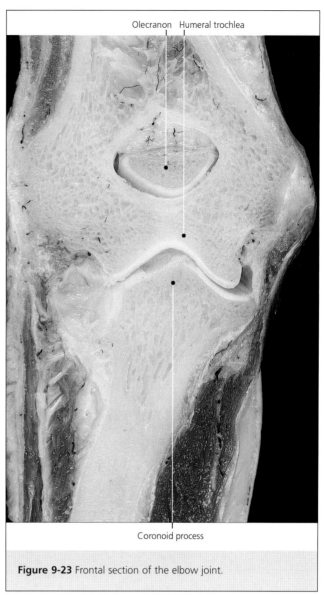

Figure 9-23 Frontal section of the elbow joint.

Ulnohumeral Joint

The ulnohumeral joint is a trochlear type of articulation, linking the greater sigmoid notch of the ulna to the humeral trochlea (Figure 9-22). With elbow flexion, the coronoid process, which is located anterior to this sagittal crest, enters the coronoid fossa. With elbow extension, the beak of the olecranon enters the olecranon fossa, limiting elbow extension (Figure 9-23). This joint is shown in different positions in Figures 9-24 through 9-26.

Radiohumeral Joint

The two articular surfaces of this joint are the hemispherical humeral capitellum and the articular fossa of the radial head (Figures 9-27 through 9-29). Although this joint is anatomically an enarthrosis, it is functionally linked to ulnar motion through ligamentous connections. During flexion and extension, the articular fossa of the radial head articulates with the capitellar hemisphere, whereas the articular circumference of the radius slides along the humeral capitellotrochlear sulcus until it achieves elbow flexion in the radial fossa of the humerus.

Proximal Radioulnar Joint

The articular circumference of the capitellum adapts and rotates on the concavity of the lesser sigmoid

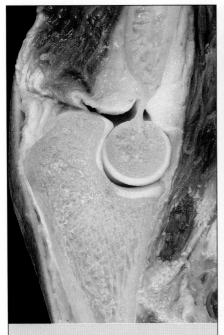

Figure 9-24 Elbow in extension.

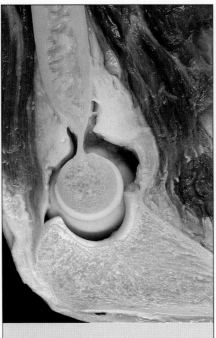

Figure 9-25 Elbow in flexion (80°).

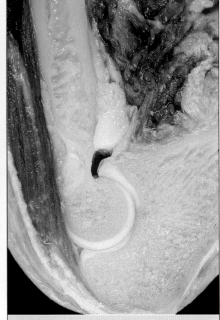

Figure 9-26 Elbow in flexion (120°).

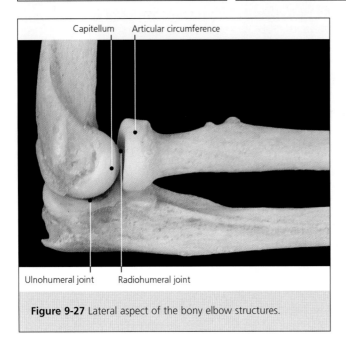

Figure 9-27 Lateral aspect of the bony elbow structures.

notch of the ulna; it forms a trochoid joint that cannot be dissociated from the distal radioulnar joint (Figures 9-30 and 9-31). These rotational movements, known as pronation and supination, can be performed throughout the whole arc of elbow flexion and extension.

The proximal radioulnar joint is stabilized by two ligaments. The annular ligament of the radius (Figures 9-32 and 9-33) attaches on the anterior and posterior aspects of the lesser sigmoid notch, around the neck and head of the articular surface of the radius. The internal face of the annular ligament is covered by a small layer of hyaline cartilage, which facilitates the rotation of the radius inside this fibrous ring. On the other hand, the quadrate ligament (Denucé's ligament) (Figures 9-34 and 9-35) is a thin fibrous band that connects the inferior edge of the lesser sigmoid notch with the radial neck, with expansions extending to the menisci located on both sides of the lesser sigmoid notch of the ulna. When the radius rotates in pronation and supination, this ligament becomes taut, limiting these movements.

Capsule and Ligaments

A fibrous articular capsule covers the elbow joint (Figure 9-36). Anteriorly, the capsule originates above the radial and coronoid fossae and attaches to the annular ligament, the anterior aspect of the coronoid process, and the two ulnar notches. The epicondyles are extracapsular. The capsule surrounds the joint, attaching on the posterior aspect above the olecranon fossa (Figure 9-37).

The elbow capsule is reinforced by a relatively thin anterior ligament that is attached proximally on the anterior humeral fossae, the anterior aspect of the medial epicondyle, and the external aspect of the lateral epicondyle. From this attachment, the fibers coalesce in a triangle with its vertex or distal insertion located on the anterior aspect of the coronoid process and the proximal radioulnar joint line. Some authors have described two small ligamentous fascicles—the medial oblique and lateral oblique—originating from the medial and lateral epicondyles, respectively.

Arthrology

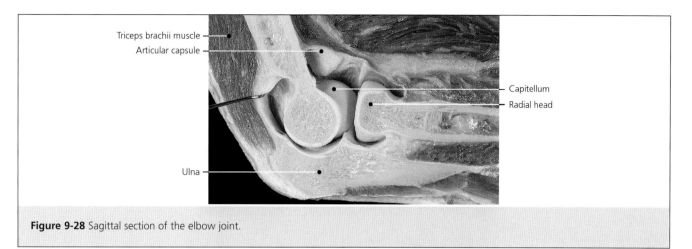

Figure 9-28 Sagittal section of the elbow joint.

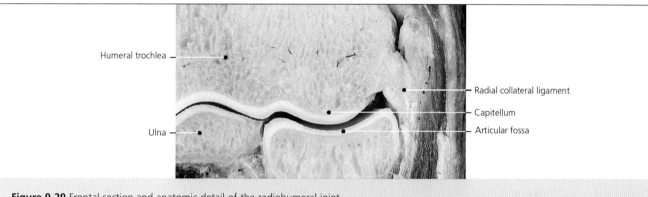

Figure 9-29 Frontal section and anatomic detail of the radiohumeral joint.

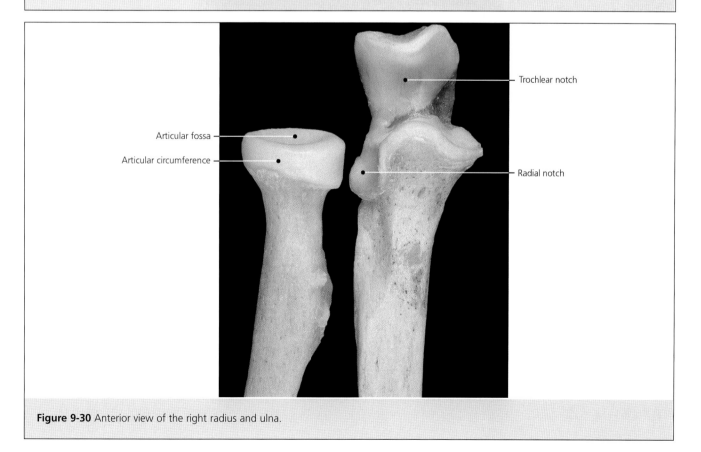

Figure 9-30 Anterior view of the right radius and ulna.

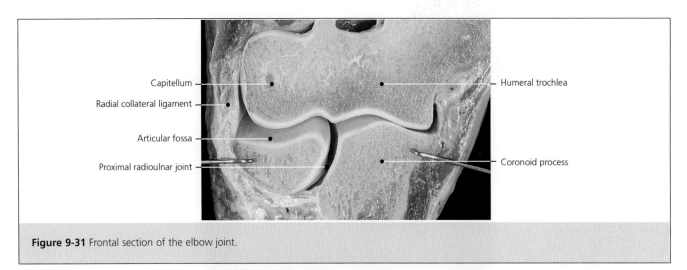

Figure 9-31 Frontal section of the elbow joint.

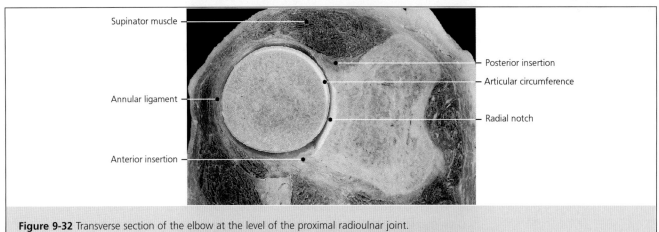

Figure 9-32 Transverse section of the elbow at the level of the proximal radioulnar joint.

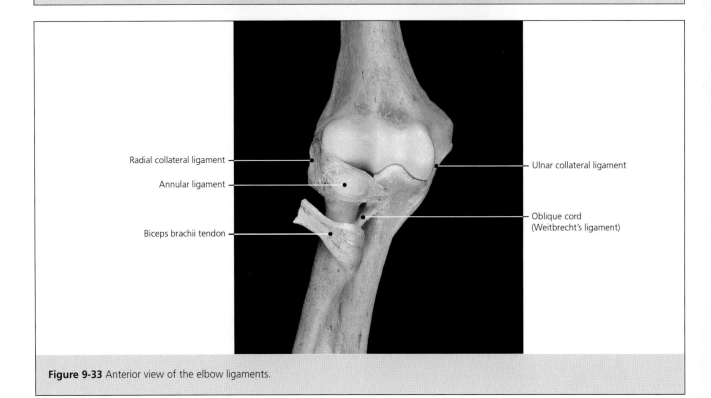

Figure 9-33 Anterior view of the elbow ligaments.

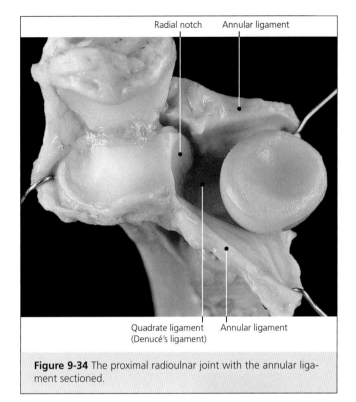

Figure 9-34 The proximal radioulnar joint with the annular ligament sectioned.

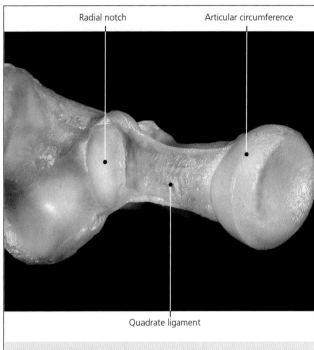

Figure 9-35 Superior aspect of the radius and ulna separated to place the quadrate ligament under tension.

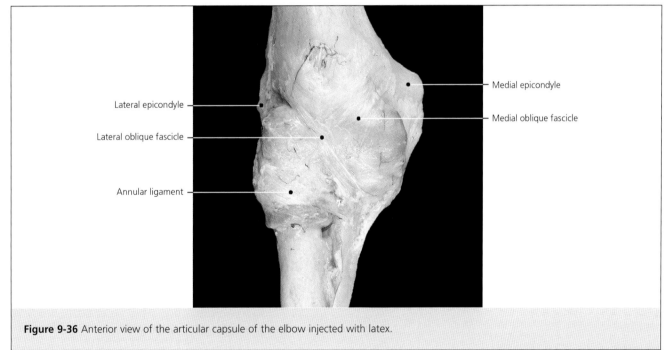

Figure 9-36 Anterior view of the articular capsule of the elbow injected with latex.

Posteriorly, the capsule is reinforced with several fascicles that connect the olecranon and humerus (Figure 9-38), directed obliquely from the vicinity of the olecranon fossa to the olecranon tip. Above these are the humerohumeral fibers, which cross the olecranon fossa transversely.

The medial side of the elbow is reinforced by the ulnar collateral ligament, which includes three fascicles (Figure 9-39). The anterior fascicle extends from the anteromedial aspect of the medial epicondyle to the coronoid process, with some fibers reinforcing the annular ligament; the middle fascicle, which is quite resistant, originates at the inferior aspect of the medial epicondyle and attaches on the medial aspect of the coronoid process and the medial ulna. These two fascicles coalesce into a fan-shaped single ante-

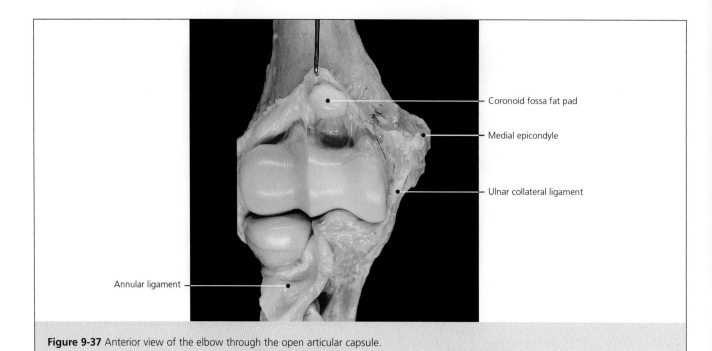

Figure 9-37 Anterior view of the elbow through the open articular capsule.

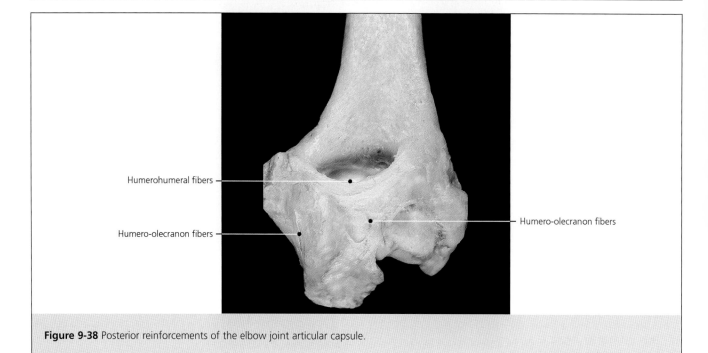

Figure 9-38 Posterior reinforcements of the elbow joint articular capsule.

rior band. The posterior fascicle (Bardinet's ligament) is also fan-shaped, originates on the posteroinferior aspect of the medial epicondyle, and attaches on the medial aspect of the olecranon. In addition, some transverse fascicle (the transverse ligament or oblique medial ligament [Cooper's ligament]) extend from the base of the olecranon to the base of the coronoid process. The ulnar nerve is in close proximity to the ulnar collateral ligament, where it enters the space between the two heads of the flexor carpi ulnaris at the olecranon-epitrochlear tunnel, or cubital tunnel.

The lateral reinforcement is formed by the lateral collateral ligament (Figure 9-40). The lateral collateral ligament is fan-shaped and usually is divided into three fascicles: an anterior fascicle extends from the anteroinferior aspect of the lateral epicondyle and over the radial head as a reinforcement of the annular ligament; a middle fascicle extends from the inferior aspect of the lat-

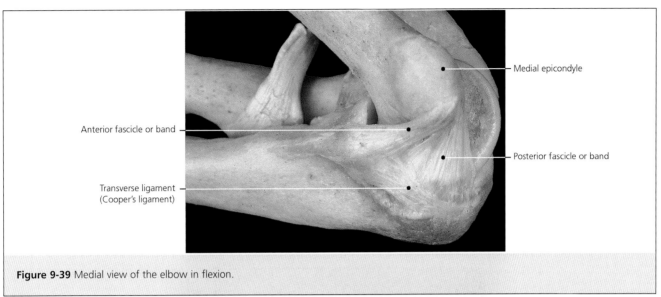

Figure 9-39 Medial view of the elbow in flexion.

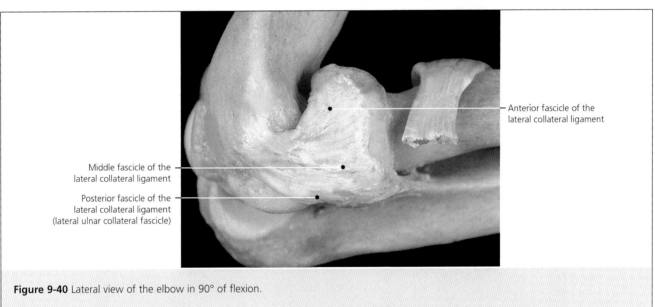

Figure 9-40 Lateral view of the elbow in 90° of flexion.

eral epicondyle and to the lesser sigmoid notch of the ulna, reinforcing the posterior aspect of the annular ligament; and the lateral ulnar collateral fascicle, which extends from the posterior aspect of the lateral epicondyle to a tubercle on the lateral aspect of the olecranon.

The humeroradial meniscus, located between the humerus and radius (Figures 9-41 and 9-42), is a synovial fold shaped like a half moon, with its internal concavity extending to the lateral aspect of the trochlea.

The vascularity of the elbow joint is formed from an arterial network of anastomoses of the collateral arteries of the arm and the radial and ulnar recurrent arteries.

Figures 9-43 and 9-44 are imaging studies of the elbow with the major bony features of the joint labeled.

RADIOULNAR SYNDESMOSIS

The radius and ulna are connected at their midportions through the forearm interosseous membrane (Figure 9-45), which forms a fibrous joint that covers the interosseous space between these two bones. This membrane is formed by a network of fibrous fascicles obliquely oriented from proximal and lateral to distal and medial on the proximal half and in the opposite direction on the distal half. All these fascicles attach to the interosseous edges of the radius and the ulna.

The forearm interosseous membrane has several functions. It increases the surface of the forearm for flexor and extensor muscle attachments. It also limits supination, a role shared with the quadrate ligament and the contact of several musculoskeletal structures in the forearm. Finally, some authors believe

Figure 9-41 Sagittal section of the elbow at the level of the radiohumeral joint.

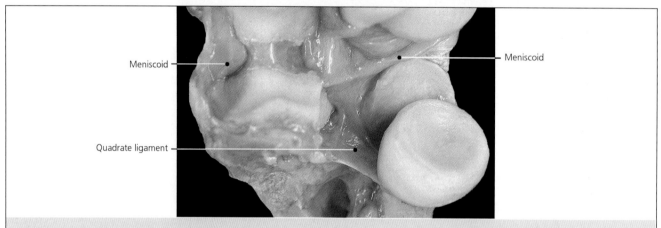

Figure 9-42 The articular capsule of the elbow joint, opened to show the elbow menisci.

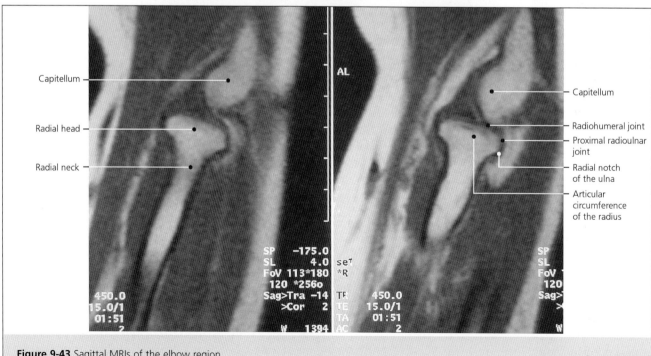

Figure 9-43 Sagittal MRIs of the elbow region.

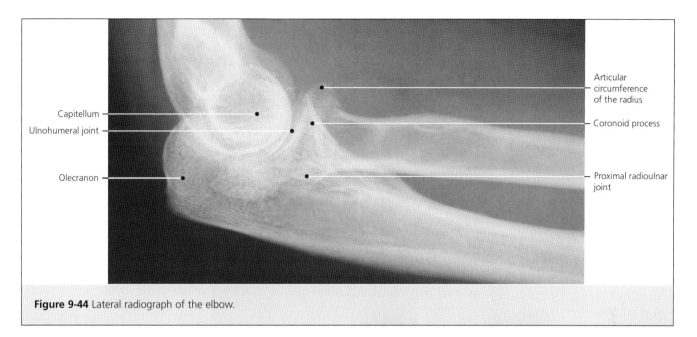

Figure 9-44 Lateral radiograph of the elbow.

it plays some role in load transmission between the radius and the ulna to avoid longitudinal displacements, although individuals without an interosseous membrane do not seem to have marked instability.

On the proximal aspect of the interosseous space is a fibrous band known as the oblique cord, or Weitbrecht's ligament (Figure 9-33). It originates at the coronoid process and tuberosity and is obliquely oriented distal to its attachment close to the radial tuberosity. Most of these ligament fibers are opposed to the proximal fibers of the interosseous membrane. Although this ligament is found consistently, its strength varies widely. Its role is not completely clear, and some authors believe that this ligament is a remnant of a muscular fascicle of the flexor pollicis longus (Gantzer's muscle).

In addition to the structures described above, proximally, a gap exists between the interosseous membrane and the oblique cord through which the distal biceps tendon and the anterior interosseous vessels course as they become posterior. Distally, there is a second space, where the anterior interosseous artery becomes posterior underneath the pronator quadratus.

DISTAL RADIOULNAR JOINT

The distal radioulnar joint is a trochoid joint formed by the concavity of the ulnar notch of the radius and the articular circumference of the ulna, which rotates during pronation and supination (Figures 9-46 and 9-47). This joint is similar to the proximal radioulnar joint; both work together.

These structures are connected by several ligaments. Distally, there is an articular disk or triangular fibrocartilage located horizontally between the ulnar head and the first carpal row. As its name implies, the tri-

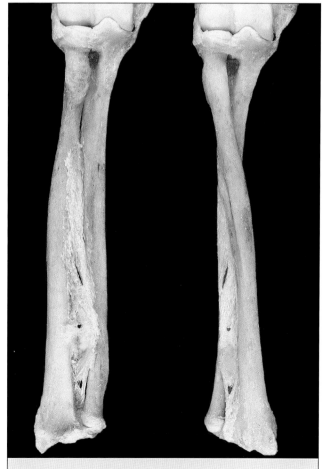

Figure 9-45 Left, forearm in supination. Right, forearm in pronation.

angular fibrocartilage is shaped like a triangle with its base attached at the inferior edge of the ulnar notch of the radius and its vertex at the base of the styloid

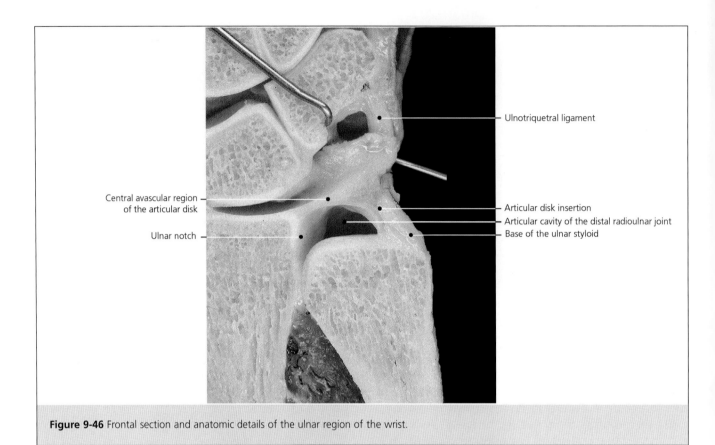

Figure 9-46 Frontal section and anatomic details of the ulnar region of the wrist.

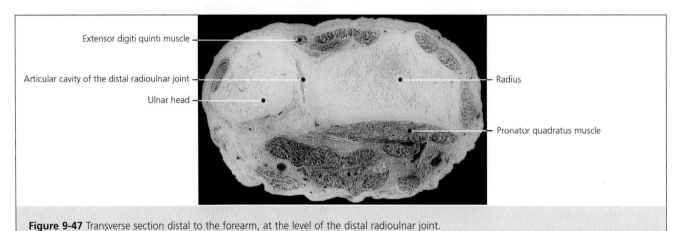

Figure 9-47 Transverse section distal to the forearm, at the level of the distal radioulnar joint.

process of the ulna. The joint fibrous capsule is attached to the superior edge of the ulnar notch, the articular circumference of the ulna, and the anterior and posterior edges of the radiocarpal ligaments. The capsule is somewhat lax because it must allow pronation and supination, which are facilitated by the presence of a proximal recess (the sacciform recess). The capsule is reinforced by the anterior and posterior radioulnar ligaments. The former is formed by some fascicle irregularly oriented on the anterior face of the fibrous capsule and extending from the anterior aspect of the ulnar notch to the head and styloid process of the ulna. The latter is similar and connects the same structures posteriorly. The tendon of the extensor digiti quinti is found on the fibrous posterior aspect of the distal radioulnar joint (fifth extensor compartment).

The synovial membrane of this joint attaches to the peripheral aspect of the articular surfaces mentioned above and to the radiocarpal ligaments. A hole connecting this joint with the radiocarpal joint is present in some wrists.

The distal radioulnar joint is supplied by the anterior and posterior interosseous arteries. The ulnar,

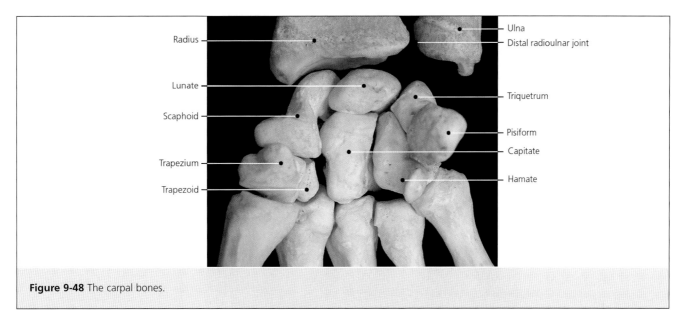

Figure 9-48 The carpal bones.

anterior interosseous, and posterior interosseous nerves innervate this joint.

RADIOCARPAL JOINT

The radiocarpal joint is formed by the two distal articular facets of the radius and the articular disk, as the ulna does not articulate directly with the carpus. These structures are combined into a proximally concave glenoid cavity that fits the convexity of the three articular facets for the bones of the first carpal row. The scaphoid and lunate articulate with their respective radial facets, whereas the triquetrum articulates with the most external aspect of the articular disk and the ulnar capsule and ligamentous structures (Figures 9-48 through 9-50). The radiocarpal joint is a condylar articulation with two axes of motion, an anteroposterior axis through the head of the capitate, which allows abduction and adduction, and a transverse axis through the radial styloid and the ulnar head, which allows flexion and extension.

Ligaments

The radiocarpal capsule (Figure 9-49) extends from the contour of the articular surface of the radius and the articular disk to the peripheral aspect of the articular facets of the scaphoid, the lunate, and the triquetrum. Anteriorly, this capsule is reinforced with several ligaments named according to their radial or ulnar insertions (Figures 9-51 through 9-54). The radial and ulnar ligaments form two Vs at the carpus—a smaller proximal one and a distal one. Some authors call this complex the anterior arcuate ligament. The ligaments attached to the radius, which form a continuous thickening of the articular capsule, are globally referred to as the palmar radiocarpal ligament. This ligament originates at the anterior edge of the articular facet of the radius and the styloid process; its fibers are obliquely oriented to the bones of the first carpal row, except for the radioscaphocapitate ligament.

The palmar radiocarpal ligament includes the following ligaments: the radioscaphocapitate ligament, the long radiolunate ligament, the radioscapholunate ligament, and the short radiolunate ligament.

Radioscaphocapitate Ligament

The most radially located ligament, it originates from the styloid process of the radius and the most lateral aspect of the anterior face of the radius. Its fibers are obliquely oriented in an ulnar direction to the scaphoid. Most fibers insert into this bone, but some continue and attached onto the capitate. These fibers interdigitate with fibers of the ulnocapitate ligament as they form the distal carpal V.

Long Radiolunate Ligament

This ligament is located on the medial side of the radioscaphocapitate ligament. It originates on the anterior aspect of the articular edge of the radius, and its fibers are obliquely directed in an ulnar direction to the lunate, on top of the intercarpal scapholunate ligament. This ligament finally attaches to the palmar face of the lunate. Most times it does not have a substantial extension into the triquetrum, although some authors have used the term radiotriquetral or radiolunatotriquetral ligament.

Radioscapholunate Ligament (of Testut and Kuenz)

This structure originates on the anterior aspect of a small crest located between the radial articular facets for the scaphoid and the lunate. Oriented in the sagittal plane, it is directed to the scapholunate articula-

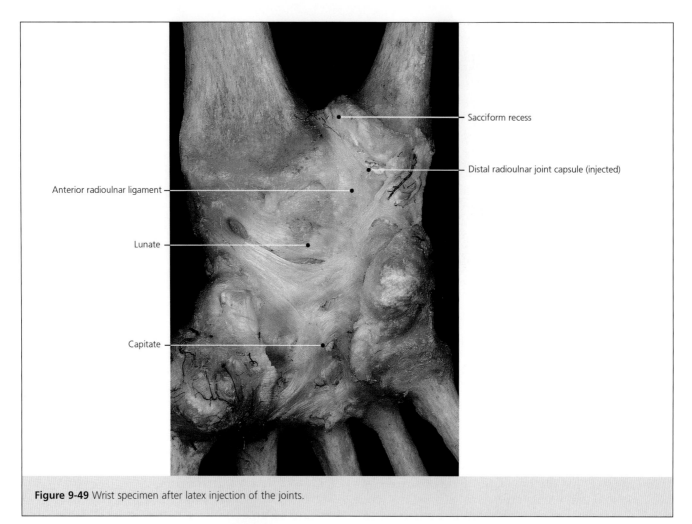

Figure 9-49 Wrist specimen after latex injection of the joints.

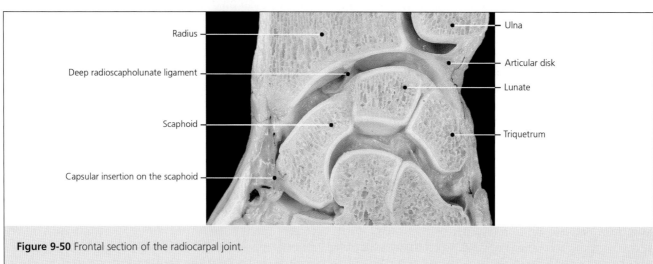

Figure 9-50 Frontal section of the radiocarpal joint.

tion, where it divides into one fascicle, attached deep into the anterior reinforcements, and a second fascicle, which reinforces the interosseous scapholunate ligament. Recent studies have shown this is a vascular structure with no true ligamentous fibers.

Short Radiolunate Ligament
Forming the anterior wall of the radiolunate joint, this ligament is shaped like a rectangle extending from the anterior edge of the articular region of the radius to the lunate. Its fibers interdigitate with the long radiolunate ligament.

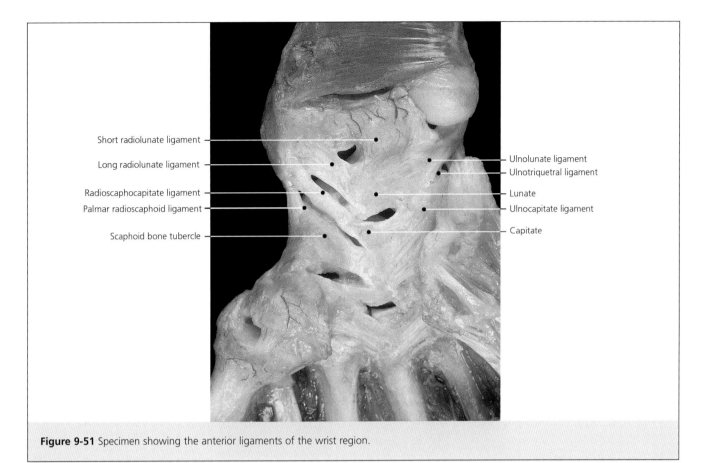

Figure 9-51 Specimen showing the anterior ligaments of the wrist region.

The ligaments originating on the ulnar side together are known as the palmar ulnocarpal ligament (Figure 9-51). These ligaments originate on the anterior aspect of the ulnar head, the styloid process, the articular disk, and the palmar radioulnar ligament. The palmar ulnocarpal ligament includes the following ligaments: the ulnolunate ligament, the ulnotriquetral ligament, and the ulnocapitate ligament.

Ulnolunate Ligament
Located medial to and in continuity with the short radiolunate ligament, this ligament originates on the ulnar head and the palmar radioulnar ligament and is obliquely oriented in a radial direction to its lunate attachment.

Ulnotriquetral Ligament
Located on the medial aspect of the former ligament, this ligament originates on the palmar radioulnar ligament and is directed to the proximal and palmar regions of the triquetrum. It usually has a foramen that communicates the radiocarpal joint with the joint between the pisiform and the triquetrum.

Ulnocapitate Ligament
This ligament originates at the base of the ulnar styloid and the palmar radioulnar ligament. Its fibers are located between the ulnolunate and ulnotriquetral ligaments, with some fiber attachments into the triquetrum and pisiform and into the intercarpal ligaments between pisiform/triquetrum and lunate/triquetrum. The ulnocapitate ligament finally reaches the capitate, interdigitating with fibers of the radioscaphocapitate ligament.

On the dorsal side, the ligaments include the following (Figure 9-53): the dorsal radiocarpal ligament, the dorsal radioscaphoid ligament, and the dorsal ulnocarpal ligament.

Dorsal Radiocarpal Ligament
This ligament originates on the posterior edge of the articular surface of the radius. The fibers run obliquely to the insertion of the ligament into the dorsal aspect of the lunate and triquetrum, although some fibers may reach the capitate.

Dorsal Radioscaphoid Ligament
This ligament is not very consistent, and, as indicated by its name, it connects the posterior aspect of the radioscaphoid articular region with the scaphoid bone.

Dorsal Ulnocarpal Ligament
This ligament originates on the ulnar styloid and articular disk and attaches into the posterior aspect of the triquetrum.

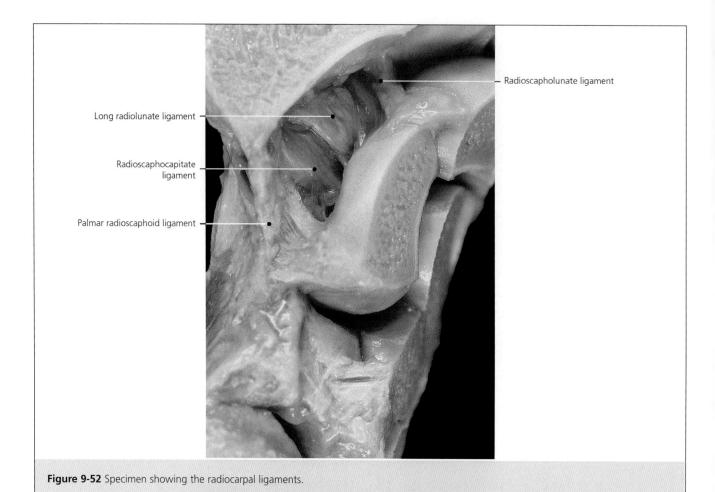

Figure 9-52 Specimen showing the radiocarpal ligaments.

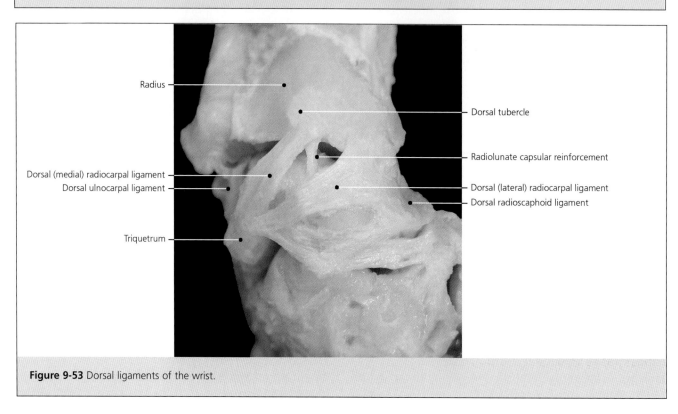

Figure 9-53 Dorsal ligaments of the wrist.

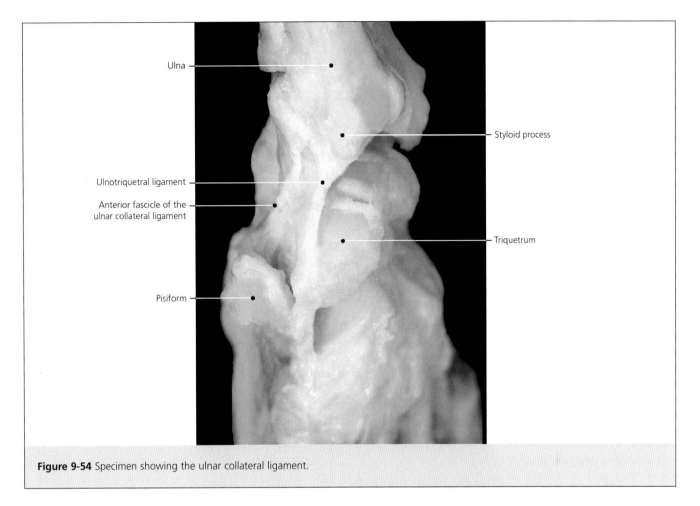

Figure 9-54 Specimen showing the ulnar collateral ligament.

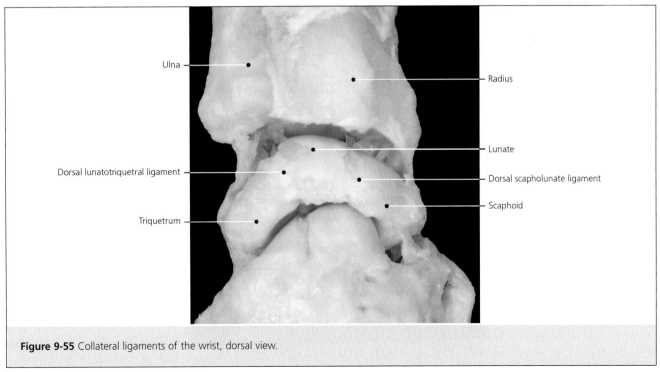

Figure 9-55 Collateral ligaments of the wrist, dorsal view.

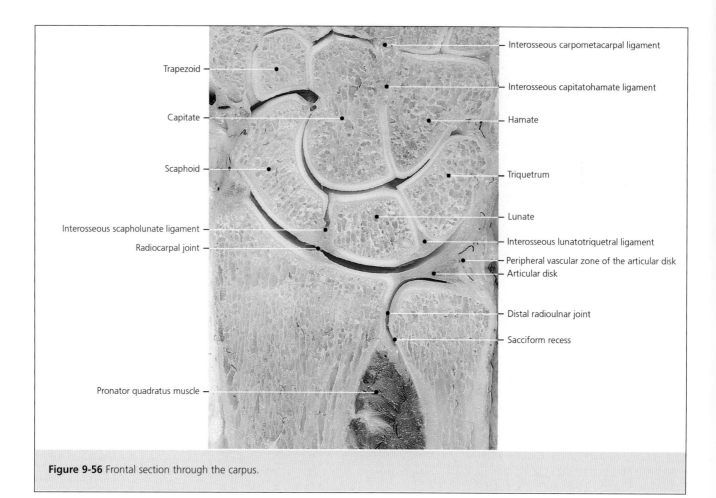

Figure 9-56 Frontal section through the carpus.

Two structures reinforce the lateral aspects of the joint: the carpal ulnar collateral ligament and the carpal radial collateral ligament.

Carpal Ulnar Collateral Ligament
This ligament (Figures 9-54 and 9-55) originates on the ulnar styloid and quickly divides into two fascicles. The anterior fascicle attaches to the pisiform, whereas the posterior attaches to the triquetrum. It also forms part of the floor of the sixth extensor compartment.

Carpal Radial Collateral Ligament
This ligament (Figure 9-55) originates on the radial styloid, and its two fascicles attach to the scaphoid tubercle. It forms part of the floor of the first extensor compartment, and it is intimately related to the radial artery in the anatomic snuffbox.

CARPAL JOINTS
The carpal joints include the intercarpal joints and the midcarpal joint.

Intercarpal joints
The intercarpal joints are the arthrodial joints between the bones of each of the two carpal rows (Figure 9-56).

Proximal Carpal Row
In the proximal carpal row, there are three intercarpal joints. From lateral to medial, the scaphoid articulates with the lunate (scapholunate joint), and the lunate articulates with the triquetrum (lunatotriquetral joint). The articular surfaces are flat and vertical. Finally, the anterior face of the triquetrum articulates with the pisiform at the joint of the pisiform bone.

The first two joints are reinforced by the interosseous ligaments, located on the superior aspect of the articular surface of the scaphoid and lunate to help form the carpal condyle; they include the palmar and dorsal intercarpal ligaments. They are formed by a series of short transverse fascicles extending from bone to bone. In addition, long intercarpal fascicles, the posterior arcuate carpal ligament, connect the scaphoid and triquetrum over the lunate. The joint of the pisiform lacks an intercarpal ligament; it is connected to the triquetrum through a dorsal ligament and to other bones through several carpal ligaments.

The joints mentioned above have their own synovium that communicates with the midcarpal synovium. The exception is the pisiform synovial membrane, which often communicates with the radiocarpal cavity.

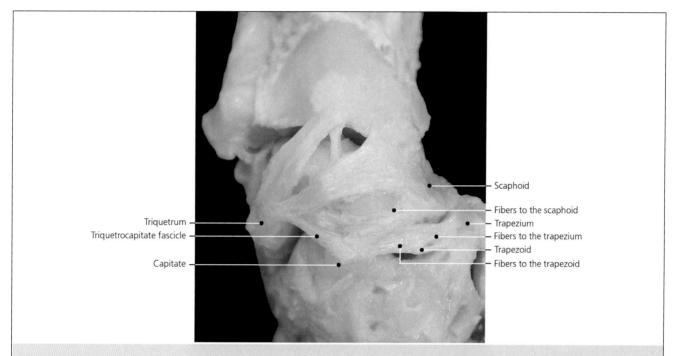

Figure 9-57 Specimen showing the dorsal ligaments of the wrist, with anatomic details of the dorsal intercarpal ligament labeled.

Distal Carpal Row
In the distal carpal row, there are three articulations. From lateral to medial, the triquetrum articulates with the trapezoid, the trapezoid articulates with the capitate, and the capitate articulates with the hamate.

The ligaments reinforcing the joints of the distal carpal row are analogous to the ones in the proximal carpal row but allow less intercarpal motion and also are known as dorsal and palmar intercarpal interosseous ligaments. There is no interosseous ligament between the trapezoid and the trapezius, which allows additional mobility. The synovial membranes of these joints are extensions of the midcarpal joint.

Midcarpal Joint
The midcarpal joint comprises the articulation between the proximal and distal carpal rows, except for the pisiform (Figures 9-48 and 9-56). The joint line is S shaped and has been described as a double condylar joint. On the lateral aspect, the scaphoid forms the convexity that fits in the concavity formed by the trapezius and trapezoid. In the center and on the medial side, the lunate, the capitate, the triquetrum, and part of the scaphoid form a proximal concavity that accommodates the convexity of the capitate and the hamate.

Ligaments
Several ligamentous reinforcements are described, but they vary widely. Extrinsically, fibrous extensions from the wrist ligaments extend from the distal epiphyses of the forearm bones to the distal carpal row (radiocarpal and ulnocarpal ligaments). The intrinsic ligaments are formed by the palmar and dorsal intercarpal ligaments, fibrous bands that connect both bones in the same carpal row (already mentioned) as bones in each of the rows. This latter group of ligaments (Figures 9-57 through 9-60) includes the following:

Dorsal Intercarpal Ligament
This ligament originates on the dorsal aspect of the triquetrum; its oblique fibers are radially oriented. Three small fascicles are differentiated, one for each of the bones the ligament attaches to (the scaphoid, the trapezoid, and the trapezius). The specimen shown in Figure 9-57 also shows fibers extending to the capitate.

Scaphotrapeziotrapezoid Ligament
This ligament originates on the palmar aspect of the scaphoid and fans out into a scaphotrapezoid and a scaphotrapezius fascicle.

Scaphocapitate Ligament
This ligament is located distal to the radioscaphocapitate ligament. It originates at the distal aspect of the trapezius and has an oblique attachment on the body of the capitate.

Triquetrocapitate Ligament
This ligament originates on the palmar aspect of the triquetrum and is directed obliquely to the body of the capitate.

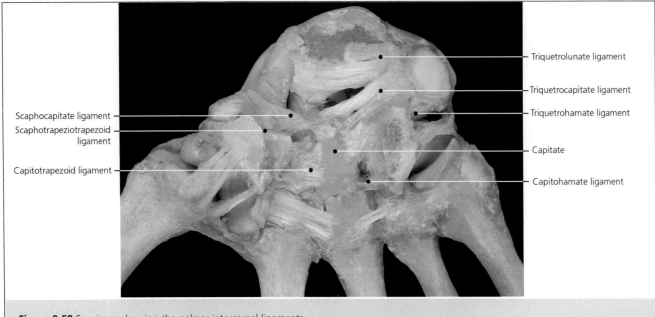

Figure 9-58 Specimen showing the palmar intercarpal ligaments.

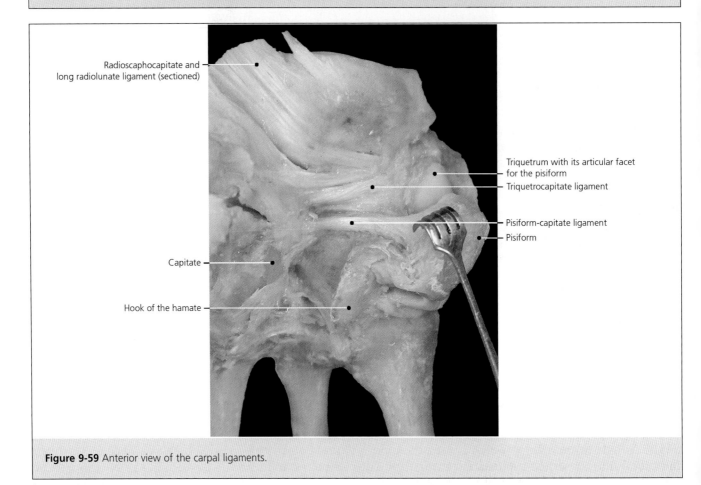

Figure 9-59 Anterior view of the carpal ligaments.

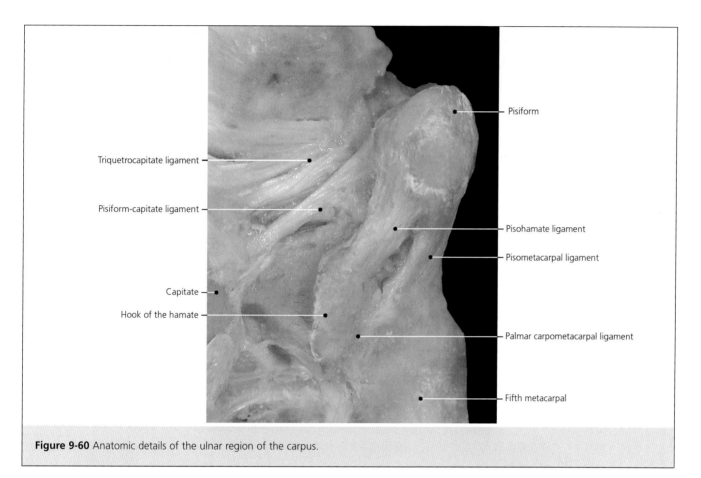

Figure 9-60 Anatomic details of the ulnar region of the carpus.

Triquetrohamate Ligament
This ligament originates at the distal aspect of the triquetrum and attaches to the base of the hook of the hamate.

Pisohamate Ligament
This ligament is considered an extension of the tendon of the flexor carpi ulnaris and extends from the pisiform to the hook of the hamate.

Pisocapitate Ligament
Note that there is a clearly differentiated fascicle from the triquetrocapitate ligament, which extends obliquely from the pisiform to the capitate.

The capitate is the central bone for insertions of the carpal ligaments (Figure 9-61); the ligaments extending from the capitate to the remaining carpal bones, except for the lunate, are together described as the radiocarpal ligament. On the dorsal aspect, fibers connect the scaphoid with the trapezius and trapezoid. On the lateral side, some fibers extend from the scaphoid tubercle to the external aspect of the trapezius. Usually, no interosseous ligaments connect the two rows.

A single articular capsule for the midcarpal joint attaches to the periphery of the articular surfaces mentioned above. In contrast, the synovial membrane may extend into the proximal carpal row, except for the pisiform, or into the distal carpal row and the carpometacarpal joints.

The carpal joints are perfused by the palmar and dorsal arterial arches of the carpus. They are innervated by the anterior and posterior interosseous antebrachial nerves, in addition to some sensory branches of the ulnar, median, and radial nerves.

The flexor retinaculum (transverse carpal ligament, or anterior annular carpal ligament; figures 9-62 through 9-64) is a clinically important structure. It is formed by a fibrous band extending transversely from the scaphoid tubercle and the trapezius on the radial side to the pisiform and hook of the hamate on the ulnar side. Two separate structures can be distinguished: a more superficial layer known as the palmar carpal ligament, which is a thickening of the antebrachial fascia formed by vertical and oblique fibers intimately related to the tendon of the palmaris longus and the intrinsic hand muscles; and a deep layer, the true transverse carpal ligament, which attaches to the bone surfaces already mentioned and is formed by transverse fibers. The palmaris longus tendon passes between these two layers with the ulnar nerve, located in Guyon's canal. The transverse carpal ligament forms the roof of a space known as the carpal tunnel. This tunnel is divided by a sagittal septum

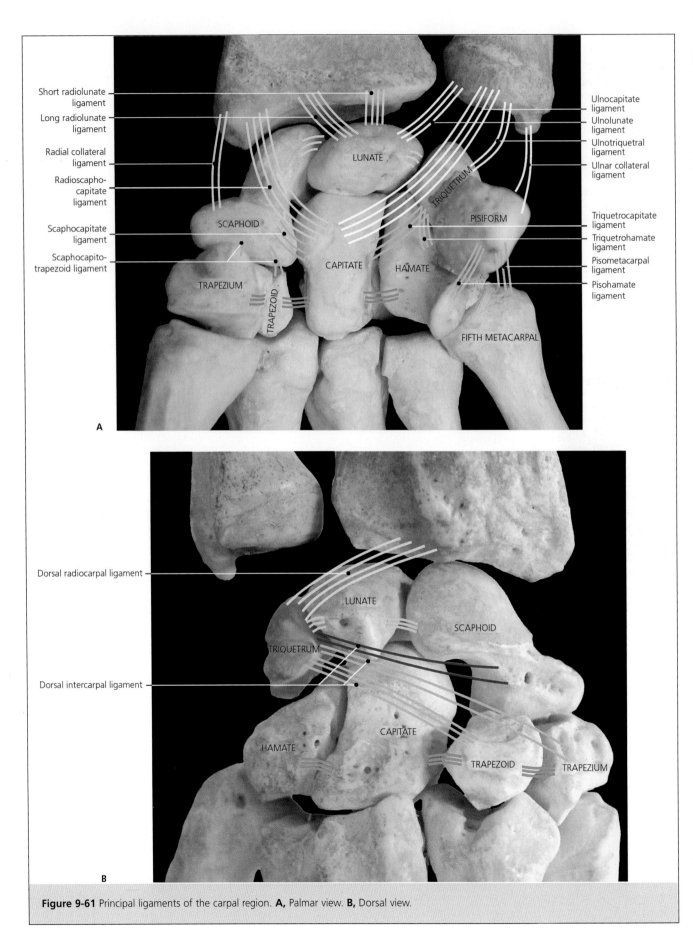

Figure 9-61 Principal ligaments of the carpal region. **A,** Palmar view. **B,** Dorsal view.

Arthrology

extending from the transverse carpal ligament to the anterior aspect of the scaphoid, trapezoid, and capitate. It delineates two spaces—the true carpal tunnel, containing the flexor tendons and the median nerve, and a more lateral space, which contains the tendon of the flexor carpi radialis.

On the posterior aspect of the forearm, at the level of the radiocarpal joint, the fascia is reinforced with transverse fibers that together form the extensor retinaculum (dorsal carpal ligament) (Figures 9-65 and 9-66). This ligament extends from the most lateral aspect of the distal epiphysis of the radius to the ulnar styloid, triquetrum, and pisiform. The extensor retinaculum is attached to the radius and the ulna, forming six osteofibrous channels or compartments for the sheaths of the extensor tendons. The first four compartments are located on the radius, the fifth over the distal radioulnar joint, and the sixth on the ulna. The first compartment is located laterally and contains the tendons of the abductor pollicis longus and the extensor pollicis brevis. The second extensor compartment is located on the lateral aspect of the dorsal tubercle of the radius and contains the tendons of the extensor carpi radialis longus and brevis. The third extensor compartment lies immediately adjacent to these two tendons, on the medial side of the dorsal tubercle, and contains the extensor pollicis longus tendon. The fourth compartment reaches the radial aspect of the distal radioulnar joint and contains the tendons of the extensor digitorum communis and the extensor indicis proprius. The fifth extensor compartment is located over the distal radioulnar joint line, its floor being formed by fibers of the posterior capsule of the joint, and contains the extensor digiti quinti tendon. The sixth compartment contains the tendon of the extensor carpi ulnaris, which passes through a groove in the ulna.

Joints of the Hand

The joints of the hand include the carpometacarpal, intermetacarpal, metacarpophalangeal, and interphalangeal joints.

Carpometacarpal Joints

The carpometacarpal joints connect the distal carpal row with the base of the five metacarpal bones. The carpometacarpal joint of the thumb (trapeziometacarpal joint) is especially interesting. Its articular surfaces include the anteroposterior convexity and transverse concavity of the first metacarpal, which forms a saddle type of joint (Figure 9-67). This joint has a very lax capsule, which extends between the circumference of the first metacarpal base to the periphery of the articular surface of the trapezius. It is reinforced by five ligaments (Figures 9-68 and 9-69): the anterior and posterior oblique ligaments, the palmar ulnar ligament, the dorsal radial ligament, and the first dorsal intermetacarpal ligament. The fibers of these ligaments run

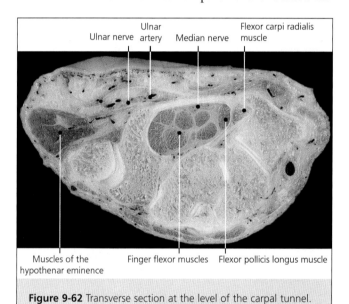

Figure 9-62 Transverse section at the level of the carpal tunnel.

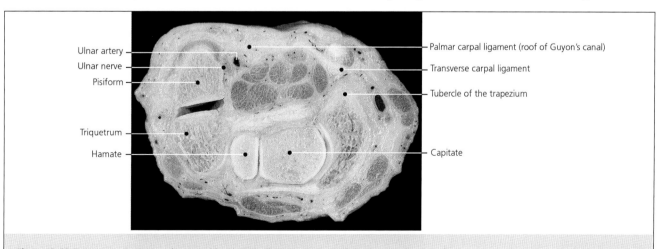

Figure 9-63 Transverse section at the carpal level. Note that Guyon's canal can be visualized perfectly.

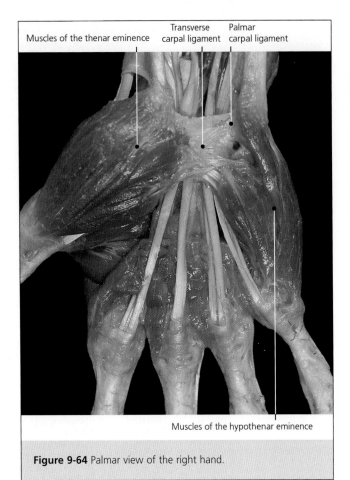

Figure 9-64 Palmar view of the right hand.

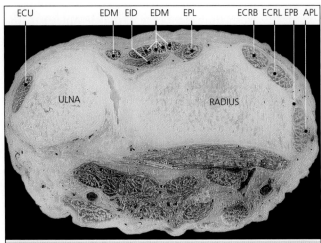

Figure 9-65 Transverse section at the distal forearm.

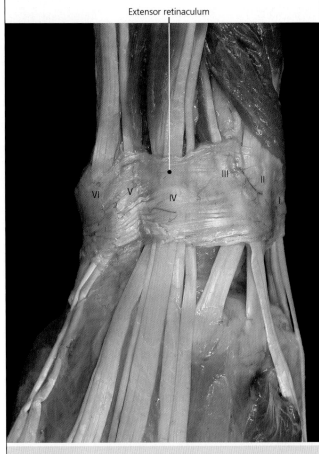

Figure 9-66 Dorsal wrist region, showing the extensor retinaculum and the six extensor compartments.

obliquely, an indication of the importance of the role of this joint in pronation and supination.

The other four carpometacarpal joints are arthrodial (Figures 9-70 and 9-71). The second metacarpal has three facets, to articulate with the trapezius, trapezoid, and capitate. The third metacarpal articulates with the capitate through a triangular facet. The fourth metacarpal articulates with the hamate and the capitate through a small articular facet. The fifth metacarpal has a convex articular facet that corresponds with the concavity of the hamate, creating a functional saddle joint.

Three separate ligamentous structures connect the joints between the carpus and the metacarpals. The main interosseous carpometacarpal ligament is attached to the capitate and hamate through two different fascicles close to the joint line of these two bones (Figures 9-71 and 9-72). From this point, this ligament is strongly attached to the medial aspect of the base of the third and fourth metacarpals.

There are several palmar carpometacarpal ligaments (Figure 9-73). One runs transversely, from the anterior aspect of the triquetrum to the second and third metacarpals. Several fibrous bands also extend from the capitate to the second, third, and, occasionally, fourth metacarpal. No palmar ligament attaches to the fifth metacarpal; instead, the pisiform/metacarpal ligament, an extension of the flexor carpi ulnaris tendon, connects the pisiform with the base of the fifth metacarpal.

The dorsal carpometacarpal ligaments (Figure 9-74) are stronger and more numerous than the palmar lig-

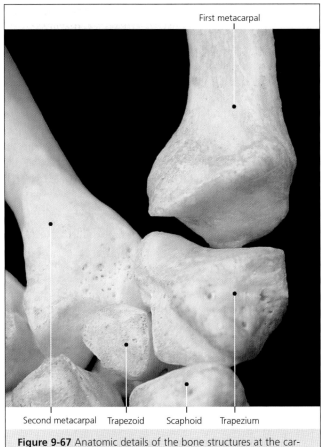

Figure 9-67 Anatomic details of the bone structures at the carpometacarpal joint of the thumb (trapeziometacarpal joint).

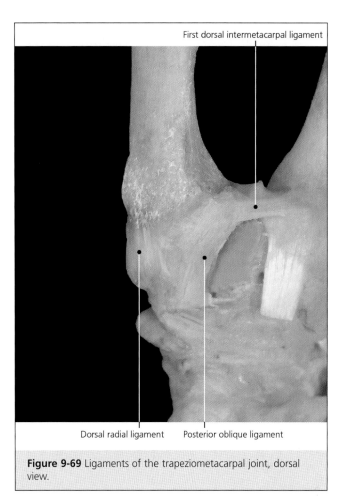

Figure 9-69 Ligaments of the trapeziometacarpal joint, dorsal view.

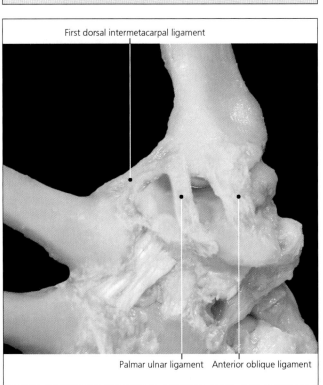

Figure 9-68 Ligaments of the trapeziometacarpal ligament, palmar view.

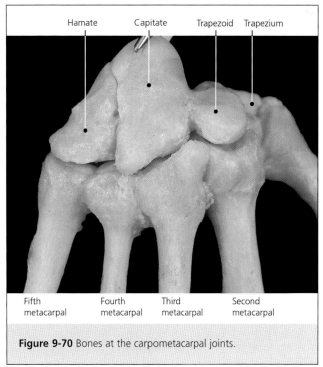

Figure 9-70 Bones at the carpometacarpal joints.

aments. The second metacarpal usually has two ligaments, one from the styloid process to the trapezoid and, less often, one for the trapezius. One ligament runs from the styloid process of the third metacarpal into the dorsal aspect of the capitate; sometimes a second ligament is present, which connects to the trapezoid. The fourth metacarpal has two dorsal ligaments, which connect the metacarpal with the capitate and the hamate. The fifth metacarpal has only one dorsal ligament, extending to the hamate. The second through fifth carpometacarpal joints share a single synovial membrane, which also communicates with the midcarpal joint.

Intermetacarpal Joints

The intermetacarpal joints are between metacarpals II through V; the first metacarpal remains separated. The intermetacarpal joints are not very mobile and are classified as arthrodial joints (Figure 9-73 and 9-74). They articulate through small, flat, oblique articular facets. The second and third metacarpals articulate through a medial facet; the third articulates with the second and the fourth through two facets; similarly, the fourth metacarpal articulates with the third and the fifth; finally, the fifth metacarpal articulates with the fourth only though a lateral facet.

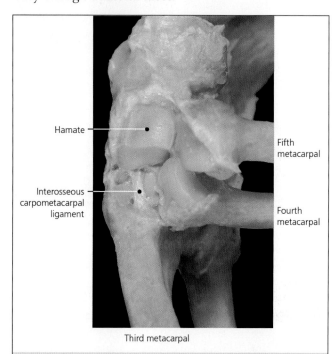

Figure 9-71 Frontal section at the level of the carpometacarpal joints.

Figure 9-72 Details of the interosseous carpometacarpal ligament after reflection of the fourth and fifth metacarpals.

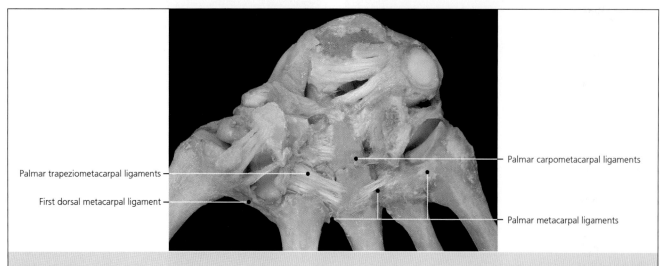

Figure 9-73 Specimen showing the palmar carpometacarpal and intercarpal ligaments.

Arthrology

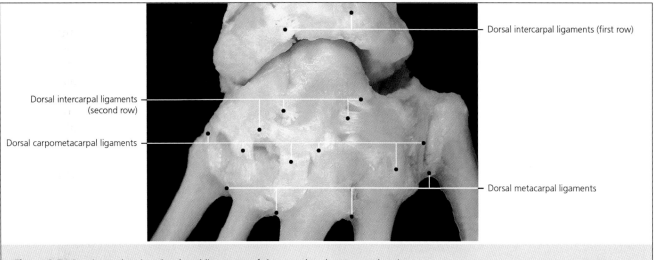

Figure 9-74 Specimen showing the dorsal ligaments of the carpal and metacarpal regions.

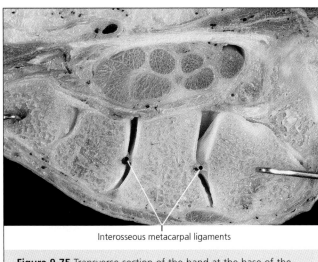

Figure 9-75 Transverse section of the hand at the base of the metacarpal bones.

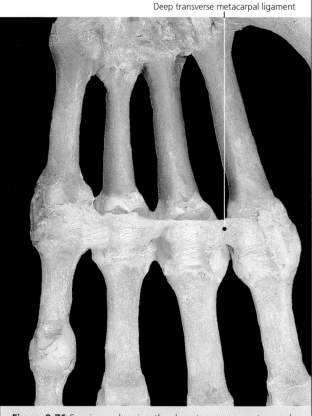

Figure 9-76 Specimen showing the deep transverse metacarpal ligament.

The spaces between the metacarpals are called interosseous metacarpal spaces. These four spaces are numbered from I to IV, from radial to ulnar.

Three ligaments that transversely connect the bases of adjacent metacarpals reinforce these joints: the palmar, dorsal, and interosseous metacarpal ligaments. The interosseous ligaments are located in the articular capsule (Figure 9-75), between the dorsal and palmar metacarpal ligaments. Distally, the heads of metacarpals II through V are connected through the deep transverse metacarpal ligament (Figure 9-76), which is described below, with the metacarpophalangeal joints. The synovial membrane of these joints continues into the carpometacarpal joint.

The intermetacarpal joints are perfused by the dorsal and palmar metacarpal arteries, from the dorsal carpal arterial arches and the deep palmar arch. They are innervated by the anterior and posterior interosseous nerves in addition to branches from the ulnar, median, and radial nerves.

Metacarpophalangeal Joints

The metacarpophalangeal joints are enarthroses, but they function as condylar joints with some degree of rotation (Figure 9-77). They are formed by the

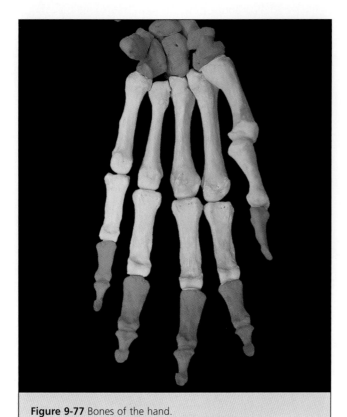

Figure 9-77 Bones of the hand.

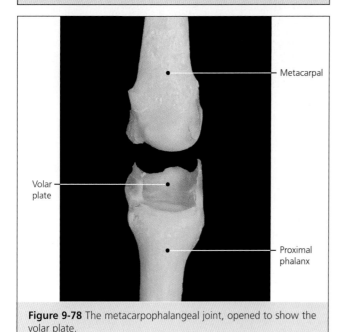

Figure 9-78 The metacarpophalangeal joint, opened to show the volar plate.

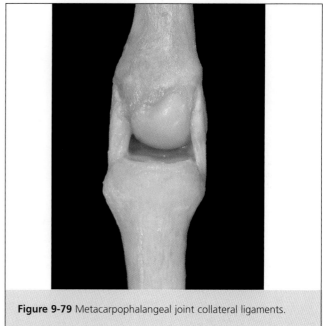

Figure 9-79 Metacarpophalangeal joint collateral ligaments.

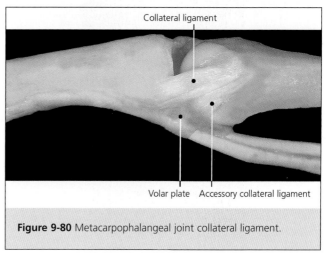

Figure 9-80 Metacarpophalangeal joint collateral ligament.

hemispherical metacarpal head and the glenoid base of each proximal phalanx. They allow flexion, extension, abduction, adduction, and small associated rotational movements. The palmar aspect of the metacarpal head is larger than the articular facet at the base of the phalanx, and there is a small glenoid fibrocartilage, the volar plate (palmar ligament), which provides a wider articular surface (Figure 9-78). This is fixed on the palmar side of the joint and extends laterally.

The articular capsule is thin and lax and is located at the periphery of the articular surfaces, slightly over the dorsal aspect of the metacarpal.

The collateral ligaments fan out on both sides of the joint from the tubercles and fossae at the metacarpal head to the tubercles at the base of the phalanx (Figure 9-79). These ligaments have two components (Figure 9-80): the collateral ligament and the accessory collateral ligament, which attaches to the palmar ligament (volar plate) and the fibrous sheaths of the fingers. They are lax in extension and tight in flexion. In this same region, the sagittal bands of the extensor mechanism are found superficially. These are described below.

At the thumb carpometacarpal joint, the more anterior fibers are attached to the sesamoids; some authors

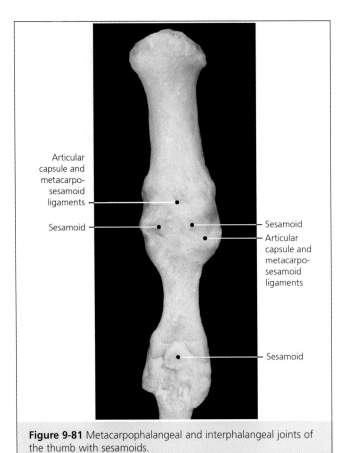

Figure 9-81 Metacarpophalangeal and interphalangeal joints of the thumb with sesamoids.

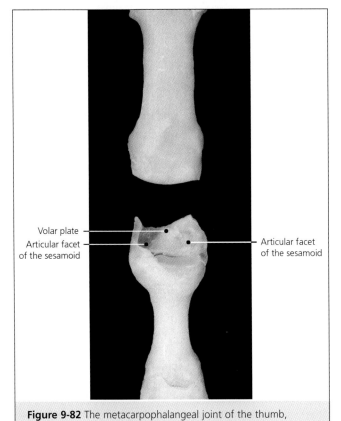

Figure 9-82 The metacarpophalangeal joint of the thumb, opened to show the articular facets of the sesamoids.

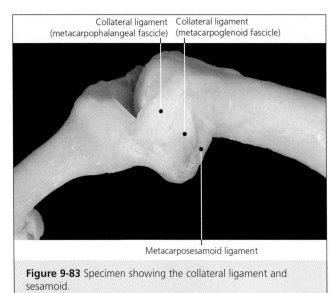

Figure 9-83 Specimen showing the collateral ligament and sesamoid.

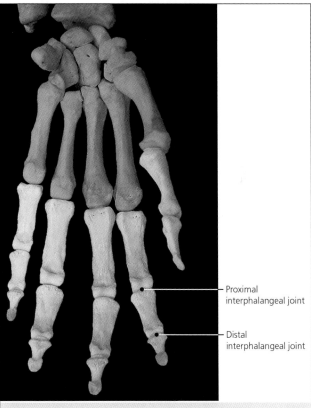

Figure 9-84 Bones of the hand with the interphalangeal joints labeled.

describe them as metacarposesamoid ligaments (Figure 9-81). On the anterior aspect are palmar ligaments (volar plates), the fibers of which are round in shape and are located dorsal to the synovial sheaths of the flexor tendons. They expand between the collateral ligaments and are strongly attached to the base of the phalanx. As mentioned earlier, the deep transverse

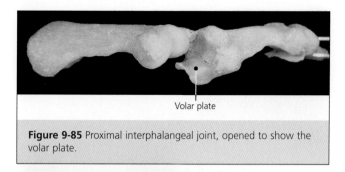

Figure 9-85 Proximal interphalangeal joint, opened to show the volar plate.

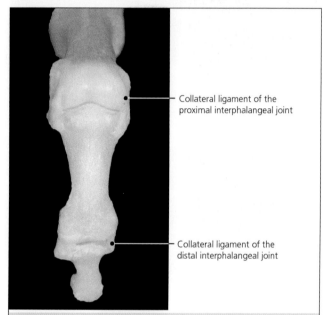

Figure 9-86 Anatomic detail of the interphalangeal joints and their collateral ligaments.

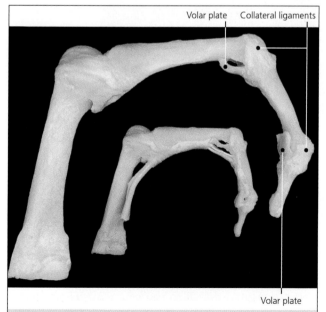

Figure 9-87 Joints of the fingers. The flexor tendons and pulleys have been preserved in miniature to show their relationships with the articular components.

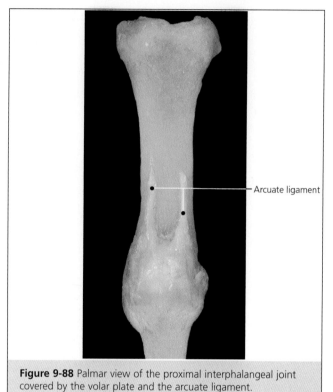

Figure 9-88 Palmar view of the proximal interphalangeal joint covered by the volar plate and the arcuate ligament.

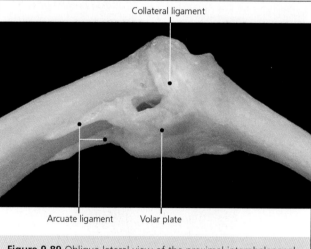

Figure 9-89 Oblique lateral view of the proximal interphalangeal joint in partial flexion. The volar plate adapts to this movement by bending at the zone of contact with the arcuate ligament.

metacarpal ligament (Figure 9-76) is located on the palmar aspect and extends from the volar plate and head (Figure 9-82) of each metacarpal to the adjacent metacarpals, except for the first.

The metacarpophalangeal joint of the thumb has some unique features. Although it has the same features as the other metacarpophalangeal joints, it is functionally a trochlear joint. On its anterior aspect, it articulates with the two sesamoids included in the thickness of the glenoid fibrocartilage; most of the collateral ligament fibers are attached to these sesamoids (Figure 9-83).

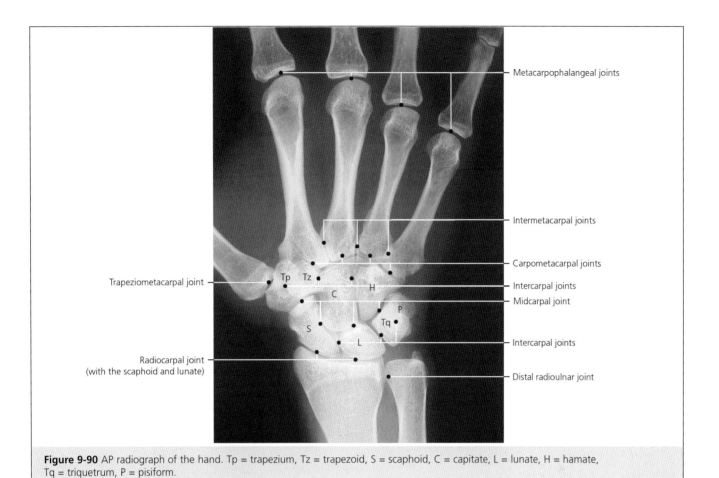

Figure 9-90 AP radiograph of the hand. Tp = trapezium, Tz = trapezoid, S = scaphoid, C = capitate, L = lunate, H = hamate, Tq = triquetrum, P = pisiform.

Figure 9-91 Coronal MRI of the wrist. Tq = triquetrum, H = hamate, C = capitate, Tz = trapezoid, L = lunate, S = scaphoid

Each metacarpophalangeal joint has a very lax synovial membrane.

The metacarpophalangeal joints are perfused by digital arteries arising from the superficial palmar arch. They are innervated by the digital nerves.

Interphalangeal Joints

The interphalangeal joints connect the three phalanges of each finger and the two phalanges of the thumb (Figure 9-84). Each finger has two interphalangeal joints: a proximal interphalangeal joint between the

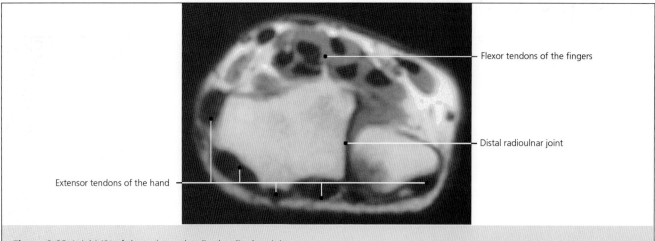

Figure 9-92 Axial MRI of the wrist at the distal radioulnar joint.

proximal and the middle phalanx, and a distal interphalangeal joint between the middle and distal phalanx. The thumb has only one interphalangeal joint.

The head of the phalanx is shaped like a pulley, with two condyles separated by a sagittal groove (Figure 9-85). These condyles articulate with the trochlear surface of the respective phalanx, divided into two cavities by a small middle crest. These joints are considered trochlear joints, although it has been shown that the flexion and extension motions have a rotation component.

The fibrous capsule is attached to the edges of the interphalangeal articular facets. Similar to the metacarpophalangeal joints, these joints have a palmar ligament (volar plate) at the base of the phalanx to increase the articular surface. They also have similar collateral ligaments (Figure 9-86), with a main and an accessory band. They fan out from the tubercle at the head of the phalanx to the tubercle at the base of the adjacent phalanx and the glenoid fibrocartilage.

Finally, there are the volar plates (Figures 9-85 and 9-87). The proximal attachment of the volar plate is commonly known as the checkrein ligament, which attaches to the phalanx margins (Figures 9-88 and 9-89). The checkrein ligament limits hyperextension at the proximal interphalangeal joint and guides the displacement of the volar plate with flexion.

Each interphalangeal joint has its own capsule with a small synovial recess at the dorsal and palmar aspect. Joint perfusion and innervation comes from the digital arteries and nerves, respectively.

Figures 9-90 through 9-93 are imaging studies of the hand, wrist, and finger.

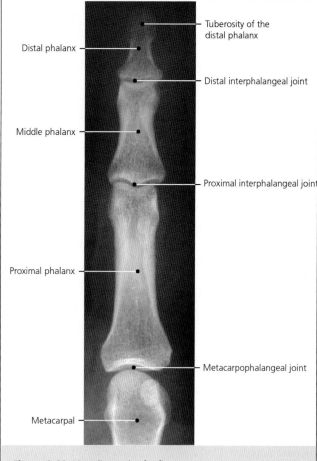

Figure 9-93 AP radiograph of a finger.

Chapter 10

MYOLOGY

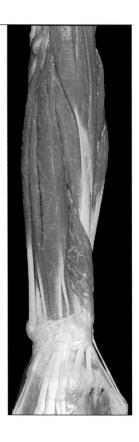

CRANIOTHORACOSCAPULAR MUSCLES

Trapezius

Early in its embryonic development the trapezius is located close to the cranium; it then migrates caudally, toward the shoulder girdle and thorax. The trapezius is found superficially in the superior dorsal thoracic region. Flat, wide, and triangular, the muscle was formerly called the cucullaris (from the Latin *culcullus*, "hood") because the two trapezius muscles together resemble a monk's hood. Three separate portions of the trapezius can be distinguished according to the direction of the muscle fibers (Figure 10-1).

The descending portion of the trapezius is the most cranial portion, and its fibers are directed obliquely and inferiorly. They arise from the external occipital protuberance and the superior nuchal line through the superior aponeurosis. They also arise from the nuchal ligament, which extends from the external occipital tuberosity to C6 or C7. All these fibers are attached to the posterior edge of the superior aspect of the acromial end of the clavicle (Figure 10-2). This portion may act on the shoulder girdle or the occipitocervical region, depending on the fixed and mobile points. When it acts on the shoulder girdle, it elevates the shoulder and tilts the scapula laterally. These actions elevate the glenoid, contributing to shoulder abduction and flexion, especially over 60°. When it acts on the occipitocervical region, contraction of both trapezius muscles extends the head; contraction of just one of the trapezius muscles extends the head as it tilts it to the same side and rotates it to the opposite side.

The transverse portion of the trapezius is formed by parallel fibers perpendicular to the spine and arising from the spinous processes and supraspinous ligament from C3 through C7 via a triangular aponeurosis. The aponeuroses of the two trapezius muscles form the cervicothoracic rhombus. These fibers attach to the acromial end of the clavicle, the posteromedial edge of the acromion, and the lateral third of the spine of the scapula, as well as the medial side of the acromioclavicular joint. The transverse portion is fixed to its origin on the cervical vertebrae, and it adducts and tilts the scapula laterally, contributing, as mentioned above, to elevation of the glenoid and shoulder flexion and abduction.

The ascending portion of the trapezius is formed by fibers directed obliquely in a superior and lateral direction. They originate on the spinous processes and supraspinous ligament from T4 through T12, through the inferior aponeurosis. The fibers ascend to the superomedial edge of the spine of the scapula, where they attach through a triangular aponeurosis that slides over the spinal trigone. From their fixed

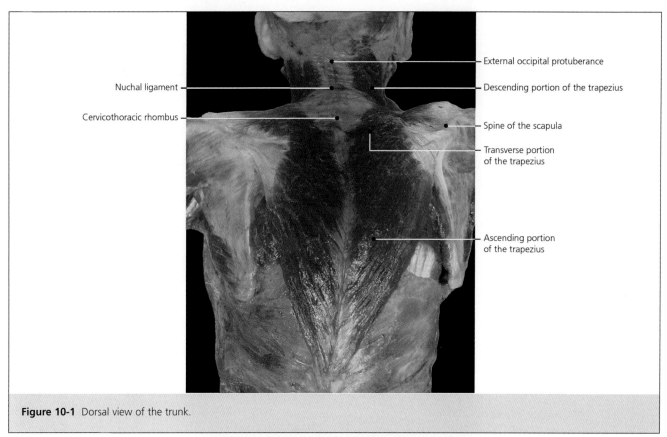

Figure 10-1 Dorsal view of the trunk.

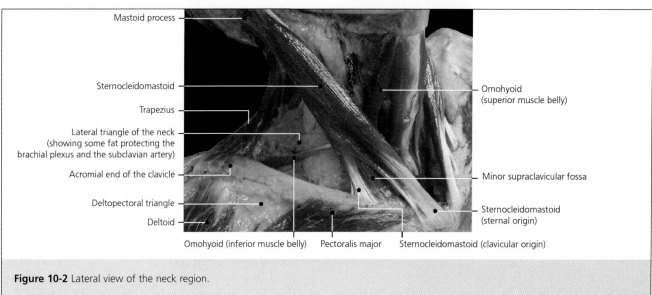

Figure 10-2 Lateral view of the neck region.

point at the thoracic vertebrae, these fibers descend and tilt the scapula laterally, contributing to shoulder flexion and abduction.

The trapezius is innervated by the accessory (spinal) nerve and branches of the deep cervical plexus to C4.

Sternocleidomastoid

The sternocleidomastoid is located on the lateral aspect of the neck (Figures 10-2 and 10-3). Two portions can be distinguished: a sternal portion and a clavicular portion.

On the sternal portion, the sterno-occipital and sternomastoid fascicles can be distinguished. They originate on the anterosuperior aspect of the sternal manubrium; the first fascicle attaches to the lateral aspect of the superior nuchal line of the occipital bone, and the second attaches to the lateral aspect of the mastoid process of the temporalis.

Myology

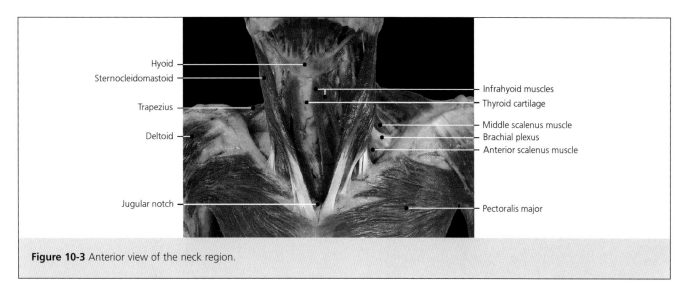

Figure 10-3 Anterior view of the neck region.

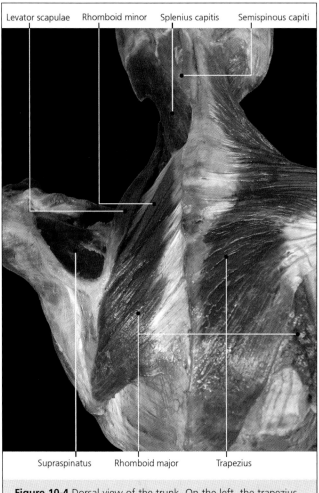

Figure 10-4 Dorsal view of the trunk. On the left, the trapezius has been resected to show the underlying muscular plane.

The clavicular portion is divided into the cleido-occipital and cleidomastoid fascicles, which originate on the superior aspect of the sternal end of the clavicle. The cleido-occipital fascicle attaches to the lateral aspect of the superior nuchal line; the cleidomastoid fascicle attaches to the vertex and anterior edge of the mastoid process.

Contraction on one side extends the head, tilting it to the same side and rotating it to the contralateral side. Contraction of both sides extends the head and assists in inspiration through the actions on the clavicle and sternum.

The sternocleidomastoid is innervated by the accessory nerve and some contributions of the anterior branches of the cervical plexus (C3 and C4).

Rhomboid Muscles

These flat, thin, rhombus-shaped muscles (Figures 10-4 and 10-5) commonly form a single muscle mass, making it difficult to distinguish the rhomboid major and the rhomboid minor muscles. These muscles are separated by a thin layer of cellular tissue and are pierced by the dorsal vein of the scapula.

The rhomboid major originates on the spinous processes and supraspinous ligament of T1 through T4. It attaches to the medial edge of the scapula, below the end of the spine.

The rhomboid minor originates on the spinous processes and supraspinous ligament from C6 and C7 and the inferior aspect of the nuchal ligament. It attaches to the medial edge of the scapula, around or superior to the spine of the scapula.

The rhomboids elevate and adduct the scapula. They also tilt the scapula medially, lowering the glenoid, which helps in shoulder adduction. In addition, they stabilize the scapula against the thoracic wall.

The rhomboid muscles are innervated by the dorsal scapular nerve (C4-C5).

Levator Scapulae

The levator scapulae originates as four fascicles on the posterior tubercle of the transverse processes from C1

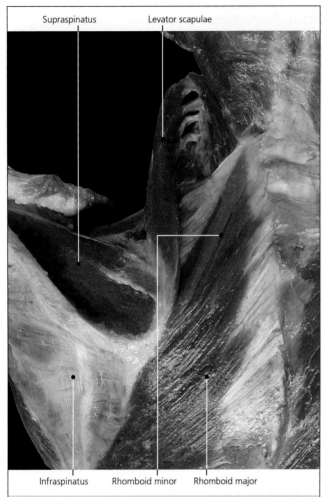

Figure 10-5 Anatomic details of the scapular region. The levator scapulae has been displaced slightly to show the digitations at its origin.

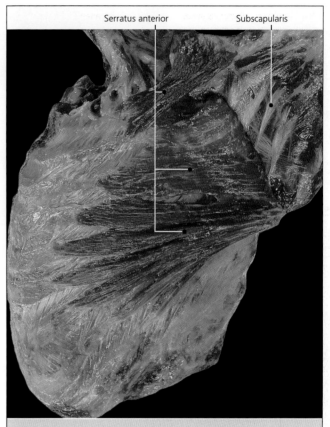

Figure 10-6 Serratus anterior. The scapula has been removed to show its insertion.

through C4 (Figures 10-4 and 10-5). These fascicles coalesce and intersect at the superior angle of the scapula—the levator scapulae is also known as the angular muscle of the scapula—and at the medial edge of the scapula, superior to the rhomboid minor.

The levator scapulae elevates and adducts the scapula. In addition, it tilts the scapula medially to lower the glenoid, which helps in shoulder adduction. Finally, it stabilizes the neck, and unilateral contraction tilts the neck laterally.

The levator scapulae is innervated by the dorsal scapular nerve (C4-C5).

Serratus Anterior

The name of this muscle derives from the Latin term *serrare* ("to saw") because the shape of the costal origins of this muscle resemble the teeth of a saw (Figure 10-6). An earlier name for this muscle was serratus lateralis, so named because it covers the lateral aspect of the thoracic wall. The serratus anterior has three portions: superior (parallel), middle (diverging), and inferior (converging).

The superior portion of the serratus anterior originates on the body of the first and second ribs and the external intercostal fascia. These fibers are oriented in a parallel direction until they attach to the anterior aspect of the medial edge of the scapula, close to the superior angle.

The middle portion of the serratus anterior originates on the second, third, and fourth ribs. Its fibers expand and attach to the entire anterior aspect of the medial edge of the scapula except for the superior and inferior angles.

The inferior portion of the serratus anterior originates on the fifth through ninth ribs. These fibers coalesce and attach to the anteroinferior aspect of the medial edge of the scapula as well as its inferior angle.

The serratus anterior fixes the scapula to the thoracic wall. In addition, it helps in forward movements of the scapulothoracic joint. These movements may elevate the glenoid, helping in shoulder abduction and flexion. It also acts on the thorax to contribute to inspiration when the scapula is the fixed point, by elevating the ribs and increasing the diameter of the thorax.

The serratus anterior is innervated by the long thoracic nerve (C5 through C7).

Myology

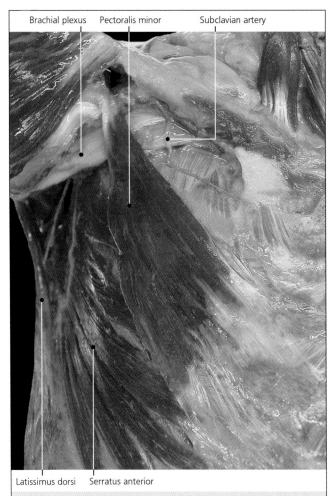

Figure 10-7 Anterior view of the trunk. The pectoralis major has been removed.

Pectoralis Minor

The pectoralis minor is a triangular muscle located beneath the pectoralis major (Figure 10-7). It originates as three separate fascicles on the superior and lateral aspects of the third, fourth, and fifth ribs. Its fibers converge in a single tendon that attaches to the medial aspect of the coracoid process.

The pectoralis minor and subclavius muscles are covered by the clavipectoral fascia, which attaches to the coracoid process, clavicle, and anterior thoracic region.

Like the serratus anterior, the pectoralis minor moves the shoulder forward. It may also act on the

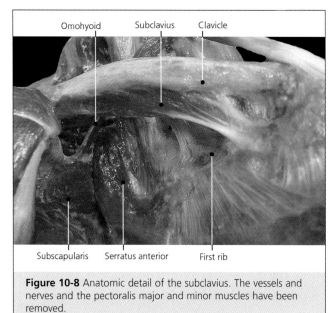

Figure 10-8 Anatomic detail of the subclavius. The vessels and nerves and the pectoralis major and minor muscles have been removed.

TABLE 10-1 THE CRANIOSCAPULAR MUSCLES

MUSCLE	Trapezius	Sternocleidomastoid
ORIGIN	External occipital protuberance, superior nuchal line, and spinous processes C7 through T12	Mastoid apophysis
INSERTION	Spine of the scapula and acromial third of the clavicle	Sternal manubrium and third of sternal end of the scapula
INNERVATION	Accessory nerve [XI]	Accessory nerve [XI] and cervical plexus
FUNCTION	Elevates the shoulder girdle and rotates the scapula, contributing to shoulder abduction; same action on the head as the sternocleidomastoid	Head extension, homolateral tilt, and contralateral rotation

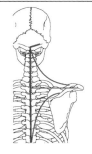
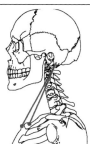

TABLE 10-2 THE THORACOSCAPULAR MUSCLES

MUSCLE	Rhomboids	Levator Scapulae	Serratus Anterior	Pectoralis Minor	Subclavius
ORIGIN	*Major:* Spinous processes C6-C7 *Minor:* Spinous processes T1 through T4	Transverse apophyses C1 through C4	Ribs I-IX	Ribs III-IV	Rib I
INSERTION	Medial edge of the scapula	Superior angle of the scapula	Medial border of the scapula	Coracoid process	Inferior aspect of the clavicle
INNERVATION	Dorsal scapular nerve	Dorsal scapular nerve	Long thoracic nerve	Pectoralis nerve	Subclavius nerve
FUNCTION	Scapular adduction and rotation, contributing to shoulder adduction	Scapular elevation	Applies the scapula against the thoracic edge	Shoulder protraction	Clavicle stabilization

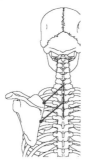 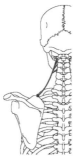 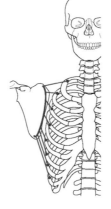 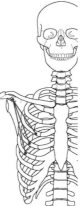 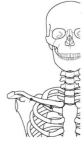

thorax when the scapula is fixed, contributing to inspiration through elevation of the ribs.

The pectoralis minor is innervated by the medial and lateral pectoral nerves (C6 through C8).

Subclavius

The subclavius is a small muscle that originates on the superior aspect of the first rib, the costochondral synchondrosis, and the rib cartilage (Figure 10-8). It attaches to the inferior aspect of the clavicle, in the subclavius sulcus.

This muscle approximates the clavicle to the sternum, stabilizing the sternoclavicular joint during motion. It also protects the brachial plexus and subclavian/axillary vessels when the clavicle fractures.

The subclavius is innervated by a small branch, the subclavius nerve (C5-C6).

Omohyoid

The omohyoid is part of the infrahyoid muscles (Figure 10-2). Although it attaches close to the scapular notch, it will be described later, with the neck muscles.

Tables 10-1 and 10-2 summarize the origin, insertion, innervation, and function of the scapular muscles.

THORACOHUMERAL MUSCLES

Table 10-3 summarizes the origins, insertions, innervation, and functions of the thoracohumeral muscles.

Latissimus Dorsi

The latissimus dorsi is a flat, wide, triangular muscle located on the superficial aspect of the dorsolumbar region (Figure 10-9). It has three fascicles—the superior (horizontal), middle (oblique), and inferior (vertical)—which coalesce into its insertion on the anterior aspect of the humerus.

The latissimus dorsi originates on the spinous processes of T7 through T12 and L1 through L5, its supraspinous ligaments, and the middle sacral crest through the thoracolumbar fascia (lumbar rhomboid). It also originates from the external lip of the iliac crest, the tenth, eleventh, and twelfth ribs, and the inferior angle of the scapula. All the fibers coalesce into a tendon attached to the depth of the intertubercular sulcus and the crest of the lesser tuberosity.

TABLE 10-3 THE THORACOHUMERAL MUSCLES

MUSCLE	Latissimus dorsi	Pectoralis major
ORIGIN	Spinous processes T7—middle sacral crest, iliac crest, last three ribs, inferior scapular angle	Clavicle, sternum, ribs I-VII
INSERTION	Crest of the lesser tuberosity	Crest of the greater tuberosity
INNERVATION	Thoracodorsal nerve	Pectoralis nerve
FUNCTION	Shoulder adduction, extension, and internal rotation	Shoulder adduction, flexion, and internal rotation

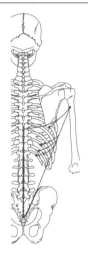
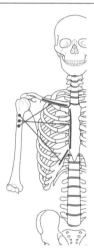

The latissimus dorsi adducts the arm through its action on the inferior angle of the scapula, which rotates and lowers the glenoid. It also extends and internally rotates the arm. When the upper limb is fixed, this muscle contributes to elevating the trunk and climbing.

Contraction of both muscles contributes to trunk extension. This muscle also contributes to forced exhalation, and its inferior fibers may be hypertrophied in patients with respiratory conditions accompanied by a frequent cough.

The latissimus dorsi is innervated by the thoracodorsal nerve (C6 through C8).

Pectoralis Major

The pectoralis major is a flat, fan-shaped muscle located on the superficial aspect of the anterior thorax (Figure 10-10). The pectoralis major is covered completely by the clavipectoral fascia, which is attached to the inferior aspect of the clavicle and the sternum. This fascia continues on the lateral side as the axillary fascia, which connects the pectoralis major with the latissimus dorsi and covers the adipose axillary tissue. The pectoralis major has three portions: clavicular (oblique descending), sternocostal (transverse), and abdominal (oblique ascending).

The clavicular portion originates on the anterior edge of the medial half of the clavicle.

The sternocostal (sternochondral) portion originates on the anterior aspect of the sternum, the lateral aspect of the first six chondral cartilages, and the seventh rib. The deep fascicles of the pectoralis major originate on the cartilage of the third, fourth, and fifth ribs.

Finally, the abdominal portion is inconsistent and originates on the sheath of the rectus abdominis and the ribs distal to the seventh rib.

All the rib fibers coalesce into a U-shaped tendon that attaches to the crest of the greater tuberosity. The clavicular portion attaches distally and forms the anterior portion of the tendon, whereas the posterior portion is formed by the sternocostal and abdominal fascicles, the latter with an oblique cranial insertion.

The clavicular and superior sternocostal fibers flex, adduct, and internally rotate the arm. The clavicular fibers also suspend the humerus. The sternocostal fibers adduct, horizontally flex, and internally rotate the arm. They also contribute to shoulder protraction. When the upper limb is fixed, these fibers also contribute to inspiration through their costal and sternal insertions. The inferior sternocostal fibers and the abdominal portion adduct the arm from the abducted position (or move the trunk during climbing activities). They also contribute to inspiration through their costal and sternal attachments.

The pectoralis major is innervated by the medial and lateral pectoralis nerves (C6 through C8).

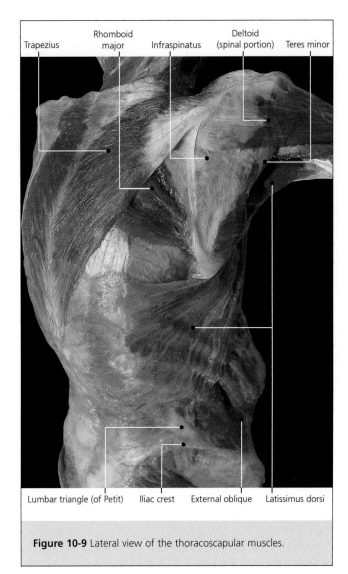

Figure 10-9 Lateral view of the thoracoscapular muscles.

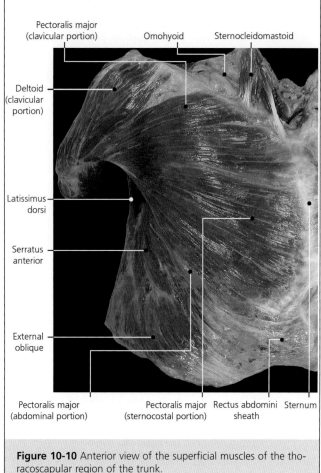

Figure 10-10 Anterior view of the superficial muscles of the thoracoscapular region of the trunk.

SCAPULOHUMERAL MUSCLES

Table 10-4 summarizes the origins, insertions, innervation, and function of the scapulohumeral muscles.

Deltoid

This multipennate muscle is triangular, with its origin at the base of the triangle and its attachment at the vertex (Figures 10-9 and 10-10). It covers the entire shoulder region, providing its typical shape. The deltoid is covered by the strong deltoid fascia.

Three portions can be distinguished, depending on the origin of the fibers: the clavicular, the acromial, and the spinal portions.

The clavicular portion originates on the anterior edge of the lateral third of the clavicle and is separated from the clavicular fibers of the pectoralis major muscle at the deltopectoral triangle. The acromial portion originates at the lateral edge of the acromion and continues with the clavicular portion without a clear separation. The spinal portion originates at the inferior edge of the spine of the scapula.

The origin is a flat aponeurosis with a smaller anterior aspect and a larger posterior aspect. The muscle fibers run obliquely in multiple directions, like other multipennate muscles. All portions converge to a V-shaped attachment at the deltoid tuberosity of the humerus.

The subdeltoid bursa is located between this muscle and the articular capsule for the glenohumeral joint, which may communicate with the subacromial bursa (Figure 10-11).

The clavicular portion of the deltoid flexes and internally rotates the arm. The acromial portion abducts the arm, although beyond 30° the clavicular and spinal fibers also contribute to abduction. The spinal portion extends and externally rotates the shoulder. In addition, the deltoid helps suspend the free end of the upper limb and stabilizes the shoulder.

The deltoid is innervated by the axillary nerve (C6 through C8).

Supraspinatus

The supraspinatus is a short muscle that originates at the supraspinous fossa of the scapula and the

Myology

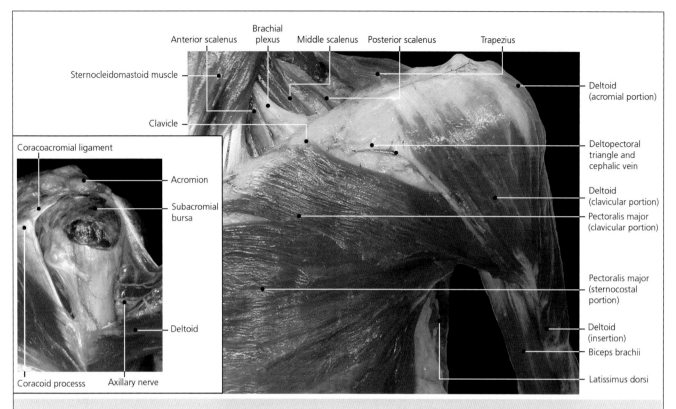

Figure 10-11 Anterior view of the superficial muscles of the thoracoscapular region of the trunk and anatomic specimen with the deltoid removed and the subacromial bursa injected with latex.

supraspinous fascia (Figures 10-4, 10-5, and 10-12). Its fibers converge and pass beneath the subacromial space formed by the acromion and the coracoacromial ligament. The subacromial bursa is located between this roof and the muscle. This is a common location for impingement.

The muscle fibers of the supraspinatus end in a tendon intimately united to the fibrous capsule of the shoulder joint. The tendon attaches to the facet of the humeral greater tuberosity for the superior supraspinatus. These insertions facilitate its role as a joint stabilizer.

The main function of the supraspinatus is to abduct the shoulder. When the shoulder is in approximately 70° of abduction, the greater tuberosity is in contact with the acromion; impingement of the supraspinatus and the subacromial bursa are among the most common problems in this location. To avoid this bony block, other muscles externally rotate the humerus to allow the arm to achieve 180° of abduction.

The supraspinatus is innervated by the suprascapular nerve (C4 through C6).

Infraspinatus

The infraspinatus is a flat muscle that originates at the infraspinatus fossa, the infraspinatus fascia, and the fibrous septum, which separates this muscle from the teres minor and major muscles (Figures 10-9 and 10-12). The muscle has three fascicles: a superior (horizontal) portion that originates beneath the spine of the scapula, a middle portion formed by oblique fibers, and an inferior portion with ascending fibers similar to the fibers of the teres minor.

The muscle fibers coalesce into a tendon intimately related with the posterior aspect of the fibrous capsule of the glenohumeral joint. This provides posterior stability to the joint, acting as an active stabilizer in internal rotation and tightening the articular capsule in external rotation. Close to the glenohumeral joint, the subtendinous bursa for the infraspinatus muscle is located beneath the tendon. The tendon is attached to the middle facet in the greater tuberosity of the humerus.

The main function of the infraspinatus is to externally rotate the arm.

The infraspinatus is innervated by the suprascapular nerve (C4 through C6).

Teres Minor

The teres minor originates on the superior aspect of the lateral edge of the scapula, the local fibrous intermuscular septum, and the infraspinous fascia (Figures 10-9 and 10-12). Its fibers are oriented in an ascending oblique direction to form a tendon that is inti-

Table 10-4 The Scapulohumeral Muscles

MUSCLE	Deltoid	Supraspinatus	Infraspinatus
ORIGIN	Clavicle, acromion, and spine of the scapula	Supraspinatus fossa	Infraspinous fossa
INSERTION	Deltoid tuberosity	Greater tuberosity of the humerus	Greater tuberosity of the humerus
INNERVATION	Axillary nerve	Suprascapular nerve	Suprascapular nerve
FUNCTION	Shoulder flexion, abduction, and extension	Shoulder abduction rotator	Shoulder external rotator

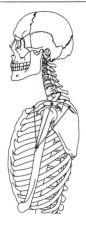

mately related to the posterior aspect of the fibrous capsule of the glenohumeral joint. The fibers attach to the inferior facet of the greater tuberosity.

When the teres minor contracts, it tightens the posterior aspect of the capsule, externally rotating the arm and avoiding impingement. It also stabilizes the joint as an active ligament.

The teres minor is innervated by the axillary nerve (C5-C6).

Teres Major

The teres major originates at the inferior facet of the lateral edge of the scapula, which is separated from the infraspinous fossa through a small oblique crest (Figures 10-12 through 10-14). In addition, it originates at the inferior edge of the scapula, the teres major fascia, and the intermuscular fibrous septum.

The fibers of the teres major ascend obliquely in an anterior direction to attach through a flat tendon into the crest of the lesser tuberosity. This tendon is separated from the humerus by the subtendinous bursa for the teres major muscle.

When the teres major acts on the humerus, it produces abduction, internal rotation, and slight arm extension. When the upper limb is fixed, the teres major tilts the scapula so that the inferior angle is positioned laterally.

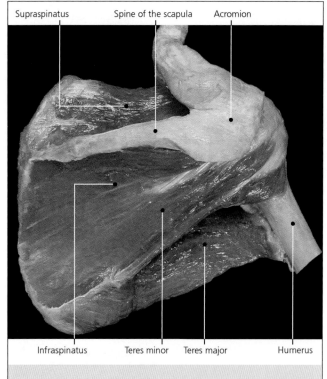

Figure 10-12 Posterior scapulohumeral muscles.

TABLE 10-4 CONTINUED

MUSCLE	Teres minor	Teres major	Subscapularis	Coracobrachialis
ORIGIN	Lateral border of the scapula	Lateral border of the scapula	Subscapularis fossa	Cocacoid process
INSERTION	Greater tuberosity of the humerus	Crest of the lesser tuberosity	Humeral lesser tuberosity	Proximal and medial half of the humerus
INNERVATION	Axillary nerve	Subscapular nerve	Subscapular nerve	Musculocutaneous
FUNCTION	Shoulder external rotation	Shoulder adduction, extension, and internal rotation	Shoulder internal rotation	Shoulder flexion and adduction

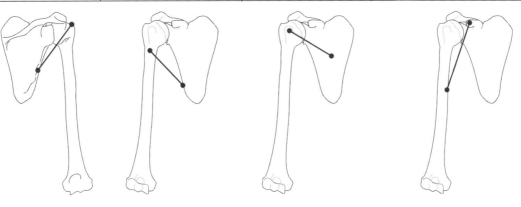

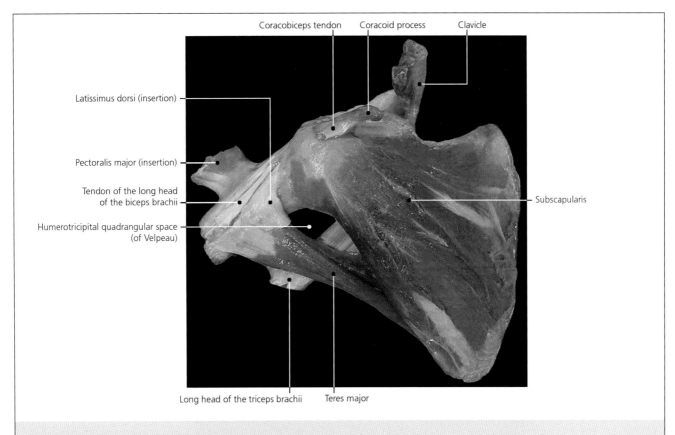

Figure 10-13 Anterior scapulohumeral muscles and humeral attachment of the pectoralis major and the latissimus dorsi.

Scapular Girdle and Upper Limb

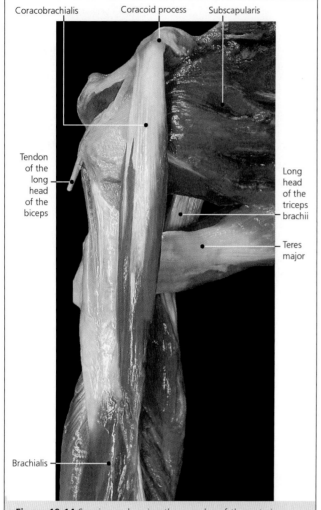

Figure 10-14 Specimen showing the muscles of the anterior region of the arm with the biceps brachii removed.

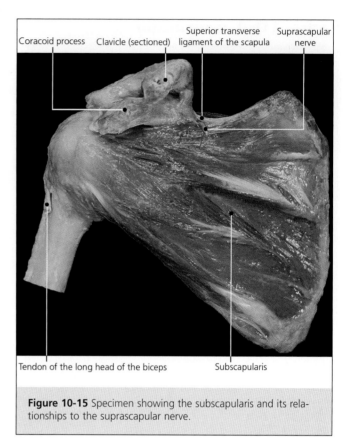

Figure 10-15 Specimen showing the subscapularis and its relationships to the suprascapular nerve.

The teres major muscle is innervated by the inferior branch of the subscapular nerve (C5 through C7) or the thoracodorsal nerve.

Subscapularis

This muscle covers the costal aspect of the scapula along several crests that form under the traction of the muscle fibers (Figures 10-13 through 10-15). The fibers ascend in an oblique and lateral direction and coalesce in a tendon that is intimately related with the fibrous capsule of the glenohumeral joint. The subtendinous bursa for the subscapularis is located close to the glenoid.

The tendon for the subscapularis is an anterior capsular reinforcement that works as an active ligament. In addition, this muscle tightens the capsule in internal rotation. Finally, its tendinous fibers attach to the lesser tuberosity of the humerus and the proximal aspect of the lesser tuberosity crest.

The main function of the subscapularis is internal rotation of the shoulder. It also adducts the shoulder slightly.

The subscapularis is innervated by two branches, the subscapular nerves (C5 through C8).

Coracobrachialis

This elongated muscle is found on the medial aspect of the arm (Figure 10-14). It originates at the apex of the coracoid process and shares its tendon with the short head of the biceps; therefore, this common origin is sometimes called the coracobiceps or conjoined tendon.

The coracobrachialis separates from the biceps at the proximal third of the humerus. It attaches on the anteromedial aspect of the upper third of the humerus.

The bursa for the coracobrachialis muscle is interposed between the coracobrachialis and the subscapularis, which lies underneath.

The main function of the coracobrachialis is to take the arm into its resting position. In addition, it is very important in the suspension of the upper limb.

The coracobrachialis is innervated by the musculocutaneous nerve (C6-C7), which perforates the muscle proximally. This is why the coracobrachialis is sometimes referred to as a perforated muscle.

Rotator Cuff

The term rotator cuff refers to a group of four scapulohumeral muscles, the main common function of which is to contribute to glenohumeral rotation and

MYOLOGY

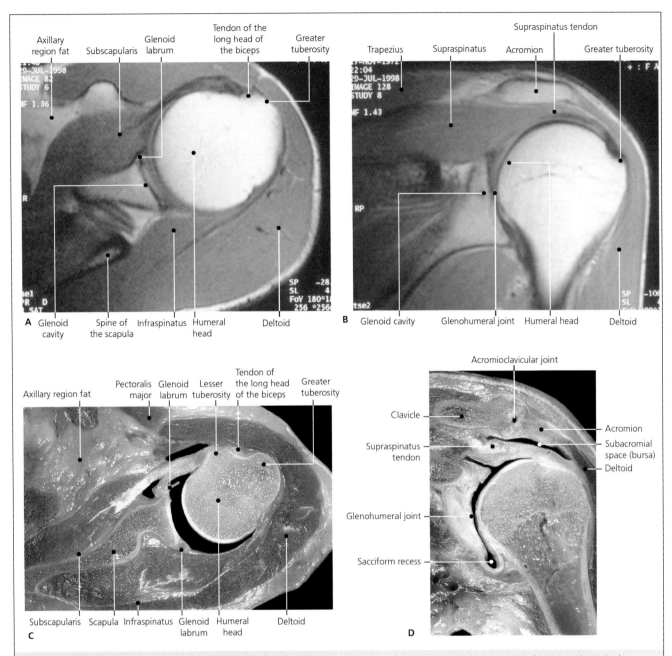

Figure 10-16 A, Transverse MRI of the shoulder. **B,** Coronal MRI of the shoulder. **C,** Anatomic specimen sectioned at approximately the same level as shown in **A**. The brachial plexus and the axillary artery and vein are included in the fat of the axillary cavity. **D,** Anatomic specimen sectioned at approximately the same level as shown in **B**.

provide stability (Figure 10-16). The rotator cuff reinforces the glenohumeral fibrous capsule, encircling it except on the inferior aspect. The muscles that comprise the rotator cuff are the subscapularis, the supraspinatus, the infraspinatus, and the teres minor.

MUSCLES OF THE ARM

There are two muscular compartments in the arm: the anterior compartment and the posterior compartment. They are bordered by the intermuscular septae. The medial intermuscular septum of the arm lies between the medial edge of the humerus and the brachial fascia. The lateral intermuscular septum of the arm lies between the lateral edge of the humerus and the brachial fascia. The anterior compartment is also known as the flexor compartment, as it includes the flexor muscles (brachialis and biceps brachii). The posterior compartment contains the main extensor (triceps brachii). All these muscles are covered by the brachial fascia, which is formed by connective tissue arranged transversely, like a dressing. Table 10-5 summarizes the origins, insertions, innervation, and functions of the muscles of the arm.

TABLE 10-5 MUSCLES OF THE ARM

MUSCLE	Biceps brachii	Brachialis	Triceps brachii
ORIGIN	*Long head:* supraglenoid tubercle *Short head:* coracoid process	Anterior region of the distal humeral half	*Long head:* infraglenoid tubercle *Lateral head:* superolateral region of the humerus *Medial head:* inferomedial region of the humerus
INSERTION	Radial tuberosity	Ulnar tuberosity	Olecranon
INNERVATION	Musculocutaneous nerve	Musculocutaneous nerve	Radial nerve
FUNCTION	Elbow flexion and forearm supination	Elbow flexion	Shoulder and elbow extension

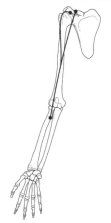

Biceps Brachii

As indicated by its name, this muscle has two heads (*bi-*, "two," and *-ceps*, "heads"). There is a short head and a long head, and the two are connected in the middle of the arm in a single muscle belly (Figure 10-17).

The long head of the biceps originates at the supraglenoid tubercle of the scapula and is directed into the intertubercular, or bicipital, groove. At this location, it passes through an osteofibrous conduit formed by the osseous structures of the sulcus and closed by the transverse humeral ligament. This is a site of tendon reflection, and as such it is protected by an intertubercular tendon sheath, which reduces friction between the tendon and the adjacent osseous structures. Once reflected, the tendon descends along the long axis of the humerus.

The short head of the biceps originates at the tip of the coracoid process, and it shares this tendinous attachment with the coracobrachialis muscle. Its fibers are directed inferiorly, toward the humerus.

Both heads send muscle fibers that coalesce approximately in the middle aspect of the arm in a single muscle belly. This continues on the long axis of the humerus over the brachialis until it reaches the elbow. At the elbow, the biceps has a very strong tendon, which descends to its attachment into the radial or bicipital tuberosity (Figure 10-18). In addition, at the point where the distal biceps tendon is formed, there is a tendinous expansion called the biceps brachii aponeurosis (lacertus fibrosus), which extends medially and coalesces with the antebrachial aponeurosis.

Two bursae are located at the level of the biceps attachment. They are the bicipitoradial bursa, located between the tendon and the radial tuberosity, and the interosseous bursa of the elbow, which separates the biceps brachii, the ulna, and the oblique cord.

The biceps brachii is a biarticular muscle with actions on both the elbow and the shoulder. At the elbow, it acts as a flexor and supinator, its main action being flexion of the supinated forearm. Its pull is transmitted ulnarly through the bicipital aponeurosis. At the shoulder it helps suspend the upper limb, and contraction of both heads produces shoulder flexion. The biceps tendon also produces abduction because the long head is reflected on the intertubercular

Myology

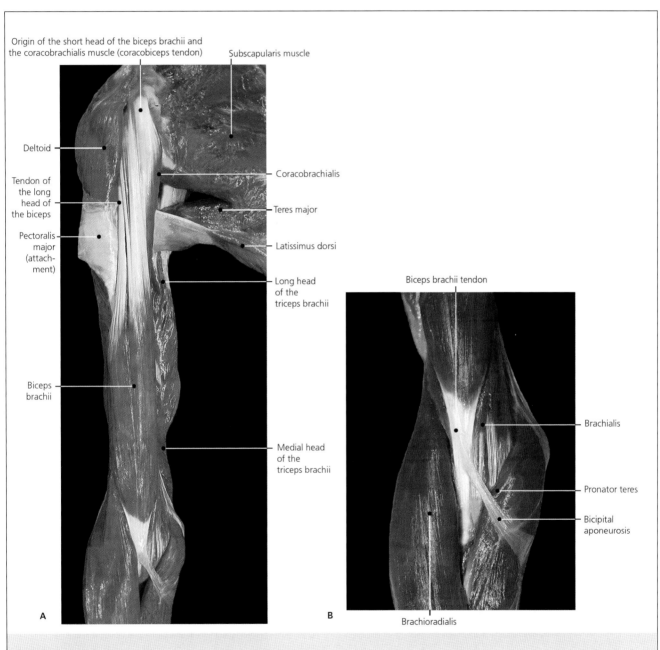

Figure 10-17 A, Superficial muscles of the arm. **B,** Anatomic detail of the elbow region.

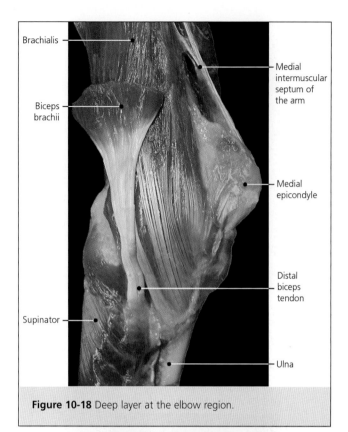

Figure 10-18 Deep layer at the elbow region.

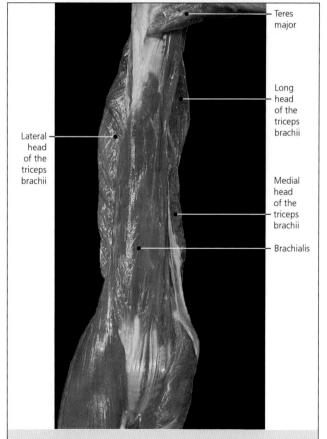

Figure 10-19 Specimen showing the brachialis after removal of the biceps brachii.

groove, whereas the short head contributes slightly to adduction.

The biceps and the brachialis are both innervated by the musculocutaneous nerve (C6-C7).

Brachialis

The brachialis is a fusiform muscle located in the anterior compartment of the arm beneath the biceps brachii, on the anterior surface of the humerus (Figures 10-17 through 10-19). It originates at the distal half of the anterior humeral surface, the inferior and medial lip of the deltoid tuberosity, and the intermuscular septum of the arm. Its fleshy origins have fibers that run distally to the elbow to attach to the tuberosity and inferior aspect of the coronoid process of the ulna, as well as the articular capsule of the elbow. The main function of this muscle is elbow flexion without concomitant forearm pronation or supination.

Triceps Brachii

The triceps brachii is a pennate muscle with three heads (*tri-*, "three," *-ceps*, "heads"): a long head, a lateral head, and a medial head (Figures 10-19 through 10-22). It is the only muscle located in the posterior compartment of the arm.

The long head of the triceps originates at the infraglenoid tubercle of the scapula. It is reflected over the teres major, forming the topographic spaces of Velpeau.

The lateral head of the triceps originates on the lateral and proximal aspect of the sulcus for the radial nerve on the humerus and the posterior aspect of the lateral intermuscular septum of the arm.

The medial head of the triceps is the largest of the three. It originates at the medial and distal aspect of the sulcus for the radial nerve as well as the intermuscular septi of the arm.

These three heads coalesce into a broad, strong tendon that attaches to the ulnar olecranon and the posterior aspect of the articular capsule of the elbow. Its medial fibers attach to the medial aspect of the tendon, the fibers of the long head attach to the superior aspect of the tendon, and the lateral fibers attach to the lateral aspect of the tendon, forming a large pennate muscle.

Beneath the triceps, and as part of this same muscle, the articular muscle of the elbow tightens the articular capsule of the elbow joint in extension.

The triceps is innervated by the radial nerve (C6 through C8).

Muscles of the Forearm

The muscles of the forearm are organized in three compartments: one anterior (flexor) compartment and two posterior (extensor) compartments.

Myology

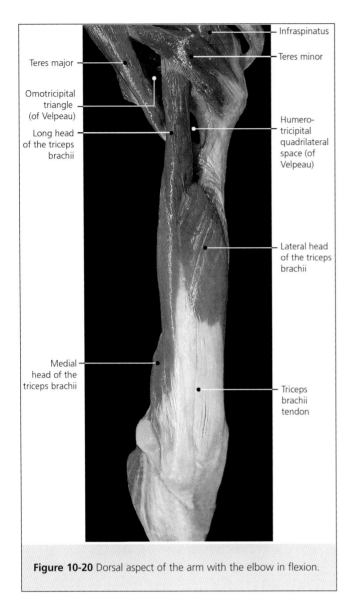

Figure 10-20 Dorsal aspect of the arm with the elbow in flexion.

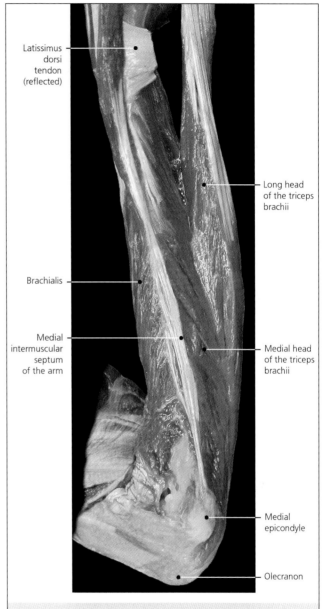

Figure 10-21 Medial view of the arm with the elbow in flexion. The biceps brachii has been removed.

Anterior Compartment
The muscles included in the anterior compartment are organized in four layers, which are described from superficial to deep.

The muscles of the anterior aspect of the forearm (Figure 10-23) are surrounded by an aponeurotic tissue, the antebrachial fascia, which resembles a dressing (Figure 10-24). This also provides insertion for both the flexors and the superficial extensors. The antebrachial fascia is strongly fixed to the medial and lateral epicondyles, the olecranon, and the posterior edge of the ulna. The forearm compartments are separated by connective tissue expansions of this fascia as well as the bones and interosseous membrane of the forearm.

First Layer
Together, the muscles of the first layer are called the epitrochlear muscles because they share a common insertion at the epitrochlea, or medial humeral epicondyle. The epitrochlear muscles are the pronator teres, the flexor carpi radialis, the palmaris longus, and the flexor carpi ulnaris (Table 10-6).

Pronator teres The pronator teres (Figure 10-25), which forms the medial border of the elbow fossa, is the most lateral muscle of the group. It has two separate origins: (1) the humeral head, which originates on the anterior aspect of the medial epicondyle, the medial intermuscular septum of the arm, and the intermuscular epitrochlear septum, which separates it from the flexor carpi radialis; and (2) the ulnar head, which originates on the medial aspect of the coronoid

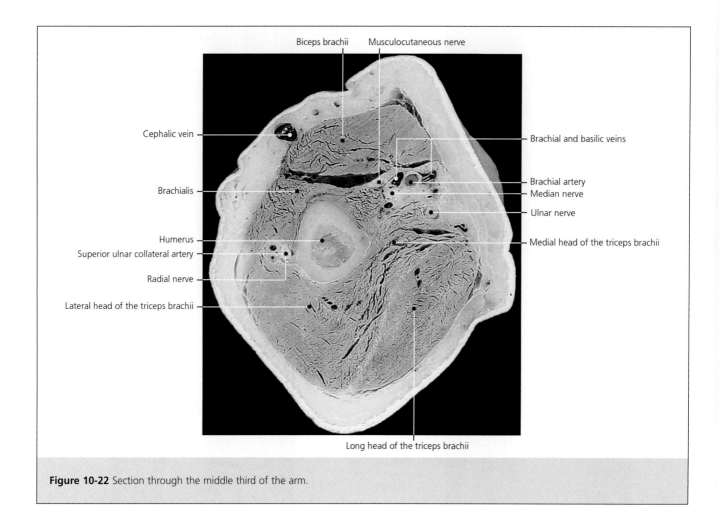

Figure 10-22 Section through the middle third of the arm.

process. Both heads are directed obliquely in an inferolateral direction until they coalesce, forming a flat tendon that wraps around the lateral aspect of the radius. Its traction forms the pronator tuberosity.

The pronator teres is innervated by the median nerve (C6-C7). The median nerve enters the forearm between the humeral (superficial) and ulnar (deep) heads of the muscle. The ulnar artery is located beneath the ulnar head.

The pronator teres has two functions. Its main function is forearm pronation. Through the epitrochlear fibers, it also contributes to elbow flexion.

Flexor carpi radialis The flexor carpi radialis (palmaris major) (Figures 10-23 and 10-26) is a fusiform muscle located on the medial side of the pronator teres. It also originates at the medial epicondyle through the common origin of the epitrochlear muscles as well as from the antebrachial fascia and the epitrochlear intermuscular septum, which separates these muscles into different compartments. All these fascicles coalesce into a muscle belly that continues in the midforearm as a long tendon attaching to the volar aspect of the second metacarpal (Figure 10-27).

This tendon has its own tunnel inside the carpal tunnel, which serves as a tendinous sheath for the flexor carpi radialis tendon and protects it from friction (Figures 10-28 and 10-29).

The flexor carpi radialis is innervated by the median nerve (C6 through C8).

The flexor carpi radialis produces carpal flexion and radial deviation (abduction). In addition, it contributes to forearm pronation and elbow flexion.

Palmaris longus The palmaris longus (palmaris minor) (Figure 10-23) is not always present. It is located on the medial aspect of the flexor carpi radialis, and it shares the same proximal origin. It has a very short muscular body, which continues as a long tendon superficially located in the anterior aspect of the forearm; its attachment into the flexor retinaculum and vertex of the palmar aponeurosis, which it continues, is fan shaped.

The palmaris longus is innervated by the median nerve (C7 through T1).

The main action of the palmaris longus is wrist flexion through traction on the flexor retinaculum. It also tightens the palmar aponeurosis, contributing to slight

Myology

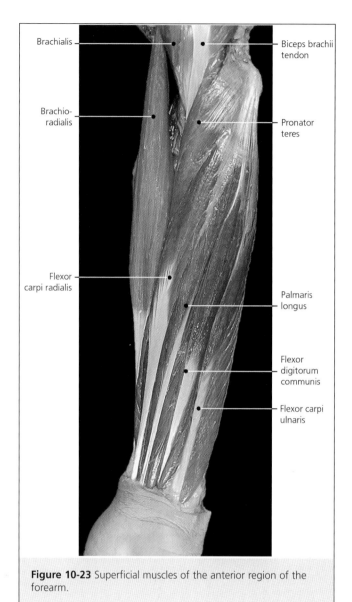

Figure 10-23 Superficial muscles of the anterior region of the forearm.

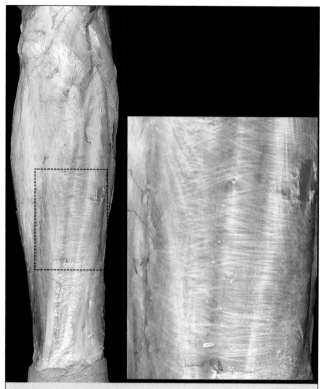

Figure 10-24 Dorsal view of the forearm. Detail shows the antebrachial fascia.

flexion of the fingers. It is minimally involved in elbow flexion.

Flexor carpi ulnaris The flexor carpi ulnaris (cubitalis anterior), located adjacent to the palmaris longus (Figure 10-23), is the most medial muscle in the first forearm layer. It has two proximal insertions, the humeral head and the ulnar head (Figure 10-30). The humeral head originates at the medial epicondyle, the medial intermuscular septum of the arm, and the epitrochlear intermuscular septum. The ulnar head originates on the posteromedial edge of the ulna, the medial aspect of the olecranon, and the antebrachial fascia. These fascicles form an arch, the arcade of Osborne, the floor of which forms the sulcus for the ulnar nerve. This epitrochlear-olecranon tunnel allows the passage of the ulnar nerve and the posterior branch of the recurrent ulnar artery. The two portions described form a muscle belly that continues medially as a tendon attached to the pisiform. (The pisiform is considered a sesamoid within this tendon.) The tendon extends to the hook of the hamate through the pisohamate ligament and to the base of the fifth metacarpal through the pisometacarpal ligament (Figures 10-31 and 10-32). Other tendon fibers extend into the palmar aponeurosis.

The flexor carpi ulnaris is innervated by the ulnar nerve (C7 through T1).

The flexor carpi ulnaris contributes to flexion and ulnar deviation (adduction) of the wrist as well as elbow flexion.

Second Layer—Flexor Digitorum Superficialis

The only muscle of the second layer is the flexor digitorum superficialis (Figure 10-33), a flat, wide muscle located beneath the pronator teres, flexor carpi radialis, palmaris longus, and flexor carpi ulnaris. The flexor digitorum superficialis has two heads: the ulnohumeral head originates at the medial epicondyle, the medial collateral ligament, and the coronoid process; and the radial head attaches to the anterior aspect of the radius. Both heads delineate a free space beneath the arch of this muscle that serves to allow passage of the median nerve and the ulnar artery. The fibers of this broad muscle are arranged in two planes; the fibers for the long (III) and ring (IV) fingers are more

TABLE 10-6 MUSCLES OF THE FIRST LAYER OF THE ANTERIOR COMPARTMENT OF THE FOREARM

MUSCLE	Pronator teres	Flexor carpi radialis	Palmaris longus (variable)	Flexor carpi ulnaris
ORIGIN	*Humeral head:* medial epicondyle *Ulnar head:* coronoid process	Medial epicondyle	Medial epicondyle	*Humeral head:* medial epicondyle *Ulnar head:* medial aspect of the olecranon
INSERTION	Pronator tuberosity of the radius	Volar aspect of the second metacarpal	Flexor retinaculum of the palmar aponeurosis	Pisiform; also extends to the hook of the hamate via the pisohamate ligament, and to the base of the fifth metacarpal via the pisometacarpal ligament
INNERVATION	Median nerve	Median nerve	Median nerve	Ulnar nerve
FUNCTION	Pronator and flexion of the elbow	Flexion and abduction of the wrist; also contributes to forearm pronation and elbow flexion	Wrist flexion via flexor retinaculum; also tightens the palmar aponeurosis and contributes to slight finger flexion	Flexion and ulnar deviation (adduction) of the wrist

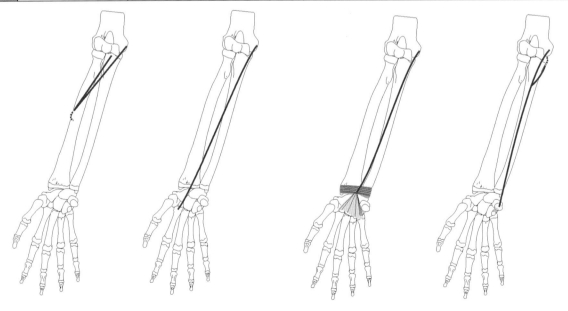

Myology

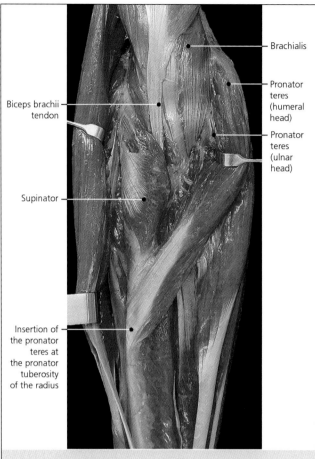

Figure 10-25 Anatomic details of the pronator teres and its attachment on the radius.

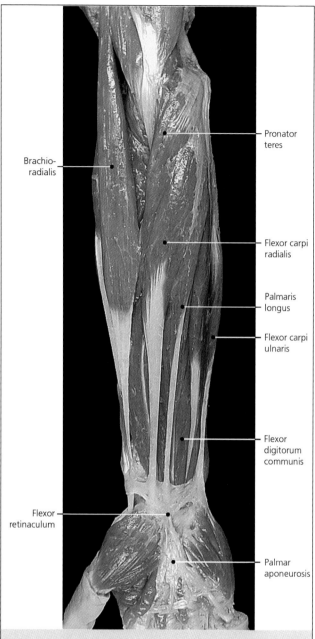

Figure 10-26 Anterior view of the superficial muscles of the forearm and hand.

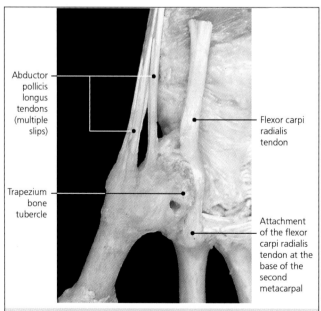

Figure 10-27 Specimen of the deep anterior aspects of the radial side of the carpus shows the attachments of the flexor carpi radialis and the abductor pollicis longus.

superficial, whereas the fibers for the index (II) and little (V) fingers are deeper. This arrangement facilitates the passage of these and other flexor tendons through the carpal tunnel. The flexor digitorum superficialis tendons are attached to the middle phalanx through two tendinous bands (Figures 10-34 through 10-36). These bands form an arch for the tendon of the flexor digitorum profundus. For this reason, the flexor digitorum superficialis is also known as the perforated flexor muscle, whereas the flexor profundus is known as the perforating flexor muscle. This relationship is explained below, in the discussion of the flexor digitorum profundus.

The flexor digitorum superficialis is innervated by the median nerve (C7 through T1).

The flexor digitorum superficialis contributes to progressive flexion of different joints, including the proximal interphalangeal joint, the metacarpophalangeal joint, the carpometacarpal joint, and the wrist. The flexor digitorum superficialis is also a very weak elbow flexor.

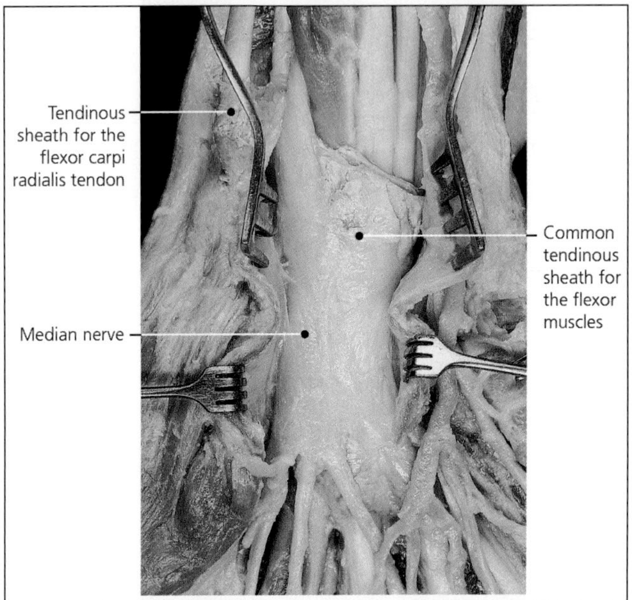

Third Layer

The third layer is located between the flexor digitorum superficialis and the forearm bones and interosseous membrane. Some authors believe that originally this was a single mass that over time became specialized through the development of a flexor tendon for the thumb. The muscles of the third layer are the flexor digitorum profundus and the flexor pollicis longus.

Flexor digitorum profundus The flexor digitorum profundus (Figure 10-37) covers the anteromedial aspect of the arm. It originates on the anterior aspect of the ulna and the interosseous membrane and occasionally reaches the anterior aspect of the radius. This muscle continues distally and divides in the midforearm into four tendons attached to the distal phalanx after passing through the carpal tunnel and the flexor

Figure 10-28 Anatomic specimen of the volar wrist with the tendinous sheaths injected with blue latex. Note the course of the flexor carpi radialis through an independent tendon sheath, open in this illustration.

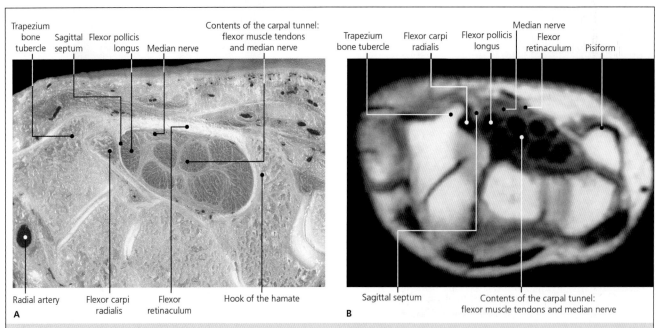

Figure 10-29 A, Transverse carpal section showing the structure and contents of the carpal tunnel. Note the course of the flexor carpi radialis tendon is separated from the remaining structures by a sagittal septum arising from the flexor retinaculum. **B,** CT at a level slightly more proximal than the specimen in **A,** showing the main contents of the carpal tunnel.

Myology

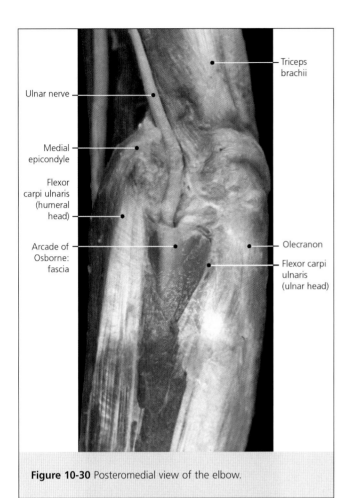

Figure 10-30 Posteromedial view of the elbow.

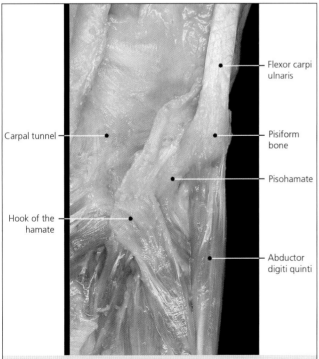

Figure 10-32 Anterior view of the distal aspect of the wrist after removing the contents of the carpal tunel. The intrinsic muscles have been maintained.

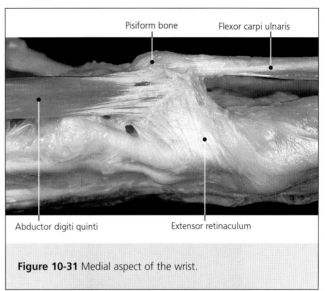

Figure 10-31 Medial aspect of the wrist.

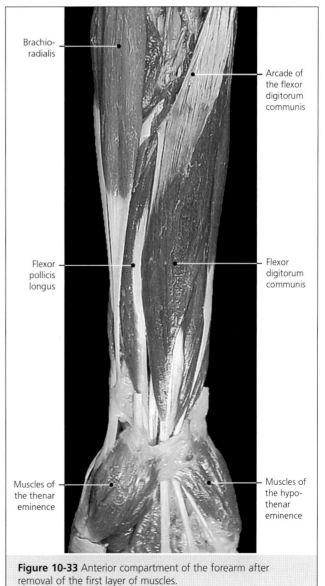

Figure 10-33 Anterior compartment of the forearm after removal of the first layer of muscles.

American Academy of Orthopaedic Surgeons

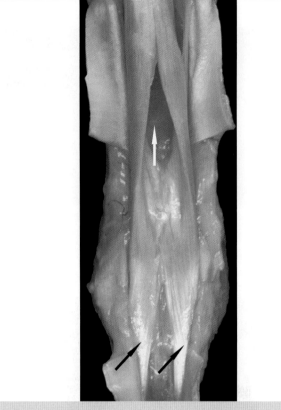

Figure 10-34 Anatomic details of the insertion of the flexor superficialis through two slips (black arrows) at the base of the middle phalanx of a triphalangeal finger. They form an opening (white arrow) for the tendon of the flexor digitorum profundus.

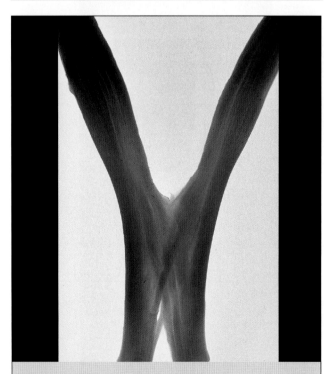

Figure 10-35 Flexor digitorum communis slips share interconnecting fibers at Camper's chiasma.

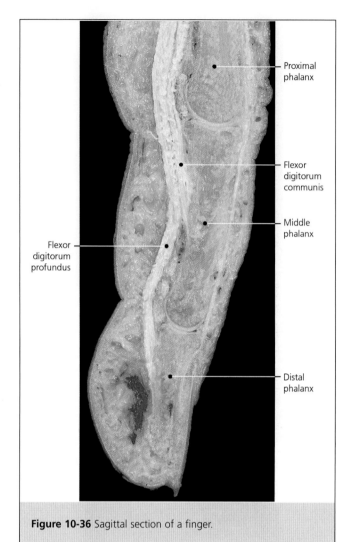

Figure 10-36 Sagittal section of a finger.

superficialis tendon (Figure 10-38). Both flexor tendons cross at the level of the proximal phalanx, at the tendinous chiasma (Figures 10-38 through 10-41). The distal attachment of the flexor digitorum profundus is a wide, flat, fan-shaped tendon.

The flexor digitorum profundus has a double innervation. The fibers that flex the index and long fingers are innervated by the anterior interosseous nerve, a branch of the median nerve. The ulnar nerve supplies the fibers for the ring and little fingers [C7-T1].

The function of the flexor digitorum profundus is similar to the function of the flexor communis, but in addition it flexes the distal phalanx. Thus, to differentiate them in the clinical examination, the flexor communis flexes only the proximal interphalangeal joint, whereas the flexor profundus flexes the distal interphalangeal joint as well.

The common sheath of the flexor tendons covers the tendons as they pass through the carpal tunnel. This sheath continues distally until it completely covers the tendon for the little finger. On the index, long,

Myology

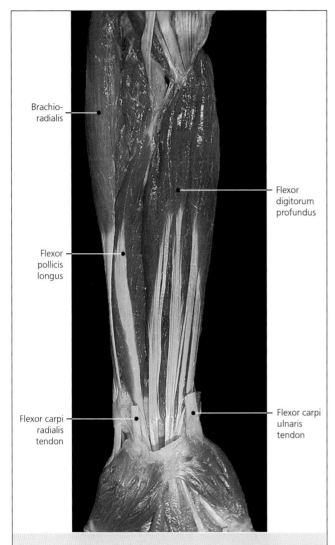

Figure 10-37 Specimen showing the third topographic layer of the anterior compartment of the forearm.

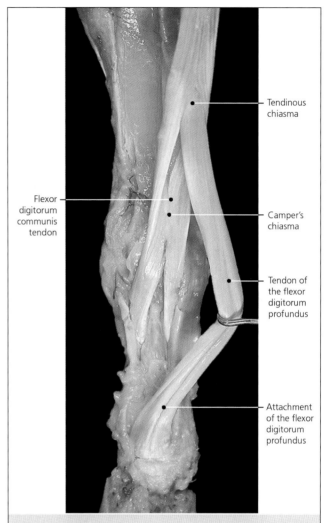

Figure 10-38 Anterior view of a finger showing the relationship between the tendons of the two flexor muscles of the fingers.

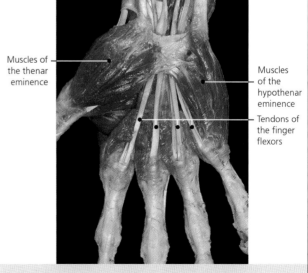

Figure 10-39 Anterior view of the muscles of the hand after removing the palmar aponeurosis.

and ring fingers, the sheath is interrupted at the metacarpal level, although it reappears distally as a synovial sheath for the fingers.

Flexor pollicis longus The flexor pollicis longus is located on the lateral side (Figure 10-42). It originates on the anterior aspect of the radius and the lateral aspect of the interosseous membrane. Occasionally it has a reinforcement fascicle, known as the accessory fascicle of Gantzer (Figure 10-43), which is quite variable. It can originate at the coronoid process, the medial epicondyle, or the epicondylar muscle mass.

The fibers of this muscle are directed vertically and distally, on the lateral aspect of the flexor digitorum profundus. It continues as a tendon, which passes through the most lateral aspect of the carpal tunnel, reflecting on the scaphoid tubercle. It perforates the two fascicles of the flexor pollicis brevis to reach its

American Academy of Orthopaedic Surgeons

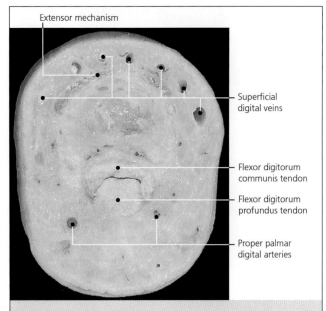

Figure 10-40 Transverse section of a finger through the proximal phalanx.

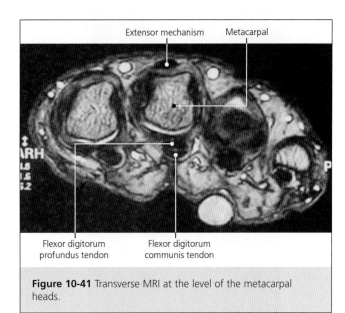

Figure 10-41 Transverse MRI at the level of the metacarpal heads.

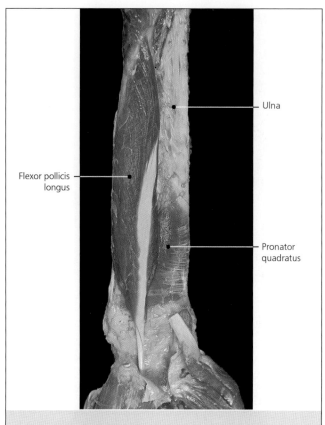

Figure 10-42 Anterior view of the deep muscles of the forearm.

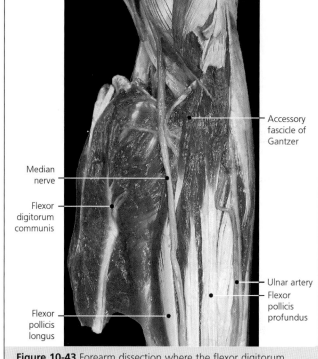

Figure 10-43 Forearm dissection where the flexor digitorum communis has been retracted to show the accessory fascicle of Gantzer.

attachment at the base of the distal phalanx of the thumb. This tendon is covered by the tendinous sheath for the flexor pollicis longus, extending from the carpal tunnel to the distal phalanx.

The flexor pollicis longus is innervated by the anterior interosseous nerve, a branch of the median nerve (C6 through C8).

The main function of the flexor pollicis longus is flexion of the distal phalanx over the proximal phalanx of the thumb, and then over the metacarpal and trapezius. It also contributes to wrist flexion.

MYOLOGY

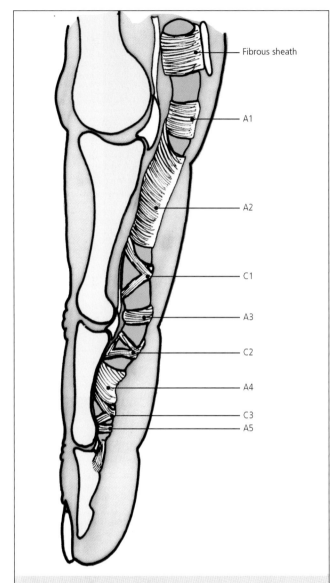

Figure 10-44 Schematic showing the fibrous sheaths (with pulleys) and synovium (in blue) in a triphalangeal finger.

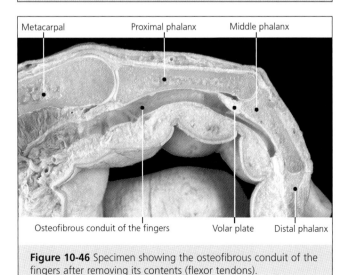

Figure 10-46 Specimen showing the osteofibrous conduit of the fingers after removing its contents (flexor tendons).

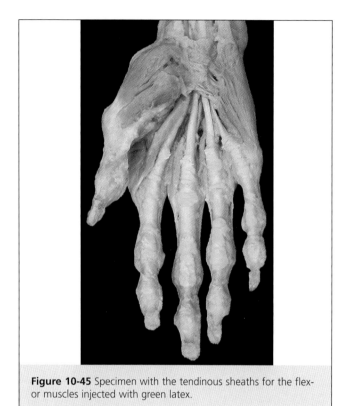

Figure 10-45 Specimen with the tendinous sheaths for the flexor muscles injected with green latex.

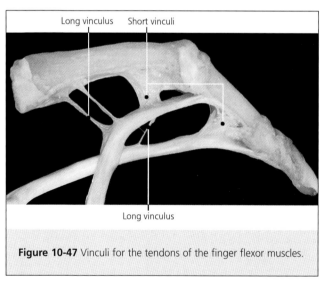

Figure 10-47 Vinculi for the tendons of the finger flexor muscles.

Relationships Among the Distal Flexor Tendons

The muscles are attached to the phalanges through the fibrous sheaths for the fingers (Figures 10-44 and 10-45), which limit the separation of the tendons from the fingers with flexion, acting as pulleys. The fibrous sheaths have five annular portions, composed of transverse fibers and numbered as A1, A2, A3, A4, and A5; and three cruciform portions, formed by cruciate fibers and numbered as C1, C2, and C3. Globally, the fibrous sheath forms an osteofibrous conduit for the flexor tendons (Figure 10-46), inside a common synovial sheath.

American Academy of Orthopaedic Surgeons 143

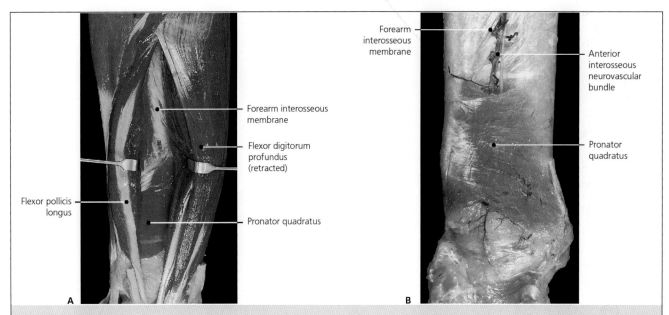

Figure 10-48 Deepest (last) layer of the forearm. **A,** The pronator quadratus is shown after removing the muscles of the third layer. **B,** Specimen showing this muscle isolated.

When the flexor tendons are separated, small fibrous condensations that lead branches of the collateral vessels inside the sheath and tendon, the so-called vincula tendina, can be observed (Figure 10-47). The vincula longa supply the flexor tendons and are located at the level of the proximal phalanx; the vincula brevia are directed to just one tendon and are located in the proximity of the distal insertion of these tendons.

Fourth Layer—Pronator Quadratus

The pronator quadratus is the only muscle in the fourth layer (Figures 10-48 and 10-49). This flat, square muscle is located in the distal quarter of the forearm. It originates as a tendon on the anterior aspect of the ulna, and its fibers are transversely oriented to its fleshy insertion on the anterior aspect of the radius.

The pronator quadratus is innervated by the anterior interosseous antebrachial nerve (C7 through T1).

The main role of the pronator quadratus is forearm pronation. It is fixed on the ulna and pulls on the radius to make it rotate, producing pronation. In addition, along with the interosseous membrane, it maintains the relationship between the radius and the ulna, especially when pushing on an object with the palm of the hand with the wrist flexed.

Table 10-7 lists the origin, insertion, innervation, and function of each muscle of the second, third, and fourth layers of the anterior compartment of the forearm.

Posterior Compartment

This region has two layers, a superficial layer and a deep layer. The superficial layer has four muscles, and the deep layer has five muscles.

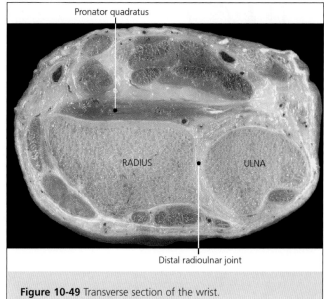

Figure 10-49 Transverse section of the wrist.

Superficial Layer

The muscles of the superficial layer share a proximal insertion on the lateral humeral epicondyle, so they are included with the epicondylar muscles, along with the muscles on the radial aspect of the posterior compartment. The muscles of this layer are the anconeus, the extensor digitorum communis, the extensor digiti quinti, and the extensor carpi ulnaris.

Table 10-8 lists the origin, insertion, innervation, and function of the muscles of the superficial layer of the posterior compartment of the forearm.

MYOLOGY

TABLE 10-7 MUSCLES OF THE SECOND, THIRD, AND FOURTH LAYERS OF THE ANTERIOR COMPARTMENT OF THE FOREARM

MUSCLE	Flexor digitorum communis	Flexor digitorum profundus	Flexor pollicis longus	Pronator quadratus
ORIGIN	Medial epicondyle, coronoid process, and anterior aspect of the radius	Proximal half of the ulna and interosseous membrane	Anterior aspect of the radius (medial epicondyle)	Anterior aspect of the distal fourth of the ulna
INSERTION	Body of the middle phalanx of the fingers	Anterior aspect of the base of the distal phalanx of the fingers	Anterior aspect of the base of the distal phalanx of the thumb	Anterior aspect of the distal fourth of the radius
INNERVATION	Median nerve	Median nerve (index and long fingers) and ulnar nerve (ring and little fingers)	Median nerve	Median nerve
FUNCTION	Flexion of the proximal interphalangeal joints and other joints where the tendons go through	Flexion of the distal interphalangeal joint and other joints where the tendons go through	Flexion of the distal interphalangeal joint of the thumb and other joints where the tendons go through	Pronation

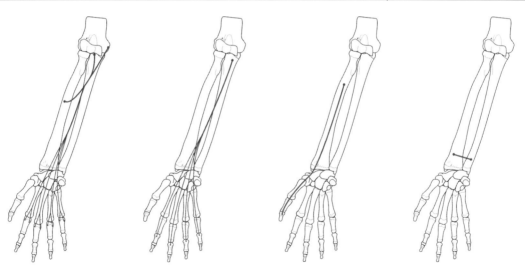

Anconeus The anconeus (Figure 10-50) is a triangular muscle that may be considered part of the elbow joint. In reality, the anconeus is the lateral and distal extension of the medial head of the triceps, as proved by their shared innervation. Its fibers fan from the lateral epicondyle and the lateral collateral ligament to the lateral edge of the olecranon and the posterior aspect of the ulna. It is separated from the elbow joint through a synovial bursa that communicates with the synovial membrane of the joint.

The anconeus is innervated by the radial nerve (C7-C8).

The anconeus functions as an accessory muscle for elbow extension, tightening the elbow joint capsule so that it is not impinged during motion.

Extensor digitorum communis The extensor digitorum communis (Figure 10-51) originates on the posterior aspect of the lateral epicondyle, the antebrachial fascia, and the intermuscular epicondylar septum. Its fibers are longitudinally oriented and are divided into three fascicles: one for the index finger, one for the long finger, and one for the ring and little fingers. The tendons for the fingers stay in the same plane and enter the fourth extensor compartment until they attach to the extensor mechanism. The extensor digitorum communis tendons have intertendinous connections (junctura tendinum) from one tendon to the adjacent tendon, at the level of the metacarpals (Figure 10-52). They are consistent between the tendons for the little and ring fingers as well as the long and ring fingers. These intertendinous connections limit independent finger extension through the extensor digitorum communis, although other tendons remain free. The extensor tendons share a sheath with the extensor indicis proprius at the fourth extensor compartment of the extensor retinaculum (Figures 10-53 through 10-55 and Table 10-9).

The extensor digitorum communis is innervated by the radial nerve (C6 through C8).

The function of the extensor digitorum communis is to extend and abduct the fingers; it also contributes to wrist adduction. The extensor digitorum communis also extends metacarpophalangeal joints II through V and can contribute to interphalangeal joint extension.

Extensor digiti quinti The extensor digiti quinti (Figure 10-51) is located on the medial aspect of the extensor digitorum communis, and it seems to be part of the extensor digitorum communis as they share the same innervation and are separated by only a thin

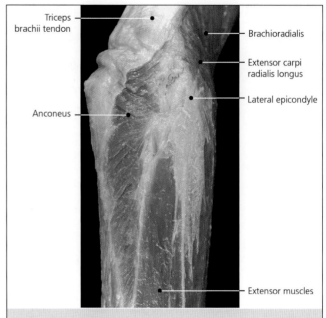

Figure 10-50 Superficial view of the lateral aspect of the forearm.

TABLE 10-8 MUSCLES OF THE SUPERFICIAL LAYER OF THE POSTERIOR COMPARTMENT OF THE FOREARM

MUSCLE	Anconeus	Extensor digitorum communis	Extensor digiti quinti	Extensor carpi ulnaris
ORIGIN	Lateral epicondyle	Lateral epicondyle	Lateral epicondyle	Lateral epicondyle; proximal posterolateral aspect of the ulna
INSERTION	Lateral edge of the olecranon; proximal posterior aspect of the ulna	Extensor mechanism of the fingers; base of the distal phalanx of the fingers	Extensor mechanism of the fingers; base of the distal phalanx of the wrist	Styloids of fifth metacarpal
INNERVATION	Radial	Radial	Radial	Radial
FUNCTION	Slight extension of the elbow	Extension of the triphalangeal fingers and the joints	Extension of the wrist and the joints	Extension and ulnar deviation of the wrist

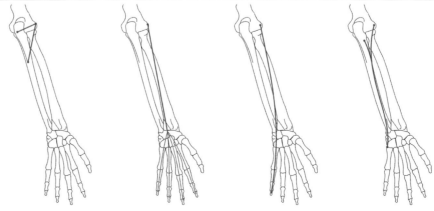

Myology

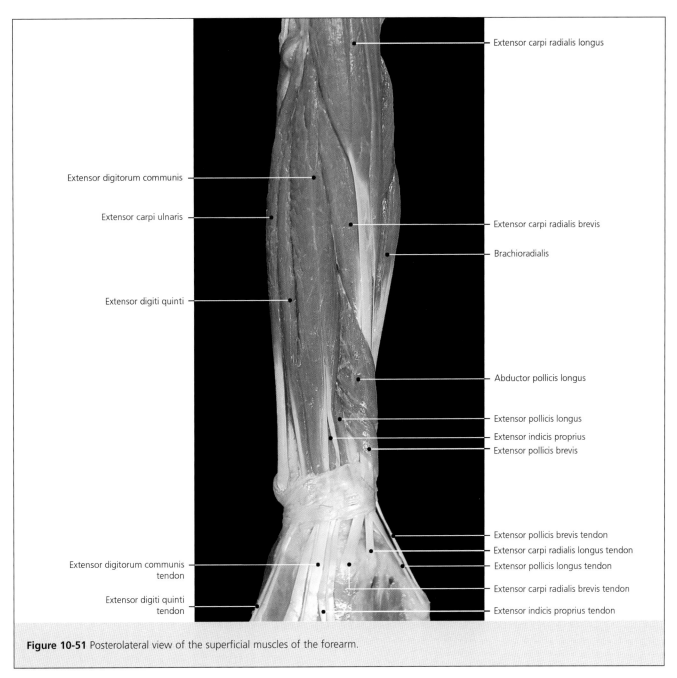

Figure 10-51 Posterolateral view of the superficial muscles of the forearm.

tendinous layer. The extensor digiti quinti originates on the lateral epicondyle, the antebrachial fascia, and the epicondylar intermuscular sheaths. Its fibers continue to the wrist, where it passes through the fifth extensor compartment protected by the tendinous sheath for the extensor digiti quinti (Figures 10-53 through 10-55). From the dorsum, it is seen on the medial side of the extensor digitorum communis and may have one or two tendons (Figure 10-56). These tendons insert through the extensor mechanism to the base of the distal phalanx.

The extensor digiti quinti is innervated by the radial nerve (C7-C8).

The extensor digiti quinti has the same function as the extensor digitorum communis; that is, reinforcement of extension of the little finger.

Extensor carpi ulnaris The extensor carpi ulnaris (cubitalis posterior) (Figure 10-51) is the most medial tendon on this plane. It has two heads, one humeral and one ulnar. The humeral head originates at the lateral epicondyle, the radial collateral ligament, and the epicondylar intermuscular sheaths. The ulnar head originates at the posterior aspect of the ulna and the antebrachial fascia. These fibers coalesce in a single muscle belly and continue into a tendon. This tendon passes through the sixth extensor compartment protected by the tendinous sheath of the extensor carpi ulnaris (Figures 10-53 through 10-55) and attaches to the ulnar styloid and the base of the fifth metacarpal.

The extensor carpi ulnaris is innervated by the radial nerve (C7-C8).

The main function if the extensor carpi ulnaris is wrist extension and adduction (ulnar deviation).

Deep Layer

To access the deep layer, it is necessary to remove the superficial or epicondylar muscles. The deep layer includes the supinator muscle as well as muscles for individual digits, which confer some independent movement, allowing actions to be performed more accurately.

Table 10-10 lists the origin, insertion, innervation, and function of the muscles of the deep layer of the posterior compartment of the forearm.

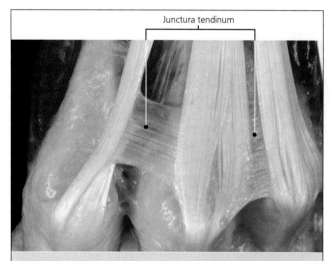

Figure 10-52 Anatomic detail of the intertendinous connections of the extensor digitorum communis muscle.

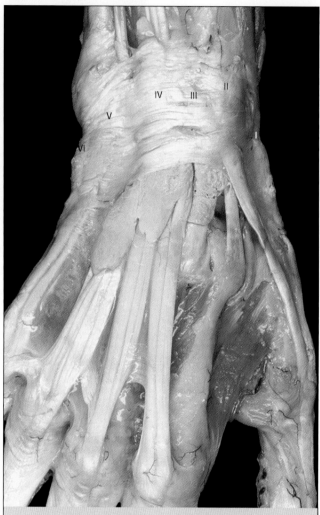

Figure 10-53 Specimen showing the sheaths for the carpal and finger extensor tendons injected with latex (see Table 10-9).

TABLE 10-9 EXTENSOR COMPARTMENTS OF THE WRIST

I Abductor pollicis longus Extensor pollicis brevis [Tendinous sheath for the abductor pollicis longus and extensor pollicis brevis]	II Extensor carpi radialis longus Extensor carpi radialis brevis [Tendinous sheath for the extensor carpi radialis]	III Extensor pollicis longus [Tendinous sheath for the extensor pollicis longus]
IV Extensor digitorum communis Extensor indicis proprius [Tendinous sheath for the extensor digitorum communis and the extensor indicis proprius]	V Extensor digiti quinti [Tendinous sheath for the extensor digiti quinti]	VI Extensor carpi ulnaris [Tendinous sheath for the extensor carpi ulnaris]

Myology

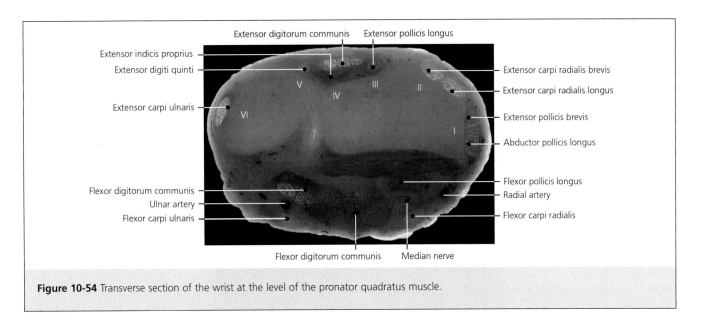

Figure 10-54 Transverse section of the wrist at the level of the pronator quadratus muscle.

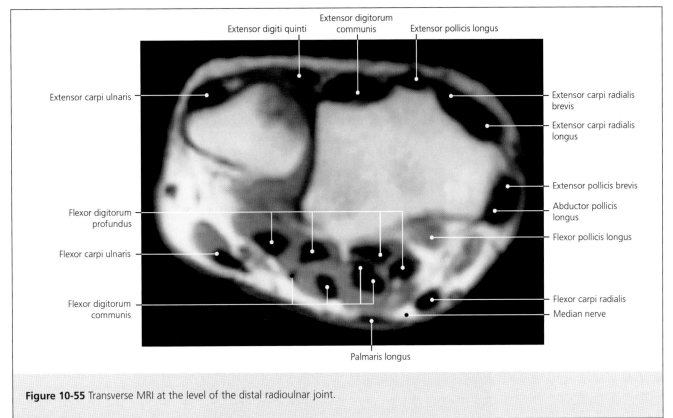

Figure 10-55 Transverse MRI at the level of the distal radioulnar joint.

Supinator The supinator muscle (Figures 10-25 and 10-57) is a flat muscle located deep in the posterolateral region of the elbow. It originates through a tendon on the lateral epicondyle, supinator crest, lateral aspect of the ulna, posterior aspect of the annular ligament, and radial collateral ligament. From these multiple origins, the fibers fan out in two layers through which the radial nerve passes (Figure 10-58). The superficial layer is formed by oblique fibers. The deep layer is wider, with fibers that are more horizontal and arise from the supinator crest of the ulna. These two layers are separated by the fatty tissue that protects the radial nerve during motion. All the fibers of the supinator surround the radius, from posterior to anterior, attaching to the posterior edge and the lateral and anterior faces of the radius. A serous bursa lies between this muscle and the extensors to facilitate motion.

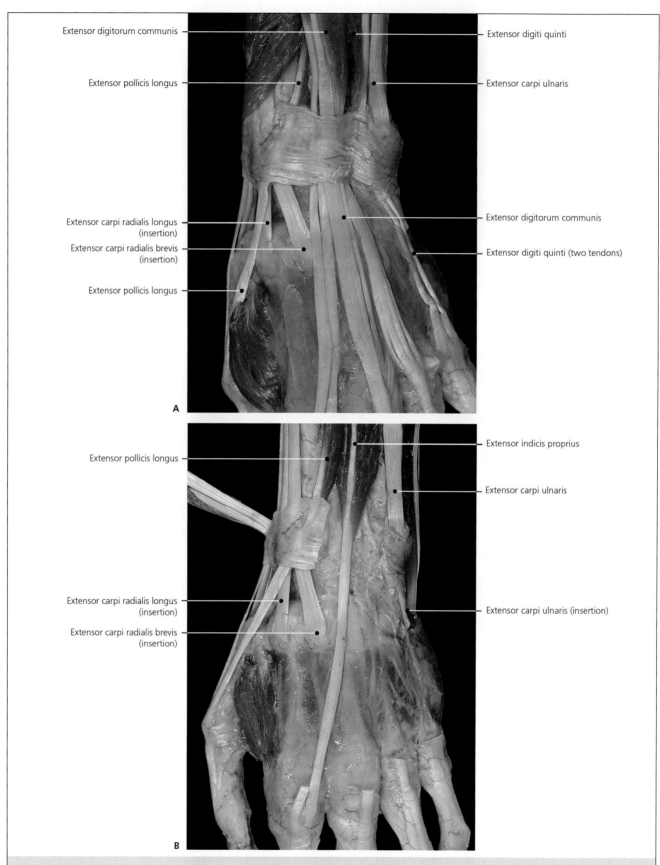

Figure 10-56 A, Anatomic details of the tendons at the dorsal aspect of the wrist. **B,** Anatomic details of the tendons of the wrist region after removal of the extensor muscles of the fingers and extensor for the little finger.

MYOLOGY

TABLE 10-10 MUSCLES OF THE DEEP LAYER OF THE POSTERIOR COMPARTMENT OF THE FOREARM

MUSCLE	Supinator	Abductor pollicis longus	Extensor pollicis brevis	Extensor pollicis longus	Extensor indicis proprius
ORIGIN	Lateral epicondyle Crista supinatoris	Posterior aspect of the interosseous membrane and adjacent areas of radius and ulna	Posterior aspect of the interosseous membrane Posterior aspect of the radius	Posterior aspect of the interosseous membrane Posterior aspect of the ulna	Posterior aspect of the interosseous membrane Posterior aspect of the ulna
INSERTION	Wrapped around the proximal fourth of the radius	Base of the first, or thumb, metacarpal	Extensor mechanism Base of the proximal phalanx of the thumb	Extensor mechanism Base of the distal phalanx of the thumb	Extensor mechanism Base of the distal phalanx of the thumb
INNERVATION	Radial nerve	Radial nerve	Radial nerve	Radial nerve	Radial nerve
FUNCTION	Supination	Thumb abduction and radial deviation (abduction) of the carpus	Thumb extension and radial deviation (abduction) of the carpus	Thumb extension and radial deviation (abduction) of the carpus	Index extension and wrist extension

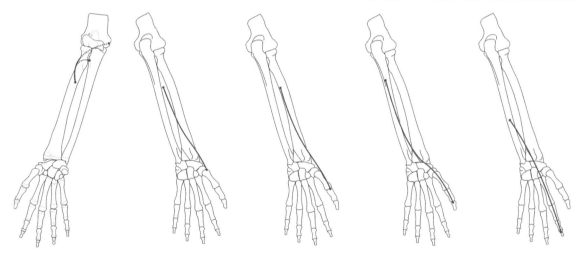

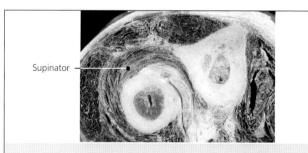

Figure 10-57 Transverse section at the level of the proximal third of the forearm showing the location of the supinator muscle.

The supinator muscle is innervated by the radial nerve (C6). After perforating this muscle, the deep branch of the radial nerve turns posteriorly and branches out in the dorsal aspect of the forearm.

The function of the supinator is, as its name implies, to supinate the forearm. During pronation and supination, the ulna acts as the fixed point; its trochlear joint with the distal humerus allows no rotation. Thus, muscle contraction uncoils the supinator, producing lateral rotation of the radius, or supination. This motion occurs at the forearm independently of elbow flexion and extension, due to the function of both radioulnar joints.

Abductor pollicis longus The abductor pollicis longus (Figure 10-59) is proximal and lateral to the extensor pollicis brevis, extensor pollicis longus, and extensor indicis proprius. It originates through fleshy insertions underneath the anconeus, on the posterior aspect of the interosseous membrane, and on the adjacent surfaces of the radius and ulna. Its fibers are obliquely oriented in a radial direction and continue as one or more tendons that cross the tendons of the extensor carpi radialis longus and brevis. A serous bursa separates the deeper radial extensors from the abductor

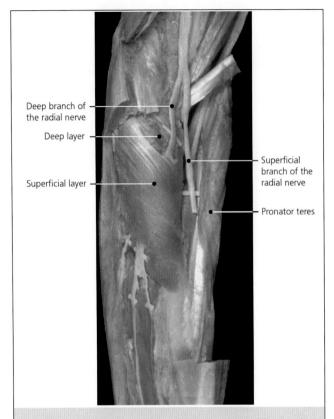

Figure 10-58 Lateral view of the elbow. The epicondylar muscles and the brachioradialis have been resected.

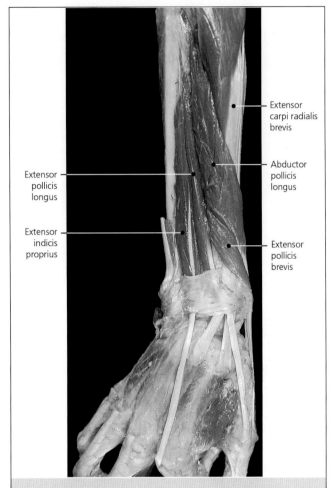

Figure 10-59 Dissection of the deep dorsal muscles of the forearm (posterolateral view).

pollicis longus and extensor pollicis brevis. These tendons pass together through the first extensor compartment, protected by the tendinous sheath for the abductor pollicis longus and the extensor pollicis brevis (Figures 10-53 through 10-55). In some individuals the abductor pollicis longus has several tendons; some insert on the posterolateral aspect of the first metacarpal base and others on the trapezius and the thenar aponeurosis (Figure 10-60).

Like all extensor muscles, the abductor pollicis longus is innervated by the radial nerve (C7-C8).

The abductor pollicis longus acts as an abductor of the trapeziometacarpal joint of the thumb and, therefore, of the wrist. It also contributes slightly to wrist flexion and supination.

Extensor pollicis brevis The extensor pollicis brevis (Figure 10-61) shares the same embryologic and functional origin as the extensor pollicis longus. Its fibers originate distal to the abductor pollicis longus and are directed obliquely to end in a tendon that crosses the radial extensor muscles. After passing through the first compartment, it inserts into the distal face of the base of the proximal phalanx of the thumb. Occasionally, the extensor pollicis brevis is absent.

The extensor pollicis brevis is innervated by the radial nerve (C7-C8).

The main functions of the extensor pollicis brevis are abduction and extension of the thumb metacarpophalangeal joint. It also abducts and flexes the wrist.

Extensor pollicis longus The extensor pollicis longus (Figures 10-59 and 10-61) is located distal and medial to the extensor pollicis longus. It originates on the posterior aspect of the middle third of the ulna and the interosseous membrane. Its oblique fibers continue on through the third extensor compartment, ending as a tendon covered by the tendinous sheath of the extensor pollicis longus (Figures 10-53 through 10-55). The tendon rotates laterally over the radial tubercle and directed to the dorsal aspect of the base of the distal phalanx of the thumb.

The extensor pollicis longus is innervated by the radial nerve (C7-C8).

The functions of the extensor pollicis longus include extension of the interphalangeal and metacarpophalangeal joints of the thumb. It also contributes to wrist extension and supination.

Myology

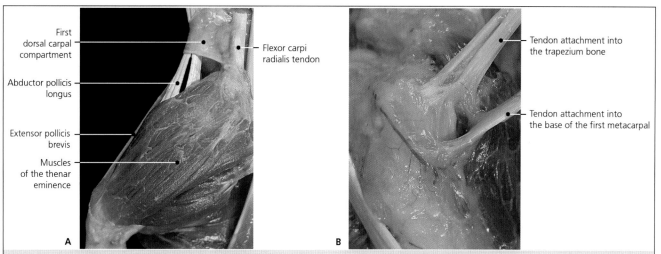

Figure 10-60 A, Palmar view of the lateral region of the wrist and proximal aspect of the hand. Note that the abductor pollicis longus has several tendons. **B,** Anatomic detail of the insertion of the tendons of the abductor pollicis longus on the trapezium bone and the base of the first metacarpal.

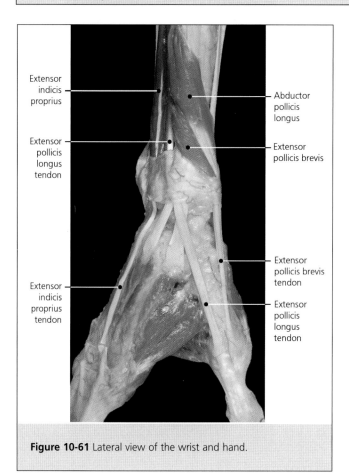

Figure 10-61 Lateral view of the wrist and hand.

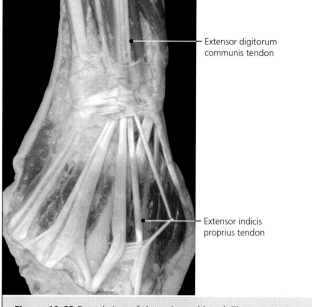

Figure 10-62 Dorsal view of the wrist and hand. The extensor digitorum communis muscle has been retracted to show how the extensor indicis proprius is independent of the intertendinous connections of the extensor digitorum communis.

Extensor indicis proprius The extensor indicis proprius (Figure 10-59) is medial and distal to the extensor pollicis longus. It originates on the posterior aspect of the ulna and the interosseous membrane. Its fibers are obliquely directed in a radial direction and continue as a tendon, which passes through the fourth extensor compartment along with the extensor digitorum communis. The extensor indicis proprius tendon continues distally into the extensor mechanism.

This tendon can be distinguished from the extensor digitorum communis tendon for the index finger in two ways. First, the tendon of the extensor indicis proprius is deeper and does not have intertendinous connections. Second, it is located on the ulnar side of the extensor digitorum communis tendon (Figure 10-62).

The extensor indicis proprius is innervated by the radial nerve (C7-C8).

TABLE 10-11 MUSCLES OF THE POSTERIOR COMPARTMENT (RADIAL ASPECT)

MUSCLE	Brachioradialis	Extensor carpi radialis longus	Extensor carpi radialis brevis
ORIGIN	Lateral edge of the humerus. Lateral intermuscular septum of the arm	Lateral supraepicondylar crest	Lateral epicondyle
INSERTION	Styloid apophysis of the radius	Base of the index finger metacarpal	Base of the long finger metacarpal (on its styloid apophysis)
INNERVATION	Radial	Radial	Radial
FUNCTION	Flexion of the elbow. Brings the forearm to the neutral position	Extension and radial deviation of the carpus	Extension of the carpus

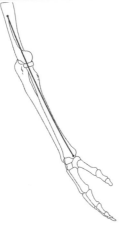

The function of the extensor indicis proprius is similar to the function of the extensor digitorum communis, reinforcing extension of the index finger and allowing action that is independent of the other fingers.

Posterior Compartment (Radial Aspect)

The radial aspect of the posterior compartment is formed by long muscles on the lateral aspect of the forearm that originate on the humerus and extend to the carpus and the radial styloid. This compartment includes three muscles: the brachioradialis, the extensor carpi radialis longus, and the extensor carpi radialis brevis (Table 10-11).

Brachioradialis

The brachioradialis (formerly called the supinator longus) (Figure 10-63) originates on the lateral aspect of the humerus, under the sulcus for the radial nerve and the lateral intermuscular septum of the arm. This flat, long muscle is directed distally on the lateral aspect of the forearm, ending medially as a broad, long tendon attached on the lateral aspect of the radial styloid. It covers the superficial (sensory) branch of the radial nerve.

The brachioradialis is innervated by the radial nerve (C5-C6) proximally to the elbow joint.

The brachioradialis is an elbow flexor. In the past, it was known as the supinator longus, but this is a misleading term because its main function is to maintain the forearm in neutral position by supinating the pronated forearm and pronating the supinated forearm.

Extensor Carpi Radialis Longus

The extensor carpi radialis longus (Figures 10-51 and 10-63) originates at the supracondylar ridge, lateral intermuscular septum, and lateral humeral epicondyle. It is located immediately distal to the brachioradialis. Its fibers are directed to the dorsal aspect of the hand, posterior to the brachioradialis. At the distal aspect of the forearm, its tendon continues along the radius and covers the tendon of the extensor carpi radialis brevis. The two tendons pass through the second extensor compartment. Both tendons share the same tendinous sheath (Figure 10-53). After leaving the sec-

Myology

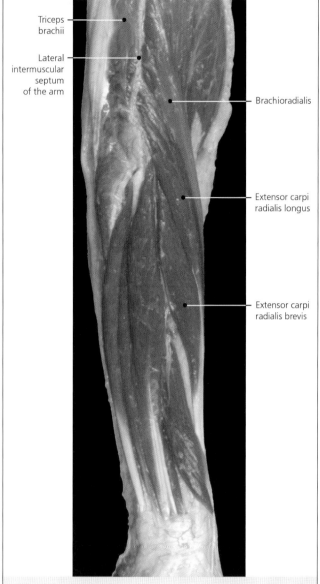

Figure 10-63 Posterolateral view of the forearm showing the superficial muscular layer.

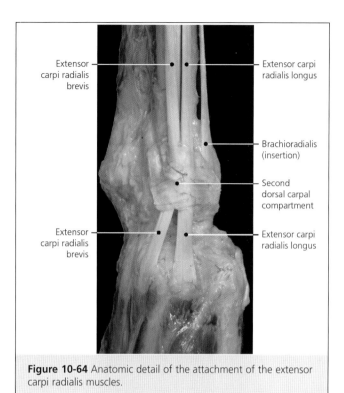

Figure 10-64 Anatomic detail of the attachment of the extensor carpi radialis muscles.

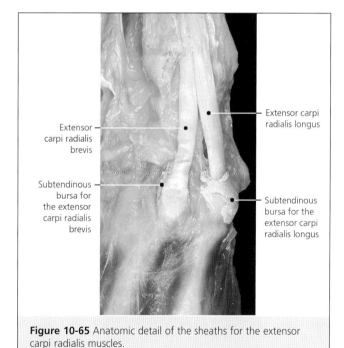

Figure 10-65 Anatomic detail of the sheaths for the extensor carpi radialis muscles.

ond extensor compartment, this tendon attaches to the base and styloid process of the second metacarpal (Figure 10-64). It is partially separated from the bone by a subtendinous bursa, which protects the tendon against friction (Figure 10-65).

The extensor carpi radialis longus is innervated by the radial nerve (C5-C6) proximal to the elbow joint.

The extensor carpi radialis longus produces extension and abduction (radial deviation) of the wrist.

Extensor Carpi Radialis Brevis

The extensor carpi radialis brevis (second external radial muscle) (Figures 10-51 and 10-63) has a common origin with the epicondylar muscles at the lateral humeral epicondyle and the intermuscular septae as well as the lateral collateral ligament. Its fibers are located under the extensor carpi radialis longus, and it continues as a long tendon to its entrance into the second extensor compartment. From there the extensor carpi radialis brevis runs obliquely, attaching to the styloid process of the third metacarpal (Figure 10-64). A subtendinous bursa that protects the tendon against friction with

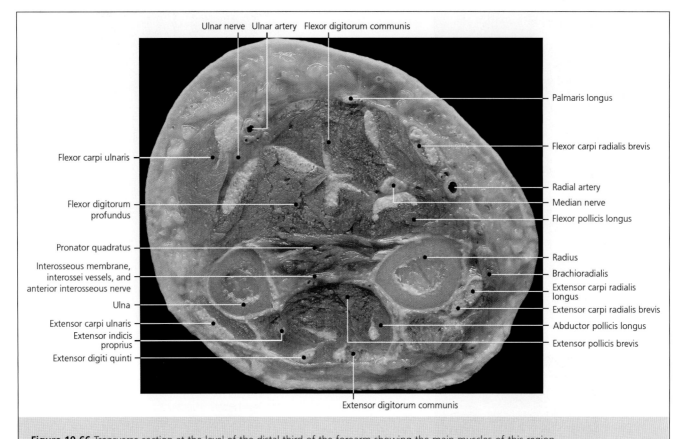

Figure 10-66 Transverse section at the level of the distal third of the forearm showing the main muscles of this region.

motion is located between the bone and the tendon (Figure 10-65).

The extensor carpi radialis brevis is innervated by the radial nerve (C6-C7).

The extensor carpi radialis brevis tendon is located close to the axis of the forearm, and that is why its main function is wrist extension. It has a minimal contribution to wrist abduction (radial deviation).

The muscle fibers of the extensor pollicis brevis and the abductor pollicis longus are located over the extensor carpi radialis tendons at the distal aspect of the forearm to reach the first extensor compartment. The extensor pollicis longus crosses the superficial aspect of these tendons after it exits the extensor compartment.

Table 10-10 lists the origin, insertion, innervation, and function of the muscles of the radial aspect of the posterior compartment. Figure 10-66 is a cross-sectional view of the muscles of this region.

Muscles of the Hand

All the muscles of the hand are located on the palmar aspect (Figure 10-67). On the dorsal aspect are only the extensor tendons and the dorsal fascia of the hand, in continuity with the antebrachial fascia on the dorsal aspect of the hand. A superficial dorsal fascia, located at the subcutaneous level, can be distinguished, as well as a deep or interosseous dorsal fascia, which covers the dorsal interossei. The extensor tendons are located between these two fascia.

The muscles of the palm of the hand are divided into three regions (Figure 10-68): the thenar eminence, the hypothenar eminence, and the middle palmar region. Remember that abduction and adduction in the hand are described with respect to an imaginary line running through the long finger, which divides the hand into two halves.

Thenar Eminence

The muscles of the thenar eminence act on the thumb, which is positioned at about a 45° angle with respect to the second metacarpal. The thenar eminence has four muscles, described here from superficial to deep.

Table 10-12 lists the origin, insertion, innervation, and function of the muscles of the thenar eminence.

Abductor Pollicis Brevis

The abductor pollicis brevis (Figure 10-69) originates at the scaphoid and the anterolateral aspect of the flexor retinaculum, and it often has an expansion to the fibrous sheath of the abductor pollicis longus. Its fibers are directed to the tubercle and lateral aspect of the base of the proximal phalanx of the thumb, the

Myology

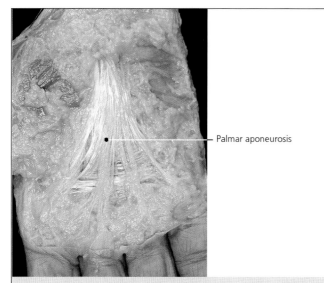

Figure 10-67 Palm of the hand showing the palmar aponeurosis and the subcutaneous adipose tissue covering the middle palmar region.

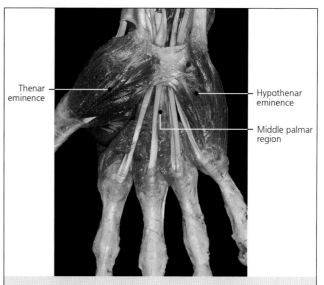

Figure 10-68 Palm of the hand with the palmar aponeurosis removed.

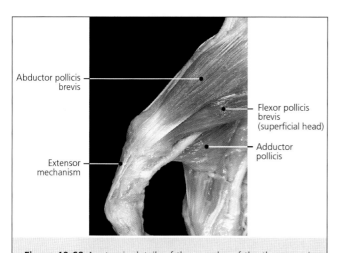

Figure 10-69 Anatomic details of the muscles of the thenar eminence.

lateral sesamoid, and the extensor mechanism; some fibers extend to the extensor pollicis brevis.

The abductor pollicis brevis is innervated by the median nerve (C6-C7).

The main function of the abductor pollicis brevis is abduction of the thumb. It places the thumb away from its anatomic position with an anterior and medial arc of motion that helps in thumb opposition. The dorsal expansion of the abductor pollicis brevis acts similarly to the interosseous muscles described below, contributing to both flexion of the metacarpophalangeal joint and extension of the interphalangeal joint.

Flexor Pollicis Brevis

The flexor pollicis brevis (Figures 10-69 and 10-70) is located in an oblique and medial position adjacent to the abductor pollicis brevis. It has two fascicles: a superficial head, which originates on the flexor retinaculum and the trapezius, and a deep head (Figure 10-71), which originates on the capitate and trapezoid bones. The flexor pollicis longus tendon is located between the two heads. Its fibers coalesce into a single muscle belly directed to the lateral sesamoid and the radial aspect of the proximal phalanx of the thumb; in some individuals, the deep head reaches the medial sesamoid. Distally, the tendon coalesces with the abductor pollicis brevis and the extensor mechanism.

The superficial head is innervated by the median nerve (C6-C7); the deep head is innervated by the ulnar nerve (C8-T1). This separate innervation explains why the superficial head is considered a true thenar muscle, whereas the deep head is considered the first interosseous muscle.

The flexor pollicis brevis produces flexion of the metacarpophalangeal joint and extension of the interphalangeal joint of the thumb through the extensor mechanism. It also contributes to thumb opposition.

Opponens Pollicis

The opponens pollicis (Figure 10-70) is a small muscle located under the abductor pollicis brevis; it is difficult to identify this muscle as a separate entity. It originates on the flexor retinaculum and the crest of the trapezius. Its fibers are oriented obliquely to its insertion into the radial aspect of the first metacarpal, slightly wrapping around it.

The opponens pollicis is innervated by the median nerve (C6-C7).

The action of the opponens pollicis is thumb opposition: it adducts, rotates, and orients the tip of the

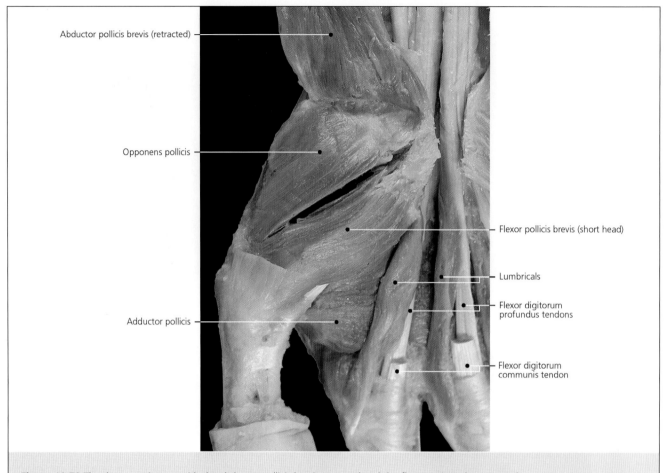

Figure 10-70 The thenar eminence with the abductor pollicis brevis removed and the flexor retinaculum opened.

thumb against the fingers, which facilitates pinch with the thumb.

The adductor pollicis (Figures 10-69 through 10-72) is the deepest and most medial muscle in the thenar eminence, covering the first two interosseous spaces on its volar side. It is formed by two heads: The transverse, or metacarpal, head originates at the anterior aspect of the third metacarpal and the base of the second metacarpal; the oblique, or carpal, head originates at the anterior aspect of the three radial bones of the first carpal row, chiefly the trapezoid and capitate, as well as on the ligament covering them. There is a hiatus between the two heads for the deep palmar arterial arch and the deep branch of the ulnar nerve. All these fibers are directed to the metacarpophalangeal joint of the thumb and attach to the medial sesamoid, the medial aspect of the base of the proximal phalanx of the thumb, and the extensor mechanism.

Adductor Pollicis
The adductor pollicis is innervated by the deep branch of the ulnar nerve (C8-T1).

The main function of the adductor pollicis is adduction and flexion of the metacarpophalangeal joint and slight internal rotation of the thumb, which approximates the thumb to the index finger.

Hypothenar Eminence
Like the thenar eminence, the hypothenar eminence contains four muscles. Three of these muscles act on the little finger, and one is subcutaneous.

Table 10-13 lists the origin, insertion, innervation, and function of the muscles of the hypothenar eminence.

Palmaris Brevis
The palmaris brevis (Figure 10-73) is formed by a series of transverse parallel fascicles that originate on the ulnar aspect of the palmar aponeurosis and the flexor retinaculum. The fibers are attached to the deep layer of the skin overlying the hypothenar aponeurosis. This rudimentary muscle has a variable size.

The palmaris brevis is innervated by the superficial branch of the ulnar nerve (C8-T1).

The function of the palmaris brevis is to contract

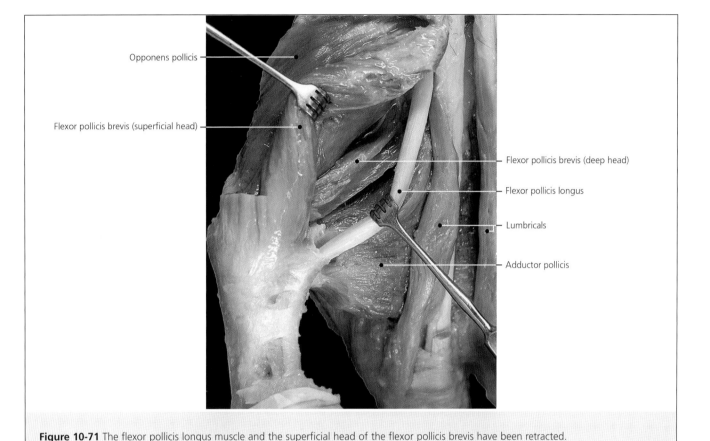

Figure 10-71 The flexor pollicis longus muscle and the superficial head of the flexor pollicis brevis have been retracted.

the skin where it is attached and protect the ulnar neurovascular bundle.

Abductor Digiti Quinti
The abductor digiti quinti (Figure 10-74) is the largest muscle in the hypothenar eminence. This flat muscle originates on the pisiform, the pisohamate ligament, and expansions of the tendon for the flexor carpi ulnaris and the flexor retinaculum. Its fibers are oriented distally, and they attach to the metacarpophalangeal joint, the ulnar aspect of the base of the proximal phalanx, and the extensor mechanism of the little finger.

The abductor digiti quinti is innervated by the deep branch of the ulnar nerve (C8-T1).

The abductor digiti quinti flexes the metacarpophalangeal joint of the little finger. It also extends the interphalangeal joints, as it shares its tendon with the abductor digiti quinti.

Opponens Digiti Quinti
The opponens digiti quinti is the deepest muscle in the hypothenar eminence. It is located under the flexor digiti quinti brevis, and it is difficult to separate the two muscles. The opponens digiti quinti arises from the hook of the hamate, the pisohamate liga-

ment, and the flexor retinaculum. Its fibers are deep to the flexor brevis and wrap around the ulnar aspect of the head and shaft of the fifth metacarpal.

The opponens digiti quinti is innervated by the deep branch of the ulnar nerve (C8-T1).

The opponens digiti quinti abducts and flexes the little finger with a small rotational component that allows opposition of the little finger to the thumb.

Middle Palmar Region
The middle palmar region includes the palmar aponeurosis and the muscles of the region, which are located deep underneath the palmar aponeurosis and the neurovascular structures of the palm. They all attach to the extensor mechanism and the extensor tendons for the fingers.

Table 10-14 lists the origin, insertion, innervation, and function of the muscles of the middle palmar region.

Palmar Aponeurosis
The antebrachial fascia continues into the palm of the hand. The fascia is thin over the thenar and hypothenar eminences and thick in the middle palmar region, where it forms the palmar aponeurosis (Figure 10-67).

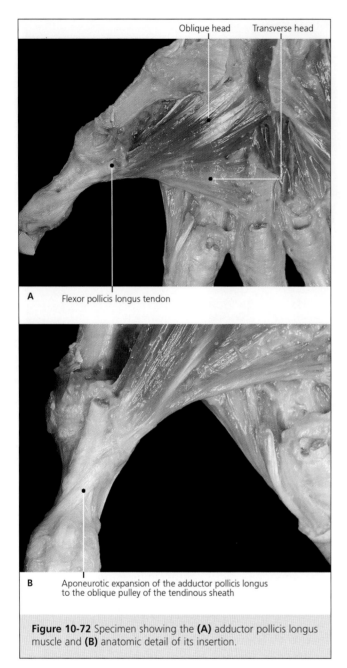

Figure 10-72 Specimen showing the **(A)** adductor pollicis longus muscle and **(B)** anatomic detail of its insertion.

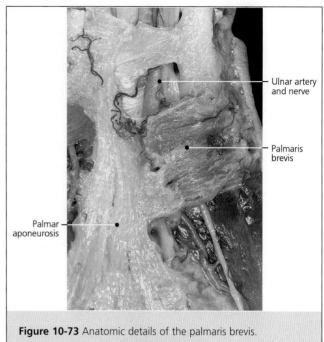

Figure 10-73 Anatomic details of the palmaris brevis.

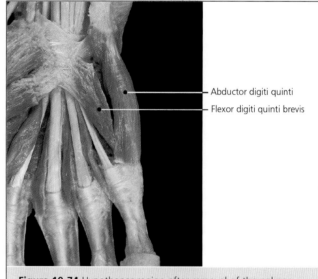

Figure 10-74 Hypothenar region after removal of the palmar aponeurosis and the palmaris brevis muscle.

The palmar aponeurosis (Figures 10-67 and 10-75) occupies the space between the two eminences of the hand. Its longitudinal fibers extend in a triangular or fan-shaped manner, with the vertex of the triangle at the attachment site of the palmaris longus. From there the fibers extend to the metacarpal heads, the fibrous sheaths for the flexor tendons, and under the skin to the bases of the proximal phalanges of the fingers. These fibers are denser at the location of the flexor muscles, forming one pretendinous band for each finger and adhering to the palmodigital skin crease. At this point, the pretendinous band divides into two fascicles around the finger and attaches to the dorsal aspect of the proximal phalanx, forming the sagittal fibrous septi. Other finer fibers located at the intertendinous spaces are known as intertendinous fibers. These fibers are directed distally and terminate in the depths of the skin, approximately at the level of the interosseous intermetacarpal spaces. There are two kinds of transverse fibers, the proximal ones and the ones located at the level of the metacarpal heads, where they are called superficial transverse intermetacarpal ligaments. These ligaments have two kinds of fibrous bundles: the long fibers are superficial and extend from the second to the fifth metacarpal, whereas the short fibers connect adjacent metacarpals. The transverse fibers cross the longitudinal fibers to

TABLE 10-12 MUSCLES OF THE THENAR EMINENCE

MUSCLE	Abductor pollicis brevis	Flexor pollicis brevis	Opponens pollicis	Adductor pollicis
ORIGIN	Flexor retinaculum Scaphoid	*Superficial head:* flexor retinaculum *Deep head:* capitate, trapezius, trapezoid, and first metacarpal	Flexor retinaculum Trapezius	*Transverse head:* II and III metacarpals *Oblique head:* capitate and radiate ligament of the carpus
INSERTION	Lateral sesamoid Radial aspect of the base of the proximal phalanx Extensor mechanism	Lateral (superficial head) and medial (deep head) sesamoids Radial aspect of the base of the proximal phalanx Extensor mechanism	First metacarpal	Medial sesamoid Ulnar aspect of the base of the proximal phalanx
INNERVATION	Median nerve	Median nerve (deep head) Ulnar nerve (superficial head)	Median nerve	Ulnar nerve
FUNCTION	Thumb abduction	Thumb flexion	Thumb opposition	Thumb adduction

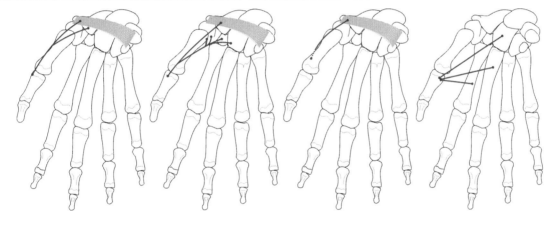

reinforce this region. Proximally located finer transverse fibers are called the transverse fascicles of the palmar aponeurosis.

The interdigital arches are spaces between the longitudinal and the transverse fibers, proximal to the transverse fascicles and distal to the superficial transverse metacarpal ligament. The lumbricals and palmar arteries and nerves pass through these arches.

The palmar aponeurosis includes sagittal septi (Figures 10-76 and 10-77), which separate the three zones of the hand. The medial fibrous septum originates at the medial edge of the palmar aponeurosis and attaches to the fifth metacarpal; the hypothenar eminence compartment is located on its medial side. Similarly, a fibrous lateral septum originates at the lateral edge of the aponeurosis and inserts into the third metacarpal; the thenar eminence compartment is located on its lateral side. The intermediate compartment is located between these two spaces.

The aponeurosis protects the structures of the hand and functions mainly when the hand is used for a firm grasp.

Lumbricals

The lumbricals (from the Latin *lumbricus*, "worm") (Figures 10-78 and 10-79) originate on the tendons of the flexor digitorum profundus, approximately at the level of the flexor retinaculum, and their tension increases with contraction of this muscle. There are four lumbricals, numbered from radial to ulnar. The first and second lumbricals originate on the radial aspect of the tendons for the index and long fingers, respectively. The third lumbrical originates on the radial aspect of the tendon for the ring finger and the medial for the long finger. The fourth lumbrical originates on the radial aspect of the tendon for the little finger and the medial aspect of the tendon for the ring finger. They all pass under the superficial transverse

TABLE 10-13 MUSCLES OF THE HYPOTHENAR EMINENCE

MUSCLE	Abductor digiti quinti	Flexor digiti quinti brevis	Opponens digiti quinti
ORIGIN	Flexor retinaculum Pisiform	Flexor retinaculum Hook of hamate	Flexor retinaculum Hook of hamate
INSERTION	Medial aspect of the base of the proximal phalanx	Base of the proximal phalanx of the little finger	Body of the little finger metacarpal
INNERVATION	Ulnar nerve	Ulnar nerve	Ulnar nerve
FUNCTION	Little finger abduction and opposition	Little finger metacarpophalangeal flexion	Opposition

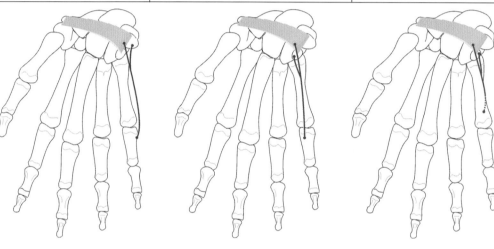

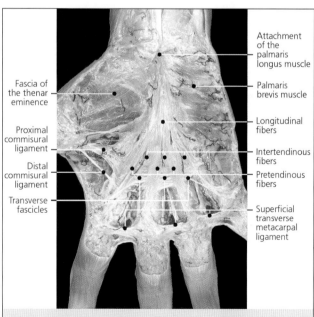

Figure 10-75 Anterior superficial view showing the palmar aponeurosis.

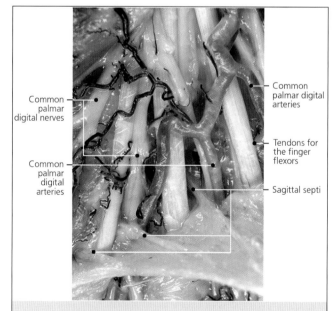

Figure 10-76 Anatomic details of the neurovascular bundle and tendons of the finger flexors entering the spaces formed by the sagittal septi of the palmar aponeurosis.

Myology

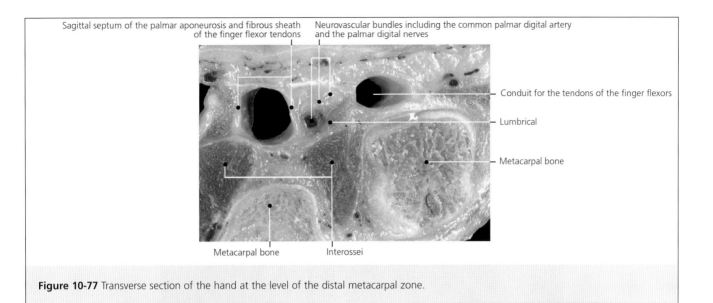

Figure 10-77 Transverse section of the hand at the level of the distal metacarpal zone.

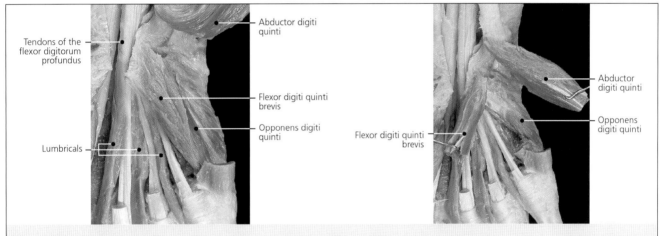

Figure 10-78 Specimens showing the hypothenar eminence after retraction of the abductor digiti quinti **(A)** and the flexor digiti quinti **(B)**.

intermetacarpal ligament and over the deep transverse ligament to the radial aspect of the extensor mechanism, where they are attached.

The first and second lumbricals are innervated by the median nerve (C7 through T1); the third and fourth lumbricals are typically innervated by the ulnar nerve (C8-T1). The median nerve also sometimes innervates the third lumbrical.

These unique muscles connect the flexor tendons with the extensor mechanism, which confers on them a complex action. They are agonists with the interosseous muscles, flexing the metacarpophalangeal joints and extending the interphalangeal joints; they also contribute to radial deviation of the fingers.

Interosseous Muscles

The muscles located at the intermetacarpal spaces are called interosseous muscles. They are organized into dorsal and palmar interosseous muscles.

Dorsal interossei The four dorsal interossei (Figures 10-80 through 10-82) are pennate. The first dorsal interosseous muscle, also known as the abductor indicis, originates at the borders of the metacarpals that limit the first interosseous space—the whole radial aspect of the second metacarpal and the posterior ulnar aspect of the first metacarpal. Its tendon is attached to the radial aspect of the proximal phalanx and the metacarpophalangeal capsule for the index finger. The second interosseous muscle originates at the second space—on the whole radial aspect of the third metacarpal and partially on the ulnar aspect of the second metacarpal. It is attached to the radial aspect of the base of the proximal phalanx and the metacarpophalangeal joint for the long finger. The third interosseous muscle originates in the third space—on the whole ulnar aspect of the third metacarpal and on part of the radial aspect of the fourth metacarpal. This muscle is also directed to the middle finger, with

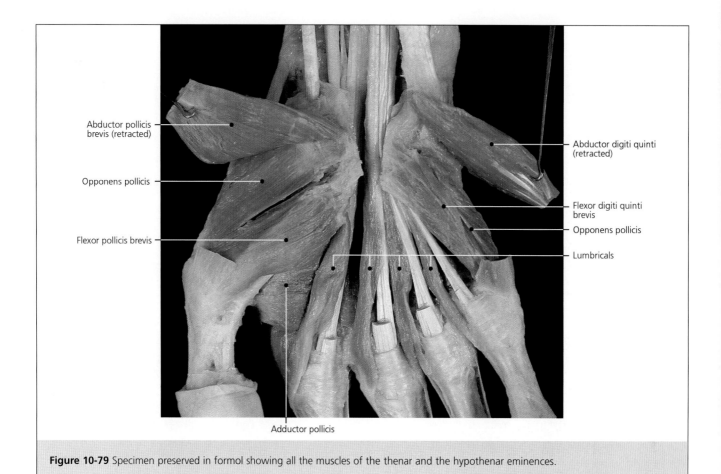

Figure 10-79 Specimen preserved in formol showing all the muscles of the thenar and the hypothenar eminences.

attachments similar to the ones of the second interosseous muscle but on the medial side. The fourth interosseous muscle originates at the fourth space—the whole ulnar aspect of the fourth metacarpal and partially on the radial side of the fifth metacarpal. It is directed to the lateral aspect of the base of the proximal phalanx of the ring finger as well as to the capsule of the metacarpophalangeal joint on the same side. These muscles are also attached to the extensor mechanism through their corresponding tendons.

The dorsal interossei are innervated by the deep branch of the ulnar nerve (C8-T1).

The functions of these muscles include finger abduction (the long finger has muscles on both sides and may actually be abducted both radially and ulnarly) and mainly flexion of the metacarpophalangeal joints and extension of the interphalangeal joints, with separation of the fingers in this position.

Palmar interossei The three palmar interossei (Figures 10-81 and 10-82) are smaller than the dorsal interossei; like the dorsal interossei, they are numbered from radial to ulnar. Some authors have suggested that the first palmar interosseous muscle originates on the ulnar side of the base of the first metacarpal and attaches to the medial sesamoid and the extensor mechanism for this finger, and consider these fibers as part of the adductor pollicis, but this description is not used currently. This explains why some textbooks describe four instead of three palmar interossei. The first palmar interosseous muscle as described here originates on the ulnar side of the second metacarpal and attaches through its tendon into the extensor mechanism of this finger. The second interosseous muscle originates on the palmar aspect of the fourth metacarpal, and the third one originates on the radial aspect of the fifth metacarpal; they both attach on the same side of their respective extensor mechanism, along with the second through fourth lumbricals. The long finger has no palmar interossei.

These muscles are innervated by the deep branch of the ulnar nerve (C8-T1).

The function of the palmar interossei is to move the fingers toward the middle line of the hand; the first palmar interosseous muscle accomplishes this with the adductor pollicis. The main role of the palmar interossei is to flex the metacarpophalangeal joints and extend the interphalangeal joints while keeping the fingers approximated to each other.

THE EXTENSOR MECHANISM

The extensor mechanism (Figures 10-83 through 10-89) includes a complex arrangement of tendons from the

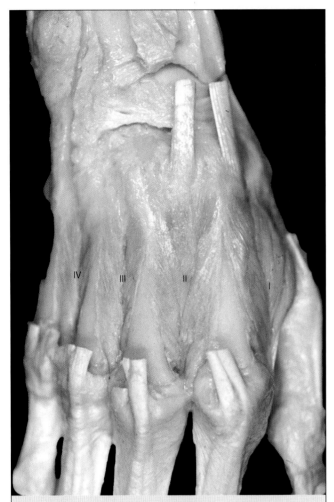

Figure 10-80 Dorsal interossei numbered (I-IV). The tendons of the dorsal muscles of the forearm have been resected.

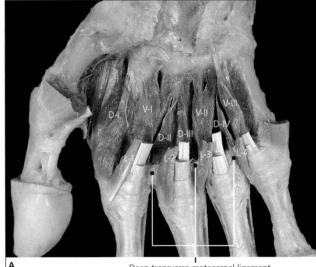

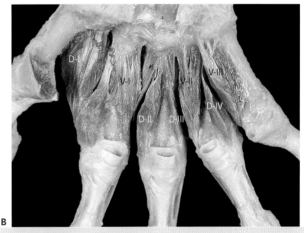

Figure 10-81 A and **B,** Specimens showing the interosseous muscles and their relationships at the palm of the hand. In B, the deep transverse metacarpal ligament has been resected and the fingers have been placed in abduction.

intrinsic muscles of the hand; the extensor digitorum communis; and the extensors for the index finger, little finger, and thumb. These tendons are interconnected through a sophisticated network of taut ligamentous and aponeurotic structures. The attachments of the various muscles on this mechanism allow two basic movements: the long or extrinsic muscles allow finger extension, and the short or intrinsic muscles allow flexion of the metacarpophalangeal joints with extension of the interphalangeal joints.

At the metacarpophalangeal joint level (Figure 10-84), the extensor mechanism includes a short middle insertion slip, oriented in an oblique and distal direction, that attaches to the articular capsule and the base of the proximal phalanx. In addition, sagittal bands attach to the volar plate and reach the fibrous sheath of the flexor tendons.

At the level of the middle third of the proximal phalanx, the main extensor tendon is divided into three bands, or slips, two lateral and one central (Figures 10-83 and 10-84).

The central slip attaches to the base of the middle phalanx. In some individuals, the central slip also receives some oblique fibers from the tendons of the lumbricals and interossei close to their attachments. In contrast, the extrinsic lateral slips (Figure 10-85), from the extensor mechanism, are obliquely oriented and directed to the proximal interphalangeal joint and lateral to the head of the proximal phalanx. At this level, they join the lateral bands of the intrinsic muscles, attaching to the base of the distal phalanx.

The interossei have a dual attachment to the lateral tubercle of the proximal phalanx and the extensor mechanism, thus participating in the extension of the interphalangeal joints, as mentioned earlier. On the dorsal aspect of the extensor mechanism is an aponeurotic sheath called the supratendinous transverse sheath or cuff of the interossei (Figure 10-86).

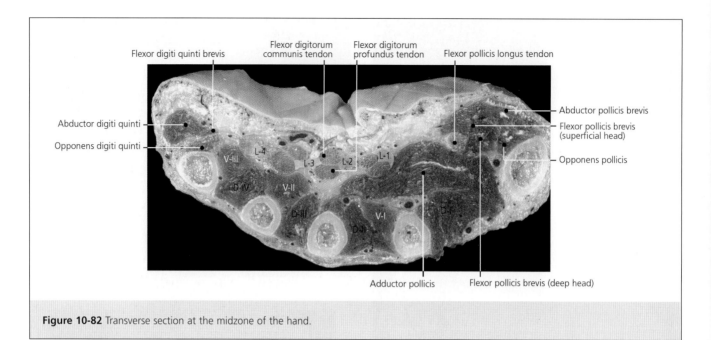

Figure 10-82 Transverse section at the midzone of the hand.

TABLE 10-14 MUSCLES OF THE MIDDLE PALMAR REGION

MUSCLE	Lumbricals	Dorsal interossei	Palmar interossei
ORIGIN	Radial side of the tendons of the flexor digitorum profundus	Interossei spaces 1-4	Interossei spaces 2-4
INSERTION	Extensor mechanism for the fingers	Extensor mechanism for the index finger (radial side), long finger (radial and ulnar sides), and ring finger (ulnar side)	Extensor mechanism of the index finger (ulnar side) and the ring and little fingers (radial side)
INNERVATION	Median nerve (index and long fingers) Ulnar nerve (ring and little fingers)	Ulnar nerve	Ulnar nerve
FUNCTION	Flexion of the metacarpophalangeal joints and extension of the interphalangeal joints Radial deviation of fingers	Flexion of the metacarpophalangeal joints and extension of the interphalangeal joints Finger abduction	Flexion of the metacarpophalangeal joints and extension of the interphalangeal joints Finger adduction

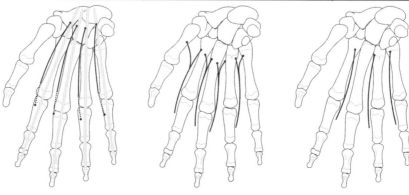

Myology

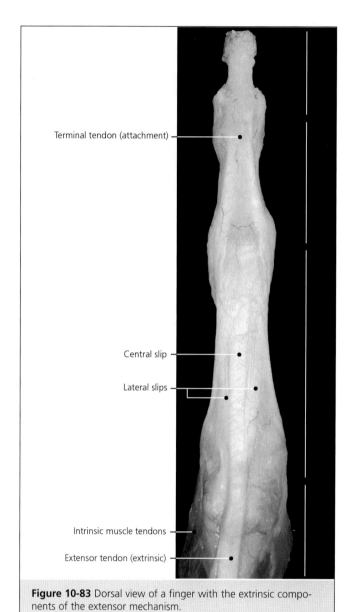

Figure 10-83 Dorsal view of a finger with the extrinsic components of the extensor mechanism.

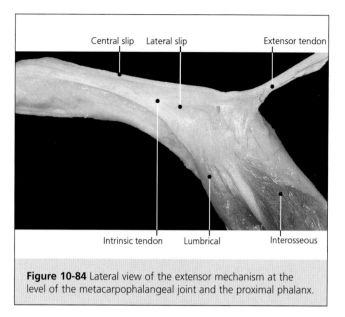

Figure 10-84 Lateral view of the extensor mechanism at the level of the metacarpophalangeal joint and the proximal phalanx.

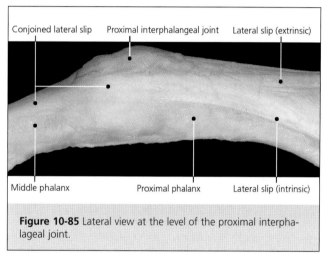

Figure 10-85 Lateral view at the level of the proximal interphalageal joint.

The interosseous tendon is divided into two fascicles, called the intrinsic central slip and the intrinsic lateral band. The first is directed to the dorsal aspect of the finger, where it coalesces with the extrinsic central slip of the extensor mechanism and attaches to the base of the middle phalanx. The second is directed obliquely to the proximal interphalangeal joint to form the conjoined lateral band, which attaches to the dorsal aspect of the distal phalanx.

The retinacular system is formed by the transverse sheath, the oblique and transverse retinacular ligaments, and the triangular lamina.

The oblique retinacular ligament (of Landsmeer) (Figure 10-87) originates on the lateral aspect of the flexor tendon sheaths at the level of the middle phalanx. It is directed distally and dorsally to its attachment at the dorsal aspect of the base of the distal phalanx.

The transverse retinacular ligament (Figure 10-88) originates at the level of the proximal interphalangeal joint and the edges of the conjoined lateral band; it attaches on the volar side at the volar plate and the subcutaneous tissue.

The triangular lamina (Figure 10-89) is a fine aponeurotic structure that connects the medial edges of the lateral slips at the distal aspect of the middle phalanx to the tendon attachment on the distal phalanx.

Ligaments of the Skin of the Fingers

The ligaments of the skin of the fingers have a very important function—to protect the neurovascular bundle of the fingers and transmit tension between the movements of the fingers and the skin. The most important are the natatory ligament, Cleland's and Grayson's ligaments, and the dorsal cutaneotendinous fibers.

The natatory ligament is located on the web spaces of the fingers, around the neurovascular bundle of the

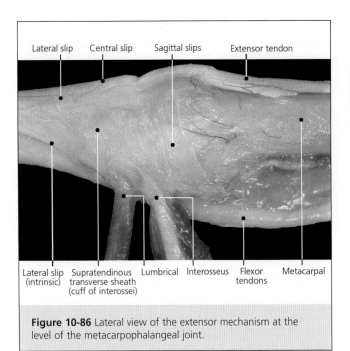

Figure 10-86 Lateral view of the extensor mechanism at the level of the metacarpophalangeal joint.

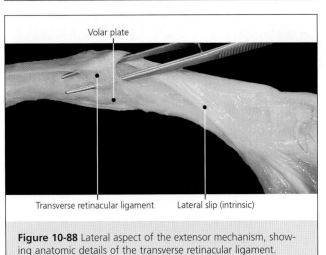

Figure 10-88 Lateral aspect of the extensor mechanism, showing anatomic details of the transverse retinacular ligament.

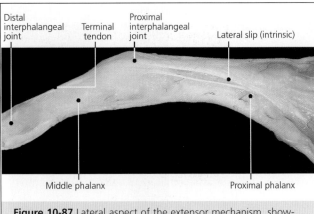

Figure 10-87 Lateral aspect of the extensor mechanism, showing anatomic details of the oblique retinacular ligament of Landsmeer.

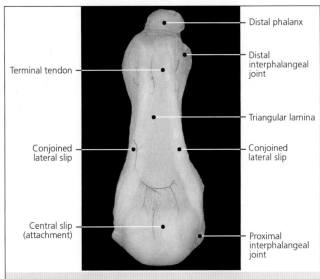

Figure 10-89 Dorsal view of the middle phalanx with the interphalangeal joints, showing anatomic details of the triangular lamina.

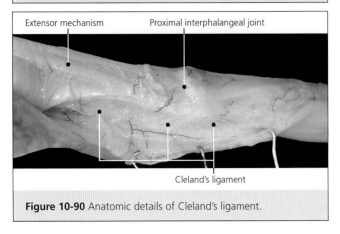

Figure 10-90 Anatomic details of Cleland's ligament.

fingers. Its transverse fibers, known as the superficial transverse metacarpal ligament (Figure 10-75), protect the neurovascular structures and limit finger abduction (separation of the fingers); these fibers and the skin overlying them just distal to the metacarpophalangeal joints become taut with finger abduction.

Cleland's ligament (Figure 10-90) is formed by four lateral fibrous septi. The first one originates between the distal half of the proximal phalanx and the proximal third of the middle phalanx. Its fibers are located on the dorsal aspect of the neurovascular bundle of the fingers and are oriented obliquely toward the overlying skin fascia. The fibers of the transverse retinacular ligament of Landsmeer are found distally, between the first and the second septi. The second septum originates from the lateral band of the extensor mechanism, from the base of the middle phalanx and the volar plate, as well as from the fibrous sheath of the flexor tendons. These fibers are oriented distally and obliquely to the overlying skin. The third septum originates on the lateral portion of the head of the middle phalanx and the articular capsule of the distal

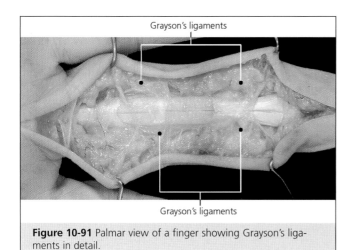

Figure 10-91 Palmar view of a finger showing Grayson's ligaments in detail.

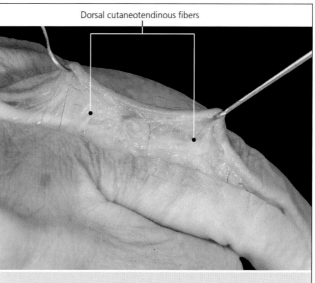

Figure 10-92 Superficial incision at the dorsal aspect of a finger.

interphalangeal joint, as well as the fibrous sheath of the flexor tendons. Its fibers are oblique and directed proximally to attach on the deep fascia along with the second septum. Finally, the fourth septum originates on the distal interphalangeal joint and the base of the distal phalanx, and its fibers are directed to the deep fascia of the fingertips.

Grayson's ligament (Figure 10-91) is located on the volar aspect and originates on the fibrous sheath of the flexor muscles, specifically, at the level of the second and third annular pulleys of the fingers. Grayson's ligament is anterior to the neurovascular bundle of the fingers. The fibers are obliquely oriented, the proximal portion directed distally and the distal portion directed proximally, although they basically extend along the whole volar aspect of the finger. These fibers form the volar creases of the interphalangeal joints.

A variable number of dorsal cutaneotendinous fibers (Figure 10-92) connect the extensor mechanism to the skin, especially at the articular regions of the fingers.

Chapter 11

Neurology

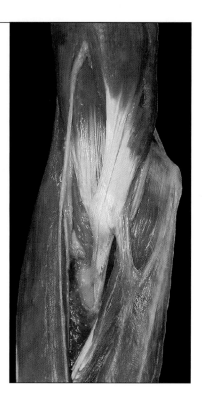

Introduction

The upper limb and shoulder girdle is innervated by the brachial plexus, which is formed from the anterior branches of spinal roots C5 through T1 (Figures 11-1 and 11-2); in some individuals, it also receives nerve fibers from C4 and T2. The brachial plexus can be visualized as it exits into the depths of the posterior cervical triangle (Figure 11-3). From there it courses through the interscalenic triangle (formed by the anterior and middle scalenic muscles) (Figure 11-4) and continues obliquely in a lateral and distal direction to the first rib and the arm. The brachial plexus is organized into roots, trunks, divisions, cords, and branches.

The superior trunk is formed by roots C5 and C6; it also commonly receives fibers from C4. The middle trunk is formed by C7. The inferior trunk is formed by C8 and T1; it commonly also receives fibers from T2 (Figure 11-5).

The trunks are divided into anterior and posterior divisions, which form the cords. The cords are named according to their topographic relationship with the axillary artery. Thus, there is a posterior cord, formed by the coalescence of the posterior divisions of the trunks; a lateral cord, formed by the anterior divisions of the superior and middle trunks; and a medial cord, formed by the anterior division of the inferior trunk (Figure 11-5).

Collateral Branches of the Brachial Plexus

The collateral branches of the brachial plexus innervate the muscles of the shoulder girdle. For discussion purposes, the branches are divided into the anterior branches, which innervate the anterior muscles, and the posterior branches, which innervate the posterior muscles (Figure 11-6).

Anterior Branches
Nerve for the Subclavius Muscle
This nerve originates from the anterior and proximal aspect of the superior trunk (Figure 11-7) and then descends medially on the anterior aspect of the brachial plexus, the anterior scalenus muscle, and the subclavian vein to reach the middle portion of the subclavian muscle. In some individuals, this nerve emanates from the accessory phrenic nerves.

Medial Pectoral Nerve
This nerve usually arises from the medial cord at the level of the clavicle (Figures 11-8 through 11-10). It is called the medial pectoral nerve because it arises from the medial cord of the brachial plexus, but it is located lateral to the lateral pectoral nerve. It runs obliquely, in an inferior and medial direction, in front of the axillary artery and through the clavipectoral fascia and the pectoralis minor muscle, dividing into sev-

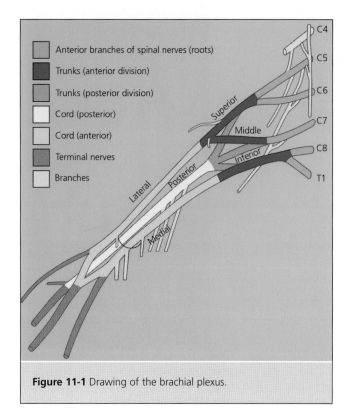

Figure 11-1 Drawing of the brachial plexus.

eral branches for the pectoralis major muscle. It innervates both the pectoralis major and minor muscles.

Lateral Pectoral Nerve
This nerve arises from the lateral cord at almost the same level as the medial pectoral nerve (Figures 11-8 and 11-10). As mentioned earlier, this nerve is located on the medial aspect of the medial pectoral nerve. The lateral pectoral nerve gives a branch that anastomoses with a similar branch from the medial pectoral nerve, forming the handle of the pectoral nerves. The deep branches of the lateral pectoral nerve reach the pectoralis minor muscle, and some of them perforate this muscle to innervate the pectoralis major.

Posterior Branches
Suprascapular Nerve
This nerve arises from the superior trunk (Figures 11-6 and 11-11), although it is not uncommon to find it arising from C5 with some fibers from C4. It courses in a lateral and dorsal direction to the scapular notch, where it passes under the superior transverse ligament of the scapula to reach the supraspinous fossa and innervate the supraspinatus muscle (Figure 11-12). From there, it continues through the spinoglenoid notch beneath the inferior transverse ligament of the scapular to reach the infraspinous fossa, innervating the infraspinatus muscle

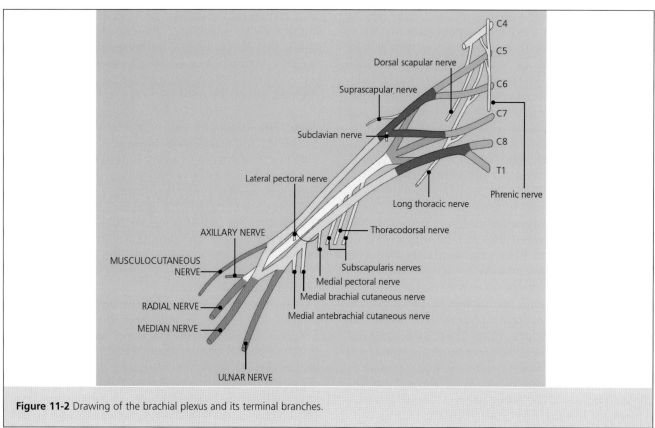

Figure 11-2 Drawing of the brachial plexus and its terminal branches.

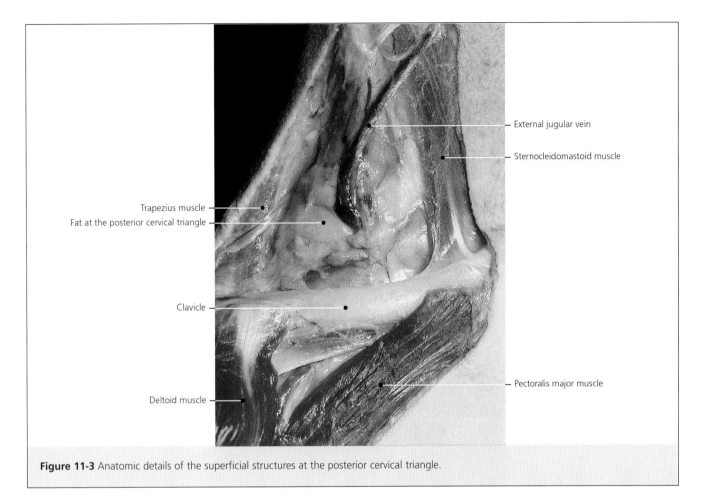

Figure 11-3 Anatomic details of the superficial structures at the posterior cervical triangle.

(Figures 11-13 and 11-14). Some branches reach the posterior articular capsule.

Dorsal Scapular Nerve
This nerve arises from C5, perforates the middle scalenus muscle, and lies on the posterior scalenus muscle (Figure 11-15). It continues in a dorsal direction beneath the levator scapulae and the rhomboids, which it innervates (Figure 11-16). The dorsal scapular nerve is close enough to the serratus anterior to occasionally provide some branches.

Subscapular Nerves
These nerves are usually formed by two branches, one proximal and one distal; the thoracodorsal nerve originates between these two branches (Figures 11-17 and 11-18). The proximal branch lies on the subscapularis muscle, which it innervates, whereas the distal branch is directed vertically to innervate the inferior aspect of the subscapularis muscle and the teres major. Some branches innervate the anterior portion of the articular capsule.

Thoracodorsal Nerve
This nerve arises from the posterior cord, between the branches for the subscapular nerves (Figures 11-17 and 11-18), and it is directed in an inferior direction over the subscapularis muscle, along with the thoracodorsal artery. It ends by innervating the latissimus dorsi.

Long Thoracic Nerve
This nerve is formed by the anterior branches of C5, C6, and C7, just after they leave their respective intervetebral foramens. The branches of C5 and C6 perforate the anterior scalenus muscle, whereas C7 is located on its anterior surface. This nerve continues vertically between the posterior scalenus and the brachial plexus (Figure 11-6). It is located on the lateral aspect of the thoracic wall, at the concavity between the subscapularis and the serratus anterior, which it innervates (Figures 11-11 and 11-18).

Terminal Nerves (Arm Region)
Musculocutaneous Nerve
The musculocutaneous nerve is the terminal branch of the lateral cord of the brachial plexus and contains fibers from C5, C6, and C7 (Figure 11-19). It is a motor nerve at the arm and a sensory nerve at the forearm. It originates at the axillary region, approximately at the level of the pectoralis minor muscle and lateral to the median nerve. It continues laterally, per-

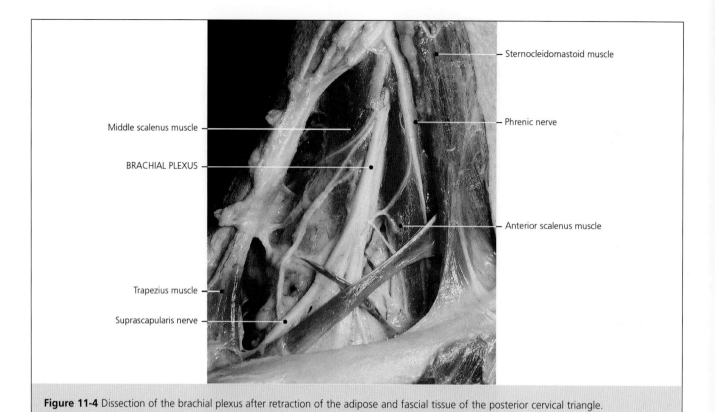

Figure 11-4 Dissection of the brachial plexus after retraction of the adipose and fascial tissue of the posterior cervical triangle.

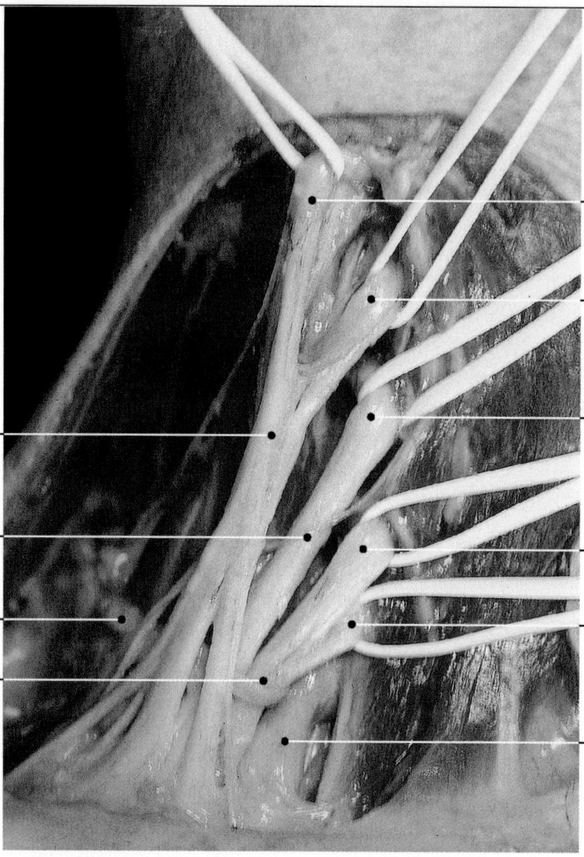

Figure 11-5 Branches and trunks of the brachial plexus.

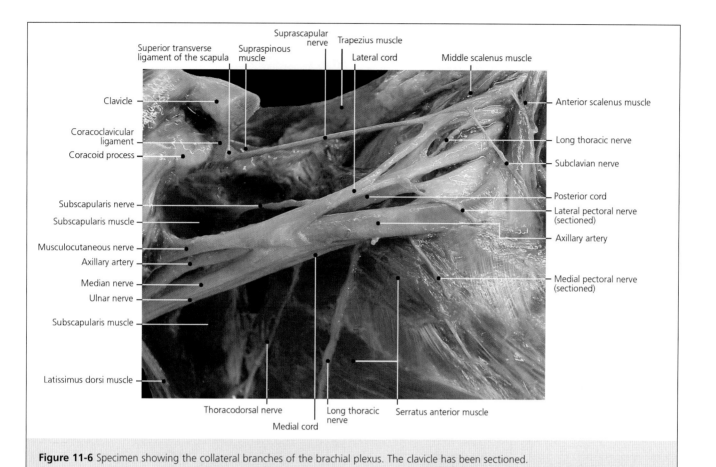

Figure 11-6 Specimen showing the collateral branches of the brachial plexus. The clavicle has been sectioned.

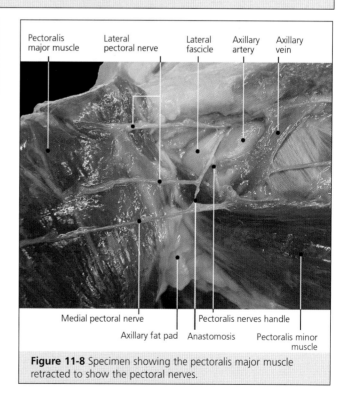

Figure 11-7 Anatomic details of the nerve for the subclavian muscle.

Figure 11-8 Specimen showing the pectoralis major muscle retracted to show the pectoral nerves.

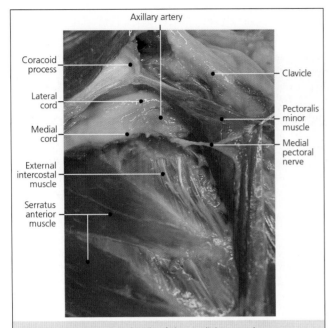

Figure 11-9 Anatomic details of the medial pectoral nerve providing innervation to the pectoralis minor muscle, which has been retracted slightly.

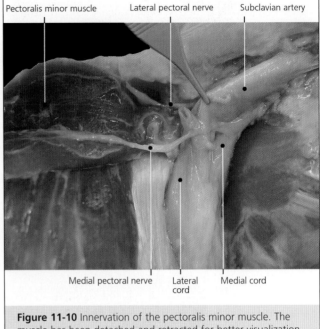

Figure 11-10 Innervation of the pectoralis minor muscle. The muscle has been detached and retracted for better visualization of the nerves.

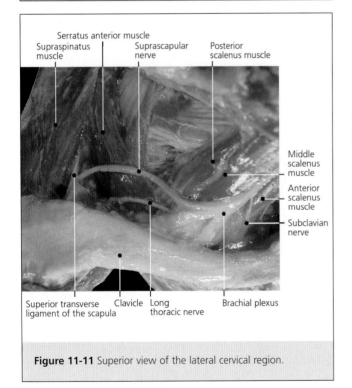

Figure 11-11 Superior view of the lateral cervical region.

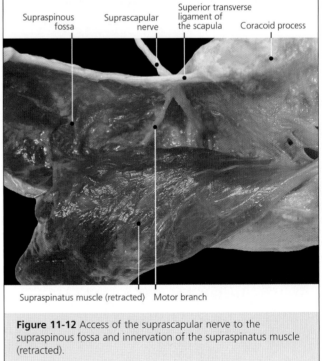

Figure 11-12 Access of the suprascapular nerve to the supraspinous fossa and innervation of the supraspinatus muscle (retracted).

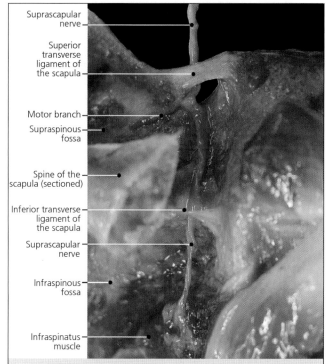

Figure 11-13 Dorsal view of the scapula. The spine has been sectioned to show the course of the suprascapular nerve and the innervation of the infraspinatus muscle.

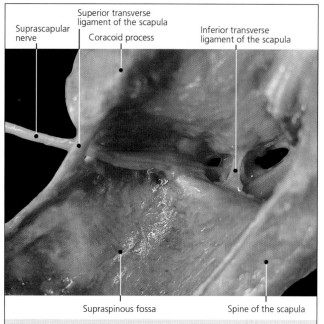

Figure 11-14 Superior view of the scapula with the supraspinatus muscle removed. The course of the suprascapular nerve may be observed.

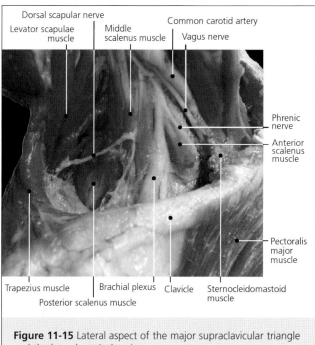

Figure 11-15 Lateral aspect of the major supraclavicular triangle and the lateral cervical region.

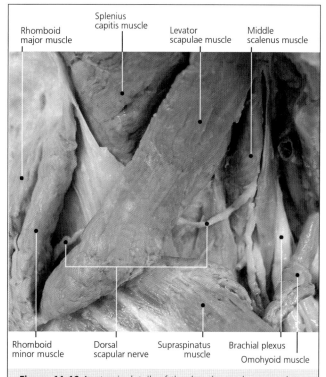

Figure 11-16 Anatomic details of the dorsal scapular nerve, lateral view.

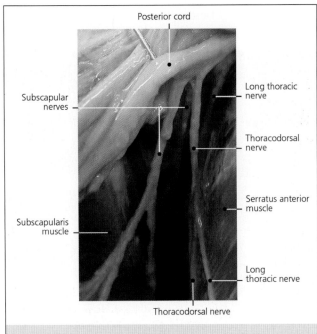

Figure 11-17 Anatomic details of the subscapular and thoracodorsal nerves.

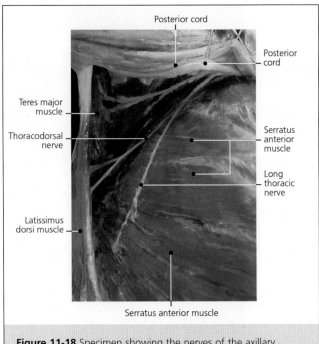

Figure 11-18 Specimen showing the nerves of the axillary region.

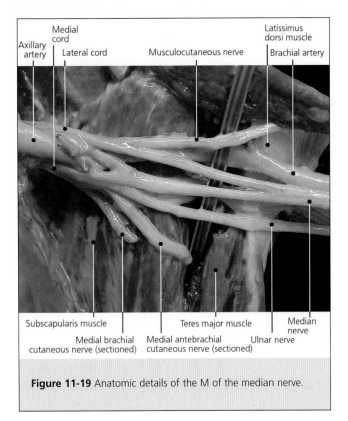

Figure 11-19 Anatomic details of the M of the median nerve.

forating the coracobrachialis muscle, to lie between the biceps brachii and the brachialis (Figure 11-20). It sends branches to these muscles, the humerus, and the anterior aspect of the elbow joint, being responsible for the innervation of the elbow flexors. In the midportion of the arm, this nerve sometimes forms an inconsistent anastomosis with the median nerve. Finally, the musculocutaneous nerve emerges lateral to the distal biceps tendon and transforms into a cutaneous nerve, the lateral antebrachial cutaneous nerve.

Median Nerve
The median nerve arises in the axilla, between the musculocutaneous and the ulnar nerves. The median nerve, together with components from the lateral and medial cords, forms an M shape (Figure 11-19). It descends vertically in the medial neurovascular compartment of the arm, ventral to the brachial artery.

Ulnar Nerve
The ulnar nerve originates at the axilla, arising from the medial cord (Figure 11-19), and descends on the medial aspect of the median nerve. It descends vertically in the medial neurovascular compartment of the arm, dorsal to the brachial artery (Figure 11-20).

Medial Brachial Cutaneous Nerve
Arising from the medial cord of the brachial plexus and usually receiving anastomotic fibers from T2 (intercostobrachial nerve), the medial brachial cutaneous nerve perforates the axillary fat and the brachial fascia (Figure 11-21) to supply the skin region, as indicated by its name.

Radial Nerve
The radial nerve originates from the posterior cord (Figures 11-20 and 11-22). From the axilla, it contin-

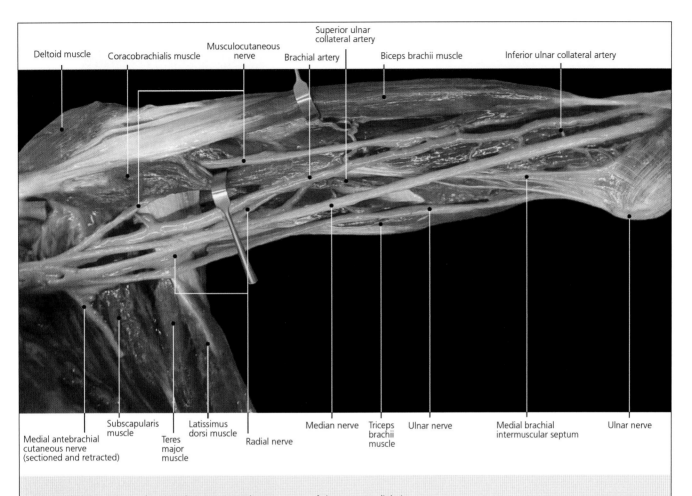

Figure 11-20 Specimen showing the neurovascular structures of the arm, medial view.

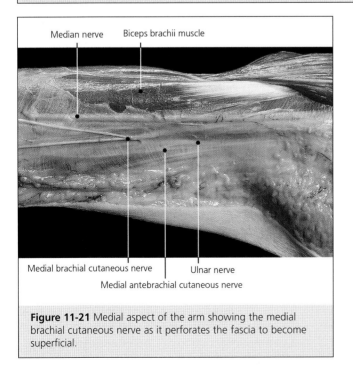

Figure 11-21 Medial aspect of the arm showing the medial brachial cutaneous nerve as it perforates the fascia to become superficial.

Figure 11-22 Terminal branches of the posterior cord. The other two cords have been retracted.

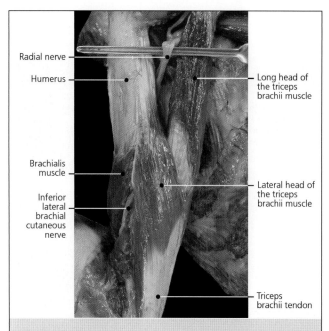

Figure 11-23 Course of the radial nerve (marked with a probe) between the muscle bellies of the triceps brachii.

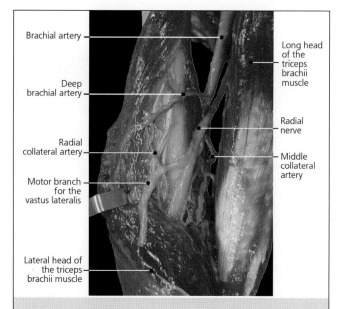

Figure 11-24 Anatomic details of the radial nerve on the spiral groove for this nerve on the humerus.

ues in a caudal and lateral direction to the posterior region of the upper limb. It crosses the arm diagonally, between the medial and lateral bellies of the triceps brachii, which it innervates (Figure 11-23), and intimately related with the sulcus for the radial nerve in the humerus, called the spiral groove (Figure 11-24). At the distal third of the arm, the radial nerve is situated laterally, and it perforates the lateral intermuscular brachial septum to lie anteriorly. At the elbow region, the radial nerve is found deep in the lateral bicipital groove.

During its course through the arm, the radial nerve sends branches for each of the bellies of the triceps brachii as well as for the anconeus muscle. In this region the radial nerve also sends sensory branches to form the posterior and lateral inferior brachial cutaneous nerves.

Axillary Nerve

The axillary nerve arises from the posterior cord of the brachial plexus in the axilla, in front of the subscapularis muscle (Figure 11-22), and continues into the topographic region known as the quadrilateral space (of Velpeau), where it turns posteriorly. In this area, it gives the motor branch for the teres minor. Next, it runs on the surgical neck of the humerus, intimately close to the glenohumeral joint capsule (Figure 11-25). Through this course, the axillary nerve innervates the deltoid muscle and conveys the sensory information from this region through the superior lateral brachial cutaneous nerve.

Figures 11-26 through 11-28 show the cutaneous innervation of the shoulder and arm.

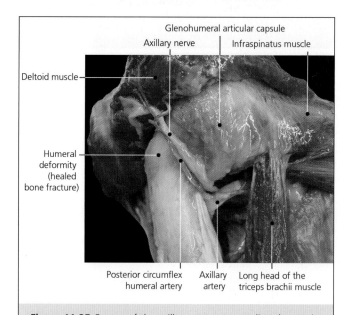

Figure 11-25 Course of the axillary nerve surrounding the surgical neck of the humerus. Note the humeral deformity secondary to a fracture healed in varus, which could have injured the nerve. The deltoid muscle has been retracted superiorly.

TERMINAL NERVES (FOREARM REGION)

Lateral Antebrachial Cutaneous Nerve

The lateral antebrachial cutaneous nerve is the terminal branch of the musculocutaneous nerve (C5 through C7). It exits between the brachialis and the biceps brachii, at the lateral bicipital canal, where it perforates the brachial fascia to lie subcutaneously along with the superficial veins (Figure 11-29). It courses in a slightly oblique direction, directed to

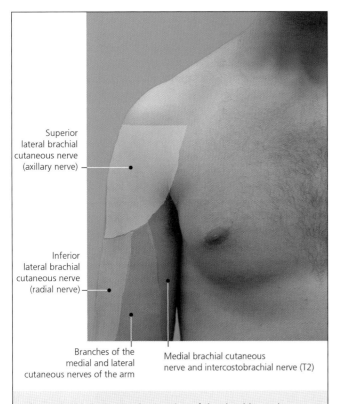

Figure 11-26 Cutaneous innervation of the shoulder and arm, anterior view.

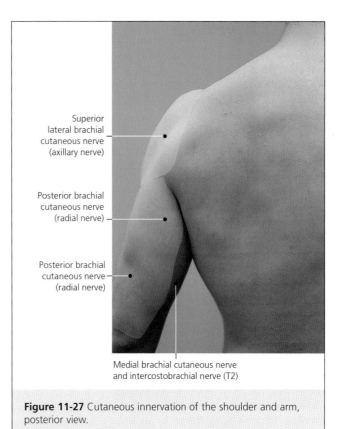

Figure 11-27 Cutaneous innervation of the shoulder and arm, posterior view.

the lateral aspect of the forearm, to innervate its skin.

The territory innervated by this nerve includes the posterolateral aspect of the forearm to the radial styloid and the dorsal aspect of the wrist. At the level of the radial styloid, this nerve commonly anastomoses with the anterior branch of the radial nerve. On the anterolateral side, it continues to the thenar eminence, where it anastomoses again with the radial nerve.

Medial Antebrachial Cutaneous Nerve

The medial antebrachial cutaneous nerve is a peripheral nerve that is also known as the accessory nerve to the medial brachial cutaneous nerve (even though it is distributed in the forearm). This is a terminal branch of the brachial plexus, specifically the medial cord (C8-T1) (Figure 11-20). It follows the course of the basilic vein on the medial aspect of the arm (Figure 11-21). At the midlevel of the arm, it perforates the brachial fascia and continues to a subcutaneous location along the basilic vein. It divides into several branches, which innervate the cutaneous territory of the medial aspect of the forearm (Figure 11-30).

The anterior branch of the medial antebrachial cutaneous nerve innervates the anteromedial aspect of the forearm to the wrist. The posterior, or ulnar, branch innervates the skin on the posteromedial aspect of the proximal two thirds of the forearm.

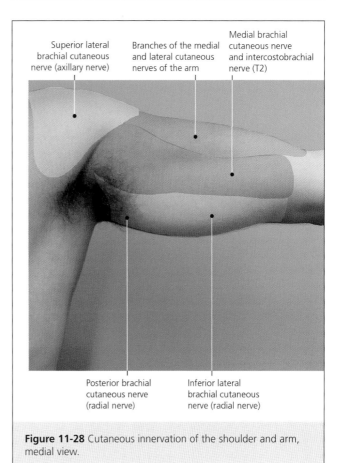

Figure 11-28 Cutaneous innervation of the shoulder and arm, medial view.

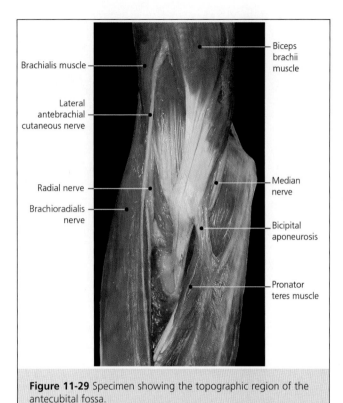

Figure 11-29 Specimen showing the topographic region of the antecubital fossa.

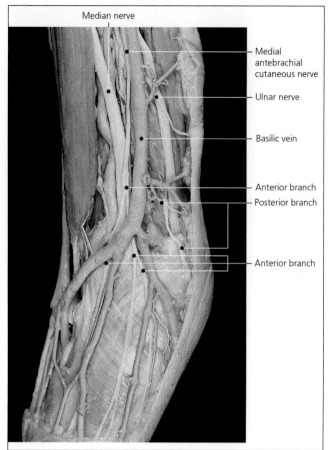

Figure 11-30 Anteromedial view of the elbow region. The veins have been injected with blue latex. Note the course of the medial antebrachial cutaneous nerve along the venous vessels.

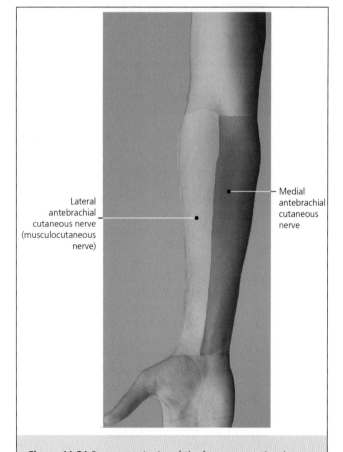

Figure 11-31 Sensory territories of the forearm, anterior view.

Figures 11-31 and 11-32 illustrate the sensory areas of the forearm.

Median Nerve

The median nerve is a terminal nerve of the brachial plexus formed by fibers from both the medial and lateral cords (Figure 11-9), which contain branches of C6 to T1. The median nerve has a medial contribution from the medial cord, which contains fibers from C8 and T1, which innervate intrinsic hand muscles. It also has a lateral root from the lateral cord, which contains fibers from C6 and C7, including motor fibers for the muscles of the forearm and sensory fibers as well.

The median nerve reaches the forearm through the medial bicipital canal, coursing along with the brachial artery and underneath the aponeurotic expansion of the distal biceps tendon (Figure 11-29). At the proximal aspect of the forearm, the median nerve sends muscular branches for the pronator teres, the flexor carpi radialis, and the palmaris longus (Figure 11-33).

It is important to note that the median nerve courses between the two heads of the pronator teres (Figure 11-34). After coursing under the arch formed by the

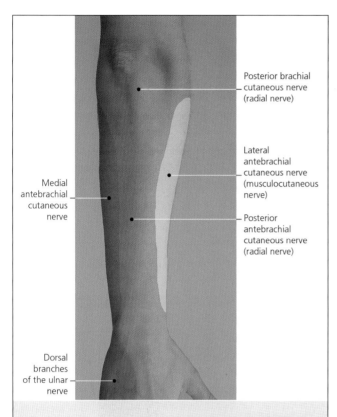

Figure 11-32 Sensory territories of the forearm, posterior view.

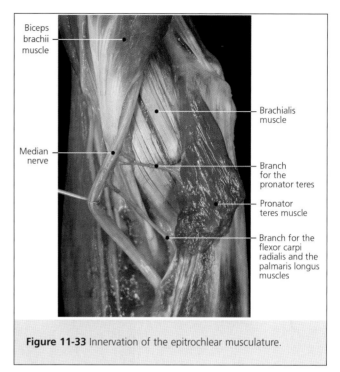

Figure 11-33 Innervation of the epitrochlear musculature.

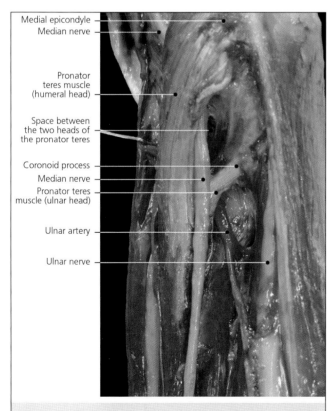

Figure 11-34 Dissection showing the course of the nerves of the forearm. The median nerve is seen passing through the pronator teres muscle.

ulnohumeral and radial fibers of the flexor digitorum superficialis, the median nerve lies between this muscle (inside its fascia) and the flexor digitorum profundus. In this region, it sends motor branches for the flexor digitorum. Just below the arch, the median nerve sends the anterior interosseous nerve, which is located between the flexor digitorum profundus and the flexor pollicis longus, both of which are innervated by the median nerve (Figures 11-35 and 11-36); it lies on the interosseous membrane (Figure 11-37). This nerve innervates the cords of the flexor digitorum profundus only for the index and long fingers; the ring and little fingers are innervated by the ulnar nerve. Its course follows the axis of the forearm until it lies underneath the pronator quadratus, which it innervates, and then continues into the articular capsule of the wrist. In addition, the median nerve innervates the diaphyses of the forearm bones.

The remaining fibers of the median nerve continue between the flexor digitorum superficialis and profundus. The anastomosis of Martin-Gruber (inconsistent) is a small branch connecting the median and ulnar nerves under the flexor digitorum profundus. At the distal third of the forearm, the median nerve is located on the medial side of the flexor carpi radialis and flexor pollicis longus tendons, and it is lateral to the flexor digitorum superficialis and the palmaris longus. In this area, it sends the palmar cutaneous branch of the median nerve (Figure 11-38), which arises from the radial side of the nerve about 3 to 4 cm proximal to the wrist. It perforates the antebrachial fascia and divides into two branches that

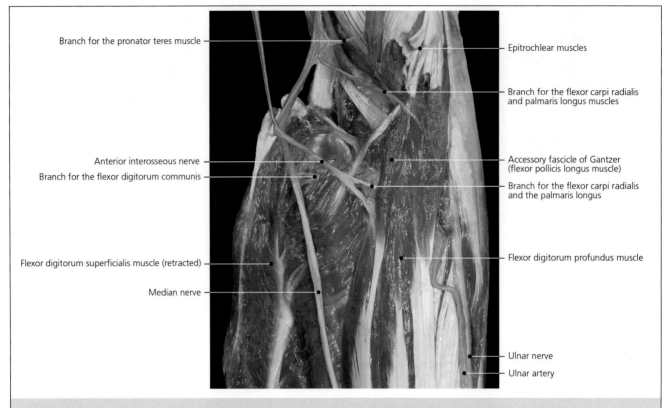

Figure 11-35 Motor branches of the median nerve above the flexor digitorum profundus; the flexor digitorum superficialis muscle has been retracted laterally.

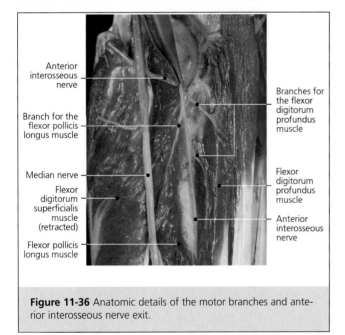

Figure 11-36 Anatomic details of the motor branches and anterior interosseous nerve exit.

Figure 11-37 Course of the anterior interosseous nerve.

innervate the skin: The first branch innervates the anterolateral aspect of the palm of the hand (the thenar eminence) (Figure 11-39), and the other branch innervates the middle region of the palm.

The median nerve enters the carpal tunnel at the wrist (Figure 11-40). Often, the median nerve sends muscular branches for the thenar eminence inside the tunnel, but more commonly a recurrent branch from the radial side of the nerve sends out a muscular branch (Figure 11-41) at the exit from the carpal tunnel; this recurrent branch is located between the superficial head of the flexor pollicis brevis and the

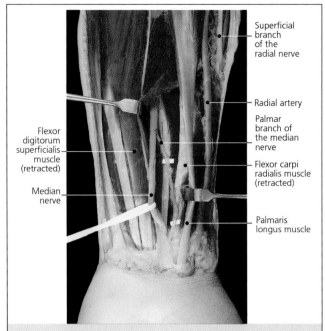

Figure 11-38 Superficial dissection at the distal third of the forearm showing the palmar branch of the median nerve and its relationships.

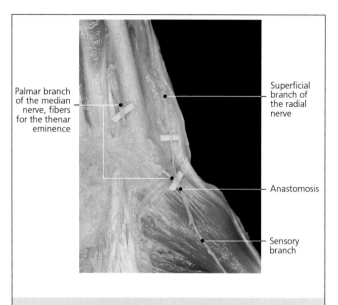

Figure 11-39 Superficial dissection of the anterior region of the wrist. Note the anastomosis between the median and the radial nerves.

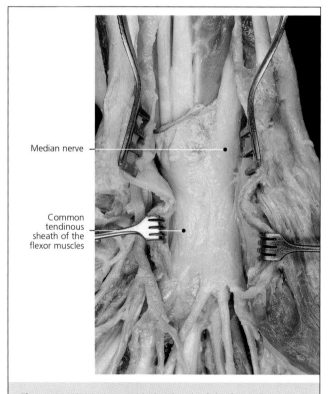

Figure 11-40 Specimen with the sheath of the finger flexor muscles injected with latex and the carpal tunnel opened to show their relationships to the median nerve.

abductor pollicis brevis, which the median nerve innervates. The recurrent branch courses deep to innervate the opponens pollicis. In addition, it sends branches for the first and second lumbricals through the common digital nerves. Occasionally, the anastomosis of Riche and Cannieu connects the median and ulnar nerves in the thenar eminence.

The common palmar digital nerves are formed at the exit of the carpal tunnel. These sensory nerves are located at the first, second, and third interosseous spaces and divide at the level of the metacarpal heads into the proper digital nerves, which are the terminal branches that end at the fingertips (Figures 11-42 and 11-43). Each common digital nerve is divided into two branches, each of which is directed to a different adjacent finger; these nerves lie on each side of the finger flexors. They innervate the palmar aspect of the thumb, index, and long fingers and the lateral half of the ring finger. In addition, through dorsal branches, they innervate the middle and distal phalanges of the index and long fingers as well as the posterolateral aspect of the ring finger (Figure 11-44).

The ramus communicans with the ulnar nerve (a sensory branch) (Figure 11-41) courses over the flexor tendons at the exit of the carpal tunnel to anastomose with the ulnar nerve (the inconsistent anastomosis of Berentini).

Figure 11-45 illustrates the motor branches of the median nerve.

Figures 11-46 and 11-47 show the cutaneous distribution of the hand.

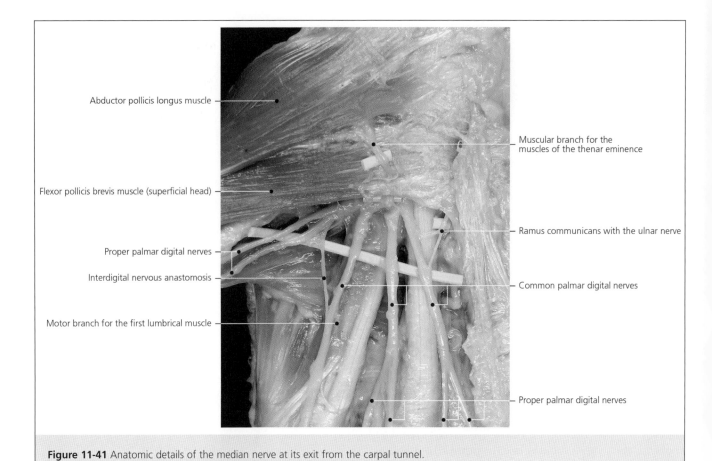

Figure 11-41 Anatomic details of the median nerve at its exit from the carpal tunnel.

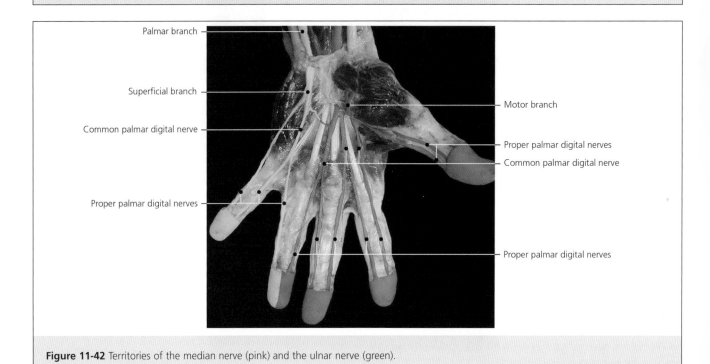

Figure 11-42 Territories of the median nerve (pink) and the ulnar nerve (green).

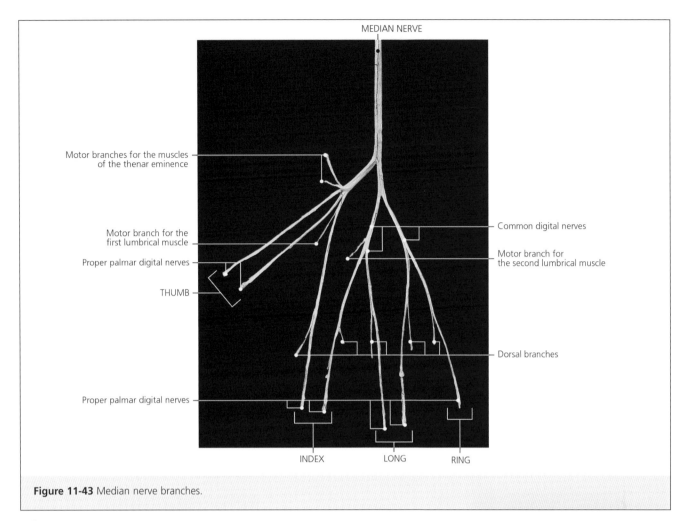

Figure 11-43 Median nerve branches.

Ulnar Nerve

The ulnar nerve is a terminal branch of the medial cord of the brachial plexus; it contains fibers from C8 and T1. It enters the forearm at the cubital canal, slides through the sulcus for the ulnar nerve, and courses under the arcade of Osborne, formed by the origin of the two heads of the flexor carpi ulnaris (Figure 11-48). It is separated from the ulnar collateral ligament by a small serous bursa. At this level, it sends an articular branch for the elbow joint.

The ulnar nerve courses parallel to the forearm axis, under the flexor carpi ulnaris. At the proximal third of the forearm, the ulnar nerve sends muscular branches for the flexor carpi ulnaris. At the midforearm, the ulnar nerve innervates the cords of the flexor digitorum profundus for the ring and little fingers (Figure 11-49). This nerve also sends small branches for the adventitial layer of the ulnar artery, the so-called nerves of Henle.

At the medial aspect of the forearm, about 5 cm proximal to the wrist, the ulnar nerve sends two branches. The first, the dorsal branch of the ulnar nerve, is directed dorsally under the flexor carpi ulnaris to innervate the skin on the posterior aspect of the ulnar side of the hand. This branch divides into several dor-

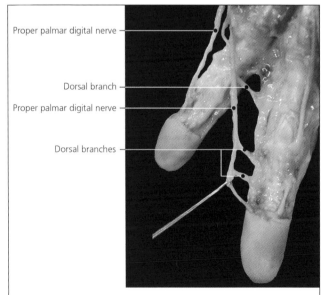

Figure 11-44 The proper palmar digital nerve placed under tension with its dorsal branches, which collects sensation from the dorsum of the fingers.

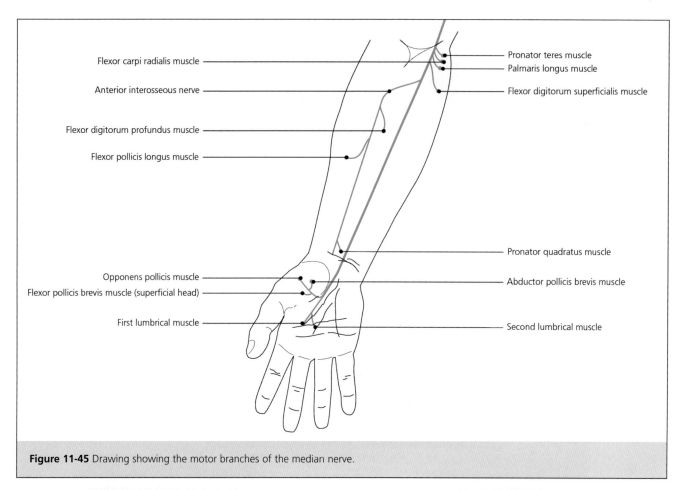

Figure 11-45 Drawing showing the motor branches of the median nerve.

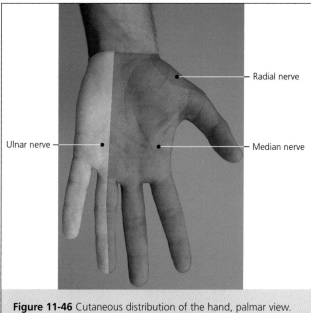

Figure 11-46 Cutaneous distribution of the hand, palmar view.

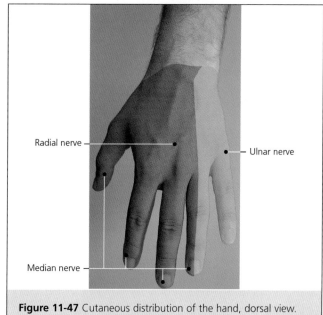

Figure 11-47 Cutaneous distribution of the hand, dorsal view.

sal digital nerves for the ring and little fingers and the ulnar side of the long finger (Figure 11-50). The main branch of the ulnar nerve is located on the palmar side.

At the wrist, the ulnar nerve lies at the medial side of the ulnar vessels on the deep layer of the flexor retinaculum, topographically known as Guyon's canal. The roof of this canal is formed by the superficial cord of the flexor retinaculum and the palmaris brevis muscle (Figure 11-51). In this region, the ulnar nerve is divided into a motor branch and a

NEUROLOGY

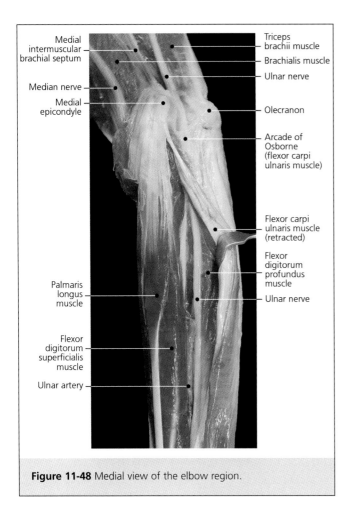

Figure 11-48 Medial view of the elbow region.

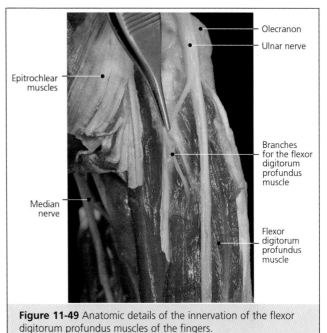

Figure 11-49 Anatomic details of the innervation of the flexor digitorum profundus muscles of the fingers.

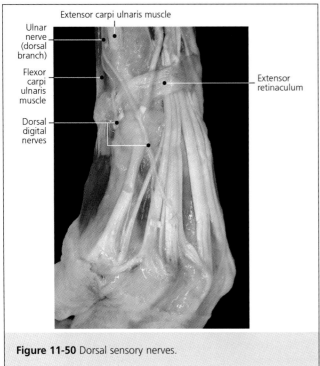

Figure 11-50 Dorsal sensory nerves.

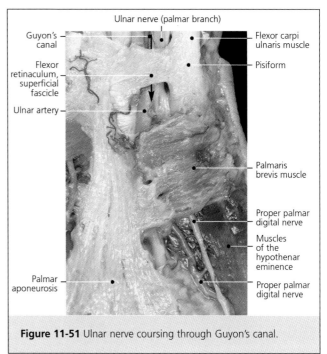

Figure 11-51 Ulnar nerve coursing through Guyon's canal.

mixed motor/sensory branch. The superficial branch anastomoses proximally with the median nerve through the ramus communicans for the median nerve. Distally, this branch is located under the superficial layer of the flexor retinaculum and the palmaris brevis, which is innervated by a small branch (Figure 11-52). At this location, the ulnar nerve sends the common digital nerve for the fourth web space, a branch that usually divides at the level of the metacarpal head into the proper digital nerves. These sensory nerves reach the skin of the dorsal

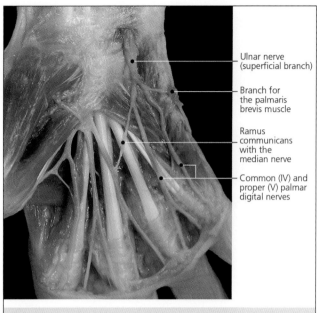

Figure 11-52 Dissection of the cutaneous nerves at the palm of the hand.

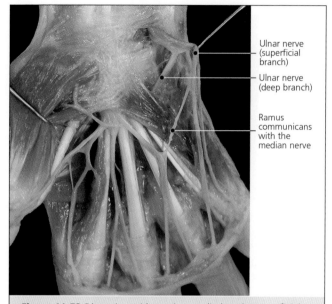

Figure 11-53 Dissection with tension applied to the superficial branch of the ulnar nerve to better show the deep branch where it courses through the arcade of the muscles of the hypothenar eminence.

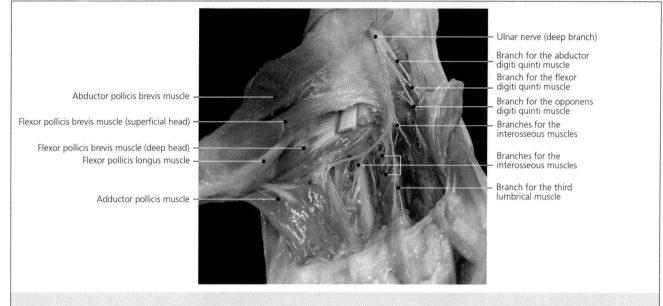

Figure 11-54 Specimen showing the motor branches of the ulnar nerve in the palm of the hand.

and palmar aspects of the little finger and the ulnar half of the ring finger. The proper digital nerve for the ulnar side of the little finger arises directly from the superficial branch.

The deep branch of the ulnar nerve is a motor branch that originates at Guyon's canal and enters an arcade formed by the origins of the muscles at the hypothenar eminence (Figure 11-53). This branch lies on the hook of the hamate and enters the interosseous plane (Figure 11-54). First it innervates all the muscles of the hypothenar eminence. Then it courses transversely along the deep structures of the hand with the deep arterial palmar arch. In this location, it sends small muscular branches to the palmar and dorsal interossei as well as the third and fourth lumbricals. It passes through the hiatus of the adductor pollicis to reach the deep head of the flexor pollicis brevis and ends at the first interosseous space, innervating all these muscles (Figure 11-55).

Figure 11-56 illustrates the motor branches of the ulnar nerve.

NEUROLOGY

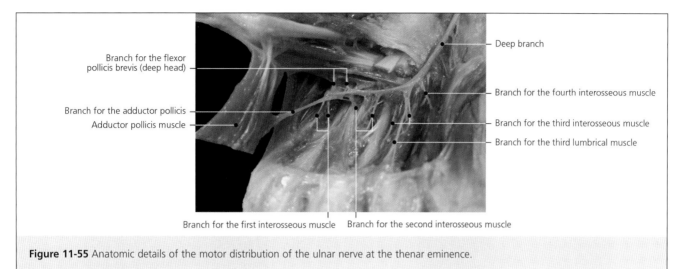

Figure 11-55 Anatomic details of the motor distribution of the ulnar nerve at the thenar eminence.

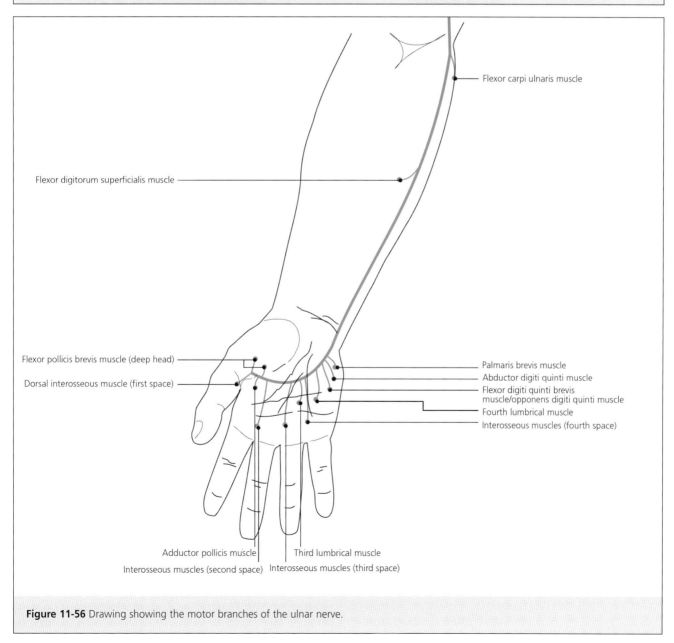

Figure 11-56 Drawing showing the motor branches of the ulnar nerve.

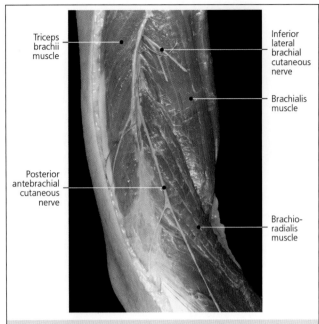

Figure 11-57 Dissection of the superficial nerves at the lateral region of the elbow.

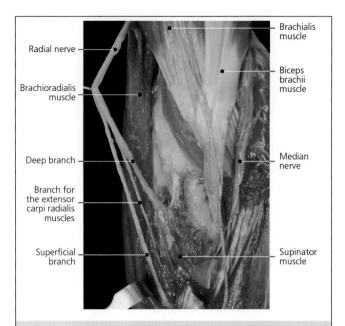

Figure 11-58 Dissection with the radial nerve placed under tension to show its superficial and deep branches as well as its motor branches.

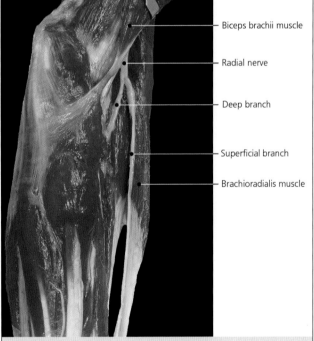

Figure 11-59 Course of the superficial branch of the radial nerve with the brachioradialis muscle partly retracted.

Radial Nerve

The radial nerve is the terminal branch of the posterior cord of the brachial plexus. It contains fibers from C5 to T1 and innervates the extensor/supinator muscles.

The radial nerve reaches the forearm on its lateral aspect, between the brachialis and the brachioradialis. Proximal to the elbow joint, the radial nerve sends muscular branches for the brachioradialis, the extensor carpi radialis longus, the anconeus, and the extensor carpi radialis brevis. It also sends a cutaneous branch, the posterior antebrachial cutaneous nerve (Figure 11-57), which innervates the dorsal skin between the regions of the lateral and medial cutaneous nerves of the forearm. At the lateral aspect of the arm, sensory fibers of the radial nerve form the inferior lateral brachial cutaneous nerve.

The radial nerve continues at the lateral bicipital groove (Figure 11-29) and divides at this location into a superficial and a deep branch (Figures 11-58 and 11-59). The sensory superficial branch continues on the deep aspect of the brachioradialis muscle inside its aponeurosis (Figure 11-59), along with the radial artery. Approximately 5 cm proximal to the wrist, this nerve perforates the antebrachial fascia and turns superficially and dorsally. It exits between the posterior aspect of the brachioradialis and the tendons for the extensor carpi radialis muscles. Finally, this nerve divides into several sensory terminal branches, the dorsal digital nerves (Figure 11-60), which innervate the dorsal aspect of the skin of the thumb and index finger and half of the long finger to the level of the proximal interphalangeal joint. At this location also is found the ramus communicans for the ulnar nerve, which anastomoses with the ulnar nerve at the dorsal aspect of the hand.

The deep branch perforates the two heads of the supinator muscle (Figures 11-61 and 11-62) through the arcade of Frohse and then turns posteriorly around the radius. At its exit from the arcade, it

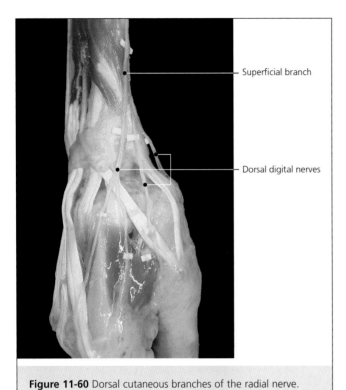

Figure 11-60 Dorsal cutaneous branches of the radial nerve.

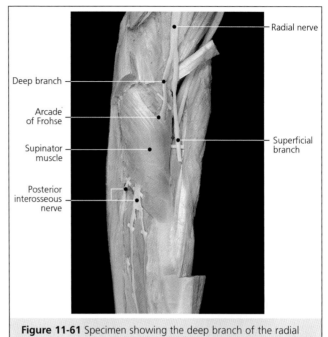

Figure 11-61 Specimen showing the deep branch of the radial nerve through the supinator muscle.

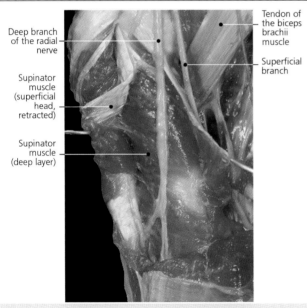

Figure 11-62 Anatomic details of the deep branch of the radial nerve between the two layers of the supinator muscle.

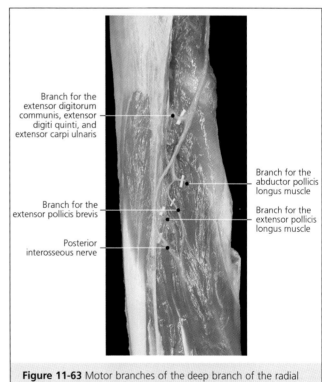

Figure 11-63 Motor branches of the deep branch of the radial nerve.

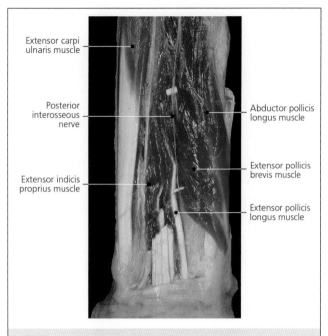

Figure 11-64 Anatomic details of the posterior interosseous nerve.

sends motor branches to innervate all the extensor muscles of the posterior compartment of the forearm. This branch innervates the superficial muscles as well as the deep muscles through the posterior interosseous nerve (Figures 11-63 and 11-64). At its final course, it is located on the interosseous membrane of the forearm and fourth extensor compartment to reach the wrist joint.

Figure 11-65 illustrates the motor branches of the radial nerve.

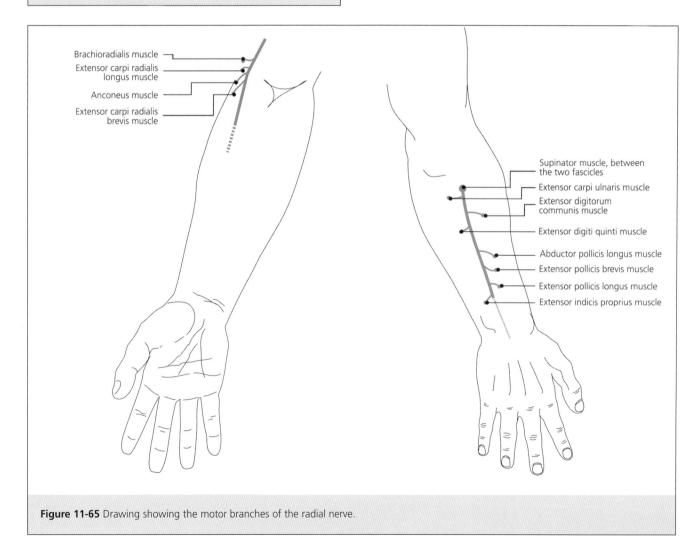

Figure 11-65 Drawing showing the motor branches of the radial nerve.

Chapter 12

Angiology

Arteries

Subclavian Artery

The arteries of the shoulder girdle and the upper limb originate from the subclavian artery. This artery originates at the anterosuperior mediastinum. It arises from the brachiocephalic trunk on the right side, behind the sternoclavicular joint, and from the aortic arch on the left side (Figure 12-1). The medial aspect of this vessel is retroclavicular, whereas the lateral aspect is subclavicular; this is due to the fact that this vessel forms an arch, the vertex of which is located between the scalenus muscles. The left and right subclavian arteries follow similar courses, although the left is located deeper than the right, is longer, and has a small thoracic portion due to its direct origin. The subclavian artery is usually divided into three portions: prescalenic, interscalenic, and postscalenic.

Prescalenic Subclavian Artery

The prescalenic portion (Figure 12-2) ascends to the vertex located at the interscalenic triangle—the space between the middle and anterior scalenus muscles. It lies on the pleura of the lung apex and the sulcus for the subclavian artery on the superior aspect of the first rib. At this position, the arch turns and is directed caudally beneath the clavicle, where it is called the axillary artery.

The main branches of the subclavian artery originate at the prescalenic portion and are directed to various areas of the trunk, neck, and brain. These branches are as follows:

Internal Thoracic Artery
Descends at the anterior aspect of the thorax, parallel to the internal face of the sternum (Figure 12-3).

Vertebral Artery
Directed to the transverse foramina of the cervical vertebrae, usually from C6 (Figure 12-4).

Thyrocervical Trunk
Formerly called the thyrocervicoscapular trunk, it originates from the subclavian artery before the level of the anterior scalenic muscle and gives four important branches:

Inferior thyroid artery This artery divides into several branches. It will be described in the chapter dedicated to the neck (chapter 18), as it supplies part of this region as well as some viscerae and the thyroid gland (Figure 12-4).

Ascending cervical artery In some individuals, this artery arises from the subclavian artery of the inferior thyroid artery. As indicated by its name, the artery continues in a cranial direction on the anterior aspect

American Academy of Orthopaedic Surgeons

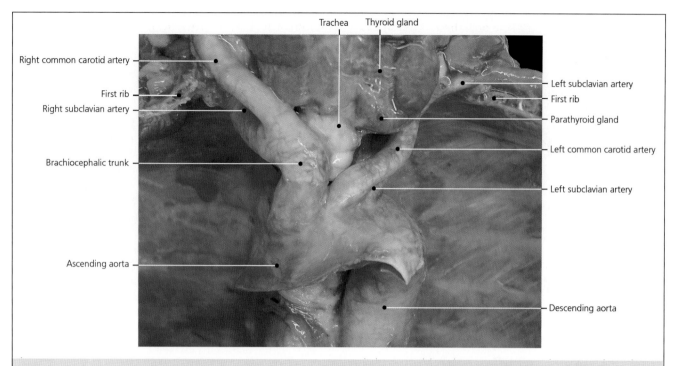

Figure 12-1 Anterior view of the mediastinum after removing the heart and lungs, showing anatomic details of the aorta and its first branches.

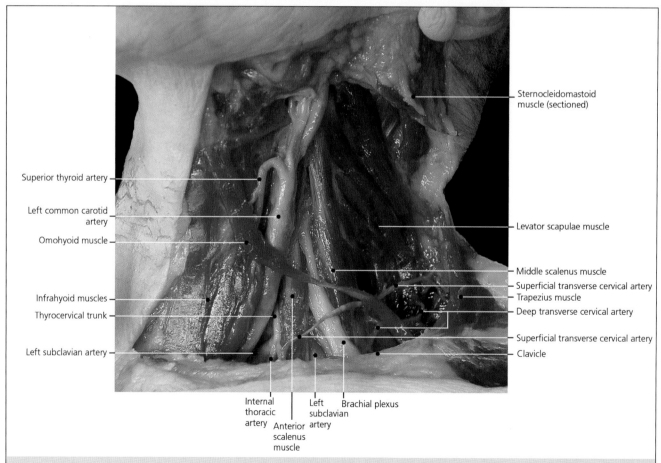

Figure 12-2 Arteries of the neck region after sectioning the sternocleidomastoid muscle. Note the prescalenic exit of the thyrocervical trunk and the interscalenic triangle (between the anterior and middle scalenus muscles), where the subclavian artery and the brachial plexus exit.

of the anterior scalenus muscle and the rectus capitis muscle, medial to the phrenic nerve, and occasionally reaching the base of the cranium. It supplies the scalenus muscles and the origin of the levator scapulae (Figure 12-4).

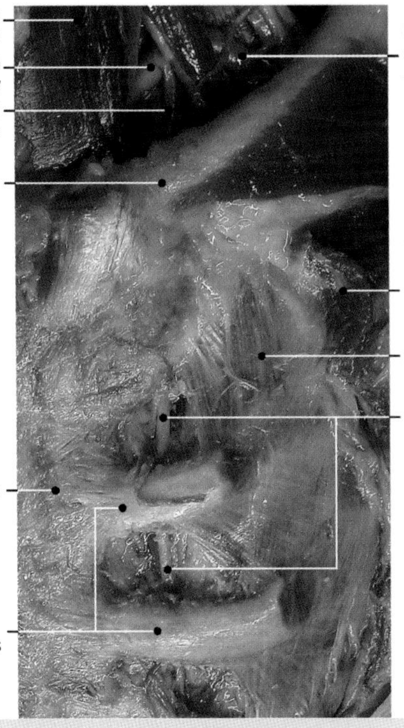

Transverse cervical artery Has a superficial branch and a deep branch. (1) The superficial branch, which in some individuals originates as a superficial transverse artery from the thyrocervical trunk, originates close to the ascending cervical artery and courses in a horizontal and lateral direction behind the sternocleidomastoid muscle and in front of the scalenus and the omohyoid muscles. It forms an arch over the brachial plexus, perfused by this vessel; it supplies the trapezius, levator scapulae, and splenius cervicis muscles. The superficial branch of the transverse cervical artery is divided into an ascending and a descending branch. (2) The deep branch of the transverse cervical artery courses behind the anterior scalenus muscle and crosses the brachial plexus between the superior and the middle trunk or between the fifth and the sixth cervical root, reaching the superior angle of the scapula over the middle scalenus muscle. At this level, it sends branches to muscles in the vicinity, including the supraspinatus (through an anastomosis with the suprascapular artery), the trapezius, the muscles of the neck, and the acromial network. The deep branch has an ascending branch between the levator scapulae and the splenius cervicis, which anastomoses with the superficial branch of the transverse cervical artery

Figure 12-3 Anterior thoracic wall with windows to observe the course of the internal thoracic artery.

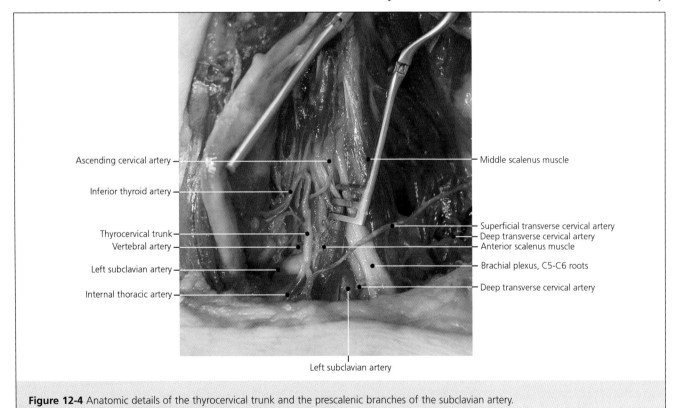

Figure 12-4 Anatomic details of the thyrocervical trunk and the prescalenic branches of the subclavian artery.

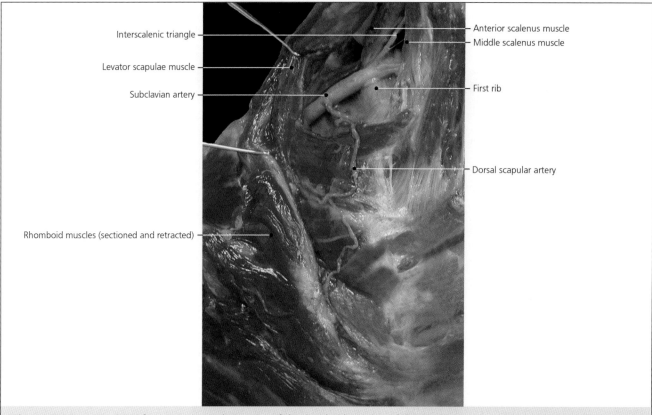

Figure 12-5 Posterior view, after sectioning and retraction of the rhomboid muscles, of the dorsal scapular artery arising directly from the subclavian artery.

and supplies the deep nuchal muscles, and a descending branch, which continues distally, along with the dorsal scapular nerve and deep with respect to the levator scapulae and the rhomboids, until it reaches the latissimus dorsi. The deep branch usually sends branches to the posterior muscles of the scapula, and it anastomoses with the suprascapular artery and the circumflex scapular artery, forming a true arterial network. In some individuals, this descending branch originates from the subclavian artery, where it is known as the dorsal scapular artery (Figure 12-5).

Suprascapular artery This artery courses horizontally in a lateral direction to the scapula, in front of the anterior scalenus and the phrenic nerve. At the level of the acromial end of the clavicle, this artery sends acromial and thoracic branches beneath the clavicle and to the anterior thoracic wall. Some authors also have described small branches for the sternoclavicular, acromioclavicular, and glenohumeral joints. The main vessel, however, continues to the scapular notch, over the superior transverse ligament; the suprascapular nerve lies beneath the ligament. The suprascapular artery enters the infraspinous fossa to supply the infraspinatus muscle. This terminal portion anastomoses with the circumflex scapular artery and some branches of the dorsal scapular artery (Figure 12-6).

Costocervical Trunk

Located in the interscalenic portion of the subclavian artery, the costocervical trunk originates from its posterior wall behind the anterior scalenus muscle. This trunk bends posteriorly over the pleural apex and sends two arteries to supply the neck and thorax. The deep cervical artery courses posteriorly to form an arch with an inferior convexity between the transverse apophyses of C7 and T1, over the eighth cervical root and the neck of the first rib. From this location, the artery ascends along the cervical musculature, between the semispinales capitus and colli and the multifidus, until it reaches the axis. In this region, it anastomoses with the occipital artery, the vertebral artery, the ascending cervical artery, and the transverse cervical artery. The supreme intercostal artery is a common branch that originates from the first and second posterior intercostal arteries, which courses in the first and second intercostal spaces. It has dorsal branches for the posterior muscles and the skin, and spinal branches that enter the intervertebral spaces of T1 and T2.

Axillary Artery

The axillary artery is the continuation of the subclavian artery; it is usually called the axillary artery at the point where it passes over the first rib, although some authors change the name from subclavian to axillary at the point

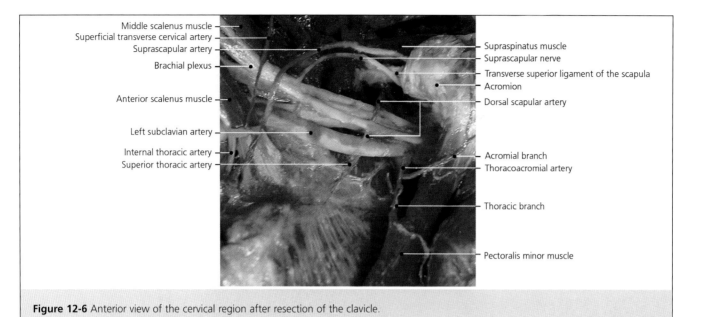

Figure 12-6 Anterior view of the cervical region after resection of the clavicle.

where it passes beneath the clavicle. The axillary artery is related posteriorly with the external intercostal muscle of the first intercostal space, the first digitation of the serratus anterior and the posterior wall of the axilla. It continues to the inferior border of the pectoralis major muscle. All the branches from this vessel originate from its anteroinferior surface. These branches are described here according to their location on the medial, posterior, or lateral aspect of the pectoralis minor muscle.

Branches Medial to the Pectoralis Minor

Superior thoracic artery This small vessel originates behind the subclavian muscle and descends along the anterior thoracic wall, piercing the clavipectoral fascia. Its muscular branches supply the subclavius muscle, both pectoralis muscles, and the superior portion of the serratus anterior. The artery then continues along the first two intercostal spaces, and some branches may reach the skin. In women, this vessel supplies the mammary gland through the lateral mammary branches. It anastomoses with the various vessels that reach the anterior thoracic wall (Figure 12-6).

Thoracoacromial artery This artery originates at the level where the axillary artery reaches the upper edge of the pectoralis minor. After a short lateral and cranial course, the thoracoacromial artery pierces the clavipectoral fascia and sends several branches in all directions; selected branches are discussed here: (1) the pectoralis branches, which descend between the pectoralis muscles, the superior aspect of the serratus anterior, and the anterior aspect of the thorax; (2) an acromial branch, directed laterally parallel to the clavicle and over the coracoid process, to the acromion, covered by the deltoid muscle; (3) a deltoid branch, oriented laterally, which follows the course of the cephalic vein in the deltopectoral interval and sends muscular branches for the deltoid and the pectoralis muscles; and (4) a small clavicular branch, which extends toward the subclavius muscle, the clavicle, and the sternoclavicular joint (Figures 12-6 and 12-7).

Branches Posterior to the Pectoralis Minor

When the axillary artery passes behind the pectoralis major, it is located behind the axillary fat, between the cords of the brachial plexus. These fascicles are called the lateral, posterior, and medial cords based on their relationship to the axillary artery. Branches of the axillary artery are directed toward the shoulder and the lateral aspect of the thorax.

Lateral thoracic artery This artery (Figure 12-8) originates behind the pectoralis minor muscle, beneath its inferior edge, and sends muscular branches to the pectoralis muscles. It descends to the fifth to sixth intercostal space, where it lies close to the long thoracic nerve between the pectoralis minor and the serratus anterior. In addition, it sends branches to the adjacent muscles, turning anterior at the inferior edge of the pectoralis major and continuing until it reaches the mammary gland through the lateral mammary branches.

Subscapular artery This artery (Figure 12-8) originates at the lateral edge of the subscapularis muscle, although in some individuals it originates from a common trunk with the humeral circumflex artery. It continues in a descending and medial direction, behind the radial nerve and in front of the median nerve. After

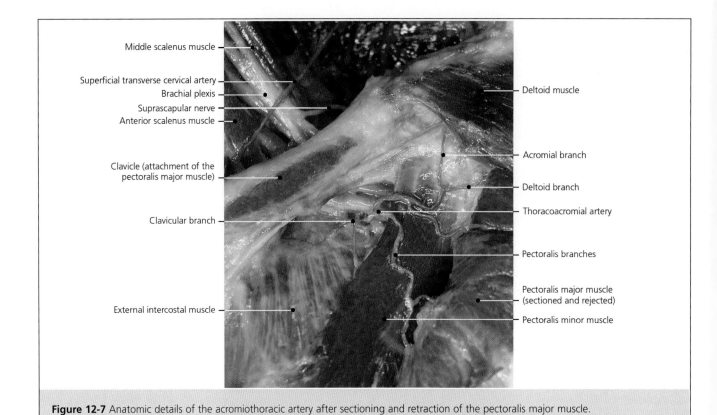

Figure 12-7 Anatomic details of the acromiothoracic artery after sectioning and retraction of the pectoralis major muscle.

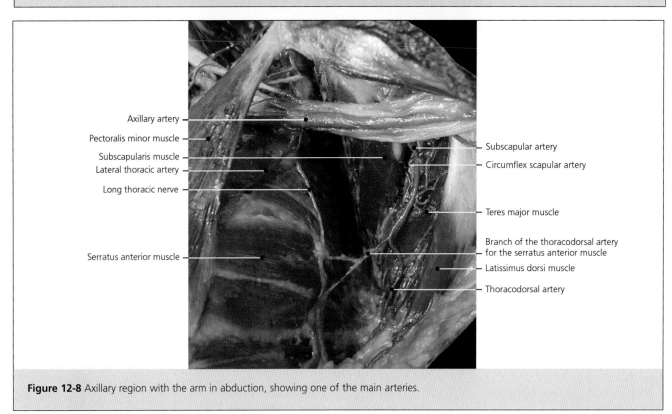

Figure 12-8 Axillary region with the arm in abduction, showing one of the main arteries.

a short course, it continues in front of the subscapularis muscle, sending muscular branches and additional branches to the adjacent fat and lymph nodes. The subscapular artery finally divides into two terminal branches, the thoracodorsal artery and the circumflex scapular artery. The thoracodorsal artery follows the course of the subscapular artery in a caudal direction behind the posterior axillary fold. It joins the tho-

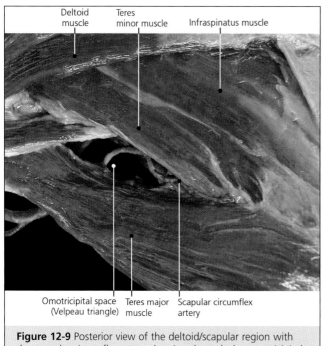

Figure 12-9 Posterior view of the deltoid/scapular region with the scapular circumflex artery showing through the omotricipital triangle.

racodorsal nerve along the lateral edge of the scapula, extending to and irrigating the teres major, the latissimus dorsi, the intercostal muscles, and the serratus anterior. It anastomoses with intercostal branches and the dorsal scapular artery to reach the skin on the lateral aspect of the thorax. The circumflex scapular artery (Figures 12-8 and 12-9) describes a posteriorly directed arch to the internal aspect of the axilla and sends three branches. The first branch is distributed on the anterior aspect of the subscapularis muscle; a second, larger, branch is directed posteriorly, passing through the omotricipital triangle and around the lateral edge of the scapula to the infraspinatus fossa to anastomose with the suprascapular artery in this region; and the third branch, the marginal artery, follows along the lateral edge of the scapula to lie between the two teres muscles to reach the inferior angle and anastomose with the dorsal scapular artery and the thoracodorsal artery.

Branches Lateral to the Pectoralis Minor
The lateral portion of the axillary artery lies close to the surgical neck of the humerus and sends branches (the circumflex arteries) to surround it. In some individuals, these arteries arise from a common trunk, although more commonlly they originate as separate arteries.

Anterior circumflex humeral artery In some individuals, this vessel (Figure 12-10) divides into several branches. The larger vessel courses over the latissimus dorsi and surrounds the humeral neck very close to the bone beneath the coracobrachialis and the short head of the biceps. As this vessel turns lateral, it courses beneath the tendon of the long head of the biceps brachii. The ascending branches follow the intertubercular groove to supply the intertubercular synovial groove and the periosteum until they reach the articular capsule of the glenohumeral joint. The descending branches are directed to the deltoid and anastomose with the posterior circumflex humeral artery.

Posterior circumflex humeral artery This artery (Figures 12-10 and 11-25) arises at approximately the same level as the subscapular artery, on the posterior wall of the axillary artery. With the radial nerve, it forms a posteriorly directed arch around the humeral neck. This branch crosses the quadrangular space close to the humerus. In this region, it sends branches to the muscles that border this space as well as to the teres minor and major, triceps, and latissimus dorsi. Finally, this vessel sends branches to the deltoid and the glenohumeral articular capsule. In most individuals it anastomoses with the anterior circumflex artery and the acromiothoracic artery, but in some it anastomoses with the deltoid branches of the deep brachial artery.

Brachial Artery

The axillary artery is called the brachial or humeral artery (Figure 12-11) from the inferior edge of the pectoralis major or the surgical neck of the humerus to its bifurcation into the radial and ulnar arteries at the elbow.

The brachial artery leaves the axillary region, and it lies on the medial aspect of the coracobrachialis muscle first and the biceps brachii next. It continues in a parallel direction along the long axis of the arm through the medial region of the anterior compartment of the arm, in a sulcus between the brachialis and the biceps brachii muscles. Finally, it enters the medial bicipital canal at the anterior aspect of the elbow, where it divides into two main branches, the radial and the ulnar arteries.

In some individuals, the superficial brachial artery, a variant of the brachial artery, is found anterior to the median nerve. This is in contrast to the usual position of the brachial artery, posterior to the median nerve.

The brachial artery sends several muscular branches, the arrangement of which varies tremendously. In general, the branches oriented radially supply the flexor muscles, and those oriented ulnarly supply the extensor muscles. The deltoid branch surrounds the humerus posteriorly, and ascending and descending branches supply the lateral aspect of the deltoid.

At the level of the deltoid branch and beneath the teres major, the brachial artery gives off the deep brachial artery (Figures 12-12 and 12-13). This branch follows the course of the radial nerve in the spiral

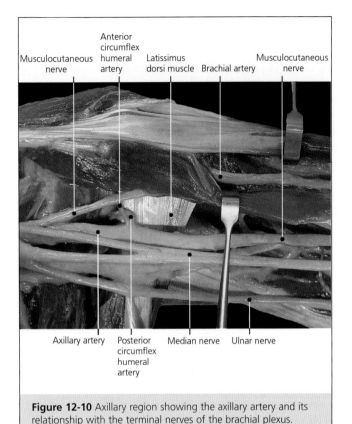

Figure 12-10 Axillary region showing the axillary artery and its relationship with the terminal nerves of the brachial plexus.

groove in a diagonal and lateral direction. Throughout its course, this vessel sends branches to adjacent muscles, including the triceps brachii, as well as retrograde branches to the deltoid and the coracobrachialis; it also gives off several nutrient branches for the humerus. When this vessel passes between the bellies of the triceps, it sends numerous muscular branches, the largest of them being the deep artery for the medial head of the triceps (Figure 12-12). The middle collateral artery originates at the lateral edge of the humerus and descends through the posteromedial aspect of the arm to reach the elbow network, where it anastomoses with the recurrent interosseous artery. The deep brachial artery continues as the radial collateral artery (Figures 12-12 through 12-14), which anastomoses with the recurrent radial artery to contribute to the elbow arterial network. A small branch forms the satellite artery for the radial nerve.

During its course, the brachial artery sends humeral nutrient arteries, which enter the humeral shaft through several orifices.

Two collateral arteries arise from the brachial artery on its ulnar side.

The superior ulnar collateral artery (Figures 12-11 and 12-15) arises close to the deep brachial artery. It sends muscular branches to the brachialis and the triceps and then continues along the forearm axis with the ulnar

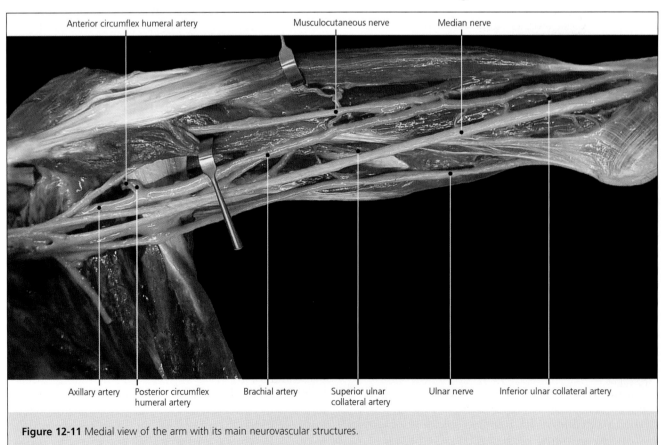

Figure 12-11 Medial view of the arm with its main neurovascular structures.

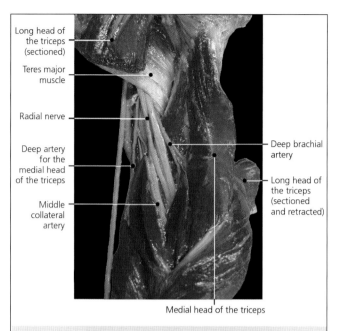

Figure 12-12 Posterior view of the tricipital region showing the course of the radial nerve and the deep brachial artery. In this specimen, the middle collateral artery is already differentiated proximally.

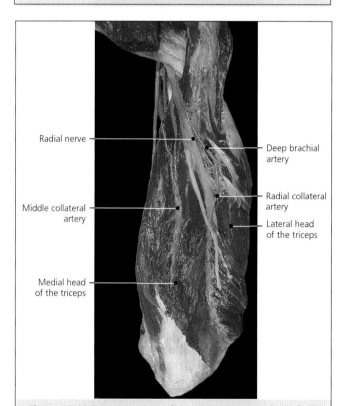

Figure 12-13 Posterior aspect of the tricipital region after reflection of the long head of the triceps brachii. This specimen shows the course of the radial collateral artery to the lateral region of the elbow.

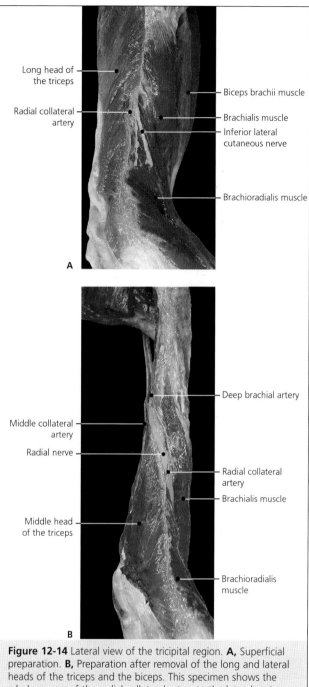

Figure 12-14 Lateral view of the tricipital region. **A,** Superficial preparation. **B,** Preparation after removal of the long and lateral heads of the triceps and the biceps. This specimen shows the whole course of the radial collateral artery to the lateral region of the elbow.

nerve until it reaches the distal forearm, where it perforates the medial intermuscular septum. It anastomoses with the inferior ulnar collateral artery, the middle collateral artery, and the recurrent collateral artery to contribute to the elbow arterial network.

The inferior ulnar collateral artery (Figures 12-11 and 12-15) arises at the distal third of the humerus, above the medial epicondyle. It is located between the median nerve and the medial antebrachial nerve, sending some branches to the brachialis and the teres major. It anastomoses with the recurrent ulnar artery and the superior ulnar collateral artery at the elbow arterial network, sending a posterior branch that perforates the medial inter-

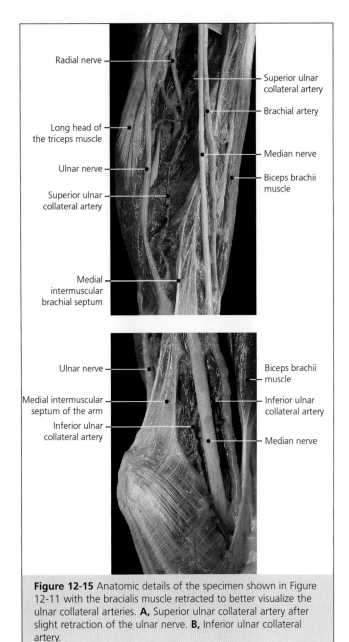

Figure 12-15 Anatomic details of the specimen shown in Figure 12-11 with the bracialis muscle retracted to better visualize the ulnar collateral arteries. **A**, Superior ulnar collateral artery after slight retraction of the ulnar nerve. **B**, Inferior ulnar collateral artery.

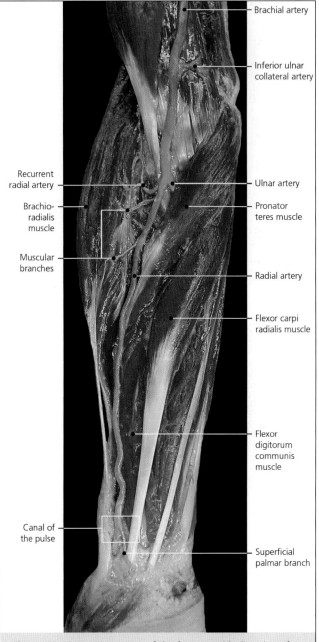

Figure 12-16 Anterior aspect of the forearm with dissecion of the main branches of the radial artery.

muscular septum of the arm. At this level, this artery sends branches for the triceps brachii and the brachioradialis, and it ends above the olecranon fossa through several anastomoses with the arterial network of the elbow. A transverse branch passes from one epicondyle to the other to connect this artery with branches of the deep brachial artery, especially the middle collateral artery.

Radial and Ulnar Arteries

The brachial artery (Figure 12-16) enters the elbow at the medial bicipital canal, beneath the lacertus fibrosus. Deep to the humeral head of the pronator teres, it divides into the radial artery and the ulnar artery.

The radial artery (Figures 12-16 and 12-17) sends a small branch, the recurrent radial artery, close to the radial neck. This branch then ascends to the lateral epicondyle until it anastomoses with the radial collateral artery. The recurrent radial artery supplies the muscles of the lateral compartment of the forearm.

The ulnar artery (Figures 12-17 and 12-18) arises from the division of the brachial artery beneath the pronator teres. It also sends a recurrent branch, which sometimes originates directly from the brachial artery. The recurrent ulnar artery originates from the ulnar artery close to the bicipital tuberosity, and it is directed in a proximal and medial direction. It divides into two branches, one anterior and one posterior. The anterior branch is located

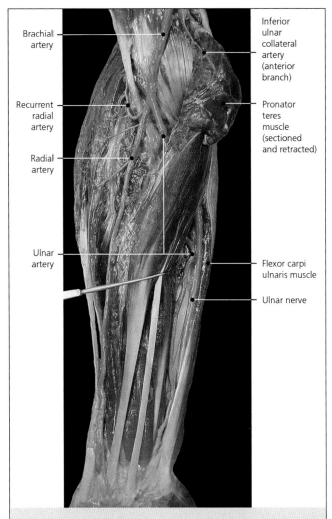

Figure 12-17 Anterior view of the forearm with the pronator teres muscle removed, showing the origin of the ulnar artery.

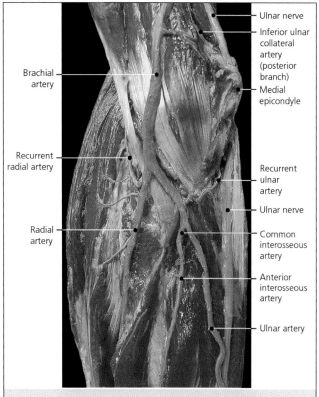

Figure 12-18 Anterior view of the forearm after removal of the epicondylar muscles, showing anatomic details of the arteries in the elbow and proximal forearm regions.

medial to the brachialis muscle and ascends to anastomose with the inferior ulnar collateral artery, sending branches to the pronator teres and the brachialis. The posterior branch follows the course of the ulnar nerve, ascending proximally and dorsally toward the medial epicondyle, where it forms part of the articular network of the elbow, anastomosing with the superior ulnar collateral artery.

This complex anastomotic network of collateral, recurrent, and small branches around the elbow joint, especially on the posterior aspect, is called the elbow arterial network.

After sending its recurrent radial branch, the radial artery courses along the anterolateral aspect of the forearm, between the flexor carpi radialis tendon on the medial side and the brachioradialis on the lateral side (Figure 12-16). It continues in a purely longitudinal direction to the wrist.

The ulnar artery (Figure 12-17) passes deep to the pronator teres and proceeds along the anteromedial aspect of the forearm, beneath the flexor carpi ulnaris, until it reaches the wrist.

Proximally, the common interosseous artery arises and divides into two branches, the anterior interosseous artery and the posterior interosseous artery (Figures 12-18 through 12-20). The anterior interosseous artery is located on the anterior aspect of the interosseous membrane, and it descends distally to the wrist, passing beneath the pronator quadratus muscle to perforate the interosseous membrane in a dorsal direction and contribute to the dorsal carpal network. This artery sends numerous branches to the deep anterior muscles of the arm, perforates the interosseous membrane at several points, and supplies the deep musculature of the posterior region. At the same level, a small arteriole, the satellite artery for the median nerve, joins the median nerve in the forearm. The second branch of the common interosseous artery is the posterior interosseous artery. This branch becomes posterior as it passes through the hiatus of the interosseous membrane and the oblique cord. It lies on the posterior aspect of the interosseous membrane with the posterior interosseous nerve and continues longitudinally to the dorsal aspect of the wrist, where it anastomoses with the dorsal carpal network. This artery sends branches to the adjacent muscles as well as to the epicondylar muscles, and it also gives off the recurrent interosseous artery,

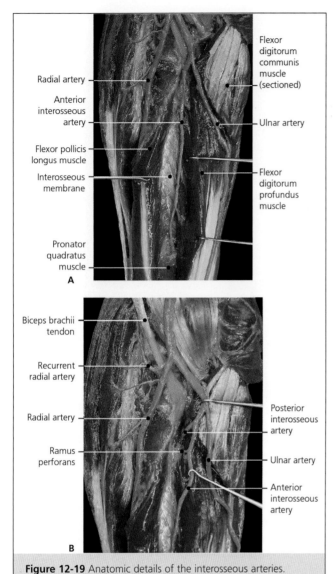

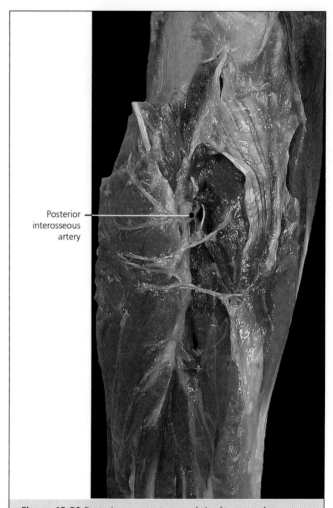

Figure 12-19 Anatomic details of the interosseous arteries. A, Anterior interosseous artery. B, The ulnar artery has been placed under tension to show the posterior interosseous artery.

Figure 12-20 Posterior compartment of the forearm after removal of the first layer to show the posterior interosseous artery.

which is located beneath the anconeus muscle. The recurrent interosseous artery ascends until it anastomoses with the middle collateral artery at the elbow articular network.

At the distal aspect of the pronator quadratus, the radial artery sends a small branch directed to the anterior carpal network. This branch is called the palmar carpal branch or transverse anterior carpal artery. At the level of the scaphoid, a small superficial palmar branch courses through the thenar eminence and anastomoses with the ulnar artery to form the superficial palmar arch (Figure 12-21). This arch is located between the palmar aponeurosis and the tendons for the finger flexors. The main trunk of the radial artery passes proximal to the trapezius and becomes dorsal at the topographic region called the anatomic snuffbox. In this area, the artery courses through the first posterior interosseous space and enters the palm of the hand again to send two branches: the dorsal carpal branch (which anastomoses with the dorsal branch from the ulnar artery and with the posterior interosseous artery to form the dorsal carpal network) and the dorsal artery of the thumb, which descends on the dorsal aspect of that digit.

After coursing through the first interosseous muscle and entering the palm of the hand, the radial artery anastomoses with the deep branch of the ulnar artery to form the deep palmar arch, which is located beneath the finger flexor tendons.

The ulnar artery sends two branches at the level of the distal ulnar epiphysis. The palmar carpal branch arises at the distal edge of the pronator quadratus to anastomose with the palmar carpal branch from the radial artery to form the palmar carpal network. In this same area, the ulnar artery sends a dorsal carpal branch, which anastomoses with the dorsal branch of the radial artery to form the dorsal carpal network, similar to the palmar branches. In a similar manner, the deep palmar arch is

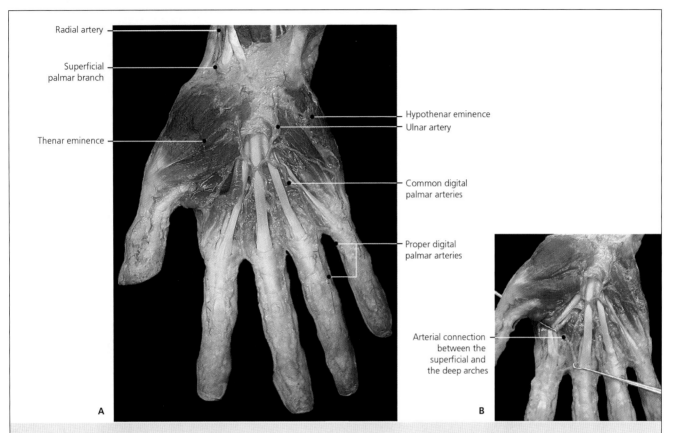

Figure 12-21 A, Specimen showing the superficial palmar arch after removal of the palmar aponeurosis. **B,** Detail of the arterial connections between the superficial and deep arches.

formed by the radial artery and a deep palmar branch (Figure 12-22), which sends an ulnar branch close to the hook of the hamate.

The superficial palmar arch (Figure 12-21) sends the common digital palmar arteries, which are located in each interosseous space. These vessels divide to form the proper palmar digital arteries, which supply each side of the finger to the fingertips.

The deep palmar arch sends several branches, the palmar metacarpal arteries (Figure 12-23), which are distributed in the intermetacarpal spaces. The thumb has its own artery from this deep arch, the main artery of the thumb, or princeps pollicis, which begins where the radial artery enters the palm of the hand through the first dorsal interosseous muscle. This artery is distributed along the flexor aspect of the thumb.

There are two important connections between the two arches. On one side, the princeps pollicis gives off the radial artery for the index finger, which anastomoses with the proper palmar digital artery on the radial side of the index finger. Numerous additional connections exist between the palmar metacarpal arteries of the deep arch and the palmar arteries of the superficial arch.

At the dorsal aspect of the hand, the dorsal carpal network originates the dorsal metacarpal arteries (Fig-

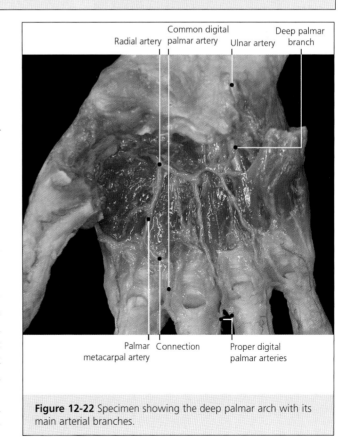

Figure 12-22 Specimen showing the deep palmar arch with its main arterial branches.

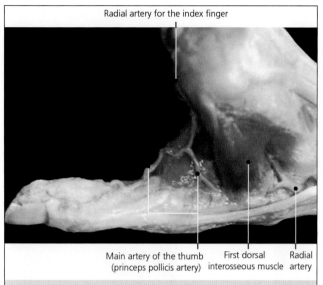

Figure 12-23 The radial artery perforates the first dorsal interosseous muscle to form the deep palmar arch at the palm of the hand. The main artery of the thumb and radial artery of the index finger are also shown.

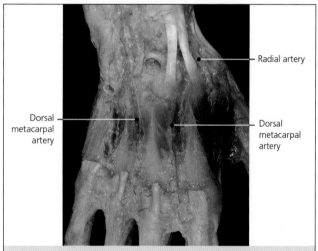

Figure 12-24 Dorsal view of the hand showing the dorsal metacarpal arteries.

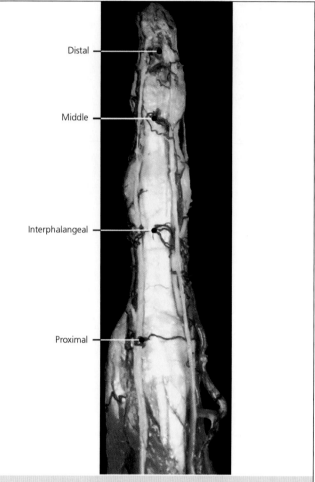

Figure 12-25 Specimen with the arteries injected with blue latex showing anatomic details of the transverse digital arteries.

ure 12-24), which are directed to the dorsal interdigital spaces. At the level of the metacarpophalangeal joints, usually after an anterior anastomosis between the common digital arteries and the palmar metacarpal arteries, there are ramus perforans, which anastamose with the dorsal metacarpal arteries.

There are four communications between the proper palmar digital arteries at the palmar aspect of the fingers, the four transverse digital arteries—one proximal, one interphalangeal, one middle, and one distal (Figure 12-25). The proximal transverse digital artery originates at the level of the metacarpal head, the interphalangeal artery originates close to the proximal interphalangeal joint, the middle artery originates at the level of the neck of the middle phalanx in triphalangeal fingers, and the distal artery corresponds to the end of the proper digital palmar arteries at the fingertips.

At the base of the proximal phalanx is a branch from the proper palmar digital artery for the mesotenon of the flexor sheath. These vessels send branches located at the mesotenon for the flexor tendons and the vinculi (Figure 12-26).

Veins

The veins of the upper limbs (Figures 12-27 through 12-29) drain to the subclavian vein and are arranged in two systems, one deep and one superficial. The deep veins, usually two per artery (except for the single axillary vein), have the same names and courses as the arteries described earlier.

Subclavian Vein

The subclavian vein extends from the edge of the first rib to its confluence into the corresponding brachiocephalic vein. It collects blood from the upper limb, shoulder, and anterior thoracic wall. The branches of the axillary and subclavian veins corre-

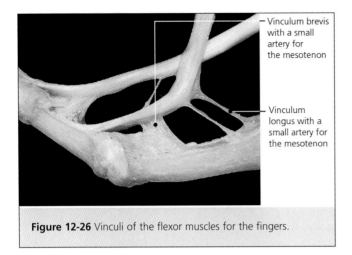

Figure 12-26 Vinculi of the flexor muscles for the fingers.

spond to the arteries with the same names described earlier.

Four veins drain directly into the subclavian vein: the pectoral veins, from the pectoralis muscles and the anterior thoracic wall; the dorsal scapular vein and the thoracoacromial vein, both of which follow the course of their homologous arteries, sometimes directly into the subclavian vein; and the axillary vein, described below.

Axillary and Brachial Veins

The axillary vein has the same boundaries as the axillary artery. It collects blood from the brachial veins and several other branches to drain into the subclavian vein. Most of the veins of the lateral thoracic wall, shoulder, and upper limb drain into this vein. The most notable satellite veins include the subscapular, circumflex scapular, thoracodorsal, circumflex humeral, and lateral thoracic. The brachial veins collect blood from the radial and ulnar veins, which follow the course of the homonymous arteries. The anterior and posterior interosseous veins lie in the deep planes.

Two venous arches are found in the palm of the hand. The superficial venous palmar arch receives the palmar digital veins, which collect the venous blood from the anterior aspect of the fingers. The deep palmar venous arch receives the palmar metacarpal veins.

Cephalic and Basilic Veins

The superficial veins of the fingers are arranged in a prominent dorsal plexus that continues through the metacarpal veins into a variable dorsal plexus of the hand. It forms a dorsal network of the hand, which originates the cephalic and basilic veins at the lateral and medial aspect of the hand, respectively.

The cephalic vein ascends on the lateral aspect of the forearm and the arm to drain into the axillary vein after it perforates the fascia of the infraclavicular fossa. Occa-

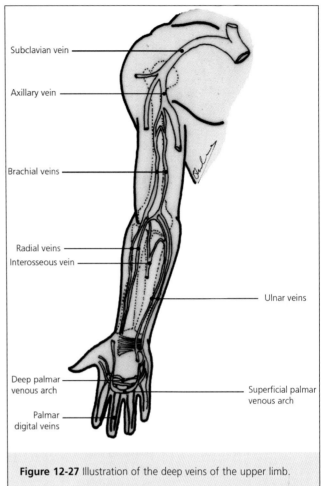

Figure 12-27 Illustration of the deep veins of the upper limb.

sionally it receives blood from an additional dorsal vein, the accessory cephalic vein.

The basilic vein ascends on the medial aspect of the forearm and the arm. In the middle of the arm, it perforates the fascia to turn deep and drain into the axillary vein.

In the anterior distal third of the arm, the antebrachial median vein divides into two branches. The lateral branch is the median cephalic vein, which drains into the cephalic vein. The medial branch is the median basilic vein, which drains into the basilic vein. This is the location most commonly used for intravenous access.

Lymphatic System

The main lymph nodes of the upper limb, which collect lymph from the thoracoabdominal wall, the mammary region, and the free portion of the upper limb, are arranged in six main groups: pectorals, subscapular, humeral, central, infraclavicular, and apical. They drain into the subclavian lymphatic trunk, which drains into the lymphatic conduit and the superior vena cava on the right side, and the thoracic conduit and brachiocephalic vein on the left.

Additional lymph nodes, including the supratrochlear nodes, are found in multiple locations of the upper limb.

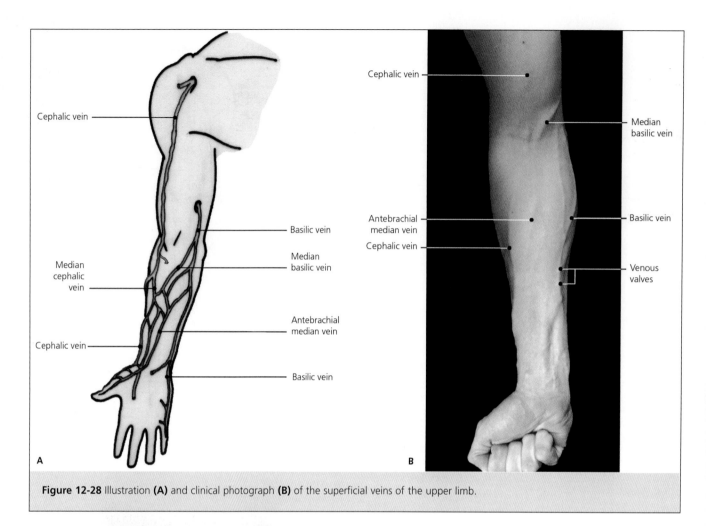

Figure 12-28 Illustration **(A)** and clinical photograph **(B)** of the superficial veins of the upper limb.

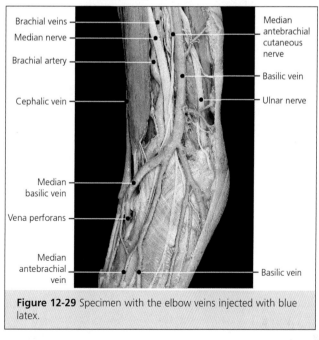

Figure 12-29 Specimen with the elbow veins injected with blue latex.

Section 3

Head and Trunk

Chapter 13

Topographic Regions and Surface Anatomy

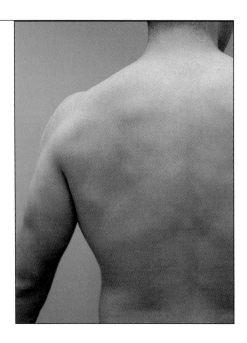

Introduction
The following regions are described in this chapter: cephalic, cervical, dorsal, and abdominal. For a discussion of the anterior and lateral thoracic regions (the scapular girdle), see chapter 7.

Cephalic Region (The Head)
The cephalic region is divided into several areas, some of which are named according to the underlying bones (Figure 13-1). The frontal region (forehead), located anteriorly, overlies the frontal bone. The parietal region is located on the sides of the head and overlies the parietal bones. The temporal region, located around the ear (the auricular region), overlies the squamous portion of the temporal bone. The mastoid region overlies the mastoid process of the temporal bone. Finally, the occipital region, found on the posterior aspect of the head, overlies the occipital bone.

The facial region overlies the craniofacial bones and the muscles of facial expression. The orbital region is the area surrounding the eyes. It is bordered superiorly and inferiorly by the suprapalpebral and infrapalpebral grooves. Beneath the orbit is the infraorbital region. The buccal regions are found at the sides of the labial region of the mouth. The cheeks of the face correspond to the zygomatic region, which overlies the zygomatic bone. Inferior and posterior to this is the parotid-masseteric region, which overlies the parotid gland and the masseter muscle. The nasal region includes the nose. Between the nose and the labial region are the nasolabial grooves. The chin, or mental region, is separated from the labial region by the mentolabial groove.

Cervical Region
The cervical region comprises an anterior region and a posterior region.

Anterior Region
The anterior region may be described by dividing it into several triangular regions (Figure 13-2).

Anterior Cervical Triangle
The anterior cervical triangle is limited by the midline, the mandible, and the sternocleidomastoid muscle. This triangle, in turn, is subdivided into four other triangular areas:
- Submandibular triangle—area limited by the mandible and the two bellies of the digastric muscle. Its contents include the submandibular salivary gland, the hypoglossal nerve, the mylohyoid muscle, and the facial artery.
- Carotid triangle—area limited by the sternocleidomastoid, the posterior belly of the digastric, and the superior belly of the omohyoid muscle. It contains the underlying carotid artery as well as the internal jugular vein and vagus nerve.
- Omotracheal (muscular) triangle—area limited by the larynx-trachea (midline), the hyoid bone, the superior belly of the omohyoid muscle, and the sternocleidomastoid muscle. It includes the infrahyoid musculature and the thyroid glands with the parathyroid glands
- Submental triangle—area located beneath the chin, limited by the mandible, the hyoid, and the anterior belly of the digastric muscle.

American Academy of Orthopaedic Surgeons

Head and Trunk

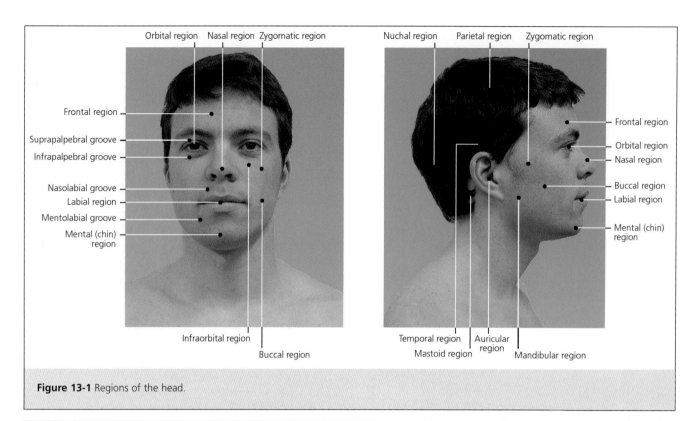

Figure 13-1 Regions of the head.

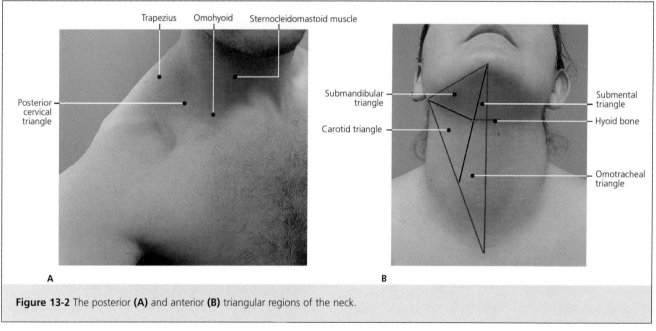

Figure 13-2 The posterior (A) and anterior (B) triangular regions of the neck.

The anterior cervical triangle is isolated from the posterior cervical triangle by the sternocleidomastoid region. Between the clavicular and sternal portion of this muscle, a small triangle is formed, the lesser supraclavicular fossa.

Posterior Cervical Triangle

The posterior cervical triangle is limited by the clavicle, the posterior border of the sternocleidomastoid muscle, and the anterior trapezius. Another triangle, the omoclavicular triangle (also called the greater supraclavicular fossa), is found within its limits. It is located between the clavicle, the sternocleidomastoid muscle, and the inferior belly of the omohyoid muscle.

Posterior Region

The nuchal region, or posterior cervical region (Figure 13-1), is bounded by the sternocleidomastoid muscle, the trapezius, and the middle third of the clavicle.

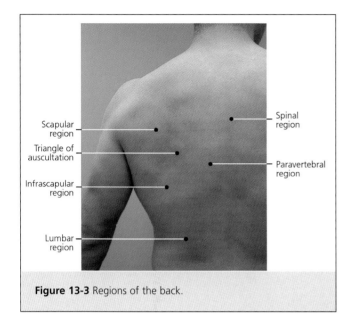

Figure 13-3 Regions of the back.

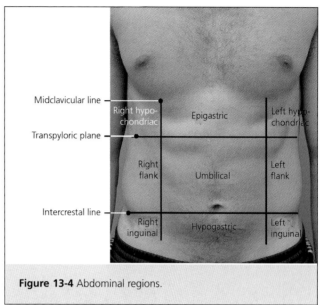

Figure 13-4 Abdominal regions.

Dorsal Region (The Back)

The spinal region is located in the midline of the back and is marked by the subcutaneous prominences of the spinous processes (Figure 13-3). The paravertebral regions are located on either side of the spinal region and consist of muscular columns parallel to the spine, formed by the erector spinae muscles.

The scapular region is located over the scapula; the infrascapular region is caudal to this. The lumbar region is localized over the iliac crest and caudal to the infrascapular region. The most caudal region of the spine is the sacral region, which overlies the sacrum.

In the dorsal region are two small triangles. The triangle of auscultation is limited by the ascending fibers of the trapezius, the latissimus dorsi and the rhomboideus major. The lumbar, or Petit's, triangle is localized between the iliac crest, the latissimus dorsi, and the external oblique.

Abdominal Region

The abdominal region is divided into nine regions (Figure 13-4). Three are to the left of the midclavicular lines, three are between the lines, and three are to the right.

The lateral regions are, from cranial to caudal: right and left hypochondriac, right and left flank, and right and left inguinal. The hypochondriac region is separated from the lateral region by the transpyloric plane (a horizontal line drawn halfway between the symphysis pubis and the superior border of the manubrium of the sternum). The lateral and inguinal regions are separated by the intercrestal line (a horizontal line drawn through the iliac crests).

The three medial regions are, from cranial to caudal: the epigastric, the umbilical (mesogastric), and the pubic (hypogastric) regions. They are defined horizontally by the transpyloric and intercrestal lines.

Chapter 14

Osteology

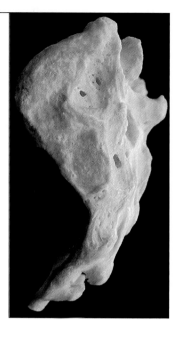

The Spine

Introduction
The spine (the axial skeleton) forms the axis of the body, which articulates with the extremities through the pelvic and scapular girdles (Figure 14-1). In most individuals, the spine comprises 33 segments called vertebrae. Between the vertebrae, articular disks (intervertebral disks) are intercalated, forming an amphiarthrosis.

The following regions of the spine are distinguished: cervical (7 vertebrae), thoracic (12 vertebrae), lumbar (5 vertebrae), sacrum (5 vertebrae), and coccyx (3 to 5 vertebrae).

In the frontal plane, the physiologic spine may present a soft sinusoidal form, with lateral curves (< 10°) toward the right and left, which is called the physiologic scoliosis. On the lateral view, four curves are clearly observed. In the cervical and lumbar regions, the curvature is anterior and is called lordosis; in the thoracic and sacral regions, the curvature is posterior and is called kyphosis. These are physiologic curves, and they provide the spine with a greater degree of stability and mobility than would be provided by a completely straight spine.

The movements allowed by the spinal column are achieved by small contributions from each of the intervertebral joints. In this way, movements of flexion, extension, lateral bending, and rotation are possible.

Typical Vertebra
Figure 14-2 shows a typical vertebra, showing the key details that are common to all vertebrae.

The vertebra is a short bone. It is composed of trabecular (cancellous) bone surrounded by a thin cortex of cortical bone.

The vertebral body forms the anterior column of the spine. Each vertebra has two end plates, which articulate with the two adjacent vertebrae through the intervertebral disks. The periphery is formed by an annular epiphysis. The vertebral bodies increase in size in the caudal direction; therefore, the bodies of the lumbar vertebrae are the largest.

The neural arch forms the posterior part of the vertebra and presents three processes: two lateral transverse processes and one dorsal spinous process. The spinous processes are located subcutaneously and can be palpated in the midline of the back. The lamina is located between the transverse and spinous processes. The slightly oblique direction of the laminae allows the superior lamina to overlap the inferior one, enhancing the protection of the spinal cord. The space between the transverse and spinous processes accommodates most of the dorsal paraspinal muscles.

The posterior arch joins the vertebral body by means of two pedicles. In this region are found two superior articular processes (to articulate with the immediately cephalic vertebra) and two inferior articular processes (to articulate with the immediately caudal vertebra), forming the zygapophyseal joints. This complex constitutes the posterior column (Figure 14-3).

The vertebral foramen is a triangular canal containing the spinal cord that is formed between the posterior arch and the posterior aspect of the vertebral body. The individual foramen, when linked in the articulated vertebral column, form the spinal canal.

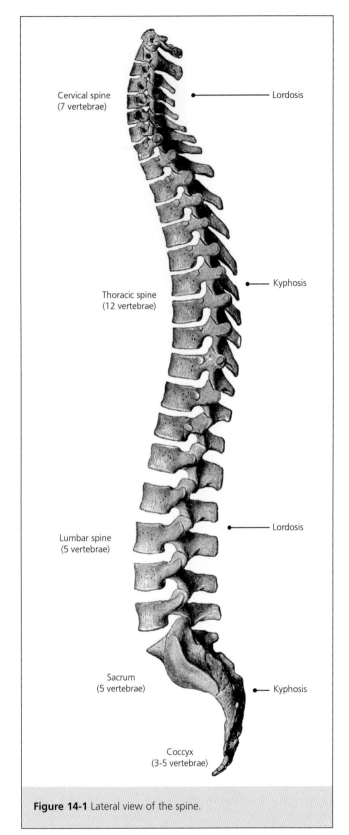

Figure 14-1 Lateral view of the spine.

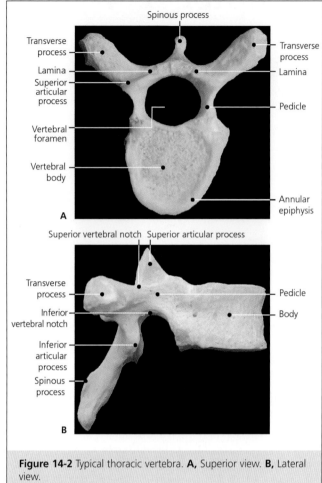

Figure 14-2 Typical thoracic vertebra. **A,** Superior view. **B,** Lateral view.

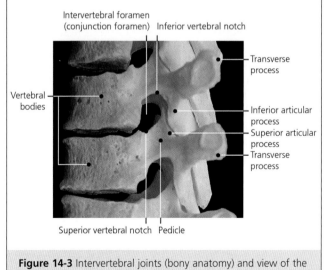

Figure 14-3 Intervertebral joints (bony anatomy) and view of the intervertebral foramen.

Osteology

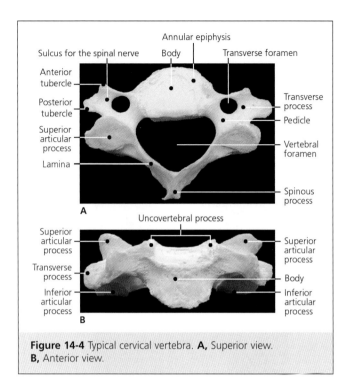

Figure 14-4 Typical cervical vertebra. **A,** Superior view. **B,** Anterior view.

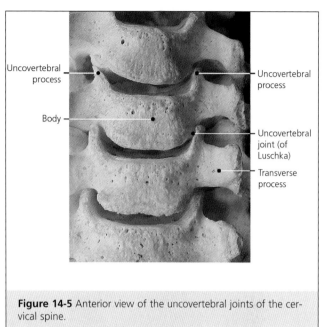

Figure 14-5 Anterior view of the uncovertebral joints of the cervical spine.

Each pedicle has a superior and an inferior vertebral notch. The articulation of these notches forms the vertebral foramen (conjunction foramen), through which the spinal nerves emerge.

Cervical Vertebrae

The seven cervical vertebrae are located between the occipital bone and the first thoracic vertebra. The first two are called the atlas (C1) and the axis (C2); they will be further described below. The vertebrae C3 through C7 (Figure 14-4) are characterized by a rectangular vertebral body with uncovertebral processes on each side. These uncovertebral processes act as hooks that allow the cervical vertebrae to fit together. This articulation forms the uncovertebral joints (of Luschka) (Figure 14-5).

The transverse processes have several distinctive features. First, the transverse foramina of C1 through C7 (with the exception of C7) contain the vertebral artery, accompanied by a satellite vein and the sympathetic nerve of François-Franck. The transverse process of vertebrae C2 through C7 present an anterior and a posterior tubercle for muscle attachments. The anterior tubercle is considered a rib remnant, and the posterior tubercle represents the proper transverse process. The carotid tubercle (or Chassaignac tubercle) represents the large, prominent anterior tubercle of C6. The common carotid artery lies anterior to this. The sulcus for the spinal nerve (C3 through C7) is located between the two tubercles and contains the corresponding spinal nerve after it leaves the intervertebral foramen. A third tubercle, the scalene tubercle, has also been described. It is located between the anterior and posterior tubercles at C3 through C5 and provides attachment for the anterior scalene muscles. Some of the cervical spinous processes have two tubercles, and therefore some authors call them bifid.

The most prominent spinous process, easily distinguished when the neck is in flexion, is that of C7. For this reason, C7 is called the vertebra prominens. This vertebra does not present an anterior tubercle in its transverse process because it does not provide muscle attachments.

In this region, the zygapophyseal joints are oriented obliquely, in an inferior and posterior direction.

The Atlas (C1)

Because it supports the skull, C1 is called the atlas (Figure 14-6), after the giant in Greek mythology who supported the heavens. The atlas has a characteristic ring shape. The body of this vertebra has developed phylogenetically into C2, forming the dens of the axis, which will be further described. The lateral masses are prominent and project a transverse process with a transverse foramen. The superior articular facets are kidney shaped and articulate with the occipital condyles; the inferior articular facets articulate with the axis. The two lateral masses are connected by two anterior and posterior arches of C1. The anterior arch presents an anterior tubercle and a dorsally oriented odontoid fossa, for articulation with the dens of the axis. The posterior arch has a posterior tubercle and a groove for the vertebral artery before it enters the skull.

The Axis (C2)

The axis (C2) (Figure 14-7) is so named because it provides the axis of rotation for the head. It is

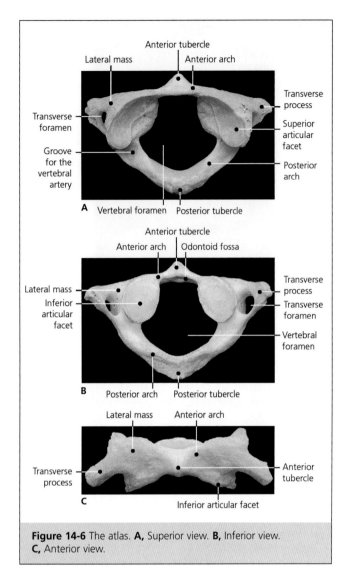

Figure 14-6 The atlas. **A,** Superior view. **B,** Inferior view. **C,** Anterior view.

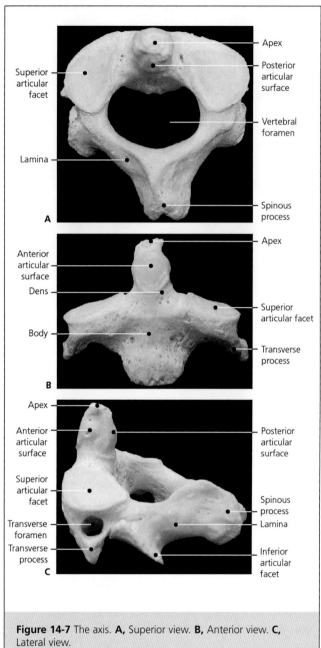

Figure 14-7 The axis. **A,** Superior view. **B,** Anterior view. **C,** Lateral view.

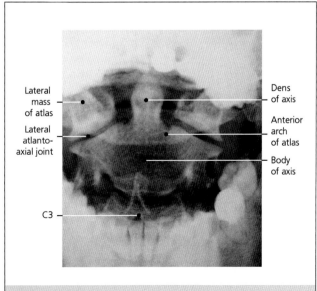

Figure 14-8 Open mouth radiographic view of the atlas and axis.

located between C1 (the atlas) and C3 and is characterized by a toothlike process called the dens (odontoid process). Its anterior articular surface articulates with the odontoid fossa of the atlas, and its posterior articular surface articulates with the transverse portion of the cruciform ligament. A ligament found on the apex of the dens is described in chapter 15, with spine arthrology.

Radiographic Images

Figures 14-8 through 14-10 are radiographic views of the atlas, axis, and cervical spine.

Osteology

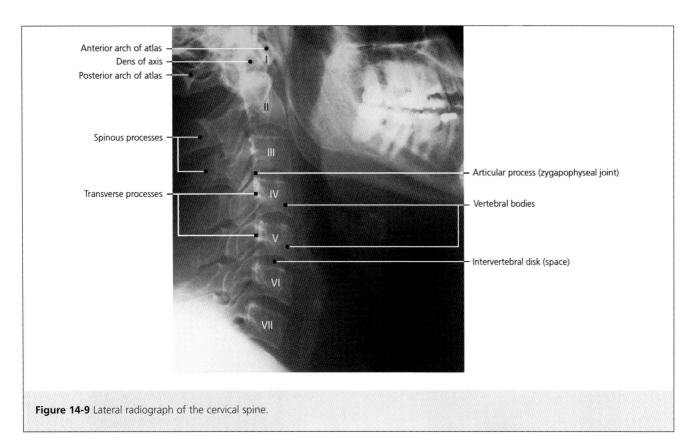

Figure 14-9 Lateral radiograph of the cervical spine.

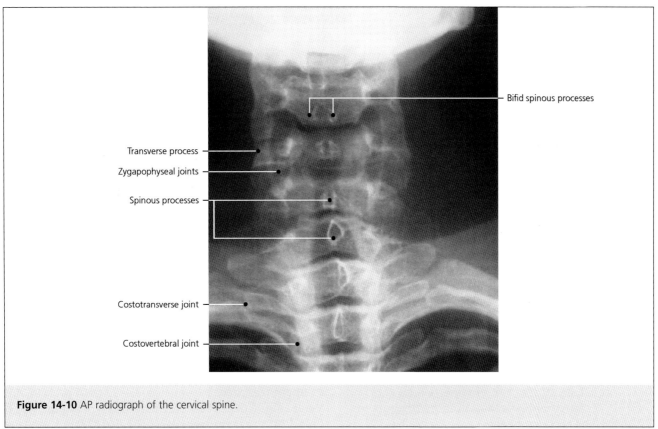

Figure 14-10 AP radiograph of the cervical spine.

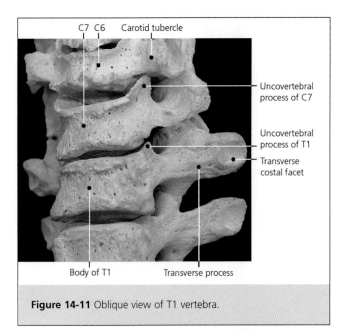

Figure 14-11 Oblique view of T1 vertebra.

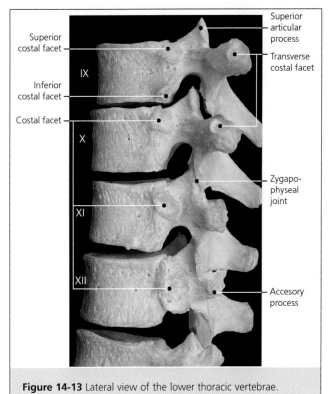

Figure 14-13 Lateral view of the lower thoracic vertebrae.

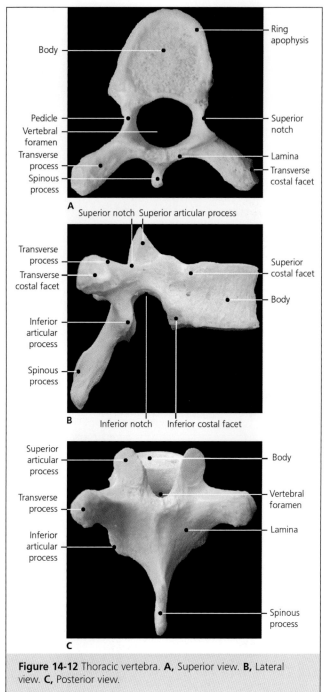

Figure 14-12 Thoracic vertebra. **A,** Superior view. **B,** Lateral view. **C,** Posterior view.

Thoracic Vertebrae

There are 12 thoracic vertebrae. T1 (Figure 14-11) is a transitional vertebra with a slightly quadrangular-shaped body and an uncovertebral process. The thoracic vertebrae provide regional kyphosis. The typical thoracic vertebra (Figure 14-12) is characterized by its small size and distinctive features that provide connection to the ribs. The vertebral bodies of T4 and T5 often present a small impression on the left lateral aspect, produced by the aorta.

The transverse costal facet articulates with the tubercle of the rib neck, whereas the rib head articulates with two adjacent vertebral bodies. For this purpose, a superior and an inferior costal facet are present on the lateral aspect of the vertebral body. The former is found on the lateral aspect of the pedicle and the latter on the base of the vertebral body. The exceptions to this are T1, T11, and T12, each of which has a sin-

OSTEOLOGY

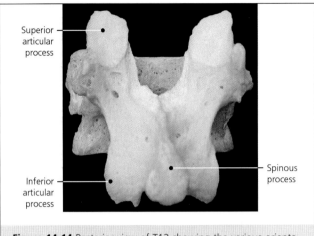

Figure 14-14 Posterior view of T12 showing the various orientations of the articular processes.

gle costal facet and hence articulates with only one rib (Figure 14-13).

In contrast to the cervical spine, the thoracic spinous processes are long and considerably caudally oriented. The articular processes are vertically oriented. The superior articular processes present a cephalad and posterior direction, while the inferior articular processes form part of the lamina and are prominent in a slightly caudal and anterior direction.

The T12 vertebra (Figure 14-14) has a laterally oriented inferior articular process allowing articulation with L1. It also has features characteristic of the lumbar vertebrae (accessory, mammillary, and costal processes). These features are described below.

Figure 14-15 is a radiographic view of the cervical and upper thoracic region.

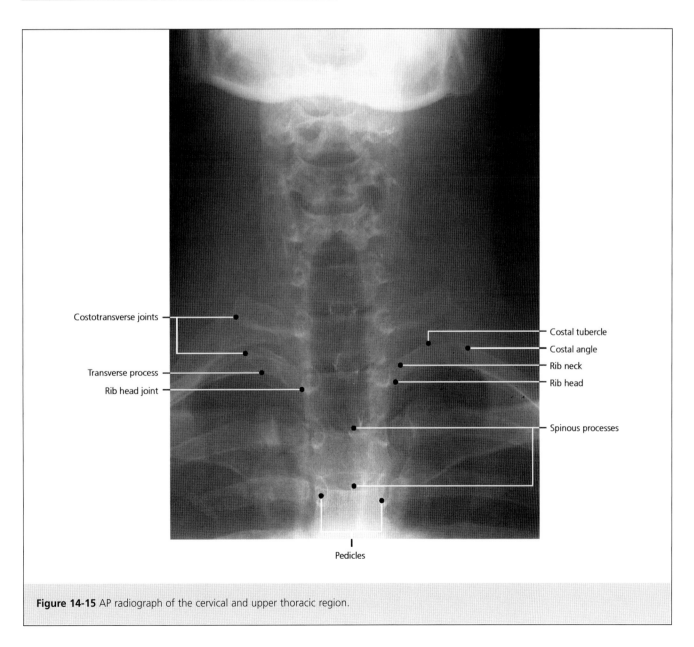

Figure 14-15 AP radiograph of the cervical and upper thoracic region.

American Academy of Orthopaedic Surgeons

Lumbar Vertebrae

Lumbar lordosis is formed by the five lumbar vertebrae and their associated intervertebral disks. In the lumbar vertebrae, the transverse process is also called the costal process (Figure 14-16) because it represents a rib remainder in this region. The "authentic" transverse process is called the accessory process. The mamillary process is found posterior to the superior articular process (Figure 14-17).

The spinous process of the lumbar vertebrae is wider in the sagittal plane than is the spinous process of the thoracic vertebrae, and it is directed perpendicularly to the lamina in the posterior direction. The superior articular processes are medially oriented, whereas the inferior articular processes are laterally oriented. As a result of the L5-S1 (lumbosacral) articulation, the inferior facet of L5 assumes a forward facing position.

The Sacrum

The sacrum is formed by the fusion of five vertebrae, and its base supports the previously described regions of the spine. On the anterior aspect of the sacrum (Figure 14-18, A) the transverse lines are clearly seen; these coincide with the ossified rudimentary intervertebral disks. The cervical, thoracic, and lumbar vertebrae are sometimes called true vertebrae, whereas the sacrum and coccyx are referred to as false vertebrae because of their fusion.

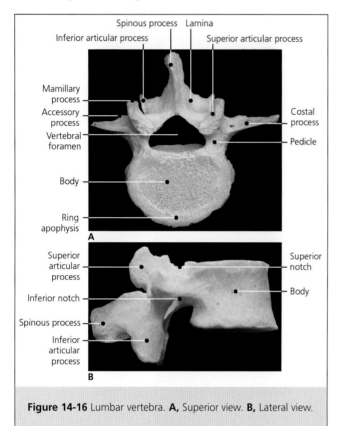

Figure 14-16 Lumbar vertebra. **A,** Superior view. **B,** Lateral view.

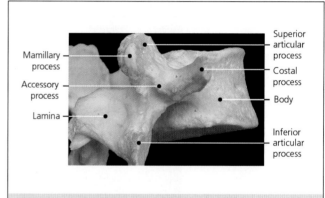

Figure 14-17 Oblique view of a lumbar vertebra showing the detail of the mamillary and accessory processes.

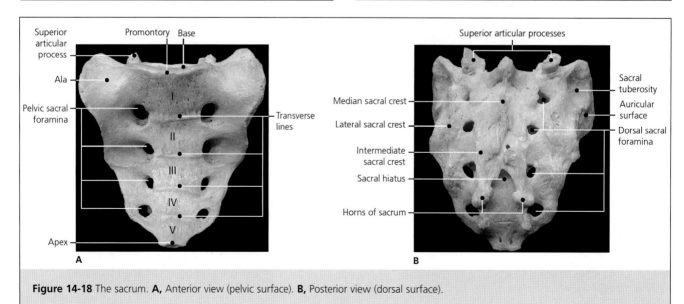

Figure 14-18 The sacrum. **A,** Anterior view (pelvic surface). **B,** Posterior view (dorsal surface).

Osteology

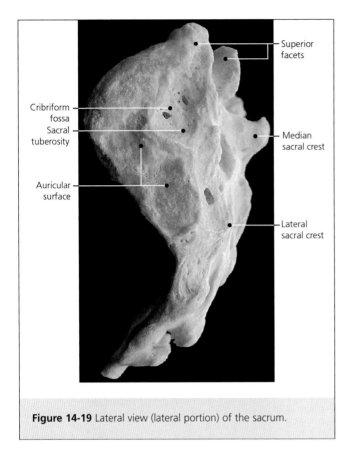

Figure 14-19 Lateral view (lateral portion) of the sacrum.

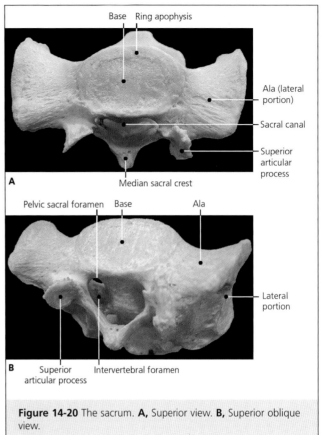

Figure 14-20 The sacrum. **A**, Superior view. **B**, Superior oblique view.

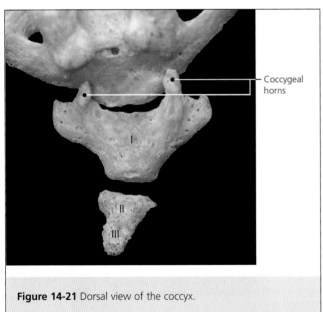

Figure 14-21 Dorsal view of the coccyx.

The sacrum is described as having a base, an apex, a dorsal surface, a pelvic surface, and two lateral parts. The anterior aspect of the base of the sacrum (the first sacral vertebra) forms the promontory, which projects slightly anteriorly and is the transition between the lumbar lordosis and the sacral kyphosis.

The superior articular processes of the sacrum articulate with the L5 inferior articular processes. On both sides of the vertebral bodies are the sacral wings. The lateral portions of the sacrum (Figure 14-19) are formed by the fusion of the transverse processes and present the auricular surfaces that articulate with their homonymous structures on the coxal bone (ilium). The sacral tuberosity is located posteriorly and consists of a ridge formed by the attachment for the posterior sacroiliac ligaments. The most cephalad part includes the cribriform fossa, where numerous blood vessels penetrate the sacrum.

The pelvic surface of the sacrum is concave, and the pelvic sacral foramina provide egress for the anterior branches of the sacral spinal nerves. The fascicles of the piriformis muscle originate between these foramina. On a transverse section, the pelvic sacral foramina may be observed, formed by the superior and inferior vertebral notches of the contiguous fused vertebrae. In the same manner, the vertebral arches fuse to form the sacral canal through which the spinal nerves travel (Figure 14-20).

The dorsal surface of the sacrum is convex and presents a medial sacral crest, a lateral sacral crest, and an intermediate sacral crest. The median sacral crest corresponds to the fusion of the spinous processes and is located in the midline of the body.

It normally ends at S3 and forms the sacral canal, the continuation of the spinal canal. The lateral sacral crest is formed by the fusion of the transverse processes. Between the medial and lateral sacral crests is the intermediate sacral crest, which is formed by the fusion of the articular processes. Also on the dorsal surface are the dorsal sacral foramina, through which the dorsal branches of the spinal nerves exit.

The distal portion of the sacrum (the apex) articulates with the coccyx. On its dorsal aspect are two horns, which limit the sides of the sacral hiatus, the terminal portion of the sacral canal (Figure 14-18, B).

The Coccyx

The coccyx comprises three, four, or five vertebrae (four is most common) that are almost always fused together (Figures 14-21 and 14-22). The first coccygeal vertebra, called the base, articulates with the sacrum. It is characterized by the presence of two posterior horns, which correspond to the superior articular processes.

The coccygeal vertebrae decrease in size in the caudal direction. Between the vertebrae are transverse lines that correspond to ossified rudimentary intervertebral disks. At the extreme caudal aspect is the apex, which is formed by a small tubercle.

Figure 14-23 is an AP radiograph of the pelvis.

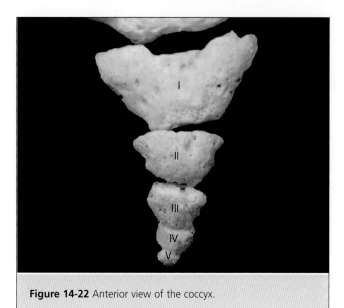

Figure 14-22 Anterior view of the coccyx.

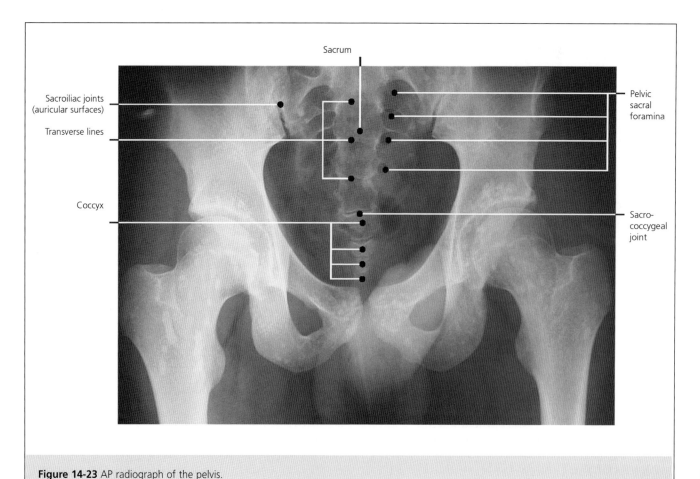

Figure 14-23 AP radiograph of the pelvis.

Osteology

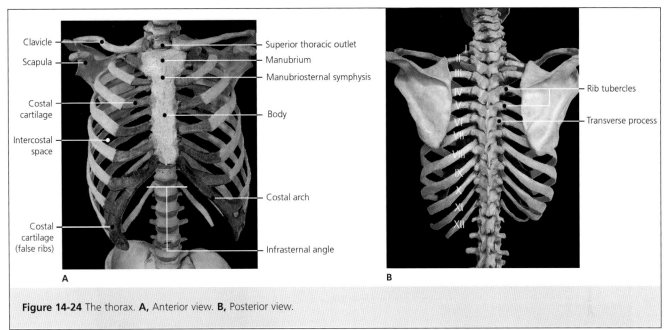

Figure 14-24 The thorax. **A,** Anterior view. **B,** Posterior view.

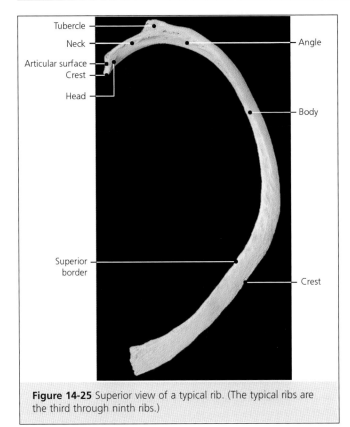

Figure 14-25 Superior view of a typical rib. (The typical ribs are the third through ninth ribs.)

The Thoracic Skeleton

Introduction

The thoracic skeleton consists of the thoracic vertebrae dorsally, the ribs laterally, and the sternum anteriorly.

The rib cage (Figure 14-24) limits the thoracic cavity, where the thoracic viscerae are located. The rib cage has a superior and an inferior thoracic inlet; these inlets allow communication with the neck and the abdominal cavity (through the diaphragm), respectively. On the dorsal aspect, the ribs and vertebrae form the pulmonary sulcus. Laterally, the intercostal spaces are occupied by the intercostal musculature, which will be described further in chapter 16.

The anterior and inferior portion of the thoracic cage create the costal arch, formed by the costal cartilage of the false ribs on both sides, which come together in the infrasternal angle.

The Ribs

The ribs are flat, elongated bones that form part of the rib cage. There are 12 ribs on each side, and they articulate dorsally with their corresponding thoracic vertebrae. The ribs are not horizontal but are directed obliquely in a caudal and medial direction. Some ribs articulate directly with the sternum anteriorly, whereas others do so indirectly through their costal cartilages.

Variations include the presence of a cervical rib (articulates with C7) or a lumbar rib.

There are true and false ribs. The true ribs are those that articulate directly, through their costal cartilage, with the sternum. The eighth, ninth, and tenth ribs are false ribs that articulate with the sternum indirectly, through the costal cartilage from other ribs. Ribs 11 and 12 are false ribs that are also called floating ribs because they are shorter and do not articulate with the sternum at all, only with their corresponding vertebrae.

The typical rib (Figure 14-25) has a wedge-shaped head with two articular surfaces, one for each of the adjacent vertebrae (except for the first, eleventh,

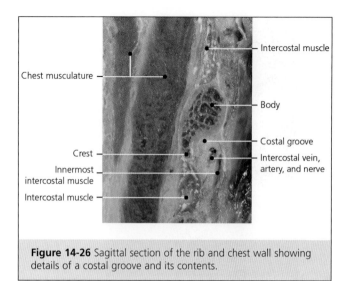

Figure 14-26 Sagittal section of the rib and chest wall showing details of a costal groove and its contents.

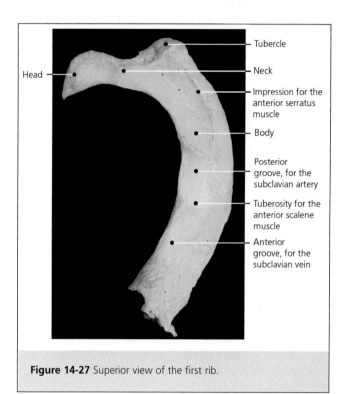

Figure 14-27 Superior view of the first rib.

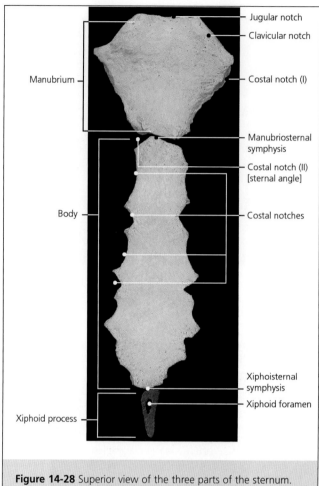

Figure 14-28 Superior view of the three parts of the sternum.

and twelfth ribs, which articulate with only one vertebra), separated by a crest. The neck continues from the head and also possesses a medially oriented crest. The body of the rib is flat and elongated. The proximal portion has a tubercle with an articular surface that accommodates the transverse process of the corresponding thoracic vertebra with which it articulates. Beyond this tubercle, the characteristic curvature of the rib starts at the costal angle and contributes to the characteristic shape of the rib cage. There are two borders, one superior and one inferior. The inferior border of the body has a crest (Figure 14-26), behind which is the costal groove for the intercostal vein, artery, and nerve.

The first and second ribs differ in some respects from the typical rib. The first rib (Figure 14-27) is broad and flat; on its superior aspect is the tuberosity for the anterior scalene muscle (of Lisfranc). This tubercle separates the anterior sulcus for the subclavian vein and the posterior sulcus for the subclavian artery, which transits through the interscalene space. An impression for the serratus anterior muscle may also be seen. The second rib is much longer than the first. It has the tuberosity for the anterior serratus muscle, which is a ridge formed by the insertion of this muscle. In some cases, the impression for the posterior scalene is also apparent.

The eleventh and twelfth ribs differ from typical ribs in that they (Figure 14-14) are shorter and straighter, articulate exclusively with their corresponding vertebrae, and do not have tuberosities.

The Sternum

The sternum (Figure 14-28) is a flat bone that is located on the midline of the chest. It comprises

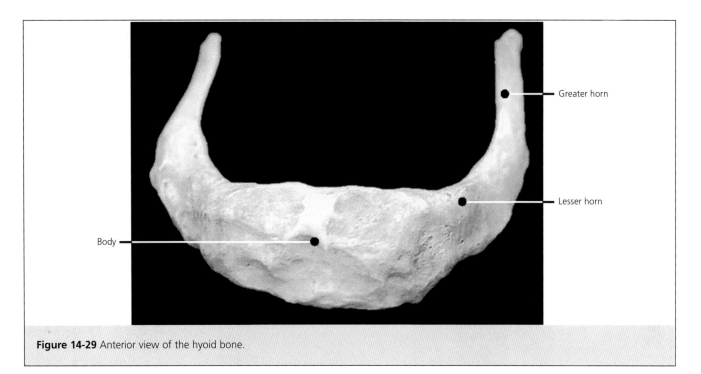

Figure 14-29 Anterior view of the hyoid bone.

three pieces: the manubrium, the body, and the xiphoid process.

The manubrium (from the Latin *manus*, "handle") is the most cephalad part. It is so named because the shape of the sternum was compared with that of a gladiator's sword, with the manubrium corresponding to the grip. The manubrium is found between the two clavicles, which it articulates with at the clavicular notches. Below these notches are the costal notches, where the manubrium articulates with the first ribs. On the superior border is the jugular notch.

The manubrium articulates with the body of the sternum through the manubriosternal symphysis, which coincides with the sternal angle (of Louis) and the costal notch for the second rib. The body of the sternum is flat. On its lateral margins are the costal notches that articulate with the second through seventh ribs. Horizontal lines that correspond to the fusion of independent ossification centers are usually visible on the body. Inferiorly, the body articulates with the xiphoid process through the xiphisternal symphysis. The xiphoid process comprises the inferior apex of the sternum. It is mostly cartilaginous, although it may ossify with advancing age. A xiphoid foramen or a bifid xiphoid process is commonly seen.

Hyoid Bone

The hyoid (Figure 14-29) is a U-shaped bone that characteristically does not articulate with any other bone. It is located between the mandible and the larynx, at the level of C3. The hyoid is a reference point for the anterior muscles of the neck, dividing the subhyoid muscles from the suprahyoid muscles.

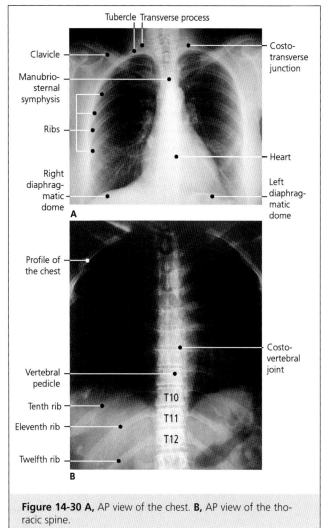

Figure 14-30 A, AP view of the chest. **B**, AP view of the thoracic spine.

The body of the hyoid is the thickest anterior part and has two horns on each side. The greater horns project posteriorly from the sides of the body. The lesser horns are located in the transitional region between the body and the greater horns and are directed cephalad. The lesser horns provide a point of attachment for the origin of the stylohyoid ligament.

Figure 14-30 is an AP radiograph of the chest.

The Skull

Because this book is intended for the study of the musculoskeletal system, anatomic specimens of the skull are presented in Figures 14-31 through 14-46 without explanatory text. A more extensive study should accompany the study of the central nervous system and cranial viscera.

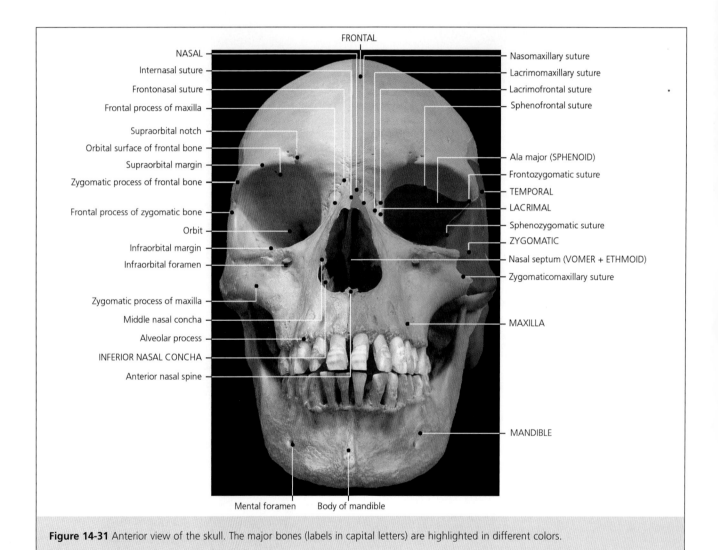

Figure 14-31 Anterior view of the skull. The major bones (labels in capital letters) are highlighted in different colors.

OSTEOLOGY

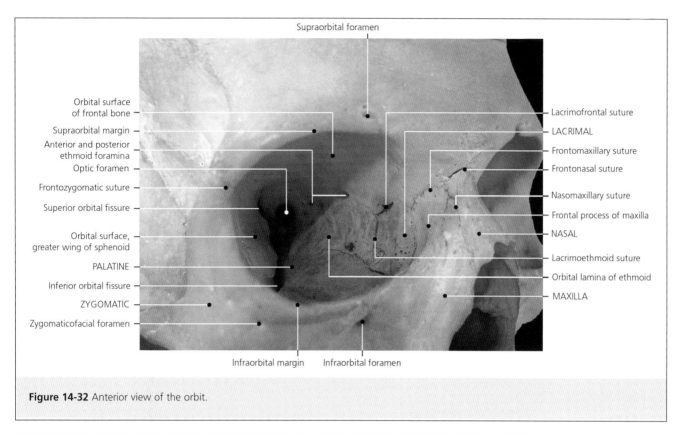

Figure 14-32 Anterior view of the orbit.

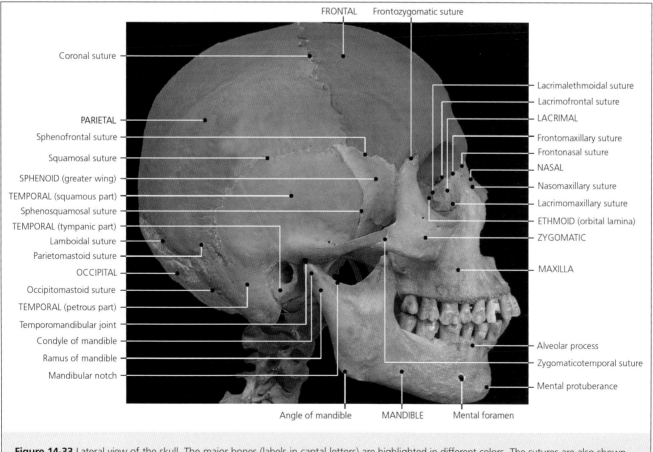

Figure 14-33 Lateral view of the skull. The major bones (labels in captal letters) are highlighted in different colors. The sutures are also shown.

American Academy of Orthopaedic Surgeons

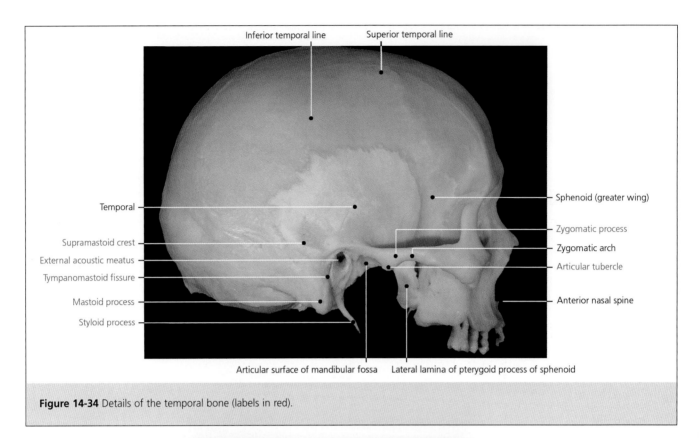

Figure 14-34 Details of the temporal bone (labels in red).

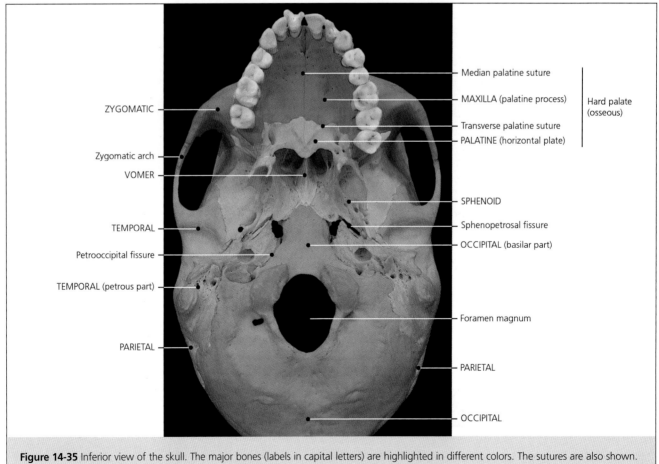

Figure 14-35 Inferior view of the skull. The major bones (labels in capital letters) are highlighted in different colors. The sutures are also shown.

OSTEOLOGY

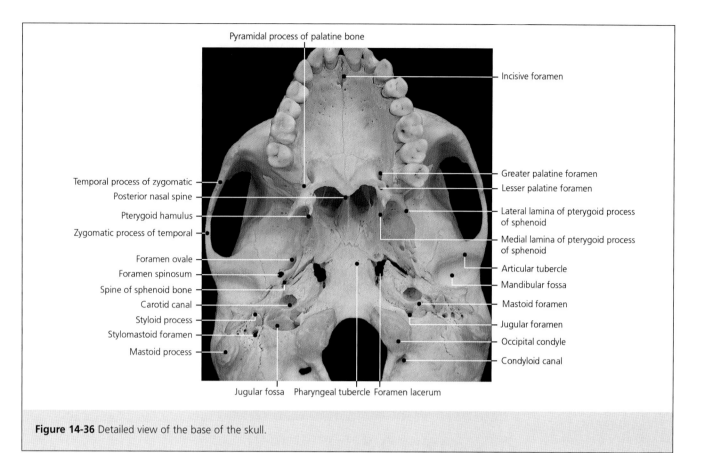

Figure 14-36 Detailed view of the base of the skull.

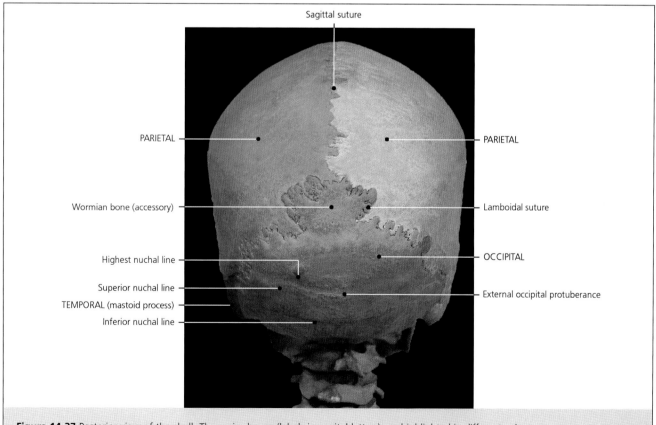

Figure 14-37 Posterior view of the skull. The major bones (labels in capital letters) are highlighted in different colors.

American Academy of Orthopaedic Surgeons

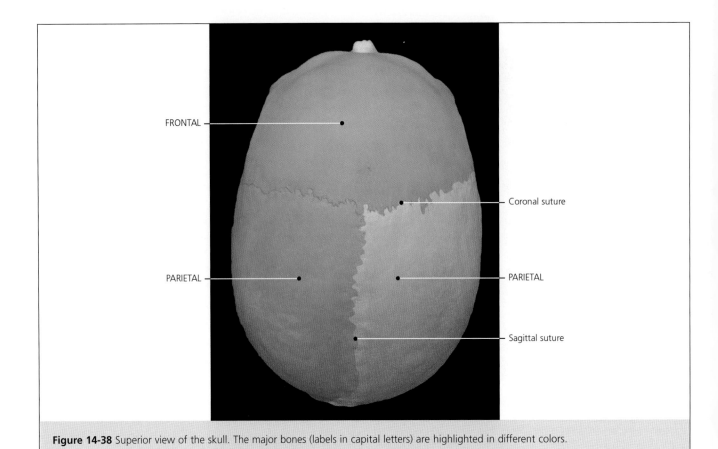

Figure 14-38 Superior view of the skull. The major bones (labels in capital letters) are highlighted in different colors.

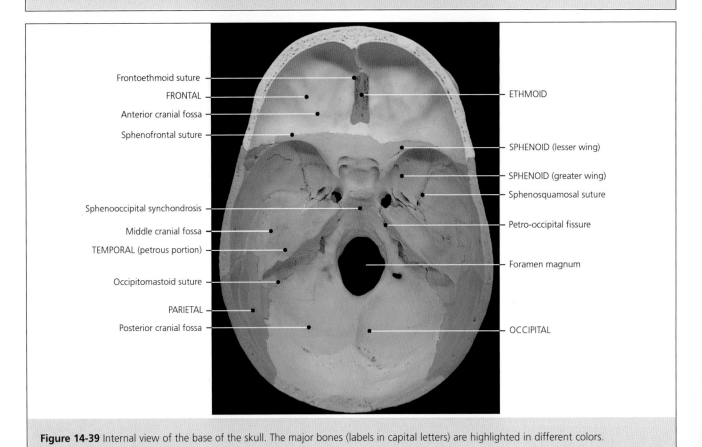

Figure 14-39 Internal view of the base of the skull. The major bones (labels in capital letters) are highlighted in different colors.

OSTEOLOGY

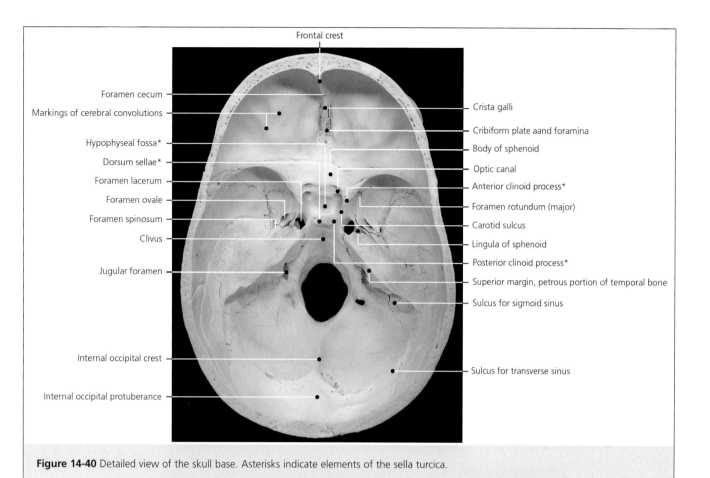

Figure 14-40 Detailed view of the skull base. Asterisks indicate elements of the sella turcica.

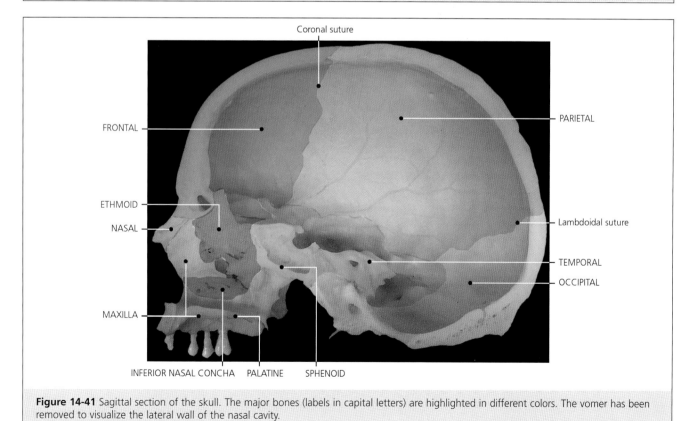

Figure 14-41 Sagittal section of the skull. The major bones (labels in capital letters) are highlighted in different colors. The vomer has been removed to visualize the lateral wall of the nasal cavity.

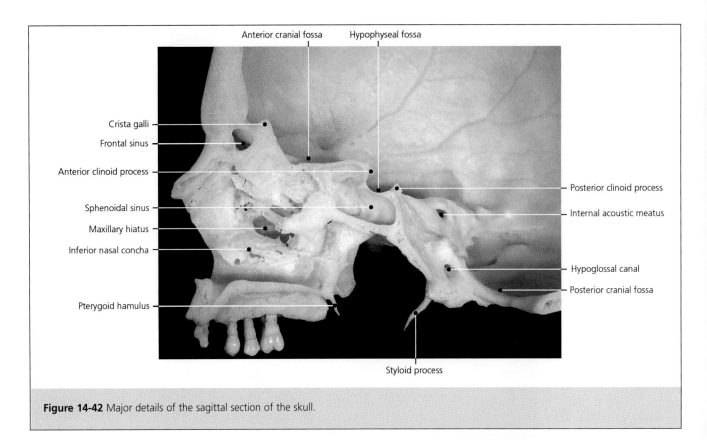

Figure 14-42 Major details of the sagittal section of the skull.

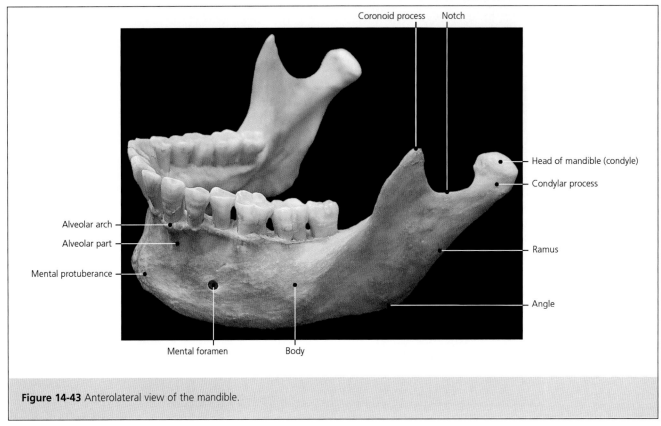

Figure 14-43 Anterolateral view of the mandible.

OSTEOLOGY

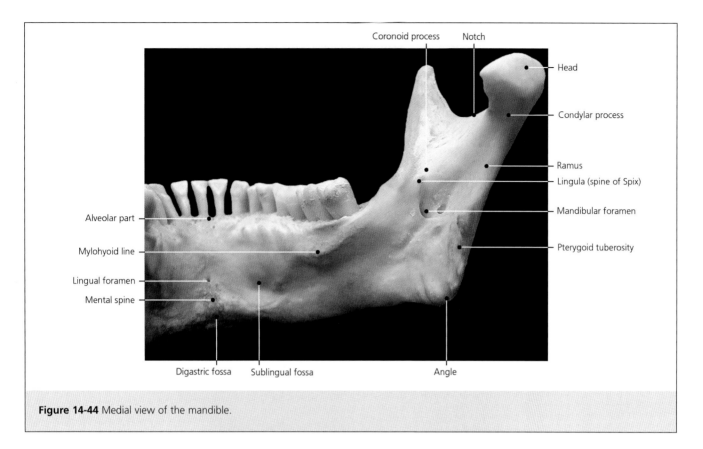

Figure 14-44 Medial view of the mandible.

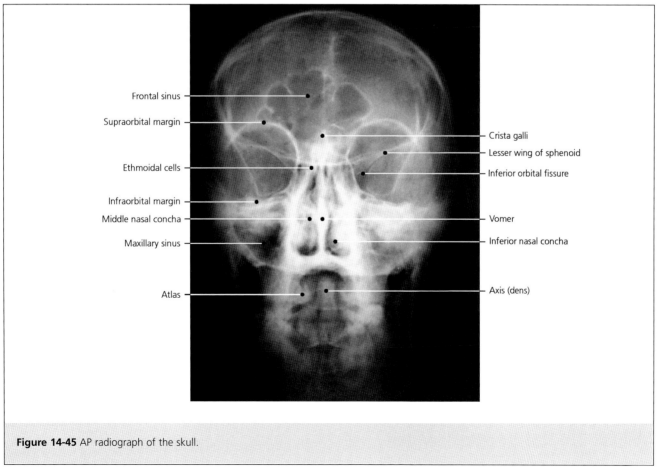

Figure 14-45 AP radiograph of the skull.

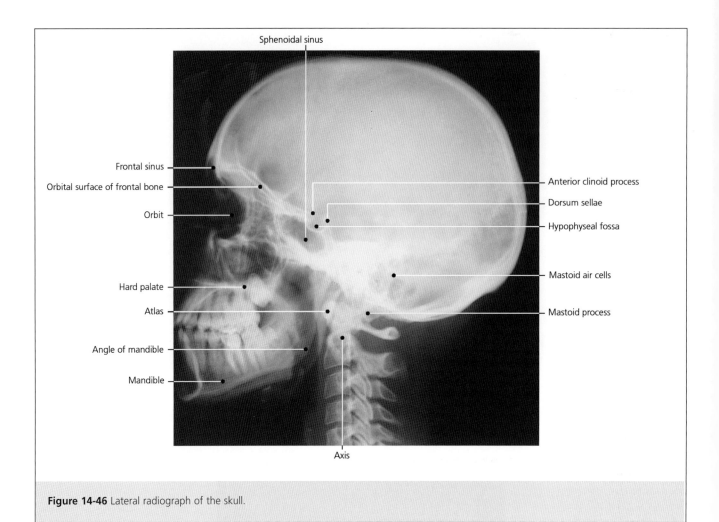

Figure 14-46 Lateral radiograph of the skull.

Chapter 15

ARTHROLOGY

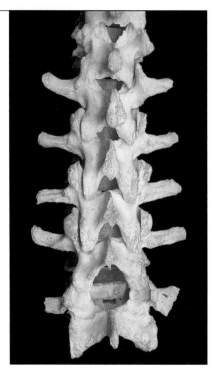

INTRODUCTION
The spine comprises 33 or 34 articulated bones that are separated by intervertebral disks. The characteristics of the individual vertebrae are described in detail in chapter 14. This chapter pays particular attention to the cervical region and the common vertebral ligaments.

CERVICAL SPINE
The cervical spine articulates with the occipital bone through the first two cervical vertebrae, called the atlas and the axis.

Atlanto-occipital Joint
The atlanto-occipital joint is a condylar joint (Figure 15-1). The articular surfaces include the two occipital condyles and the superior articular facets of the atlas. The convex shape of the occipital condyles fit with the somewhat less concave superior articular facets of the atlas (Figure 15-2). This joint is reinforced by the ligaments and membranes described below.

Capsular Ligaments
The capsular ligaments are composed of vertical bundles. Cranially, they attach around the condyle; caudally, they attach to the periarticular surfaces of the corresponding articular facets of the atlas. Laterally, the atlanto-occipital capsule is relatively thick and is reinforced by a group of fibrous bundles, the lateral atlanto-occipital ligaments (Figure 15-3). These oblique fascicles extend from the transverse process of the atlas to the jugular process of the occipital bone. Internally, the joint is lined by synovium. Anteriorly and posteriorly, the atlanto-occipital capsule is reinforced by thick ligaments that blend with the corresponding parts of the anterior and posterior atlanto-occipital ligaments. In addition, several bundles arise from the occipital bone in the retrocondylar (posterior) fossa and attach to the apex of the transverse process of the atlas. These bundles form a quadrilateral oriented obliquely from cranial to caudal and from medial to lateral, constituting a reinforcement of the fibrous capsule. This capsule relates laterally with the rectus capitis lateralis and medially with the transverse ligament of the atlas and axis.

Membranes
Anterior atlanto-occipital membrane This membrane (Figure 15-3) courses from the anterior portion of the foramen magnum to the superior border of the anterior arch of the atlas. It is reinforced by the anterior atlanto-occipital ligament, which descends from the basilar surface to the anterior tubercle of the atlas. This reinforcement is seen in the midline as a superficial bundle and is a continuation of the anterior longitudinal ligament. Anteriorly, the anterior longitudinal ligament is covered by the longus capitis and rectus capitis, separating it from the pharynx; posteriorly, it relates to the api-

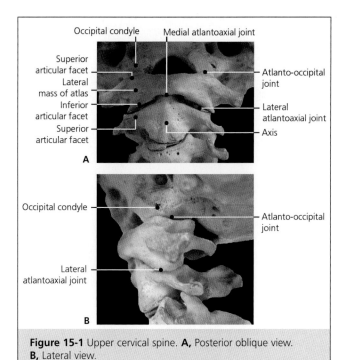

Figure 15-1 Upper cervical spine. **A,** Posterior oblique view. **B,** Lateral view.

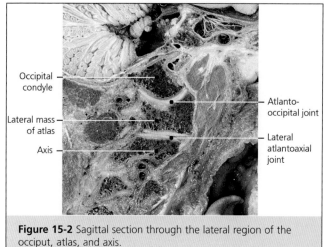

Figure 15-2 Sagittal section through the lateral region of the occiput, atlas, and axis.

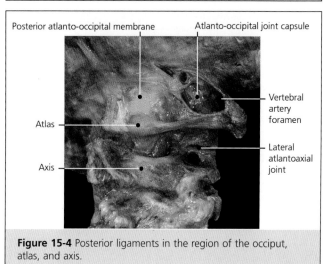

Figure 15-4 Posterior ligaments in the region of the occiput, atlas, and axis.

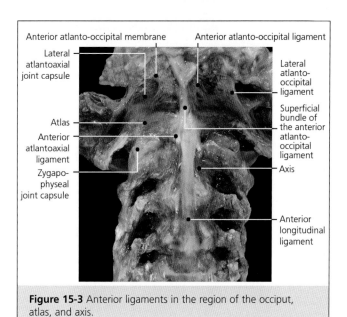

Figure 15-3 Anterior ligaments in the region of the occiput, atlas, and axis.

cal ligament of the dens and the atlantoaxial synovial membrane.

Posterior atlanto-occipital membrane This membrane (Figure 15-4) extends from the posterior border of the foramen magnum to the superior border of the posterior arch of the atlas. Posterior to the condyles, it occupies the interval between the occipital bone and the atlas. Laterally, this membrane blends with the atlanto-occipital joint capsule. After surrounding the lateral masses of the atlas, the vertebral arteries pass from outside in through an orifice in this membrane, the vertebral artery foramen, and penetrate the skull.

This membrane is covered posteriorly by the superior oblique muscles and the rectus capitis posterior major and minor.

Occipitoaxial Joint

The occipitoaxial joint does not have any direct point of contact between the occipital bone and the axis. The occipital bone and the axis are coupled by abundant and sturdy ligaments. These ligaments are divided into two groups: the tectorial membrane, and the occipito-odontoid ligaments.

Tectorial Membrane

This structure (Figure 15-5) is divided into the middle and lateral occipitoaxial ligaments. The former is attached to the basilar part of the occipital bone and to the posterior aspect of the body of the axis, passing behind the dens of the axis and the cruciform ligament as it descends. The middle occipitoaxial ligament blends with the cruciform ligament (Figure 15-6) and with the posterior longitudinal ligament, being an extension of this ligament in this region by means of

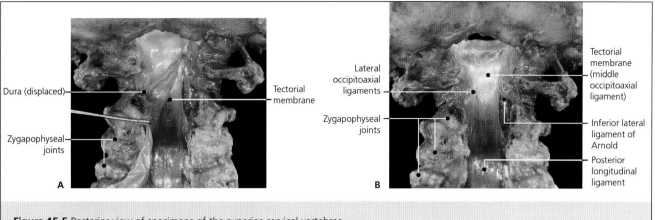

Figure 15-5 Posterior view of specimens of the superior cervical vertebrae.

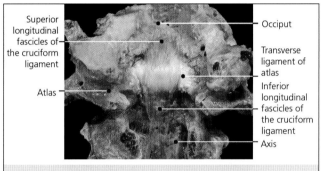

Figure 15-6 Detailed view of the cruciform ligament with its components.

a double layer. The lateral occipitoaxial ligament attaches obliquely to the lateral regions of the foramen magnum and to the posterior aspect of the body of the axis. The most external fibers, called the inferior lateral ligament of Arnold, constitute a bundle that reinforces the internal portion of the atlantoaxial capsule.

Occipito-odontoid Ligaments

These three ligaments—the apical ligament of the dens and the two alar ligaments of the dens—couple the occipital bone with the dens of the axis. The apical ligament of the dens extends from the anterior aspect of the foramen magnum to the apex of the dens (Figure 15-7). This ligament is located posterior to the anterior atlanto-occipital ligament and anterior to the tectorial membrane. The two alar ligaments of the dens (Figure 15-8) arise from the sides of the superior halves of the odontoid process and insert on the medial aspect of the occipital condyles. The most prominent bundles, which are oriented in a slightly posterior oblique direction, approach the posterior midline of the odontoid process where the two alar ligaments blend. A small, loop-shaped bundle is thus formed that extends from one side to the other of the occipital bone, passing behind the odontoid process.

This occipito-occipital bundle is the transverse occipital ligament of Lauth.

In addition, some of these ligaments have fibers that cover the articular complex in both the front and the back.

Atlantoaxial Joint

The atlantoaxial joint is composed of two lateral atlantoaxial joints and one medial atlantoaxial joint. The lateral joints are diarthrodial and correspond to the articulation between the inferior articular facet of the atlas and the superior articular facet of the axis. This is the most common type of joint caudal to the axis, in which the inferior facet of the superior vertebra articulates with the superior facet of the caudal vertebra.

The four facet joint surfaces are convex, particularly in the anterior-posterior direction. They are covered with hyaline cartilage that is thicker in the center than at the periphery. The point of contact between the two facets occurs in this convex region (Figure 15-2). In the space anterior and posterior to this, small synovial folds may be found. The lateral atlantoaxial joints are reinforced by the lateral, anterior, and posterior atlantoaxial ligaments.

Lateral Atlantoaxial Ligaments

Thick capsular fibers reinforce the atlantoaxial joint; without these fibers, the facet joint capsules would be very thin. Medially, the joint is reinforced through a fibrous bundle that arises from the posterior aspect of the body of the axis and is directed obliquely cephalad and laterally to the lateral masses of the atlas, behind the insertion of the transverse ligaments of the atlas. This is the inferior lateral ligament of Arnold, which is a portion of the lateral occipitoaxial ligament.

Anterior Atlantoaxial Ligament

This ligament consists of a fibrous layer that extends from the inferior border of the anterior arch of the

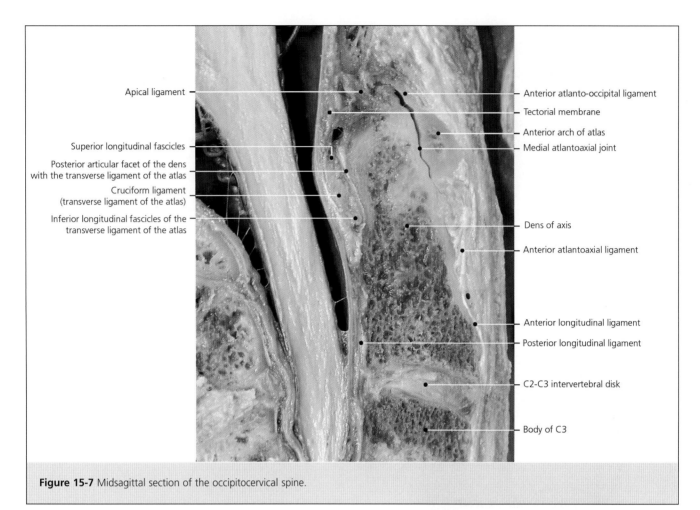

Figure 15-7 Midsagittal section of the occipitocervical spine.

atlas to the tubercle on the anterior aspect of the body of the axis. Laterally, this layer continues with the lateral atlantoaxial ligaments. On the anterior midline, a small vertical cord is observed; this represents the superior portion of the anterior longitudinal ligament.

Posterior Atlantoaxial Ligament

This ligament arises from the posterior arch of the atlas and attaches to the laminae and base of the spinous process of the axis. It is composed of two types of bundles: superficial bundles, which occupy the midline and represent the interspinous ligament; and deep bundles, rich in elastic fibers, homologous to the ligamentum flavum. Laterally, the atlantoaxial ligament is penetrated from anterior to posterior by the greater occipital nerve.

Synovium

A very loose synovium, particularly anteriorly, facilitates the gliding action of the atlantoaxial facet joint surfaces. Two small synovial folds (one anterior and one posterior) together with small adipose reinforcements act as wedges that fill up the peripheral parts of the joint line. This synovium frequently communicates with one or both synovial membranes of the medial atlantoaxial joint.

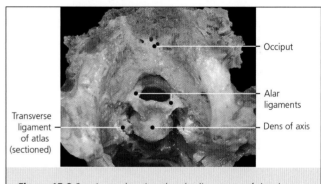

Figure 15-8 Specimen showing the alar ligaments of the dens.

Medial Atlantoaxial Joint (Atlanto-Dens Articulation)

The medial atlantoaxial joint (Figures 15-1 and 15-7) articulates the osteofibrous ring of the atlas with the odontoid process.

The osteofibrous ring of the atlas is formed anteriorly by the facet for the odontoid process on the posterior aspect of the anterior arch of the atlas (Figures 15-9 and 15-10), and posteriorly by the transverse ligament of the atlas, a fibrous band that extends horizontally between the lateral masses of the atlas. The anterior and medial

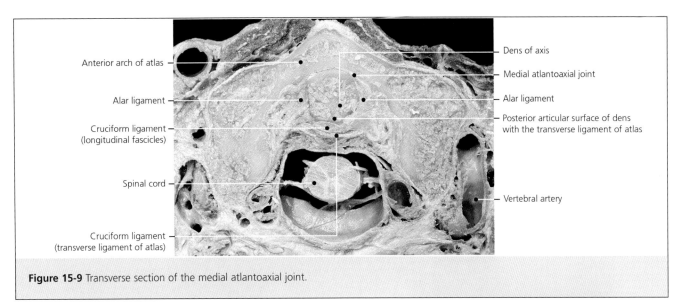

Figure 15-9 Transverse section of the medial atlantoaxial joint.

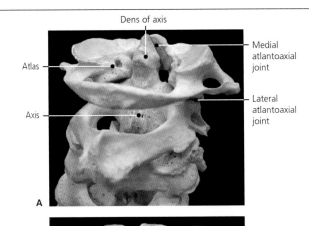

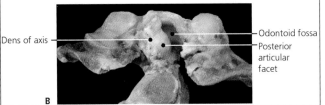

Figure 15-10 A, Posterior oblique view of an articulated specimen of the atlas and axis. **B,** Medial atlantoaxial joint after sectioning of the cruciform ligament. The posterior arch of the atlas as well as the axis have been removed to show the articular facets.

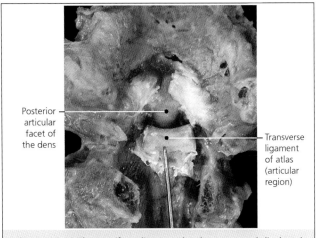

Figure 15-11 The cruciform ligament has been cut and displaced to allow visualization of the posterior articular facet of the dens and the transverse ligament.

aspects of this ligament are covered by a thin layer of hyaline cartilage (Figure 15-11). Superior and inferior longitudinal fascicles project from its superior and inferior borders. The entire complex, including the transverse ligament of the atlas and the two longitudinal fascicles, is called the cruciform ligament of the atlas (Figures 15-6 and 15-7).

The odontoid process occupies the center of the ring of the atlas and acts as its axis of rotation, articulating through two facet joint surfaces (Figures 15-9 through 15-11). The anterior articular facet articulates with the articular facet for the dens of the atlas, and the posterior articular facet articulates with the transverse ligament of the atlas. This middle atlantoaxial joint has two independent synovial membranes that correspond with each of these joints. The anterior synovial membrane is smaller than the posterior one.

Figure 15-12 shows various radiographic views of the occipitocervical region.

COMMON SPINAL JOINTS

The spinal joints caudal to the axis share several common characteristics. Each vertebra articulates with the adjacent vertebra via the body and two zygapophyseal joints. In addition, the joints are joined by ligaments attached to the laminae, spinous processes, and transverse processes.

Intervertebral Symphysis

The intervertebral symphysis is a typical amphiarthrosis that articulates two adjacent vertebral bodies (Fig-

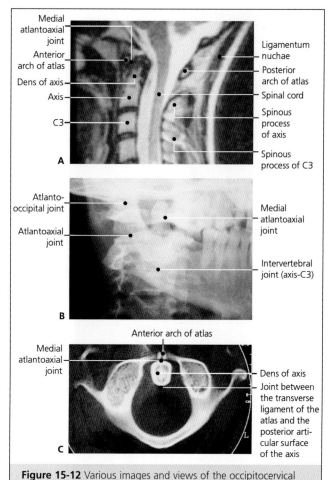

Figure 15-12 Various images and views of the occipitocervical region. **A,** Sagittal MRI. **B,** Oblique radiographic view of the occipitocervical region. **C,** CT scan of the medial atlantoaxial joint.

ure 15-13). The central part is concave, with the surface covered by hyaline cartilage that attenuates the concavity. The vertebral bodies are joined by intervertebral disks and the anterior and posterior longitudinal ligaments.

Intervertebral Disk

The intervertebral disk (Figures 15-13 and 15-14) lies between and is intimately attached to the end plates of the vertebral body. The size of the intervertebral disks increases from the cervical to the lumbar region. The anterior portion of the disk (and vertebral body) is taller in the cervical and lumbar vertebrae; the posterior portion is taller in the thoracic vertebrae. This creates the physiologic curvatures of the spine.

Two structures form the intervertebral disk: the annulus fibrosus and the nucleus pulposus. On the periphery of the disk is the annulus fibrosus, composed of fibrous bundles arranged in an oblique manner extending from the superior to the inferior vertebra in various directions. The obliquity is more pronounced closer to the center of the nucleus, where the fibers are almost horizontal. These fibers are grouped into layers, forming circular, alternating lamellae. The fibers on each ring have one orientation, and the neighboring fibers are oriented in the exact opposite direction; that is, if one layer runs obliquely from left to right, the next layer runs obliquely from right to left. The fibers of the annulus fibrosus insert on the cortex of the vertebral body via Sharpey's fibers. These fibrous lamellae are stronger and more numerous on the anterior and lateral aspects of the disk than on the posterior aspect, where they are thinner and scarcer. The various lamellae that form the peripheral portion of the disks are well limited by septae composed predominantly of elastic fibers.

The central portion of the intervertebral disk, the nucleus pulposus, is a semiliquid nuclear mass contained by the annulus. The tissue of the nucleus is separated from the bone by a well-defined layer of hyaline cartilage that extends to the inner margins of the annular insertion. Embryologically, the nucleus pulposus stems from the notochord.

The intervertebral disks are nonvascular structures that depend on diffusion from subchondral capillaries for nutrition. The restricted nutrient flow to the nucleus contributes to disk degeneration in adults.

Anterior Longitudinal Ligament

The anterior longitudinal ligament joins the anterior aspects of the vertebral bodies (Figure 15-15). This ligament is the continuation of the anterior atlanto-occipital membrane (Figure 15-3) and extends from the body of the axis down to the sacrum. The anterior longitudinal ligament has different characteristics in each region of the spine. In the cervical region, it is narrow and covers the medial zone. In the thoracic region, the ligament broadens and completely covers the vertebral body anterior to the ribs. A medial portion can be differentiated from the two thicker lateral portions. It can be distinguished by its longer fibers, a continuation of the medial cervical bands. In the lumbar region, this ligament narrows again and is reduced to a single midline band, similar to that in the cervical region. It extends into the anterior surface of S1 and usually ends at S2, where it blends in with the periosteum.

In several areas of the spine, the medial portion of the anterior longitudinal ligament is separated from the lateral portion, leaving a cleft that allows for the passage of veins into the vertebral body.

Posterior Longitudinal Ligament

The posterior longitudinal ligament is located posterior to the vertebral bodies and covers the anterior margin of the spinal canal (Figure 15-16). It extends from C3 to the sacrum and is an extension of the tectorial membrane. At the level of the sacrum, it narrows into a sim-

ARTHROLOGY

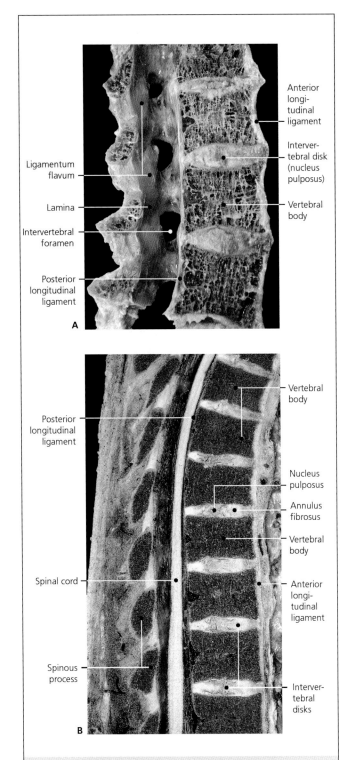

Figure 15-13 Sagittal sections of the spine where the intervertebral symphysis is evident.

ple middle cord that may extend down to the coccyx. This ligament is wide at the level of the intervertebral disks, extending to the inner cortex of the pedicles, and is thinner at the level of the vertebral bodies.

Uncovertebral Joints

The uncovertebral joint is a type of diarthrosis present only in the cervical region. This joint articulates the uncinate process of the inferior vertebra with the inferolateral notch of the superior vertebra (Figure 15-17). They are conjugated by the intervertebral disk and the uncovertebral ligament. A small synovial membrane covers this ligament internally.

Zygapophyseal Joints

The zygapophyseal joint also is a type of diarthrosis. The orientation of the joint surfaces in the zygapophyseal joints differs by region.

In the cervical region, the adjacent joint surfaces are connected by a fibrous capsule that is thin and lax and thickens only anteriorly (Figures 15-5 and 15-18). The articular surface of the superior vertebra faces caudally and anteriorly; that of the inferior vertebra faces cephalad and posteriorly. In its middle portion, the facet joint capsule is reinforced by the ligamentum flavum.

In the dorsal region, the facet joint capsule is thicker. The facet joint surface of the superior vertebra faces anteriorly, and that of the inferior vertebra faces posteriorly (Figure 15-19). The joint is covered medially by the ligamentum flavum, and its posterolateral aspect is covered by a fibrous fascicle, the posterior ligament.

In the lumbar region, the zygapophyseal joints have the same characteristics as those of the dorsal region, although the posterior ligament is thicker and even more resistant. The facet joint surface of the inferior vertebra faces laterally, and that of the superior vertebra faces medially (Figure 15-20).

The last thoracic vertebra, T12, has the function of transitioning to the lumbar spine. Its superior facet joints are oriented in a manner similar to those of the thoracic spine, while the inferior facet joints are oriented in the same direction as the lumbar spine (Figure 15-20).

Figure 15-21 shows various radiographic views of the uncovertebral and zygoapophyseal joints.

Ligamentous Structures

Ligamentum Flavum

The vertebral laminae are united by the ligamentum flavum (from Latin *flavum*, "yellow"), so-called because of its distinctly yellow color. The ligamentum flavum is the continuation of the posterior atlantooccipital membrane (Figure 15-22).

Each lamina is conjugated with the adjacent lamina through two rectangular ligaments, one on each side. The anterior surface of this ligament is intimately related to the spinal dura and is separated only by semi-

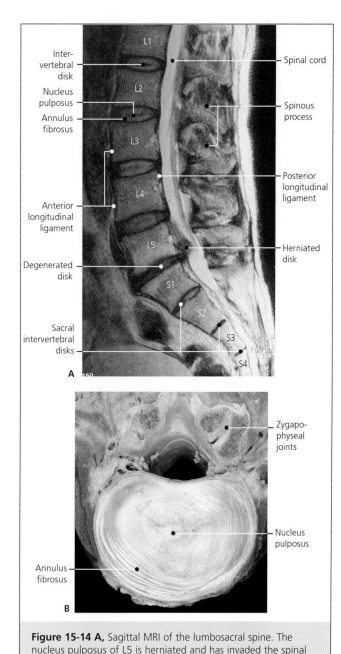

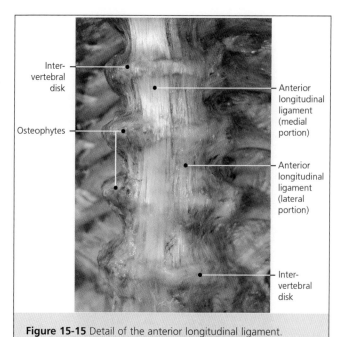

Figure 15-15 Detail of the anterior longitudinal ligament.

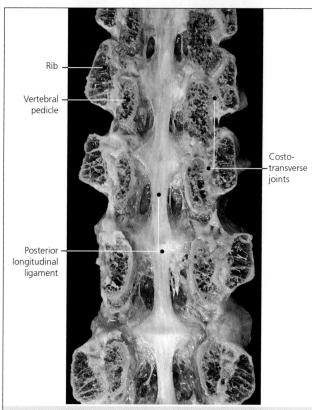

Figure 15-14 A, Sagittal MRI of the lumbosacral spine. The nucleus pulposus of L5 is herniated and has invaded the spinal canal. **B,** Axial section of a lumbar intervertebral disk.

Figure 15-16 Dissection of the posterior longitudinal ligament. (Same specimen as shown in 15-22 but with the neural arch removed.)

fluid adipose tissue and epidural veins. The posterior aspect is related to the laminae and paraspinal muscles. The lateral border corresponds to the posterior part of the intervertebral foramen. It lines the medial aspect of the articular processes and reinforces the facet joint capsule. The medial borders blend in the midline, in the vicinity of the spinous processes. In the cervical region, they are separated by minimal connective tissue.

Interspinous Ligaments

These broad ligaments (Figure 15-23) occupy the spaces between the spinous processes. Anteriorly, they project toward the ligamentum flavum. Posteriorly, they blend with the supraspinous ligament.

Supraspinous Ligaments

These long fibrous bands extend between the apexes of the spinous processes. In the lumbar region, they seem to be constituted by the crossover of numerous muscle bands in the midline. In the thoracic region, they form

Arthrology

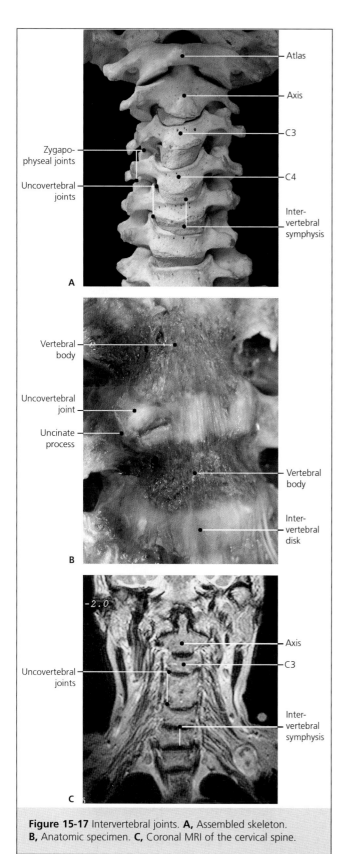

Figure 15-17 Intervertebral joints. **A,** Assembled skeleton. **B,** Anatomic specimen. **C,** Coronal MRI of the cervical spine.

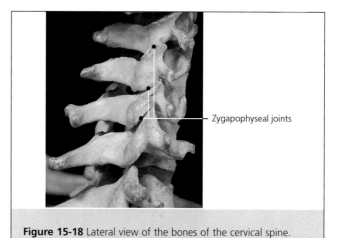

Figure 15-18 Lateral view of the bones of the cervical spine.

Figure 15-19 Detail of the thoracic zygapophyseal joints.

a cord between the spinous processes that tenses in flexion. In the cervical region, the supraspinous ligaments form a triangular, vertical septum called the ligamentum nuchae (Figure 15-12). The base of the triangle arises from the external occipital protuberance and the midline crest, the apex attaches to the spinous process of C6 or C7, and the anterior border firmly attaches to the apex of the cervical spinous processes, including the posterior tubercle of the atlas.

Intertransverse Ligaments

The transverse processes are connected by fibrous bands jointly called the intertransverse ligaments (Figures 15-23 through 15-25). In the cervical region, these ligaments are ill-defined and sometimes even absent; they can be confused with the intertransverse muscles. In the thoracic region, they form small rounded bundles that bind the apexes of the transverse processes. In the lumbar region, they are more developed and run from the base of the transverse process of the superior vertebra to the mammillary process and superior articular process of the inferior vertebra.

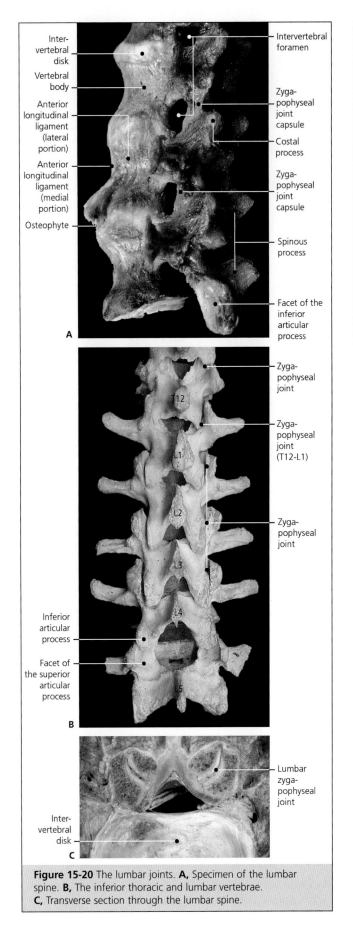

Figure 15-20 The lumbar joints. **A,** Specimen of the lumbar spine. **B,** The inferior thoracic and lumbar vertebrae. **C,** Transverse section through the lumbar spine.

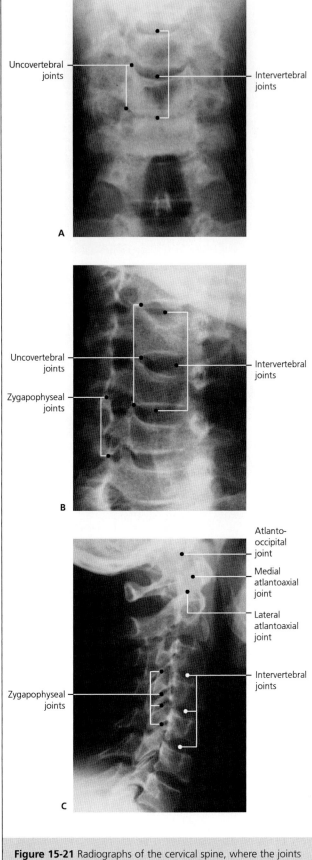

Figure 15-21 Radiographs of the cervical spine, where the joints have been labeled. **A,** AP view. **B,** Oblique view. **C,** Lateral view.

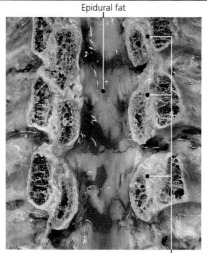

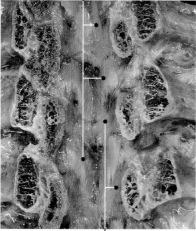

Figure 15-22 Thoracic spine specimen including an anterior view of the neural arch (facing posteriorly).

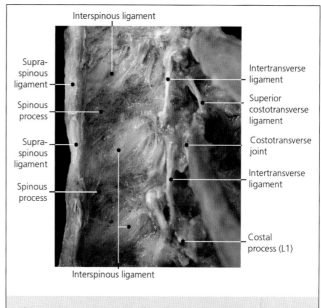

Figure 15-23 Lateral view of the ligaments of the thoracolumbar region.

Figure 15-26 shows various images of the joints of the lumbosacral region.

Arthrology of the Lumbosacral Spine

Lumbosacral Joint

The lumbosacral joint connects the sacrum with L5 (Figure 15-24). Like the other vertebrae, it articulates through the vertebral body and by two zygapophyseal joints. The reinforcements are similar to those at other levels of the spine, with the addition of the lumbosacral ligament. This ligament extends from the transverse processes of L5 to the base of the sacrum, blending with fibers of the anterior sacroiliac ligament.

Sacrococcygeal Joint

As an amphiarthrosis, the sacrococcygeal joint binds the inferior oval surface of the sacrum with the slightly concave surface of the coccyx through interposed fibrocartilage. The peripheral ligaments that reinforce the joint include the anterior sacrococcygeal ligament, the posterior superficial sacrococcygeal ligament, and the lateral sacrococcygeal ligaments.

The anterior sacrococcygeal ligament binds the anterior surfaces of both bones (Figure 15-27) and morphologically constitutes the prolongation of the anterior longitudinal ligament. It may present with longitudinal as well as horizontal bundles that converge on the midline.

Cranially, the posterior superficial sacrococcygeal ligament attaches on the distal end of the median sacral crest and on the edges of the sacral hiatus; caudally, it is attached through two lateral bands on the posterior surface of Co1 or Co2 (Figure 15-28). Below this is the deep posterior sacrococcygeal ligament, representing the prolongation of the posterior longitudinal ligament.

Three lateral sacrococcygeal ligaments can be distinguished on each side of the sacrococcygeal articulation: one internal, one median, and one external (Figure 15-28). The internal ligament extends from the sacral horn to the corresponding horn of the coccyx. The median ligament runs from the sacral horn to the posterior aspect of the coccyx. The external ligament descends from the lateral aspect of the apex of the sacrum to the corresponding aspect on the coccyx.

Intercoccygeal Joints

The intercoccygeal joints are rudimentary amphiarthroses that join the pieces of the coccyx. They have

Head and Trunk

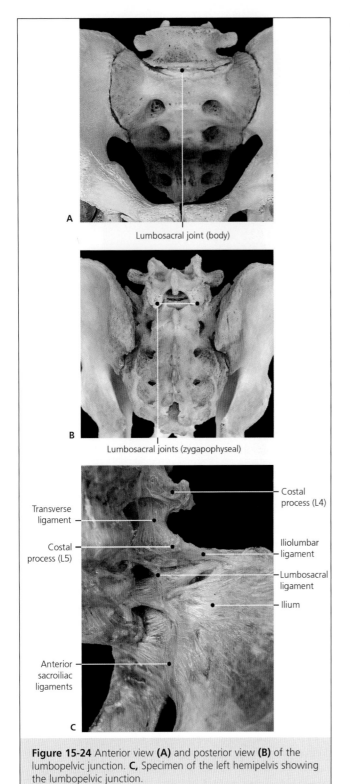

Figure 15-24 Anterior view **(A)** and posterior view **(B)** of the lumbopelvic junction. **C,** Specimen of the left hemipelvis showing the lumbopelvic junction.

small interosseous disks and are reinforced by peripheral ligaments.

From the apex of the coccyx, a small fibrous bundle emerges that attaches to the deep subcutaneous tissues. In some individuals, this coccyx-cutaneous ligament is an extension of the spinal column cau-

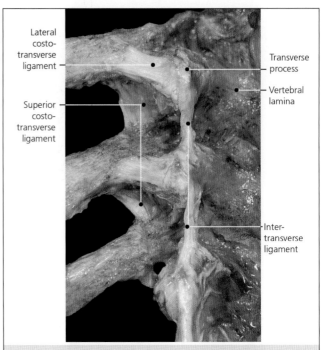

Figure 15-25 Posterior dissection of the posterior thoracic spine and ribs with details of the costotransverse and intertransverse ligaments.

dally. This ligament creates a depression in the skin immediately caudal to the coccyx, which is called the coccygeal fossa or fovea.

Arthrology of the Thorax

This section describes the joints that join the ribs to the vertebrae (costovertebral), the ribs to the costal cartilages (costochondral), the ribs to the sternum (sternocostal), and the costal cartilages to each other (interchondral). In addition, the joints of the sternum are described.

The ribs and the spine form a diarthrosis at two different points, at the head and at the tubercle of the rib.

Costovertebral Joint

Each rib articulates with two adjacent vertebrae, the one at its own level and the one superior to it, with the exception of the first, eleventh, and twelfth ribs, which articulate only at their own level (Figure 14-24). The joint surfaces consist of two flat surfaces on the rib head, one superior and one inferior. They are oriented obliquely and are covered by a thin layer of fibrocartilage (Figure 15-29). On the vertebral side, the inferior (superior vertebra) and the superior (vertebra at the same level of the rib) joint surfaces are separated by an intervertebral disk. They are obliquely oriented and are covered by a thin layer of cartilage fitted to the rib head in the form of a wedge. This joint is reinforced by the intra-articular ligament of the rib head, the radiate ligament of the rib head, and the posterior ligament.

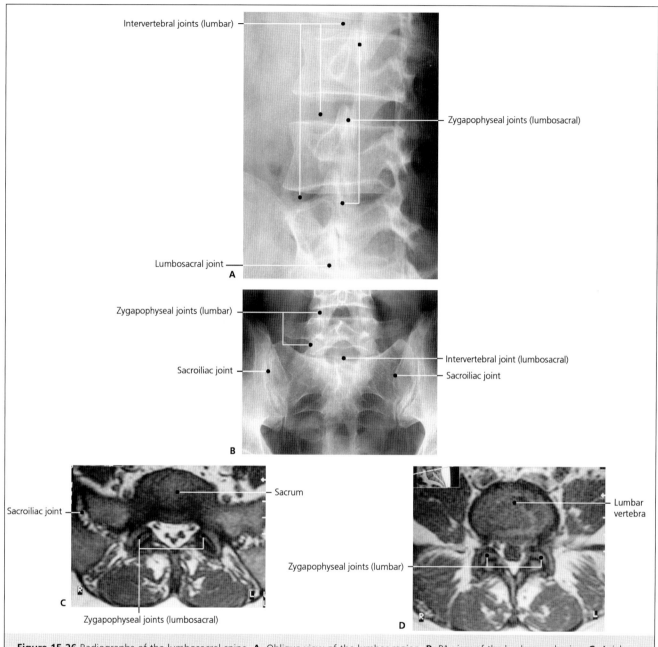

Figure 15-26 Radiographs of the lumbosacral spine. **A**, Oblique view of the lumbar region. **B**, PA view of the lumbosacral spine. **C**, Axial MRIs of the lumbar and lumbosacral spine.

The intra-articular ligament of the rib head (interosseous) extends from the crest that divides the two costal articular surfaces to the intervertebral disk (Figure 15-30). This ligament is not found on the first, eleventh, and twelfth ribs, which articulate directly with the vertebral body and not an intraarticular disk.

The radiate ligament of the rib head is located on the anterior part of the joint and extends like a fan from the rib head to the neighboring vertebral bodies and intervertebral disk (Figure 15-30).

Some authors describe a posterior ligament, which consists of resistant fascicles that run from the posterior superior aspect of the neck of the rib to the posterior aspect of the superior vertebral body.

Each costovertebral joint has two rudimentary synovial membranes, one superior and one inferior. These membranes are partially divided by the interosseous ligament, around which they are continuous.

Costotransverse Joint

The costotransverse joint connects the rib tubercle with the corresponding transverse process (Figures 15-31 through 15-33), with the exception of the eleventh and twelfth ribs, which lack these connections and are therefore called floating ribs (Figure 15-34). The joint surfaces are covered by a thin layer of fibrocartilage. The costotransverse ligament; the lateral, superior, and costotransverse ligaments; and the costlaminar ligament reinforce this joint.

The costotransverse ligament (Figure 15-32) is a group of fibrous bundles that run from the posterior and inferior surface of the neck of the rib to the anterior surface of the corresponding transverse process.

The lateral costotransverse ligament is a rectangular band that attaches to the posterolateral part of the tubercle of the rib, immediately external to the joint surface, and runs to the apex of the corresponding transverse process.

The superior costotransverse ligament (Figures 15-25 and 15-30) is a flat, squared bundle that runs from the superior border of the neck of the rib to the inferior aspect of the superior articular process. There are two

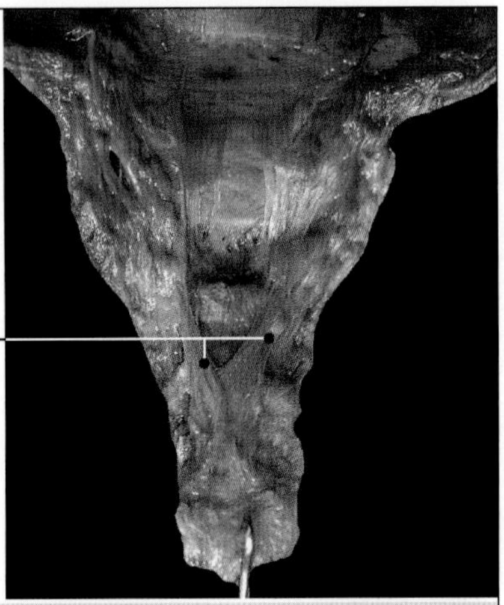

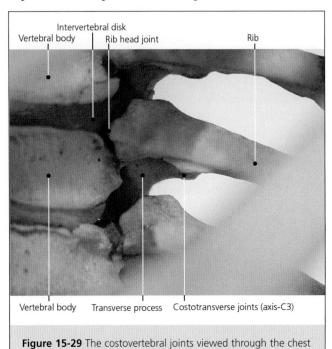

Figure 15-27 Detailed view of the anterior sacrococcygeal ligament.

Figure 15-29 The costovertebral joints viewed through the chest on an assembled skeleton.

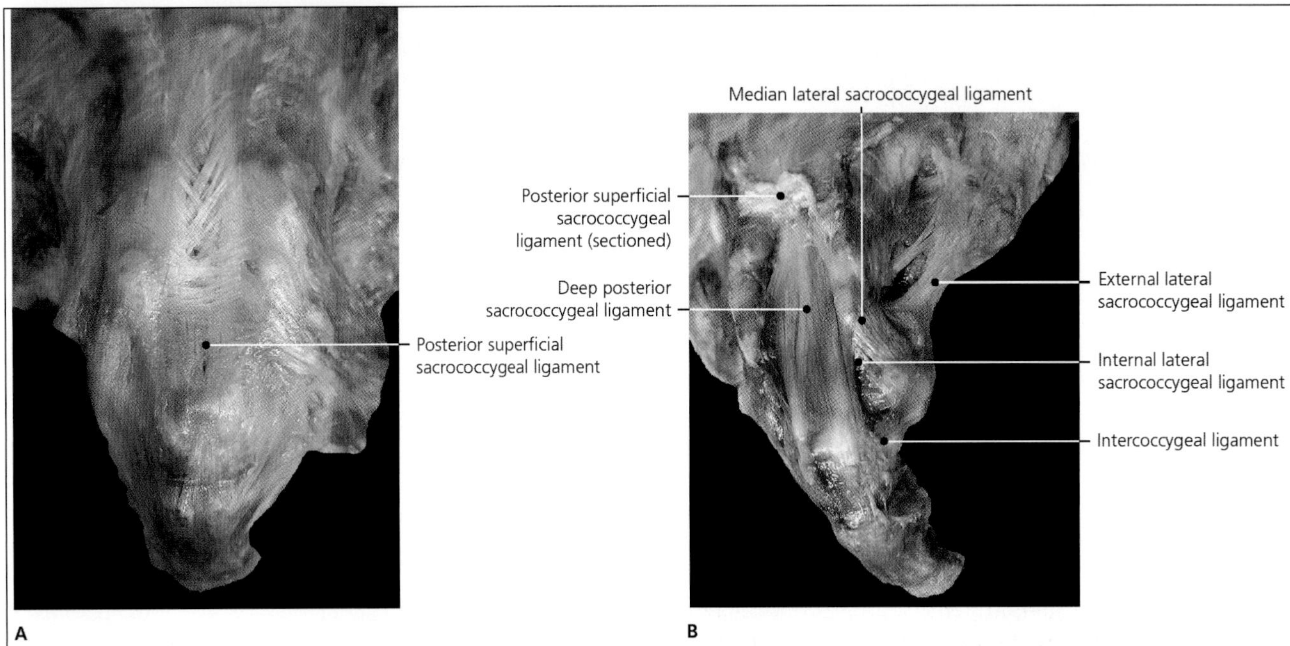

Figure 15-28 Posterior and lateral sacrococcygeal ligaments. **A,** Superficial dissection of the posterior sacrococcyx. **B,** Deep dissection of the posterior sacrococcyx from an oblique view.

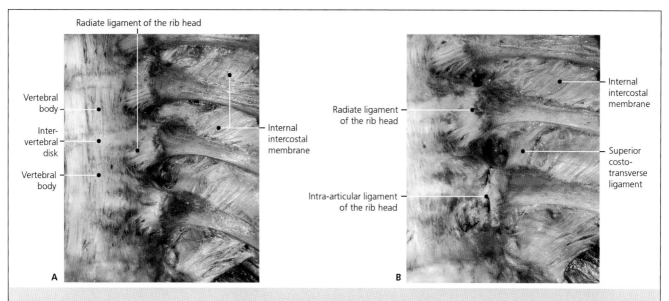

Figure 15-30 Ligaments of the rib head. In **B,** the radiate ligament has been removed to expose the intra-articular ligament.

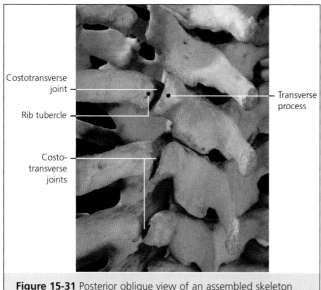

Figure 15-31 Posterior oblique view of an assembled skeleton showing the costotransverse joints.

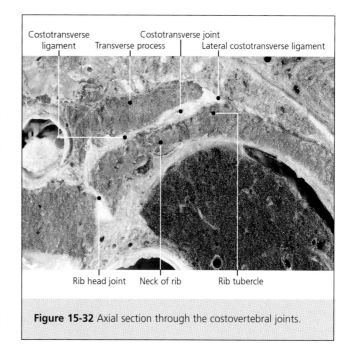

Figure 15-32 Axial section through the costovertebral joints.

accessory ligaments, the lateral and the medial accessory bundles. The lateral accessory bundle attaches to the inferior border of the apex of the transverse process, where it may be confused with the corresponding intertransverse ligament, and descends intermingled with the superior costotransverse ligament. The medial accessory bundle, which is variable, normally extends from the base of the transverse process to the head of the rib. When this bundle is present, it divides the space between the main ligament and the spine into two spaces: the anterior space, which allows passage for a tributary vein to the intercostal vein, and the posterior space, which allows passage for the intercostal nerves and arterial branches for the cord.

The inferior costotransverse ligament is visible when the rib is elevated. It is formed by a group of small bundles that insert superiorly, on the inferior rib border, and inferiorly, on the apex of the corresponding transverse process.

The costolaminar ligament arises from the superior costal border, immediately medial to the lateral costotransverse ligament. It runs obliquely in a medial and superior direction, inserting on the inferior border of the lamina, close to the transverse process.

Each costotransverse joint has a rudimentary synovial membrane that enhances gliding.

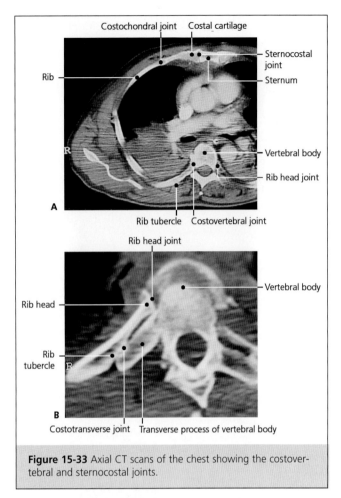

Figure 15-33 Axial CT scans of the chest showing the costovertebral and sternocostal joints.

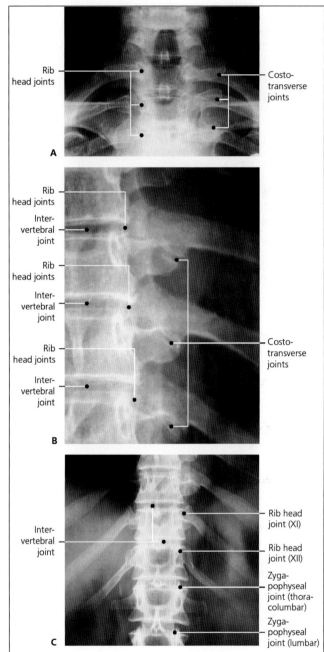

Figure 15-34 PA radiographs of the costovertebral joints. **A,** Upper thoracic region. The first rib articulates with T1 exclusively. **B,** Detail of the midthoracic spine. **C,** Lower thoracic spine. The eleventh and twelfth ribs also articulate exclusively with their corresponding vertebra.

Costochondral Joints

The osseous and the cartilaginous portions of the rib are joined by the costochondral joint, which is a suture-type joint (Figure 15-35). The joint consolidates through the continuity of the periosteum with the perichondrium. It does not have a joint cavity.

Sternocostal Joints

These diarthroses join the costal cartilage of the first seven ribs to the costal notches of the sternum through two wedge-shaped articular surfaces (Figures 14-24 and 15-35). Both joint surfaces are covered by fibrocartilage and are reinforced by the radiate and intra-articular sternocostal ligaments and the capsular ligaments, which are described below.

The radiate sternocostal ligaments reinforce the joint capsule both anteriorly and posteriorly (Figure 15-36). They originate on the anterior end of the costal cartilage and attach to the sternum in a fan like fashion. On the anterior aspect, these bundles interlace with those of the contralateral side, forming the membrane of the sternum. The posterior ligaments are thinner and are confused with the joint capsule.

The intra-articular sternocostal ligaments join the apex of the cartilaginous wedge to the costal notch, within the joint (Figure 15-37). This is especially prominent on the second rib.

The capsular ligaments correspond with the fibrous membranes that cover the costal cartilage (perichondrium) and blend with the sternal periosteum.

These joints are rudimentary synovial joints. The sternocostal joint of the first rib has distinctive features and is called the sternochondral synchondrosis of the first rib (Figure 15-38). The joint surface is

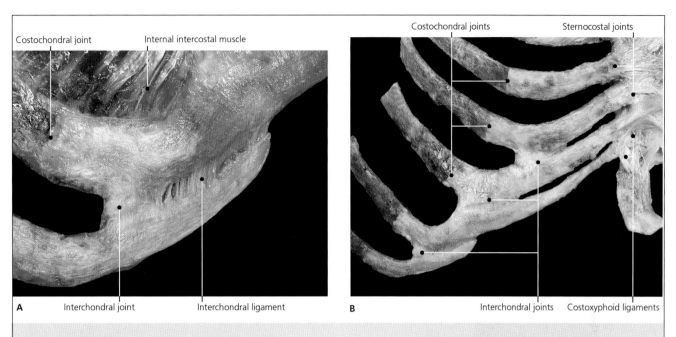

Figure 15-35 Interchondral **(A)** and costochondral **(B)** joints.

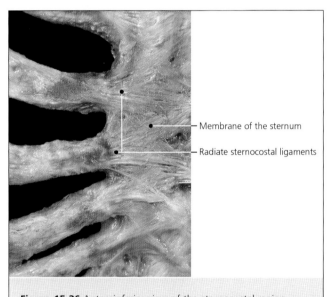

Figure 15-36 Anteroinferior view of the sternocostal region.

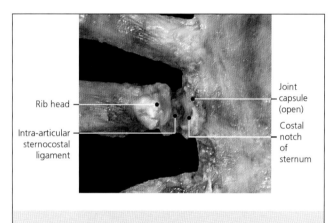

Figure 15-37 Posterior endothoracic view of a specimen showing an exposed sternocostal joint.

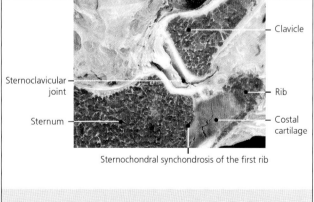

Figure 15-38 Cross-sectional anatomic specimen through the sternocostoclavicular joint.

more extensive than in the other sternocostal joints and normally is in direct continuity with the sternum. The joint surfaces are slightly separated from each other, leaving a small triangular depression for the costal surface of the clavicle. The sternocostal joint of the seventh rib is characterized by the costoxyphoid ligaments (Figure 15-35), which join the costal cartilage to the anterior surface of the xyphoid process.

Interchondral Joints

In the sixth through ninth ribs, the costal cartilage usually unites with each other through the perichon-

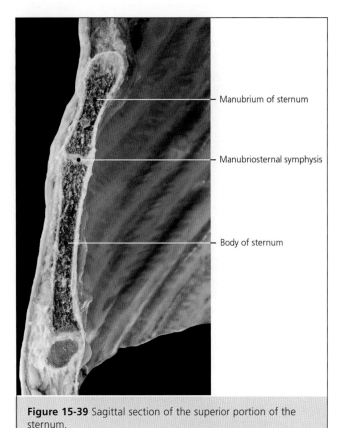

Figure 15-39 Sagittal section of the superior portion of the sternum.

drium and some fibrous bundles (Figure 15-35). The interchondral joints have a rudimentary synovium that allows small sliding movements.

Sternal Synchondroses

These articulations unite the three portions of the sternum.

Manubriosternal Symphysis

The inferior part of the manubrium articulates with the body of the sternum through a fibrocartilage (Figures 14-24 and 15-39), constituting an amphiarthrosis. It is reinforced by the uninterrupted perichondrium and by multiple fibrous bundles.

Xyphisternal Symphysis

The body of the sternum bonds with the xyphoid process (Figures 14-24 and 15-35), similar to the manubriosternal symphysis.

Chapter 16

MYOLOGY

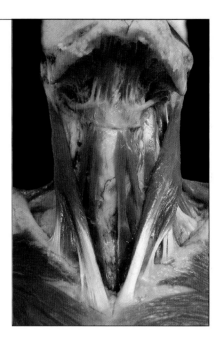

MUSCLES OF THE TRUNK

This section focuses on the muscles exclusive to the trunk. The muscle groups that arise in the trunk and insert onto the structures forming the shoulder girdle are described in chapter 10.

Dorsal Region

The anatomy of the dorsal musculature of the trunk is difficult to learn. For didactic purposes, anatomists named the group of muscles on both sides of the midline of the back the erector spinae (Figure 16-1), a name that reflects their fundamental function, of resisting gravity while keeping the trunk erect. The caudal limit of the erector spinae is the two layers of thoracolumbar fascia. The erector spinae are innervated by the dorsal branches of the spinal nerves. The individual muscles of this region are not described in detail here.

The muscles of the back are divided into a lateral and a medial compartment. The lateral compartment is located at the level of and lateral to the transverse

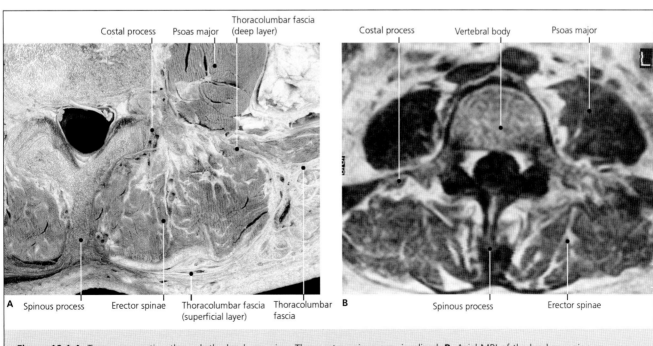

Figure 16-1 A, Transverse section through the lumbar region. The erector spinae are visualized. **B,** Axial MRI of the lumbar region.

American Academy of Orthopaedic Surgeons 257

TABLE 16-1 MUSCLES OF THE LATERAL COMPARTMENT

MUSCLE	Iliocostalis	Longissimus	Serratus posterior superior
ORIGIN	*Lumbar:* Iliac crest *Thoracic:* Lower six ribs *Cervical:* Most-cranial ribs	*Thoracic:* Aponeurosis extending from the iliac crest and sacrum through the seventh thoracic vertebra *Cervical:* Transverse processes of T1 through T6 *Cephalic:* Transverse processes of C3 through T3	Transverse processes of C6 through T2
INSERTION	*Lumbar:* Rib head angles of the fifth through twelfth ribs *Thoracic:* Six cephalad ribs *Cervical:* Transverse processes of the middle cervical vertebrae	*Thoracic:* Transverse processes of the thoracic and lumbar vertebrae and rib head angles of the 11 most caudal ribs *Cervical:* Transverse processes of C2 through C7 *Cephalic:* Mastoid process	Second through fifth ribs
INNERVATION	Posterior primary rami of the regional spinal nerves	Posterior primary rami of the regional spinal nerves	Posterior primary rami of the regional spinal nerves
FUNCTION	Extension of the spine	Extension of the spine and head	Accessory respiratory

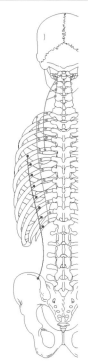
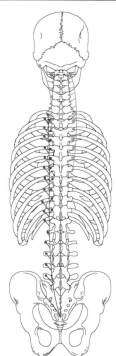
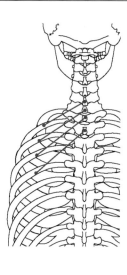

TABLE 16-1 MUSCLES OF THE LATERAL COMPARTMENT (CONT.)

Serratus posterior inferior	Splenius cervicis	Splenius capitis	Intertransverse
Transverse processes of T11 through L2	Spinous processes of T3 through T6	Spinous process of C4 through T2	Transverse processes
Ninth through twelfth ribs	Transverse processes of C1 and C2	Mastoid process	Transverse processes
Posterior primary rami of the regional spinal nerves	Posterior primary rami of the regional spinal nerves	Posterior primary rami of the regional spinal nerves	Posterior primary rami of the regional spinal nerves
Accessory respiratory	Extension and ipsilateral rotation of the neck	Extension and ipsilateral rotation of the head and neck	Lateral bending of the trunk

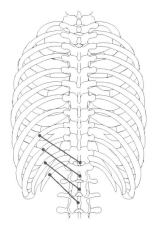 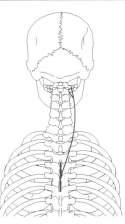 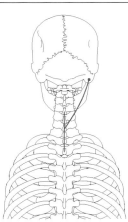 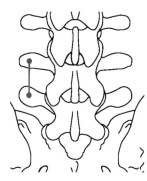

processes. The medial compartment is limited by the spinal canal, including the laminae.

The Lateral Compartment

The most lateral muscle of this compartment (Table 16-1) is the iliocostalis (Figure 16-2). It arises from the iliac crest and inserts onto all the thoracic ribs. For descriptive purposes, three segments are distinguished: lumbar, thoracic, and cervical.

The lumbar portion of the iliocostalis (iliocostalis lumborum) arises from the iliac crest and inserts onto the fifth through twelfth rib head angles. The thoracic portion (iliocostalis thoracis) originates from the six most caudal rib head angles and inserts onto the six most cranial ribs. Finally, the cervical portion (iliocostalis cervicalis) arises from the most cranial ribs and inserts onto the posterior tubercle of the transverse processes of the middle cervical vertebrae.

Medial to the iliocostalis lies the longissimus, or long dorsal muscle (Figures 16-2 and 16-3). As with the ilio-

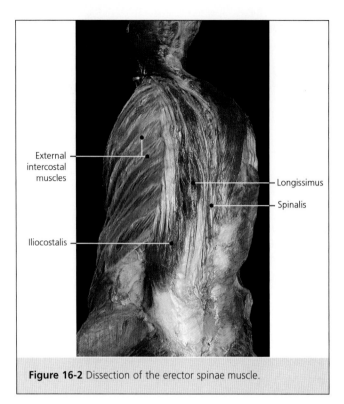

Figure 16-2 Dissection of the erector spinae muscle.

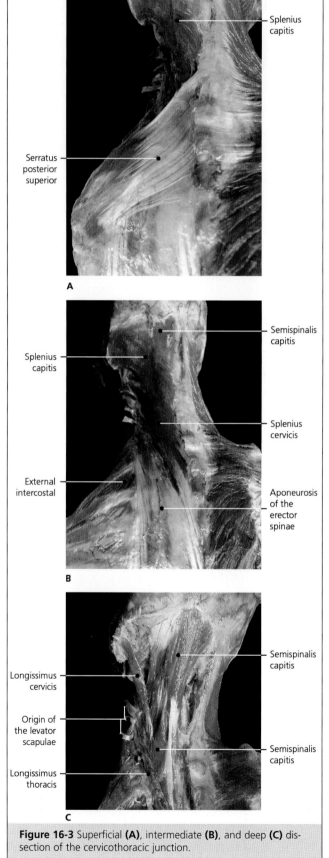

Figure 16-3 Superficial (**A**), intermediate (**B**), and deep (**C**) dissection of the cervicothoracic junction.

costalis, the longissimus is described as having thoracic, cervical, and cephalic segments.

The thoracic portion of the longissimus (longissimus thoracis) originates from the aponeurosis extending from the iliac crest and sacrum through the seventh thoracic vertebra. This origin is similar to that of the latissimus dorsi, but it differs in its depth. The longissimus inserts onto the transverse processes of the thoracic and lumbar vertebrae and onto the rib head angles of the eleven most caudal ribs. The cervical portion (longissimus cervicis) arises from the transverse processes of T1 through T6 and inserts onto the posterior tubercle of the transverse processes of C2 through C7. The cephalic portion (longissimus capitis) arises from the transverse processes of C3 through T3 and inserts onto the mastoid process of the temporal bone.

The serratus posterior superior and inferior muscles are found in the superior and inferior thoracic region, respectively, superficial to the thoracolumbar fascia. These are fine, small aponeurotic laminae where some muscle fibers may be found. The serratus posterior superior (Figure 16-3) is found beneath the rhomboideus and levator scapulae muscles. It arises from the spinous processes of C6 through T2 and inserts onto the second through fifth ribs. The serratus posterior inferior arises from the spinous processes of T11 through L2 and inserts onto the ninth through twelfth ribs. These muscles act as accessory respiratory muscles, although they are weak because of their morphology.

In the cervical region, two muscles lie deep to the serratus posterior superior. The splenius cervicis (Fig-

MYOLOGY

TABLE 16-2 MUSCLES OF THE MEDIAL COMPARTMENT

MUSCLE	Spinalis	Interspinalis	Semispinalis	Multifidus	Rotators
ORIGIN	*Thoracic:* Spinous processes T10 through L3 *Cervical:* Spinous processes C6 through T2 *Cephalic* (variable): Reinforces the cervical portion	Spinous processes	*Thoracic:* Transverse processes T7 through T12 *Cervical:* Transverse processes T1 through T6 *Cephalic:* Transverse processes C4 through T6	Transverse processes	Transverse processes
INSERTION	*Thoracic:* Spinous processes of T2 through T8 *Cervical:* Spinous processes of C2 through C5 *Cephalic* (variable)	Spinous processes	*Thoracic:* Spinous processes C7 through T6 *Cervical:* Spinous processes C2 through C5 *Cephalic:* Occiput between the superior and inferior nuchal lines	Spinous processes of the two to four levels cephalad	*Short:* Lamina of the adjacent cephalad vertebra *Long:* Lamina of the second most proximal cephalad vertebra
INNERVATION	Posterior primary rami of the regional spinal nerves	Posterior primary rami of the regional spinal nerves	Posterior primary rami of the regional spinal nerves	Posterior primary rami of the regional spinal nerves	Posterior primary rami of the regional spinal nerves
FUNCTION	Extension of the trunk	Extension of the trunk	Extension of the trunk and head; slight rotation of the trunk	Extension of the trunk	Contralateral rotation of the spine

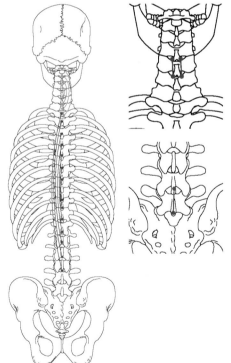
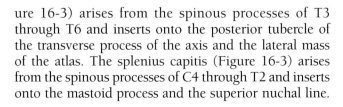
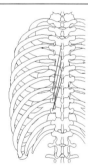
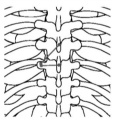

ure 16-3) arises from the spinous processes of T3 through T6 and inserts onto the posterior tubercle of the transverse process of the axis and the lateral mass of the atlas. The splenius capitis (Figure 16-3) arises from the spinous processes of C4 through T2 and inserts onto the mastoid process and the superior nuchal line.

The unilateral contraction of these muscles extends and rotates the neck and head toward the ipsilateral side.

The lateral intertransverse muscles are found between the transverse processes. They are named according to their location with respect to the spine; ie, lumbar, thoracic, and cervical. These muscles contribute to lateral bending of the trunk.

Medial Compartment

These muscles (Table 16-2) are located in the vertebral canal. The most prominent are the spinalis and the semispinalis.

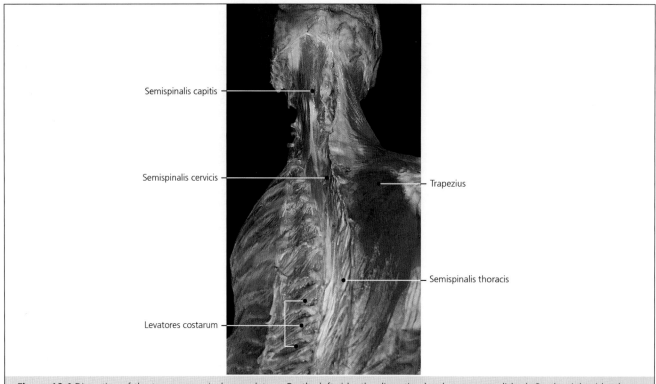

Figure 16-4 Dissection of the transverse spinal musculature. On the left side, the dissection has been accomplished. On the right side, the superficial muscles have been preserved.

The spinalis muscle (Figures 16-2 and 16-4) joins the spinous processes. Three portions are distinguished: the spinalis thoracis, the spinalis cervicis, and the spinalis capitis.

The spinalis thoracis arises from the spinous processes of T10 through L3 and inserts onto the spinous processes of T2 through T8. The spinalis cervicis arises from the spinous processes of C6 through T2 and inserts onto the spinous processes of C2 through C5. Finally, the spinalis capitis is a mere reinforcement of the spinalis cervicis and is not always present.

Between the spinous processes are the interspinalis muscles, named after the region of the spine in which they are located: the interspinalis thoracis, the interspinalis cervicis, and the interspinalis capitis. The interspinalis cervicis may be present in pairs because of the bifid nature of the cervical spinous processes.

The semispinalis, multifidi, and rotator muscles are also called transversospinal because they originate in the transverse processes and insert onto the spinous processes. The semispinalis is one of the major muscles of this complex.

The semispinalis muscle (Figures 16-3 and 16-4) is described as having three sections: thoracic, cervical, and cephalic.

The semispinalis thoracis arises from the transverse processes of T7 through T12 and inserts onto the spinous processes of C7 through T6. The semispinalis cervicis arises from the transverse processes of T1 through T6 and inserts onto the spinous processes of C2 through C5. Finally, the semispinalis capitis arises from the transverse processes of C4 through T6 and inserts onto the occiput, between the superior and inferior nuchal lines.

Deep to this muscle along the vertebral canal are the multifidi (Figure 16-5). They arise from the transverse processes and insert onto the spinous processes of the vertebrae two to four levels cephalad.

The rotator muscles (Figure 16-5) are found deep to the multifidi. They are named according to the region of the spine in which they are found: lumbar, thoracic, or cervical. The rotators arise from the transverse processes and insert onto the laminae of the adjacent one or two levels.

SUBOCCIPITAL MUSCULATURE

This musculature is located underneath the occiput (Table 16-3). There are three muscle groups: posterior, lateral, and anterior.

Posterior Group

Deep to the semispinalis capitis lie the rectus and oblique muscles.

The rectus capitis posterior major (Figure 16-6) arises from the spinous process of the axis and is directed obliquely toward the inferior nuchal line. Its function includes extension and ipsilateral rotation of the head. Unlike the long trunk musculature, the rectus capitis posterior major allows for

MYOLOGY

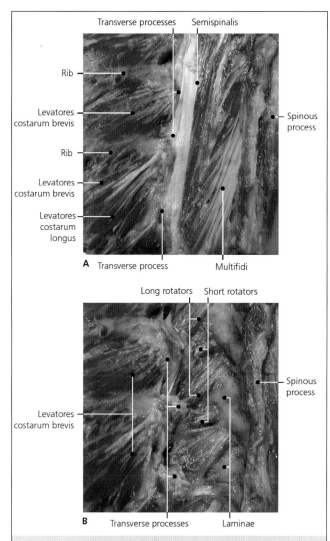

Figure 16-5 A, Specimen of the transverse spinal muscles of a spinal segment. A small portion of the semispinalis has been preserved. The multifidi can be observed deep to the semispinalis and skip two to four vertebrae to insert onto the spinous process of the vertebra two to four vertebrae cranial to their origin. **B,** Deep dissection with respect to Figure 16-5A.

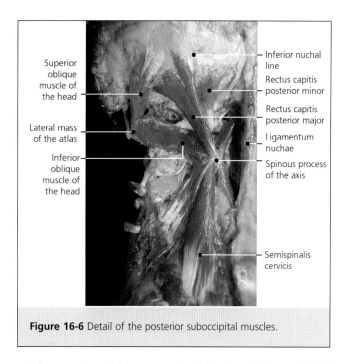

Figure 16-6 Detail of the posterior suboccipital muscles.

fine movements. It is innervated by the suboccipital nerve.

The rectus capitis posterior minor (Figure 16-6) is located medial to the rectus capitis posterior major. It arises from the posterior tubercle of the atlas and is directed cranially, inserting onto the inferior nuchal line. It has the same function as the rectus capitis posterior major, although it has a smaller rotational component. It is also innervated by the suboccipital nerve.

The oblique muscles are localized lateral to the rectus muscles. The inferior oblique muscle of the head (Figure 16-6) courses between the spinous process of the axis and the lateral mass of the atlas. Because of its lateral location, this muscle functions as an ipsilateral rotator of the head. It is innervated by the dorsal rami of the cervical nerves.

The superior oblique muscle of the head (Figure 16-6) is found cephalad to the inferior oblique muscle of the head. It arises from the lateral mass of the atlas and inserts immediately cranial to the posterior rectus capitis major. Together with the inferior oblique, it is responsible for ipsilateral rotation of the axis and, consequently, the head. This short muscle allows for fine movements to direct the head in space. It is also innervated by the dorsal rami of the cervical nerves.

Lateral Muscles
The only muscle in this region is the rectus capitis lateralis (Figure 16-7). It arises from the lateral mass of the atlas and inserts onto the jugular process of the occiput. Its function consists of lateral bending of the head. It is innervated by the ventral rami of the C1 and C2 nerves.

Anterior Muscles
Medial to the rectus capitis lateralis is the rectus capitis anterior. It originates on the lateral mass of the atlas and attaches to the basilar portion of the occiput, in front of the foramen magnum. Its main function is fine flexion of the head. It is innervated by the ventral rami of the cervical nerves.

The longus capitis covers part of the cervical vertebral bodies (Figure 16-7). It is the cephalic extension of the longus colli, which is discussed below. It arises from the anterior tubercles of the transverse processes from C3 through C6 and inserts onto the anterior aspect of the basilar portion of the occiput. Contraction of one side of the longus capitis produces anterior and lateral flexion of the head; contraction of both sides produces flexion of

Table 16-3 The Suboccipital Muscles

MUSCLE	Rectus capitis posterior	Oblique muscles of the head	Rectus capitis lateralis	Rectus capitis anterior	Longus capitis
ORIGIN	*Major:* Spinous process of the axis *Minor:* Posterior tubercle of the atlas	*Inferior:* Spinous process of the axis *Superior:* Lateral mass of the atlas	Lateral mass of the atlas	Lateral mass of the atlas	Transverse processes C3 through C6
INSERTION	Inferior nuchal line	*Superior:* Inferior nuchal line *Inferior:* Lateral mass of the atlas	Jugular process of the occiput	Basilar portion of the occiput	Basilar portion of the occiput
INNERVATION	Posterior rami of the cervical spinal nerves	Posterior rami of the cervical spinal nerves	Anterior rami of the cervical spinal nerves	Anterior rami of the cervical spinal nerves	Anterior rami of the cervical spinal nerves
FUNCTION	Extension of the head	Extension and ipsilateral rotation of the head	Lateral flexion of the head	Flexion of the head	Flexion of the head

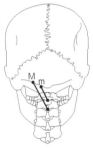
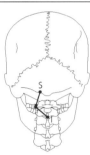
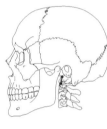
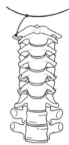
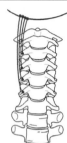

the head. It is innervated by the ventral rami of the cervical spinal nerves.

Muscles of the Neck

A lateral group and an anterior group of muscles are described. The anterior muscles are divided into superficial and deep (prevertebral) muscles.

Lateral Muscles

The scalene muscles are found in this region. The anterior scalene (Table 16-4 and Figures 16-7 and 16-8) arises from the anterior tubercles of the transverse processes of C3 through C6 and inserts onto the superior aspect of the first rib, on the tubercle for the anterior scalene muscle. The subclavian vein is situated anterior to this muscle, and the subclavian artery is situated posterior to it, coursing through the interscalene triangle formed by the anterior and middle scalene muscles. When the rib is fixed, the anterior scalene produces lateral flexion of the head. When the head is fixed, the anterior scalene aids in inspiration by elevating the first rib.

The middle scalene muscle (Figures 16-7 and 16-8) arises from the anterior tubercles of the transverse processes of C2 through C7 and inserts onto the first rib (occasionally onto the second rib, through an accessory fascicle), posterior to the anterior scalene, thus forming the interscalene triangle. Its functions are identical to those of the anterior scalene.

The posterior scalene muscle (Figure 16-7) arises from the posterior tubercles of the transverse processes of C4 through C6 and inserts onto the superior border of the second rib. Similar to the anterior and middle scalene muscles, the posterior scalene may produce lateral flexion of the neck or elevate the second rib, depending on the fixed and mobile points.

The scalene muscles are innervated by the anterior rami of the spinal nerves (C2 through C8).

The minimal scalene muscle is variably present. This muscle arises between the transverse processes of C6 and C7 and inserts onto the first rib and the apical pleura. It is localized between the subclavian artery (anterior) and the brachial plexus (posterior).

Anterior Musculature

In the deep prevertebral region is a muscle that is intimately related to the longus capitis, the longus colli (Figure 16-8). The longus colli is an arch-shaped muscle that joins the vertebral bodies and transverse processes of the upper thoracic vertebrae

MYOLOGY

with those of the inferior cervical vertebrae. One side produces flexion combined with lateral flexion of the neck, and both sides act simultaneously to produce flexion of the neck. The longus colli is innervated by the anterior rami of the cervical spinal nerves.

In the superficial region of the neck, a superficial cutaneous muscle, the platysma, may be distinguished (Figure 16-9). The platysma tenses the skin from the mandible down to the superior aspect of the chest, and it is innervated by the facial nerve. Deep to the platysma, the musculature is classified according to its relationship with the hyoid bone—suprahyoid and infrahyoid.

Suprahyoid Muscles

Four muscles are classified as suprahyoid: the digastric, the stylohyoid, the mylohyoid, and the geniohyoid (Table 16-5).

The digastric muscle (Figures 16-10 and 16-11), as the name implies, is formed by two muscle bellies separated by an intermediate tendon. The posterior belly arises from the mastoid ("digastric") notch on the mastoid process and is directed toward the hyoid, where it has an intermediate tendon. Its distal portion passes through an aponeurotic buttonhole formed by the stylohyoid. The intermediate tendon reflects under this pulley on the hyoid bone and continues as the anterior belly. It inserts onto the posterior aspect of the anterior mandible (digastric fossa). Both digastric muscles lower the mandible and reach the posterior aspect of the hyoid bone. The posterior belly is innervated by the facial nerve, and the anterior belly is innervated by the mylohyoid branch of the mandibular nerve (trigeminal nerve).

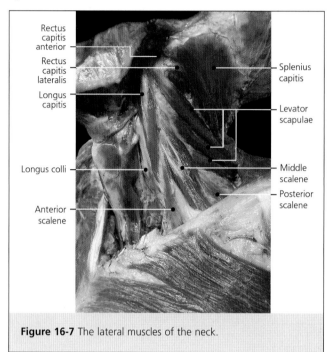

Figure 16-7 The lateral muscles of the neck.

TABLE 16-4 THE LATERAL MUSCLES OF THE NECK

MUSCLE	Anterior scalene	Middle scalene	Posterior Scalene
ORIGIN	Transverse processes C3 through C6	Transverse processes of C2 through C7	Mental spine
INSERTION	First rib	First rib	Second rib
INNERVATION	Ventral rami of the regional spinal nerves	Ventral rami of the regional spinal nerves	Ventral rami of the regional spinal nerves
FUNCTION	Elevates the first rib and produces lateral flexion of the neck	Elevates the first rib and produces lateral flexion of the neck	Elevates the second rib and produces lateral flexion of the neck

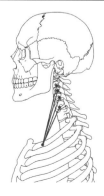 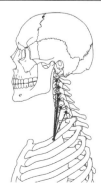 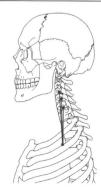

American Academy of Orthopaedic Surgeons

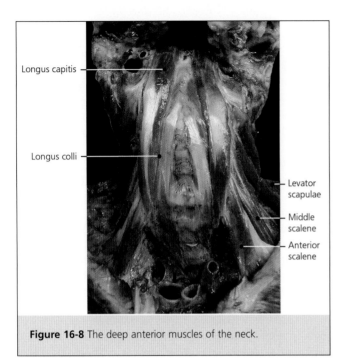

Figure 16-8 The deep anterior muscles of the neck.

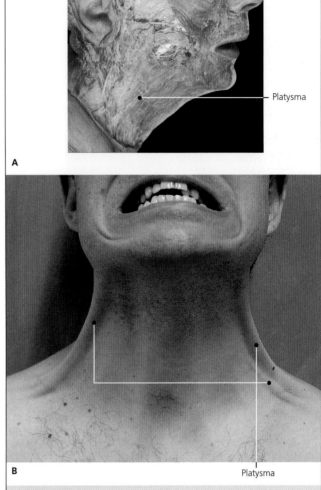

Figure 16-9 A, Superficial dissection of the neck revealing the platysma muscle. **B,** Photograph of a man tensing the platysma. Observe the lateral expansion of the skin.

The stylohyoid muscle (Figure 16-11) arises from the styloid process of the temporal bone and inserts onto the hyoid bone. It accompanies the posterior belly of the digastric muscle anteriorly and forms a buttonhole on which the digastric muscle reflects (digastric buttonhole). The function of the stylohyoid muscle is to elevate and translate the hyoid bone posteriorly. It is innervated by the facial nerve.

The mylohyoid muscle (Figures 16-10 through 16-12) is a broad muscle that forms part of the muscle floor of the mandible; some authors refer to it as the diaphragm of the mouth. It arises from the mylohyoid line of the mandible and inserts onto the body of the hyoid bone. Its function is to elevate the hyoid bone and to lower the mandible. It is innervated by the mylohyoid nerve.

Finally, the geniohyoid (Figure 16-12) is located cranial to the mylohyoid, on the floor of the mouth. It originates on the mental spine (apophysis geni) and attaches to the body of the hyoid. Its main function is to elevate the hyoid bone. It is innervated by spinal nerve C1, which is carried by the main trunk of the hypoglossal nerve.

Infrahyoid Muscles

There are also four muscles below the hyoid bone: the sternohyoid, the sternothyroid, the thyrohyoid, and the omohyoid. All are innervated by the ansa cervicalis (Table 16-6).

The sternohyoid muscle (Figure 16-10), as its name indicates, arises from the posterosuperior face of the manubrium of the sternum, the clavicle, and the posterior sternoclavicular ligament. It inserts onto the inferior part of the body of the hyoid bone. Its function is to lower the hyoid.

The sternothyroid muscle (Figure 16-13) arises from the posterior and cephalad aspects of the manubrium of the sternum and the first costal cartilage. It inserts onto the oblique line of the thyroid cartilage. Its functions include depressing the thyroid cartilage and lowering the larynx.

The thyrohyoid muscle (Figure 16-13) is a flat muscle that constitutes the continuation of the sternothyroid. It arises from the oblique line of the thyroid cartilage and inserts onto the inferior border of the body and greater horn of the hyoid bone. Its function consists of lowering or depressing the hyoid bone.

The omohyoid muscle (Figure 16-10) is also a digastric muscle. Its inferior belly arises medial to the scapular notch and progresses medially. The intermediate tendon is located posterior to the sternocleidomastoid muscle, accompanying the neurovascular bundle of the neck. The superior belly inserts onto the inferior region

TABLE 16-5 THE SUPRAHYOID MUSCLES

MUSCLE	Digastric	Stylohyoid	Mylohyoid	Geniohyoid
ORIGIN	*Posterior belly:* Mastoid process	Styloid process	Mylohyoid line of the mandible	Mental spine
INSERTION	*Anterior belly:* Mandible	Hyoid bone	Hyoid bone	Hyoid bone
INNERVATION	*Posterior belly:* Facial nerve *Anterior belly:* Mylohyoid nerve	Facial nerve	Mylohyoid nerve	C1
FUNCTION	Lowers the mandible; posterior thrust of the hyoid bone	Elevates the hyoid bone	Elevates the hyoid bone	Elevates the hyoid bone

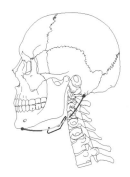
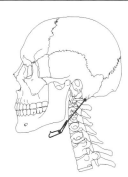
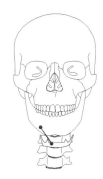
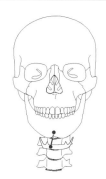

TABLE 16-6 THE INFRAHYOID MUSCLES

MUSCLE	Sternohyoid	Sternothyroid	Thyrohyoid	Omohyoid
ORIGIN	Sternum and clavicle	Sternum	Thyroid cartilage	*Inferior belly:* Superior border of scapula
INSERTION	Hyoid bone	Thyroid cartilage	Hyoid bone	*Superior belly:* Hyoid bone
ATTACHMENT	Ansa cervicalis	Ansa cervicalis	Ansa cervicalis	Ansa cervicalis
FUNCTION	Lowers the hyoid bone	Lowers the thyroid cartilage	Lowers the hyoid bone	Lowers the hyoid bone

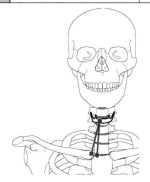

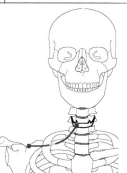

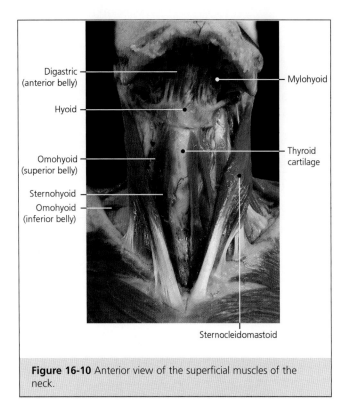

Figure 16-10 Anterior view of the superficial muscles of the neck.

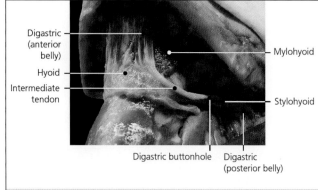

Figure 16-11 Oblique anterolateral view of the neck, with details of the stylohyoid and digastric muscles.

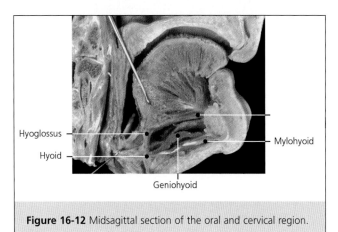

Figure 16-12 Midsagittal section of the oral and cervical region.

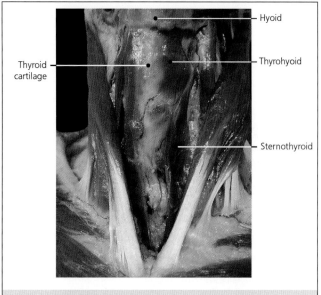

Figure 16-13 Deep dissection of the anterior neck. The sternohyoid muscles have been removed.

of the base of the hyoid bone. Its function includes lowering of the hyoid bone.

Muscles of the Chest

Some of the muscles that lie on the surface of the chest wall were described in chapter 10. This section describes muscles that are intrinsic to the trunk.

Dorsal

The levatores costarum are found in the dorsal region (Figure 16-4). These are small, flat muscles that arise from the transverse processes of C7 through T11 and insert onto the rib head angles. Two types can be differentiated: the levatores costarum longus inserts two rib segments caudal to its origin, whereas the levatores costarum brevis inserts onto the immediately caudal rib head angles. As the name indicates, the function of these muscles is to elevate the ribs, thus assisting inspiration. They are innervated by the dorsal rami of the spinal nerves.

Intercostal

In the spaces between the ribs are three flat muscle layers commonly called the intercostal muscles. The most superficial ones are the external intercostal muscles (Figures 16-14 through 16-16), which are characterized by the orientation of their fibers, which run obliquely from the superior rib toward the inferior rib and from lateral to medial. These muscles assist in inspiration and are innervated by the intercostal nerves. The external intercostal muscles do not occupy the complete intercostal space but are limited to the area between the costotransverse joint and the costochon-

MYOLOGY

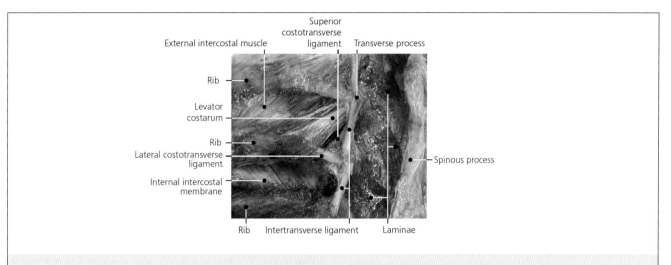

Figure 16-14 Specimen of the trunk. The external intercostal muscle has been removed from the inferior intercostal space, exposing the internal intercostal membrane.

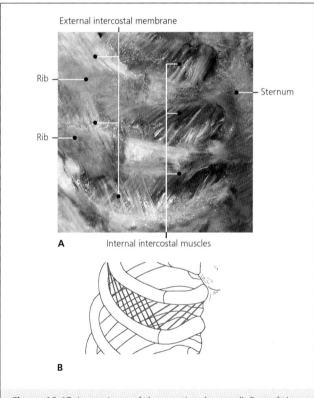

Figure 16-15 A, specimen of the anterior chest wall. Part of the external intercostal membrane has been removed to expose the underlying internal intercostal muscles. B, Diagram showing the direction of the intercostal muscle fiber orientation; the external intercostal muscles are in red, and the internal intercostal muscles are in blue.

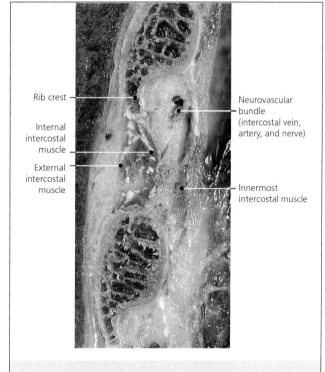

Figure 16-16 Sagittal section and detail of an intercostal space with its three muscle layers.

dral junction. Thereafter, they continue as the external intercostal membrane (Figure 16-15).

The internal intercostal muscles are situated deep to the external intercostal muscles (Figures 16-15 through 16-17). Their fibers also run obliquely but in a different orientation, running from the superior border of one rib to the next inferior rib crest, in a craniomedial orientation. Dorsally, these muscles do not occupy the intercostal space. They continue as the internal intercostal membrane (Figure 16-14) from the rib head angle toward the lateral aspect of the vertebral body. These muscles lower the ribs and aid in expiration. They are innervated by the intercostal nerves.

Finally, the innermost intercostal muscles (Figures 16-16 and 16-17) are the deepest, and some authors consider them a fold of the internal intercostal muscles. The intercostal neurovascular bundle is located between the internal and innermost intercostal muscles and includes the intercostal vein, artery, and nerve. They are innervated by the same intercostal nerve.

The subcostal muscles are found in the internal aspect of the chest and constitute parts of the internal intercostal muscles that attach to nonadjacent ribs. Finally, the transversus thoracis muscle (previously called the triangular sternal muscle), a continuation of the trasversus abdominis muscle, arises from the posterior aspect of the body of the sternum and inserts onto the costochondral junctions (Figure 16-17). Both the subcostal and the transversus thoracis muscles are innervated by the intercostal nerves.

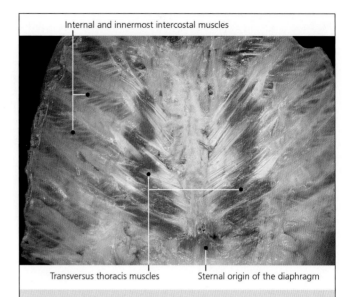

Figure 16-17 Posterior view of the anterior chest wall.

TABLE 16-7 THE ABDOMINAL MUSCLES

MUSCLE	Rectus abdominis	External oblique	Internal oblique	Transversus abdominis	Quadratus lumborum
ORIGIN	Fifth through seventh ribs	Fifth through twelfth ribs	Iliac crest, anterior superior iliac spine, and inguinal ligament	Seventh through twelfth ribs, iliac crest, and thoracolumbar fascia	Costal processes of L1 through L4 and iliac crest
INSERTION	Pubic crest	Linea alba and inguinal ligament	Tenth through twelfth ribs and linea alba	Linea alba	Twelfth rib and costal processes of the lumbar vertebrae
INNERVATION	Seventh through twelfth intercostal nerves, iliohypogastric nerve	Eighth through twelfth intercostal nerves, iliohypogastric and ilioinguinal nerves	Eighth through twelfth intercostal nerves, iliohypogastric and ilioinguinal nerves	Eighth through twelfth intercostal nerves, iliohypogastric and ilioinguinal nerves	Subcostal nerve and lumbar plexus
FUNCTION	Flexes the trunk	Flexion of the trunk and contralateral rotation	Flexion of the trunk and ipsilateral rotation	Decreases the diameter of the abdomen	Extension and ipsilateral bending of the trunk

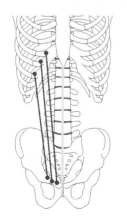 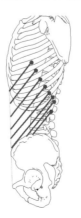  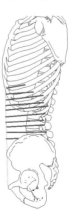 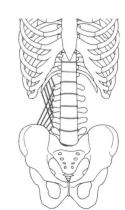

Myology

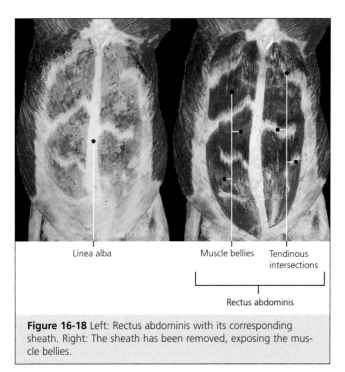

Figure 16-18 Left: Rectus abdominis with its corresponding sheath. Right: The sheath has been removed, exposing the muscle bellies.

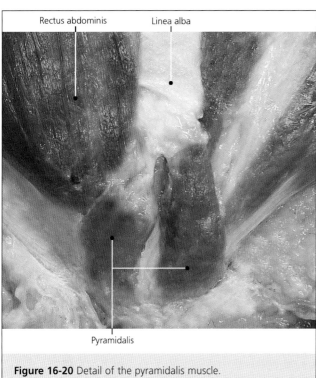

Figure 16-20 Detail of the pyramidalis muscle.

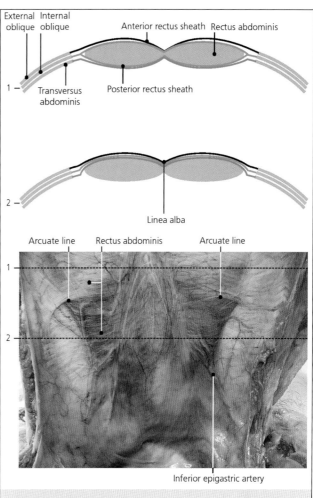

Figure 16-19 Top: Drawings of transverse sections across the points labeled in the photograph below (section 1 is cephalad to the umbilicus; section 2 is caudal). The drawings illustrate the aponeurosis of the lateral muscles. Bottom: Posterior view of the abdominal wall. The arcuate line is formed where the aponeurosis of the lateral muscles becomes anterior to the rectus abdominis.

Muscles of the Abdomen

The abdominal muscles form the abdominal wall. Two groups of muscles may be distinguished, those that form the anterolateral wall of the abdomen and those that form the posterior wall. The diaphragm, which provides the upper limit of the abdominal cavity, is described in detail, but the pelvic diaphragm, its inferior limit, will not be discussed (Table 16-7).

Anterolateral

The anterolateral muscles of the abdomen include the following: rectus abdominis, pyramidalis, external oblique, internal oblique, and transversus abdominus.

The rectus abdominis muscle is a long, flat muscle located at the sides of the linea alba (Figures 16-18 through 16-20). It has three to five bellies separated by tendinous intersections. It arises from the costal cartilages of the fifth through seventh ribs and the xyphoid process and attaches onto the pubic crest and symphysis pubis.

The rectus abdominis is covered by the rectus sheath of the abdomen (Figure 16-19), which consists of two distinct fascial laminae derived from the lateral musculature. The aponeurosis of the external oblique muscle covers the rectus abdominis anteriorly, above the umbilicus. The internal oblique muscle bifurcates into anterior and posterior aponeurotic sheaths. The anterior

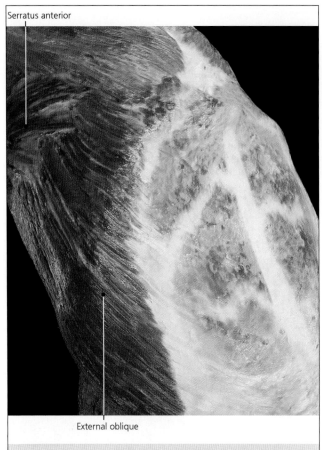

Figure 16-21 Lateral view of the abdomen, exposing the external oblique muscle.

The pyramidalis muscle (Figure 16-20) is a variable triangular muscle that is located anterior to the rectus abdominis. It arises from the pubic crest and symphysis pubis and inserts onto the linea alba. The pyramidalis is innervated by the subcostal nerve.

The lateral muscles are flat muscles that overlap in three layers. The external oblique muscle of the abdomen is the most superficial of the three (Figure 16-21). It takes its name from the oblique orientation of its fibers, which run from lateral to medial and from cephalad to caudal. It arises from the fifth through twelfth ribs, intertwining with the serratus anterior. The fibers attach through the aponeurosis that covers the anterior aspect of the rectus abdominis to the linea alba and are continuous with the contralateral abdominal musculature, forming a musculoaponeurotic loop. The most inferior fibers attach to the external border of the iliac crest, anterior superior iliac spine, and pubic tubercle. This aponeurosis reflects over itself, forming the inguinal ligament, the anterior wall, and the floor of the inguinal canal.

In the inguinal region, the reflected inguinal ligament (of Colles) arises from the pubic tubercle (Figure 16-22), forming a medial crus. Other fibers, called the lacunar ligament (of Gimbernat), arise from the lateral portion of the pubis and insert onto the pectineal crest to form a lateral crus. The pectineal ligament (of Cooper) is formed by the confluence of several aponeurotic structures, including the lacunar ligament, the transversus fascia, the ligament of Henle, and the origin of the pectineus muscle with its corresponding fascia. The two crura separate at their attachment, forming the superficial inguinal ring, which is partially covered by the intercrural fibers.

The inguinal canal is limited by the aponeurosis of the external oblique (anterior wall), the transversus abdominis fascia and reinforcements (posterior wall), the internal oblique and transversus abdominis fascia (roof), and the inguinal and lacunar ligaments (floor). It contains the spermatic cord in males and the round ligament of the uterus in females, entering through the deep inguinal ring.

The function of the external oblique is flexion and contralateral rotation of the trunk. If both sides act simultaneously, the rotation is neutralized and the trunk goes into flexion. It is innervated by the eighth through twelfth intercostal nerves and the iliohypogastric and ilioinguinal nerves.

The internal oblique of the abdomen is found deep to the external oblique (Figure 16-23). Its fibers arise from the intermediate line of the iliac crest, anterior superior iliac spine, deep layer of the thoracolumbar fascia, and inguinal ligament. Its fibers are directed obliquely from caudal to cranial and from lateral to medial. The internal oblique inserts onto the tenth

aponeurotic sheath of the internal oblique and that of the external oblique form the anterior layer of the rectus sheath. The posterior aponeurotic layer of the rectus sheath is formed by the posterior aponeurotic sheath of the internal oblique and that of the transversus abdominis muscles. Below the umbilicus, the aponeurosis of the external oblique, internal oblique, and transversus abdominis lie anterior to the rectus abdominis, forming the arcuate line (of Douglas), which can be observed from an intra-abdominal view (Figure 16-19). Caudally, the posterior rectus sheath is absent.

The function of the rectus abdominis is to flex the trunk, approximating the costal arch to the pubis (or the pubis to the costal arch, depending on the fixed and movable points). It is innervated by the seventh through twelfth intercostal nerves and the iliohypogastric nerve.

The linea alba, which separates the two rectus abdominis muscles, is created at the midline of the abdomen by the confluence of the aponeurosis of the abdominal musculature. It is nonvascular and therefore constitutes an ideal location for surgical approaches to the abdomen. Approximately in the middle, the linea alba is a reinforcement that forms the umbilical ring. The linea alba attaches to the pubis through the adminiculum.

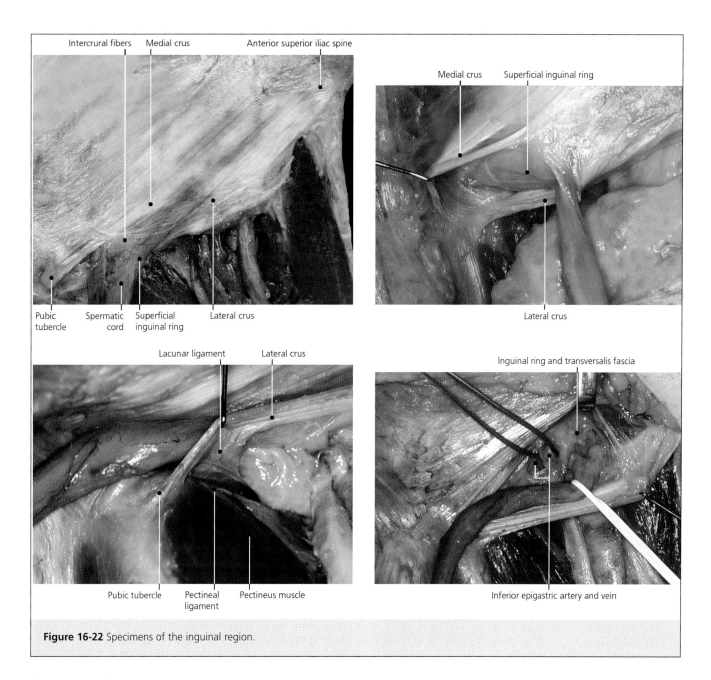

Figure 16-22 Specimens of the inguinal region.

through twelfth ribs, and in the anterior region its aponeurosis bifurcates, forming the anterior rectus sheath and the posterior rectus sheath. The most caudal fibers unite with those from the transversus abdominis to form the conjoined tendon, which attaches to the pubis. Its muscle fibers contribute to form the cremaster muscle.

The function of the internal oblique is flexion of the trunk when both sides act simultaneously and ipsilateral rotation of the trunk when only one side acts. The internal oblique is innervated by the eighth through twelfth intercostal nerves and the iliohypogastric and ilioinguinal nerves.

Posteriorly, the lumbar triangle (Petit's triangle) is found; this is limited by the iliac crest, the latissimus dorsi, and the external oblique. It allows direct access to the internal oblique muscle (Figure 10-9).

The transversus abdominis muscle (Figure 16-24) is located deep to the internal oblique. Its dorsal aspect is separated from the peritoneum by the transversalis fascia. The transversus abdominis arises from six digitations on the seventh through twelfth ribs and from the thoracolumbar fascia, the lumbocostal ligament, the internal lip of the iliac crest, the anterior superior iliac spine, and the inguinal ligament. The fibers of the transversus abdominis are transverse, acting as an anatomic swathe, and insert to the rectus sheath, linea alba, and symphysis pubis. The musculoaponeurotic union forms the semilunar line (of Spiegel), which is typically seen from an internal view of the abdominal wall (Figure 16-25).

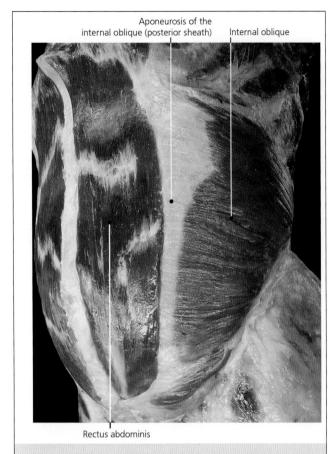

Figure 16-23 Anterolateral view of the abdomen. The external oblique has been removed to expose the internal oblique.

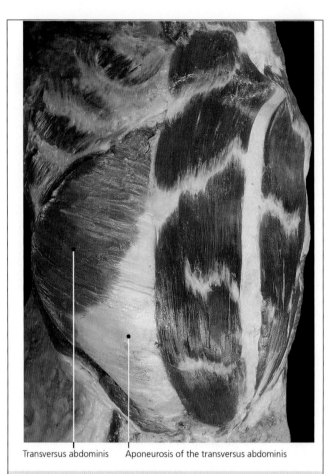

Figure 16-24 Anterolateral view of the abdomen. The internal oblique muscle has been removed to expose the transversus abdominis.

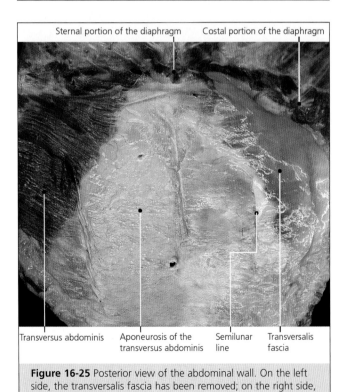

Figure 16-25 Posterior view of the abdominal wall. On the left side, the transversalis fascia has been removed; on the right side, it has been preserved.

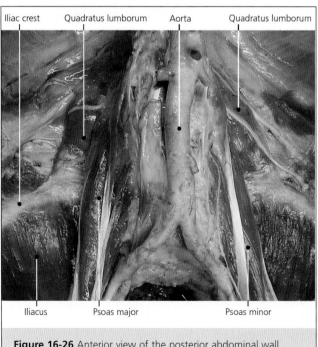

Figure 16-26 Anterior view of the posterior abdominal wall.

MYOLOGY

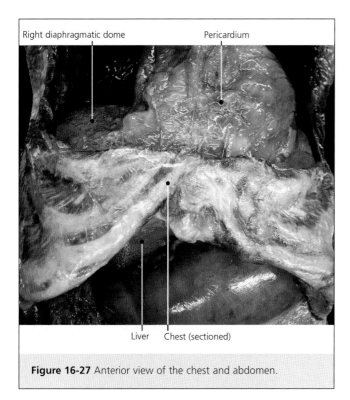

Figure 16-27 Anterior view of the chest and abdomen.

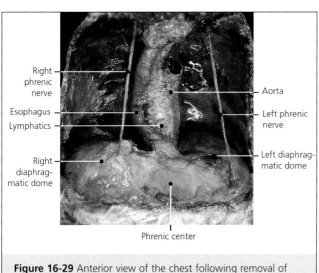

Figure 16-29 Anterior view of the chest following removal of the lungs and heart.

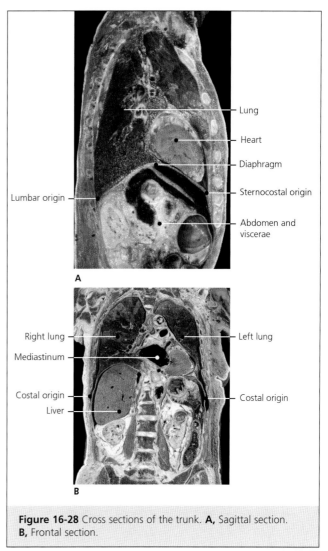

Figure 16-28 Cross sections of the trunk. **A,** Sagittal section. **B,** Frontal section.

The transversus abdominis muscle decreases the diameter of the abdomen when increased by cavitary pressure. It assists in forced expiration required for coughing, sneezing, defecation, and voiding. It is innervated by the eighth through thirteenth intercostal nerves, the iliohypogastric nerve, and the ilioinguinal nerve.

Posterior

The posterior wall of the abdomen is formed by two muscles, the quadratus lumborum and the psoas (Figure 16-26). The psoas is described in chapter 22.

The quadratus lumborum is a wide muscle that is organized into two layers. It arises from the costal processes of L1 through L4 and from the posteromedial third of the internal lip of the iliac crest. Some fibers that arise from the costal processes attach to the twelfth rib; those that originate from the iliac crest attach to the costal processes of the lumbar vertebrae. The functions of the quadratus lumborum include extension and ipsilateral bending of the trunk. It is innervated by the subcostal nerve and the lumbar plexus.

THE DIAPHRAGM

The diaphragm is a flat muscle that separates the thoracic and the abdominal cavities. It has two domes that project into the chest cavity. The right side is the larger, containing the liver in its concavity (Figures 16-27 through 16-30).

The diaphragm has three areas of origin (Figures 16-28, 16-30, and 16-31). The fibers of the sternal portion arise from the posterior aspect of the

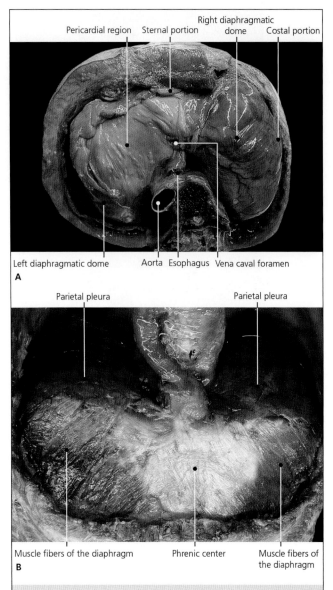

Figure 16-30 A, Superior view of the diaphragm. **B,** The parietal pleura has been removed from the diaphragm to expose the muscle fibers and central tendon.

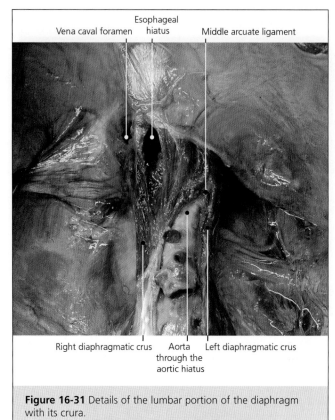

Figure 16-31 Details of the lumbar portion of the diaphragm with its crura.

xyphoid process (Figures 16-28 and 16-30). The costal portion arises from the posterior aspect and costal cartilages of the seventh through twelfth ribs. Finally, a lumbar portion arises from the sides of the vertebral bodies via the diaphragmatic crus.

The right diaphragmatic crus arises from vertebrae L1 through L3 and from the corresponding intervertebral disks. The most medial fibers form an arch that blends with those of the left diaphragmatic crus. The left diaphragmatic crus arises from vertebrae L1 and L2 and the corresponding intervertebral disk. The two crura form a tendinous arch, the middle arcuate ligament, that forms a hiatus, called the aortic hiatus, through which the aorta enters the abdomen. This hiatus also allows passage of the azygos vein and the thoracic duct. The hemiazygos vein crosses directly through the left crus. Immediately lateral to both crura, the medial arcuate ligament (psoatic arch) is formed, allowing passage of the psoas major muscle together with the sympathetic chain. Further lateral, the lateral arcuate ligament (arch of the quadratus lumborum) is formed, for the passage of the quadratus lumborum muscle.

The fibers from all the diaphragmatic origins converge toward the central tendon (phrenic center) (Figure 16-30), which constitutes the tendinous insertion. Between the three portions of the diaphragm are small "weak" spaces. Between the sternal and costal portions, the sternocostal triangle (hiatus of Larrey) is formed, where the internal thoracic artery anastomoses with the inferior epigastric artery. Between the costal and lumbar portions, the lumbocostal triangle (hiatus of Bochdaleck) is formed.

In the tendinous center, the vena caval foramen is found, where the inferior vena cava passes toward the right auricle. In the left region, the esophageal hiatus is found, where the esophagus, accompanied by the vagus nerves, passes to reach the stomach (Figure 16-31).

Contraction of the diaphragm makes the domes flatten, consequently lowering the intrathoracic pressure and expanding the lungs to produce inspiration. Relaxation of the diaphragm allows the domes to move upward, leading to expiration. The descent of the

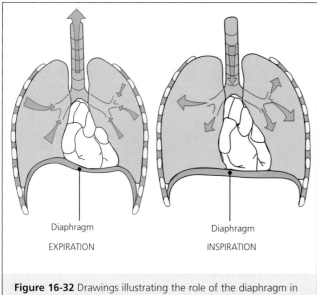

Figure 16-32 Drawings illustrating the role of the diaphragm in respiration.

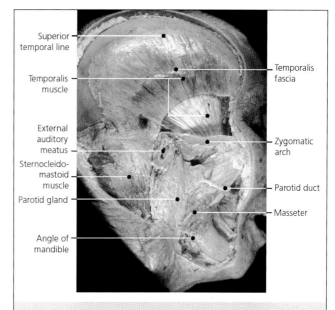

Figure 16-33 Superficial specimen of the lateral region of the face.

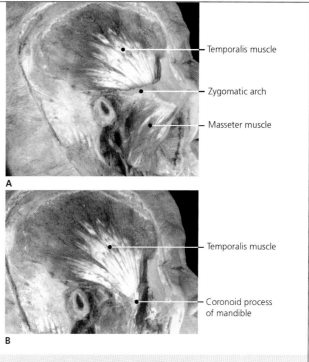

Figure 16-34 A, Specimen of the superficial muscles of mastication. **B,** Detail of the temporalis muscle after removal of the zygomatic arch.

diaphragm during inspiration also increases intra-abdominal pressure (Figure 16-32).

The diaphragm is innervated by the two phrenic nerves (which arise from spinal nerves C3 through C5), which surround the pericardium before reaching the corresponding domes of the diaphragm (Figure 16-29). The right phrenic nerve usually passes with the inferior vena cava through its foramen, reaching the abdominal surface of the diaphragm. The left phrenic nerve directly perforates the central tendon. These nerves provide motor and sensory innervation, including sensation and propioception. A myriad of abdominal conditions may induce irritation of these nerves, manifesting as motor dysfunctions (hiccup, cough, etc.).

Muscles of Mastication

In many teaching centers, the muscles of mastication are included in the study of the musculoskeletal system. For this reason, they are included here. Four muscles are included in this group: the temporalis, the masseter, the lateral pterygoid, and the medial pterygoid.

The temporalis is a flat, fan-shaped muscle (Figures 16-33 and 16-34). It is covered by a resistant temporalis fascia that extends between the superior temporal line and the zygomatic arch. This fascia is divided into two layers. The superficial layer inserts onto the lateral border, and the deep layer inserts onto the medial border of the zygomatic arch. The muscle fibers of the temporalis arise from the fossa temporalis and converge toward a powerful tendon inserted at the coronoid process of the mandible. Its function is to elevate the mandible, enabling mastication. The temporalis is innervated by the mandibular nerve (branch of the trigeminal nerve).

The masseter muscle is covered by the masseter fascia and is organized into two layers. The superficial layer arises from the anterior two thirds of the zygomatic arch, and its fibers are oriented toward the angle of the mandible. The deep layer arises from the posterior two thirds of the zygomatic arch and inserts in an oblique fashion to the angle of the mandible and the anterior adjacent area (Figures 16-33 through 16-35).

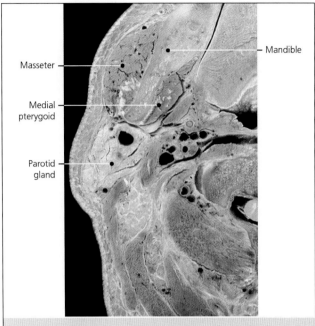

Figure 16-35 Transverse section cephalad to the angle of the mandible.

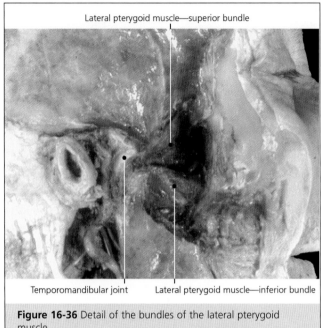

Figure 16-36 Detail of the bundles of the lateral pterygoid muscle.

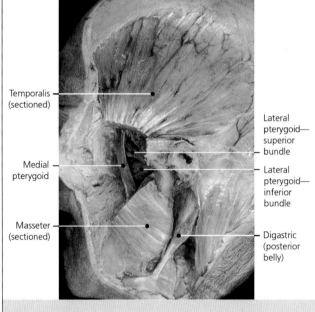

Figure 16-37 Deep layers of the muscles of mastication.

This muscle elevates the mandible, producing mastication. It is innervated by the mandibular nerve.

The lateral pterygoid muscle arises from the external aspect of the lateral lamina of the pterygoid process of the sphenoid bone and inserts by means of two bundles to the articular disk of the temporomandibular joint and to the pterygoid fossa on the condylar process of the mandible (Figures 16-36 and 16-37). This muscle aids in coordinating masticatory movements with those of the articular disk of the mandible. The simultaneous contraction of the lateral pterygoid muscles produces forward thrust of the jaw. If only one contracts, lateral translation is produced. This is accessory to the musculature that lowers the jaw. The lateral pterygoid is innervated by the mandibular nerve.

The medial pterygoid muscle arises from the pterygoid fossa and inserts in an oblique direction onto the pterygoid tuberosity on the medial aspect of the angle of the mandible (Figures 16-35 and 16-37). It contributes to mastication by coordinating the three previous muscles. It is innervated by the mandibular nerve.

Chapter 17

Neurology

Introduction
The following structures are formed from the spinal cord, in cranial to caudal order: cervical plexus, brachial plexus, intercostal nerves, and lumbosacral plexus. The brachial plexus is described in chapter 11. The lumbosacral plexus is described in chapter 23, as it relates to the pelvis and lower limbs. This chapter describes the cervical plexus and the general characteristics of the intercostal nerves.

Cervical Plexus
In the cervical region are the anterior and posterior rami of the nerves of the cervical spine. The posterior rami are directed dorsally, toward the nuchal region. The C1 nerve deserves special attention. This motor nerve is located in the triangle formed by the rectus capitis posterior and superior oblique muscle of the head, and it innervates these muscles. It also accompanies the vertebral artery. The C2 nerve (Figure 17-1) is a prominent, primarily sensory branch, called the greater occipital nerve (nerve of Arnold). C2 emerges between the posterior arch of the atlas and the lamina of the axis, crossing through the semispinalis capitis and the trapezius muscle, innervating the skin of the occipital region. The posterior branches inferior to C2 have a dorsal trajectory, innervating the paraspinal muscles and then reaching the skin over the spinous processes and erector spinae.

The cervical plexus is formed by the ventral rami of cervical nerves C1 through C4 (Figure 17-2). It is characterized by having loops formed by the anastomosis of the different segments. From the C2 vertebra a nerve emerges that bifurcates into two rami, one superior and one inferior. The superior ramus anastomoses with C1 and forms the superior root of the ansa cervicalis. Laterally, the inferior ramus anastomoses with C3 and forms the inferior root of the ansa cervicalis. A small portion of C3 anastomoses with C4. C4 also contributes to C5 for the brachial plexus. These arches are also connected with the hypoglossal nerve (which carries part of the fibers from C1) and with the vagus nerve. They are similarly connected with the superior and medial cervical ganglions.

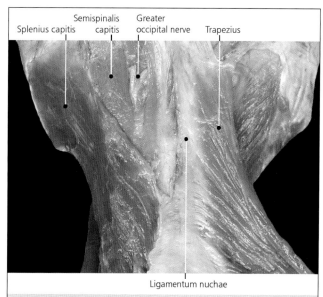

Figure 17-1 The greater occipital nerve emerging through the semispinalis capitis.

American Academy of Orthopaedic Surgeons

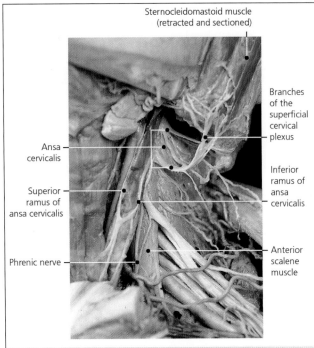

Figure 17-2 The cervical plexus. The superficial cervical plexus is observed as it exits toward the sternocleidomastoid muscle. The ansa cervicalis (C1 through C3), which innervates the infrahyoid muscles among others, is evident.

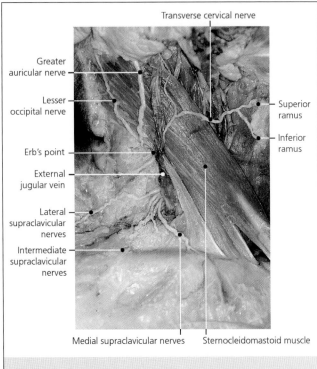

Figure 17-3 Details of the superficial cervical plexus.

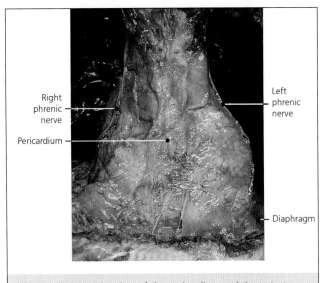

Figure 17-4 Anterior view of the pericardium and the trajectory of the phrenic nerves.

The most evident portion of the cervical plexus is the superficial one (Figure 17-3). Four important branches emerge through Erb's point (neural point), which is located in the middle of the posterior border of the sternocleidomastoid muscle. The transverse cervical nerve (C3) surrounds the sternocleidomastoid and is distributed through the anterior region of the neck, with superior rami (suprahyoid region) and inferior rami (infrahyoid region). The greater auricular nerve (C2-C3) courses obliquely toward the auricular region. The lesser occipital nerve (C2-C3) ascends toward the occipital region, following the sternocleidomastoid muscle, and anastomoses with the greater occipital nerve. Finally, the supraclavicular nerve emerges slightly distal to Erb's point. Its medial branches are distributed over the anterior chest, the sternoclavicular joint, and manubrium. The intermediate branches innervate the skin over the chest down to the fourth rib. The lateral branches travel toward the acromion and deltoid.

The phrenic nerve (Figure 17-2) is formed from the anterior ramus of C4, with contributions from C3 and C5. It is localized over the anterior scalene muscle and enters the mediastinum on both sides of the pericardium (Figure 17-4) until it reaches its corresponding hemidiaphragm; occasionally it continues to the peritoneum. The accessory phrenic nerves are contributions from C5-C6 to the phrenic nerve that travel with the subclavian nerve.

Figure 17-5 is a diagram of the cervical plexus.

Intercostal Nerves

The intercostal nerves correspond to the anterior rami of the 12 thoracic pairs, which are distributed along the intercostal grooves together with the intercostal artery and vein (Figures 17-6 and 17-7). The intercostal nerves are located between the internal and innermost intercostal muscles and are the most caudal structure of the neurovascular bundle. These nerves perforate the inter-

Neurology

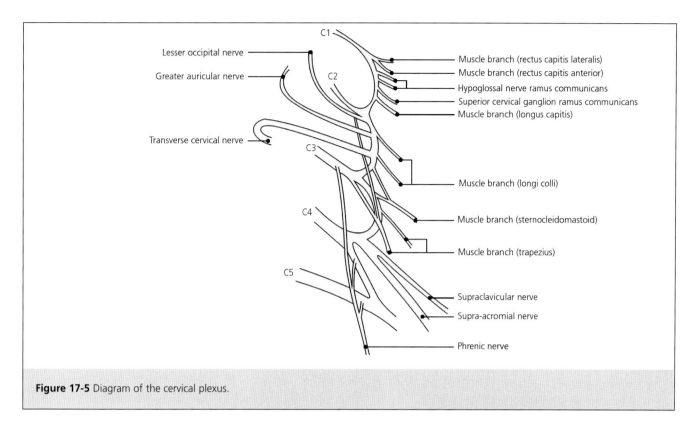

Figure 17-5 Diagram of the cervical plexus.

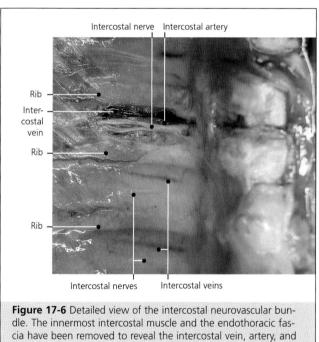

Figure 17-6 Detailed view of the intercostal neurovascular bundle. The innermost intercostal muscle and the endothoracic fascia have been removed to reveal the intercostal vein, artery, and nerve.

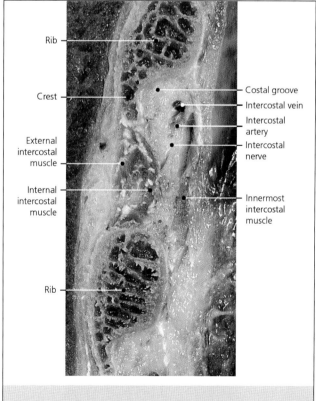

Figure 17-7 Detailed view of the intercostal space and neurovascular bundle.

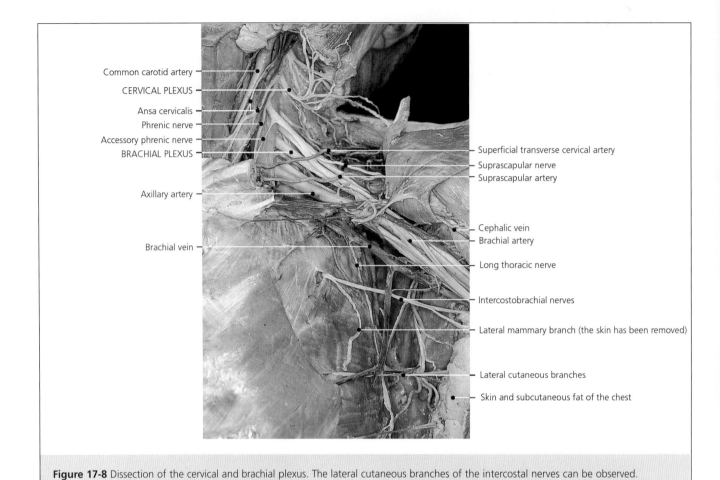

Figure 17-8 Dissection of the cervical and brachial plexus. The lateral cutaneous branches of the intercostal nerves can be observed.

costal muscles as well as other chest wall muscles through lateral cutaneous branches, which provide local sensory innervation to the skin (Figure 17-8). Between T4 and T6, these nerves extend into the mammary region and are called the lateral mammary rami.

The T2 and, occasionally, T3 nerves have special characteristics and are called intercostobrachial nerves because they are distributed into the upper extremity.

Finally, the intercostal nerves distributed beneath the twelfth rib are called the subcostal nerves.

Additional information on the innervation of the trunk can be found in chapter 5.

Chapter 18

Angiology

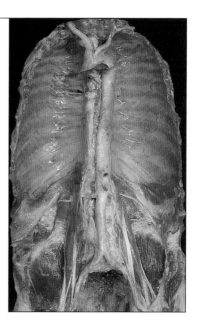

Introduction
This chapter describes the vasculature that supplies the head, neck, and trunk (superficial). Some of these vessels are described in chapter 12 (angiology of the scapular girdle and upper limb).

Arteries

The Aorta
The aorta is the main artery of the body, responsible for distributing blood from the heart to the tissues. It originates from the left ventricle and terminates into the common iliac arteries and middle sacral artery at the lower lumbar levels.

Immediately arising from the left ventricle, the ascending aorta (Figure 18-1) is short in length and is controlled by the aortic valve, consisting of three cusps. Responsible for myocardial circulation, the right and left coronary arteries originate from small foramina in the right and left cusps, respectively.

Following the short ascending portion of the aorta is the curved aortic arch (Figure 18-1) located to the left of the vertebral bodies. The aortic arch gives rise to the brachiocephalic trunk and to the left subclavian and common carotid arteries, responsible for the blood supply to the upper limbs, neck, and head, which will be described later.

Following the aortic arch at the T4 level, the descending aorta arises. It divides into the thoracic aorta (cephalad to the diaphragm) and the abdominal aorta (caudal to the diaphragm) (Figure 18-2).

The thoracic aorta is located on the left side of the vertebral bodies, on the posterior mediastinum. The bronchial, esophageal, pericardial, and mediastinal branches arise in this portion. Similarly, the posterior intercostal arteries arise from the posterolateral wall of the aorta and are distributed in the intercostal spaces (Figures 18-3 and 18-4). Other collaterals arise from the intercostal arteries. The first is the dorsal branch that goes to the dorsal aspect of the trunk to supply blood to the muscles and skin through medial and lateral cutaneous branches. From the dorsal branch, spinal branches arise that are distributed to the vertebrae and spinal cord. The posterior intercostal arteries travel in a neurovascular bundle located between the intercostal muscles. A collateral branch divides in the vicinity of the costal angle and descends to the superior border of the next caudal rib. The intercostal arteries penetrate the muscles on the lateral aspect of the chest and reach the skin through a lateral cutaneous branch. An anterior branch stems from this, running toward the pectoral and mammary region, forming the lateral mammary branches. In the anterolateral region of the chest, the posterior intercostal arteries anastomose with the internal thoracic artery (Figure 18-5).

Finally, before going through the aortic hiatus, the superior phrenic arteries branch from the anterior aspect of the aorta and distribute over the thoracic aspect of the diaphragm.

The abdominal aorta commences past the aortic hiatus and ends in its bifurcation at the L4-L5 level into the common iliac and middle sacral arteries. It localizes closer to the midline and provides multiple branches for abdominal organs. Only those branches that vascularize the walls and muscles of the abdomen are described here.

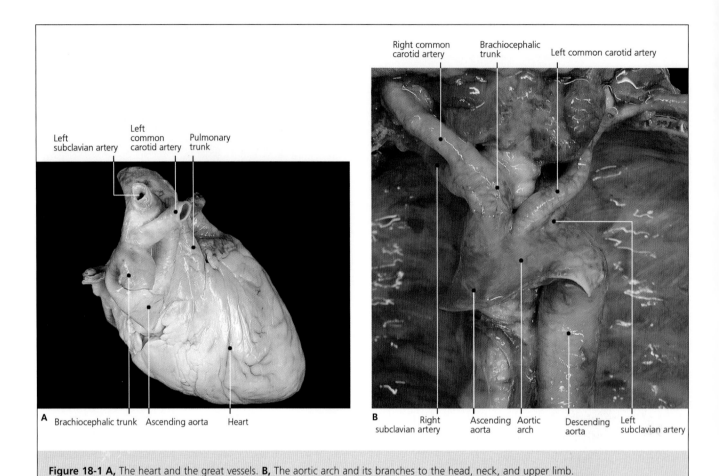

Figure 18-1 A, The heart and the great vessels. **B,** The aortic arch and its branches to the head, neck, and upper limb.

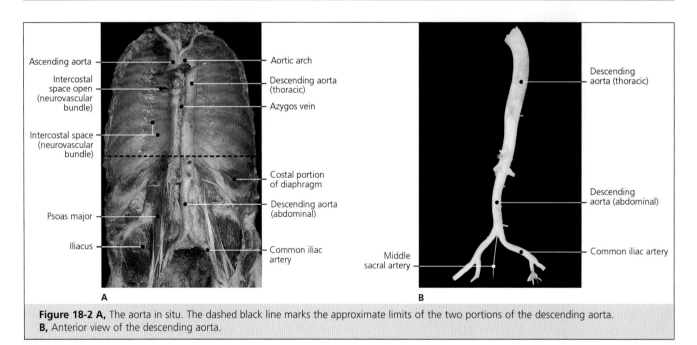

Figure 18-2 A, The aorta in situ. The dashed black line marks the approximate limits of the two portions of the descending aorta. **B,** Anterior view of the descending aorta.

The first branch of the abdominal aorta is the inferior phrenic artery, which supplies the abdominal side of the diaphragm. Under the twelfth rib lies the subcostal artery, the most distal of the intercostal arteries. Caudal to this, the lumbar arteries arise from both sides of the abdominal aorta and present a dorsal and a spinal branch, which are distributed throughout the walls of the abdomen, similar to the intercostals.

Angiology

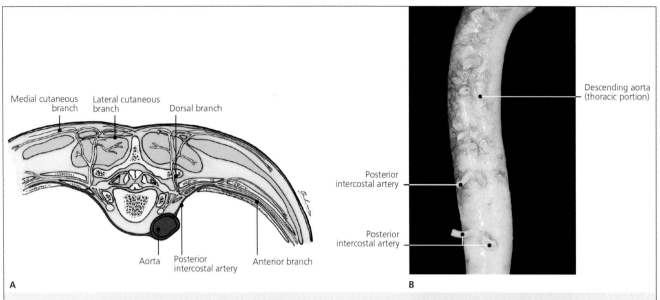

Figure 18-3 A, Diagram of a posterior intercostal artery. **B,** Posterior aspect of the aorta with the origin of the two posterior intercostal arteries for each corresponding space.

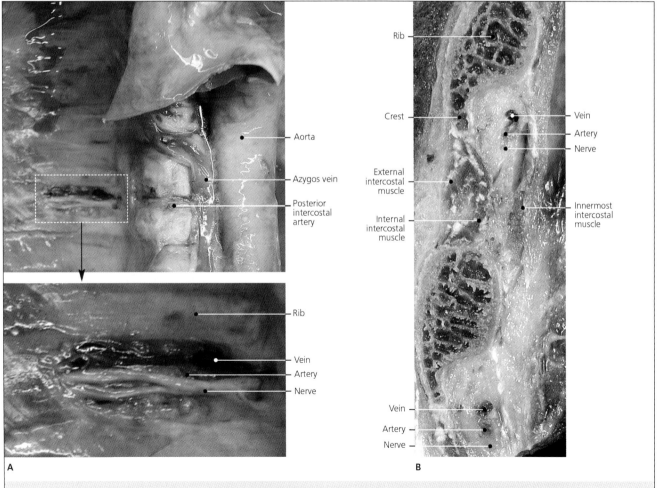

Figure 18-4 A, Dissection of the intercostal spaces. The lower panel shows a closeup of the boxed area in the upper panel. **B,** Frontal section.

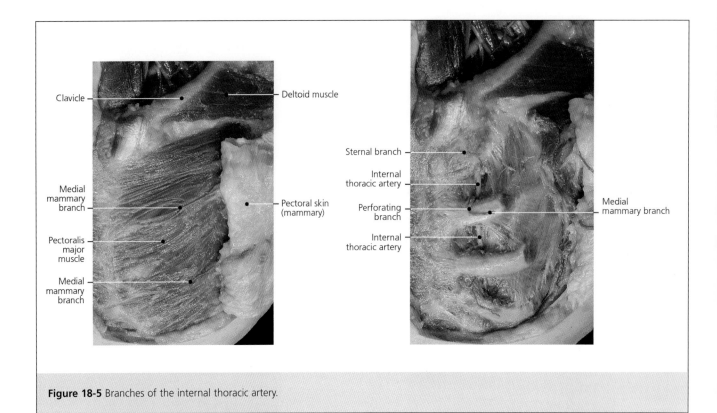

Figure 18-5 Branches of the internal thoracic artery.

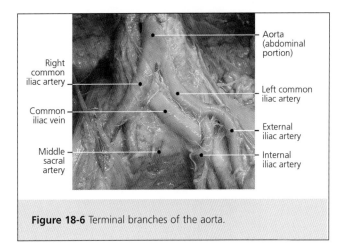

Figure 18-6 Terminal branches of the aorta.

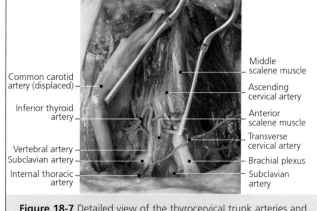

Figure 18-7 Detailed view of the thyrocervical trunk arteries and the origin of the vertebral artery.

Finally, the aorta bifurcates into two common iliac arteries at the L4 level, which further subdivide into the internal iliac artery and external iliac artery (Figure 18-6). This is described further in chapter 24. A smaller vessel, the middle sacral artery, arises as a midline extension of the aorta and is located anterior to the sacrum, ending in the coccygeal glomus. This is a small paraganglion formed by a meshwork of capillaries located at the apex of the coccyx. From the middle sacral artery, small collateral branches may arise, including the inferior lumbar arteries at the L5 level and the lateral sacral arteries, which anastomose to the internal iliac arteries.

Arteries of the Neck and Head

The arteries that supply the head, neck, scapular girdle, and superior limbs are all branches of the subclavian and common carotid arteries. The branches of the subclavian artery related to the scapular girdle and superior limb are further described in chapter 12.

The first two branches of the subclavian artery (Figure 18-7), the vertebral artery and internal thoracic artery, supply the axial structures. The vertebral artery (Figure 18-8) branches from the prescalene portion of the subclavian artery and is deeply directed until it reaches the spinal column, forming the prevertebral portion. The transverse portion of this artery pene-

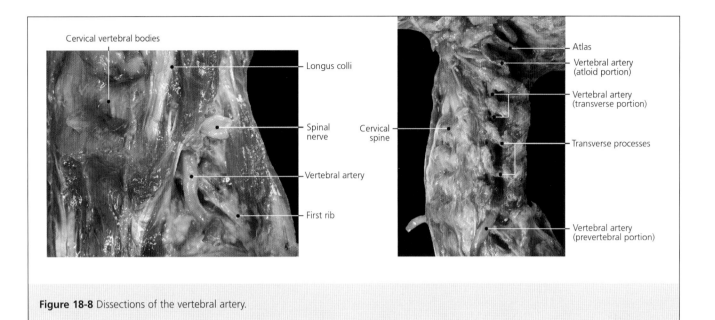

Figure 18-8 Dissections of the vertebral artery.

trates and climbs through the vertebral artery foramina from C6 to C1. The last portion is related to the atlas and, for this reason, is called the atloid portion. It presents two curves, the first to circumvent the lateral mass of the atlas and a second to penetrate into the skull. The vertebral artery forms part of the arterial blood supply to the brain.

The thyrocervical trunk branches off the subclavian artery (Figure 18-7) and provides four branches: the transverse cervical, suprascapular, inferior thyroid, and ascending cervical arteries. The first two are described in chapter 12. The inferior thyroid artery is directed toward the inferior portion of the thyroid gland, behind the common carotid artery. The ascending cervical artery originates from the thyrocervical trunk or from the inferior thyroid artery. It ascends anterior to the anterior scalene muscle, medial to the phrenic nerve, reaching the base of the skull. Another artery that contributes to the vascularization of the trunk is the transverse cervical artery. Its superficial, superior, and inferior branches are distributed over the superficial muscles of the trunk (trapezius, scalenes, and levator scapulae).

Another important collateral branch that originates from the scalene portion of the subclavian artery and supplies the neck and chest is the costocervical trunk. This trunk branches into the deep cervical artery and the highest intercostal artery. The deep cervical artery follows a deep path toward T1 and then ascends toward the nuchal region by following the semispinalis capitis. The highest intercostal artery enters the chest deep to the internal border of the first rib and distributes into the first and second intercostal spaces as the first and second posterior intercostal arteries; they have characteristics similar to the previously described intercostal arteries.

The common carotid artery arises from the brachiocephalic trunk on the right side and directly from the aortic arch on the left side (Figure 18-9). It ascends on both sides of the trachea and does not branch until it bifurcates into internal and external carotid arteries.

The internal carotid artery does not branch in the neck and ascends vertically until it penetrates the carotid canal on the base of the skull.

The external carotid artery provides several branches to the organs and muscles of the neck and head. They are described from caudal to cephalad (Figure 18-9).

The superior thyroid artery gives rise to the following branches: the sternocleidomastoid, infrahyoid, cricothyroid, glandular, and superior laryngeal and ascending pharyngeal arteries.

The linguofacial trunk is the origin of the lingual and facial arteries. They may also arise independently from the external carotid artery. The lingual branch is distributed to the tongue and suprahyoid muscles. The facial branch forms an arch close to the angle of the mandible anterior to the masseter muscle and branches off obliquely on the face until it reaches the orbit and the frontal bone.

The occipital artery originates on the posterior aspect of the external carotid artery and ascends toward the mastoid process. Medial to this, it gives rise to the mastoid, auricular, and sternocleidomastoid branches. It courses to the occipital bone, providing occipital branches and a descending branch to the occiput.

The posterior auricular artery also originates from the posterior aspect of the external carotid artery and provides the auricular, parathyroid, mastoid, stylomastoid, and posterior tympanic branches.

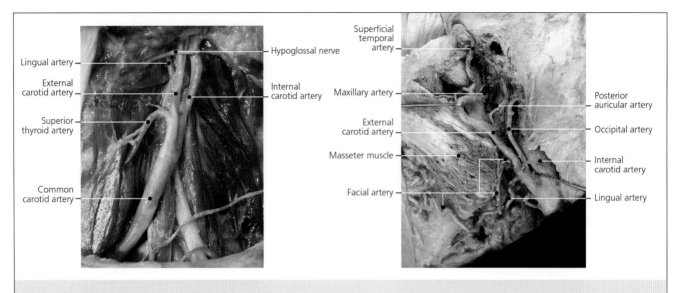

Figure 18-9 Dissection of the common carotid and external carotid arteries.

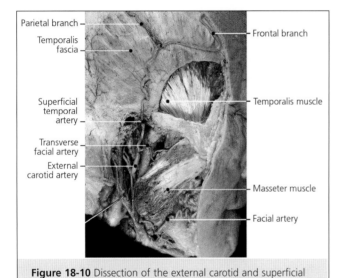

Figure 18-10 Dissection of the external carotid and superficial temporal arteries.

The external carotid artery bifurcates into two terminal branches, the superficial temporal and maxillary arteries.

The superficial temporal artery (Figure 18-10) provides small anterior auricular and parathyroid branches for the parathyroid gland. The artery's main collaterals include the transverse facial artery and the zygomatico-orbital artery. The former lies deep to the parotid gland and runs underneath the zygomatic arch until it reaches the cheek. The latter follows a similar path until it reaches the lateral orbital border. It also provides a collateral for the temporalis muscle as the medial temporal artery. The superficial temporal artery bifurcates into the frontal and parietal terminal branches, distributing and anastomosing in this region.

The maxillary artery is distributed on the deep side of the maxilla and mandible, toward the teeth and, through a medial meningeal artery, toward the brain.

Arteries of the Chest and Abdomen

The collateral branches of the aorta that supply the chest and abdominal walls are described above.

The intercostal arteries anastomose with the internal thoracic artery (Figure 18-11). This is a collateral branching of the prescalene portion of the subclavian artery at the middle third of the clavicle. It runs caudally on the internal side of the chest wall, behind the costal cartilages, on both sides of the sternum. It provides branches to the mediastinum, in particular the pericardiophrenic artery, which follows the phrenic nerve and distributes on the pericardium and the superior aspect of the diaphragm. In addition, it provides sternal branches (medial) and anterior intercostal branches (laterals), located in the intercostal spaces, anastomosing with the terminal portion of the posterior intercostal arteries. Perforating and medial mammary branches emerge from the intercostal spaces between the first through sixth ribs. The internal thoracic artery bifurcates at the level of the diaphragm into the musculophrenic and superior epigastric arteries. The first follows the costal arch, providing anterior intercostal branches below the seventh rib. The second enters the abdominal cavity through the sternocostal angle formed by the sternal and costal origins of the diaphragm. Once in the abdomen, it is named the superior epigastric artery. It anastomoses with the inferior epigastric artery (Figure 18-11).

The inferior epigastric artery is a collateral branch of the external iliac artery formed posterior to the

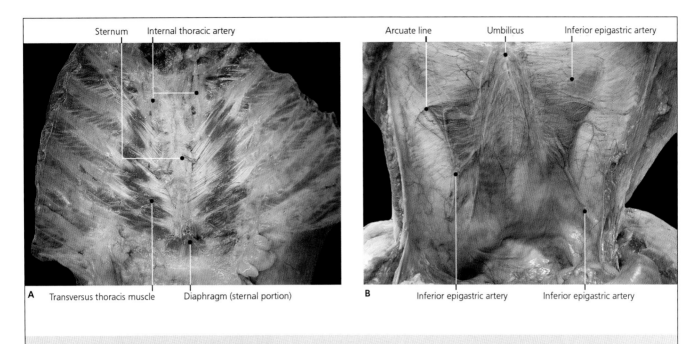

Figure 18-11 **A,** Posterior wall of the chest. **B,** Posterior wall of the abdomen.

inguinal ligament. It ascends on the posterior aspect of the rectus abdominis muscle, passing by the arcuate line forming the lateral umbilical fold and finally anastomosing with the superior epigastric artery.

VEINS

Veins of the Head and Neck

The chief veins that remove deoxygenated blood from the head and neck are the jugular veins.

The deep veins of the head and neck receive the same name as the arteries they accompany and drain into the subclavian vein or into the internal jugular vein (Figure 18-12). The veins that accompany collateral arteries arising from the subclavian artery drain into the subclavian vein. In contradistinction, veins that accompany collateral branches of the carotid arteries drain into the internal jugular vein.

The internal jugular vein stems from the inferior bulb and the subclavian vein at the level of the first rib. Both form a venous arch and join on the corresponding brachiocephalic vein, which carries blood to the superior vena cava and into the right auricle.

The two most superficial veins in this region include the external jugular and anterior jugular veins. The external jugular collects blood from the transverse cervical, suprascapular (also directly from the subclavian vein), and occipital veins, which accompany the homonymous arteries, and drain into the subclavian vein. The second originates at the level of the hyoid and drains into the subclavian vein, close to the venous arch.

Veins of the Trunk and Abdomen

Blood derived from the anterior part of the chest and abdomen drains into the internal thoracic and inferior epigastric veins, homonymous with the corresponding arteries in this region. The internal thoracic veins drain into the brachiocephalic vein, which has branches similar to those of its arterial counterpart. The inferior epigastric vein drains into the external iliac vein and presents identical distribution as the corresponding artery.

The superior chest veins differ somewhat from the arteries. The highest intercostal vein receives blood only from the first intercostal space and generally drains into the corresponding brachiocephalic vein. On the left side, a left superior intercostal vein collects blood from the second and third intercostal spaces, and occasionally from the fourth, and drains into the left brachiocephalic vein.

The azygos vein system is responsible for the remaining drainage from the chest and abdominal walls (Figures 18-13 and 18-14). The lumbar veins to L3 and L4 drain directly into the inferior vena cava, while the L1 and L2 lumbar veins drain into an ascending lumbar vein on both sides of the spine. This ascending lumbar vein continues on the right side with the azygos vein and, on the left side, with the hemiazygos vein.

The right subcostal vein is responsible for the venous drainage caudal to the twelfth rib. The posterior intercostal veins, which are responsible for the drainage of the right intercostal spaces from ribs twelve through two, drain into the hemiazygos vein.

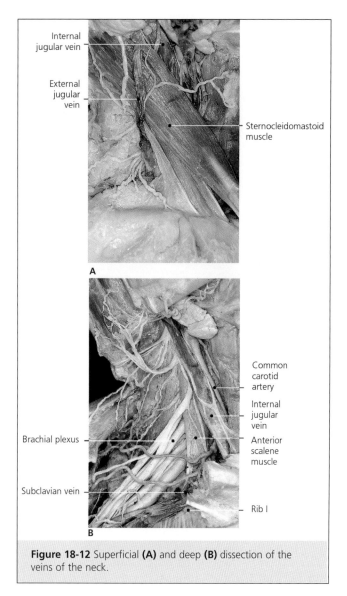

Figure 18-12 Superficial **(A)** and deep **(B)** dissection of the veins of the neck.

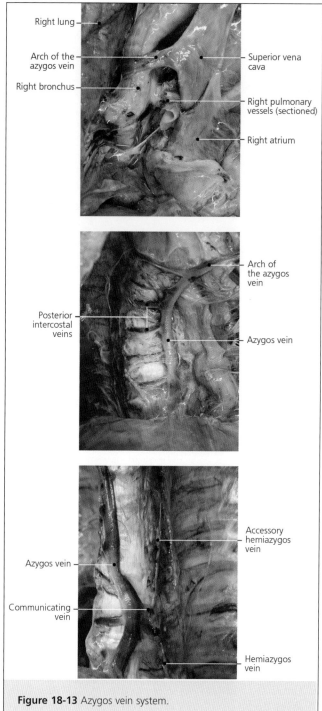

Figure 18-13 Azygos vein system.

The hemiazygos vein drains to the azygos vein at the T9 or T10 level. The posterior intercostal veins from the eighth through fourth intercostal spaces drain to the accessory hemiazygos vein. Drainage of the accessory hemiazygos vein is variable: It may drain directly to the azygos vein or indirectly if it blends with the hemiazygos vein, or it may drain directly to the left brachiocephalic vein if it anastomoses with the left superior intercostals and highest intercostal veins.

The azygos vein continues with the arch of the azygos vein and drains to the superior vena cava and the right auricle.

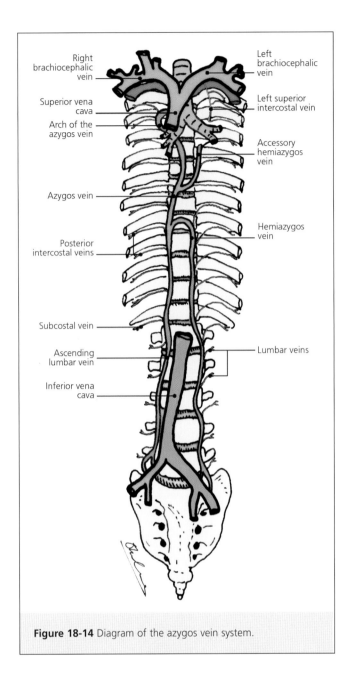

Figure 18-14 Diagram of the azygos vein system.

Section 4

Pelvic Girdle and Lower Limb

Chapter 19

Topographic Regions and Surface Anatomy

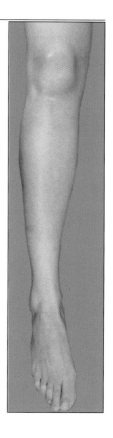

The pelvic girdle and lower limb are intimately related with the spinal column. Thus, any limb-length difference or anomaly in the hip can affect the spine and alter gait.

The dorsal and proximal region of the lower limb is called the buttock or gluteal region (Figure 19-1). This elevation is caused by the gluteus maximus muscle, which partially covers the gluteus medius and minimus muscles, except laterally. The gluteus medius can be palpated on the gluteal side of the ilium, whereas the gluteus minimus is completely covered by the medius. The two gluteal regions are separated by the intergluteal cleft. The inferior limit of the buttocks is marked by the gluteal fold.

Deep and lateral to the gluteal region is the coxal, or hip, region. Pain caused by the hip joint is often felt in the groin.

In the thigh, or femoral region (Figure 19-2), anterior and posterior regions are distinguished. Proximally and medially in the anterior femoral region is the topographic area called the femoral triangle, or Scarpa's triangle. This triangle is bordered laterally by the sartorius muscle, medially by the adductor longus, and proximally by the inguinal ligament. It contains the femoral artery, vein, and nerve, which is where the femoral pulse is felt.

The muscle bulges of the thigh allow its subdi-

vision into three compartments: the anterior, with the hip flexor and knee extensor muscles (with the exception of the sartorius, which is a knee flexor); the medial or adductor muscles; and the posterior femoral muscles, which act as knee flexors and hip extensors.

The knee region is divided topographically into anterior and posterior regions. The posterior region contains the topographic area known as the popliteal

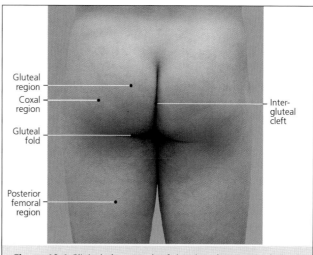

Figure 19-1 Clinical photograph of the gluteal region, with various features labeled.

American Academy of Orthopaedic Surgeons

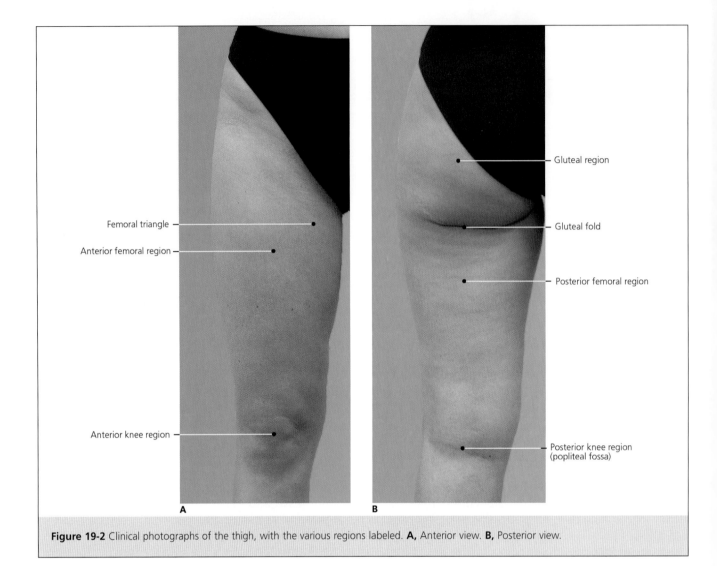

Figure 19-2 Clinical photographs of the thigh, with the various regions labeled. **A,** Anterior view. **B,** Posterior view.

fossa, which is delimited distally by the two heads of the gastrocnemius and proximally by the hamstrings and the biceps femoris. The popliteal fossa contains the popliteal artery and vein and the sciatic nerve, which divides within the fossa into the common peroneal and the tibial nerves.

The leg, called the crural region (Figure 19-3), is similarly subdivided into an anterior region and a posterior region. Within the posterior region is the sural region (the calf), where the two heads of the gastrocnemius and the soleus can be observed.

The ankle is called the talocrural region. It also is divided into anterior and posterior parts. In the posterior talocrural region are clefts through which course the tendons of the extrinsic muscles of the foot; these are the medial and lateral retromalleolar regions. The posterior tibial neurovascular bundle passes through the medial retromalleolar region. The malleoli serve as a fulcrum for the tendons of the long extrinsic muscles.

Finally, the foot (Figure 19-4) can be divided into the tarsal region, which includes the heel, or calcaneal region; and the plantar and dorsal regions of the foot. The foot has medial (tibial) and lateral (peroneal) borders. In the sole of the foot is the plantar arch or dome. The area including the metatarsal bones is the metatarsal region. In the normal foot, the heel, the lateral border, and the metatarsals touch the ground when the person is standing; the medial region is the plantar arch. The condition where the plantar arch is absent is called a flat foot; when the arch is extreme, it is called a cavus foot.

The five digits, or toes, of the foot are numbered from medial to lateral. The first toe is bigger than the others and is called the great toe or hallux. Each toe has a plantar and a dorsal aspect.

TOPOGRAPHIC REGIONS AND SURFACE ANATOMY

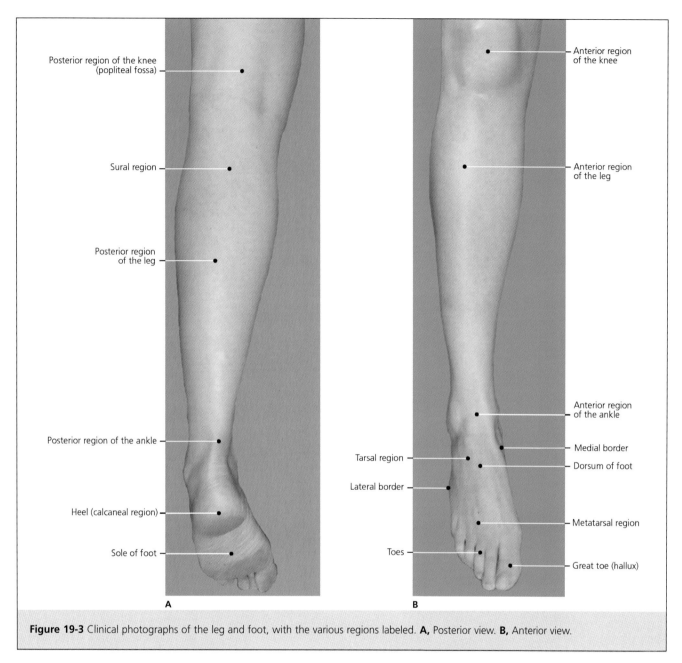

Figure 19-3 Clinical photographs of the leg and foot, with the various regions labeled. **A,** Posterior view. **B,** Anterior view.

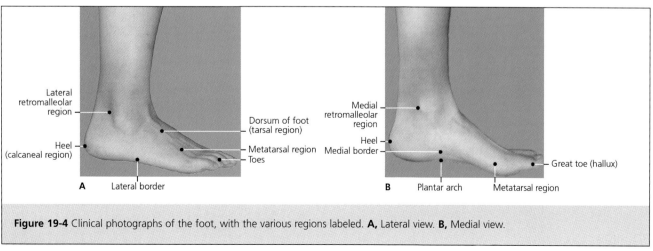

Figure 19-4 Clinical photographs of the foot, with the various regions labeled. **A,** Lateral view. **B,** Medial view.

American Academy of Orthopaedic Surgeons

Chapter 20

Osteology

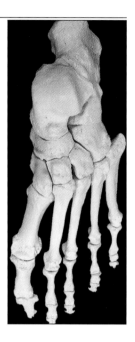

Introduction
The skeleton of the pelvic girdle and lower extremity includes the following bones: coxal or innominate, femur, patella, tibia, fibula, tarsal bones (talus; calcaneus; navicular; cuboid; and medial, intermediate, and lateral cuneiforms), metatarsals, phalanges, and sesamoids.

The pelvic girdle, like the scapular girdle, joins the free portion of the lower extremity to the axial skeleton, the vertebral column. It is composed principally of the coxal bones.

The pelvis (Figure 20-1) is formed by the innominate bone and the sacrum. The region of the pelvis containing the rectum and genitourinary viscera and attachments for the pelvic floor muscles is known as the pelvis minor, and the region containing the viscera and attachments for the abdominal muscles is known as the pelvis major. The border between the minor and major pelvis is formed by the terminal line. There are two openings in the pelvis: the superior opening, located above the line, and the inferior opening, formed by the coccyx, the pubic arch, and the sacrotuberous ligament.

Coxal (Innominate) Bone
The coxal bone is formed by the union of the ilium, the pubis, and the ischium (Figure 20-2). These three bones have their separate ossification centers, and they join to form the acetabulum. Although this bone has been named the coxal bone, the anatomic details refer to each of these three bones. The anterior superior iliac spine and the pubis are in the same frontal plane, and the symphyseal aspect of the pubis is in the midsagittal plane.

The ilium articulates with the sacrum to form the sacroiliac joint. The two pubic bones join at the symphysis pubis. Below the symphysis is the pubic arch. The acetabulum articulates with the femoral head to form the coxofemoral, or hip, joint. This is described in chapter 21.

The coxal bone contains a foramen formed by the pubic and ischial bones called the obturator foramen, although some authors call it the obturator. A foramen is really incapable of obturating, but rather it is obturated by a series of structures, such as the obturator muscles and the obturator artery, and it is covered by the obturator membrane.

The acetabulum is formed by the three pelvic bones. The acetabulum is surrounded by the acetabular rim, and it contains the femoral head, forming the coxofemoral joint. The concave articular surface inside the cavity is normally covered with hyaline cartilage, which is the articular area. This surface and the acetabular rim are open in the inferior aspect, forming the acetabular notch. In the depths of the acetabulum and bounded by the articular lunate surface is the acetabular fossa, a nonarticular portion occupied by the ligamentum teres and by fatty tissue, the pulvinar, which is described in chapter 21.

Ilium
The ilium forms the superior aspect of the coxal bone (Figure 20-3). Its body is located near the acetabulum, just above the supra-acetabular notch. The most important part of this bone is the iliac wing, crowned by the iliac crest. In this crest, there is an internal border where the transverse abdomi-

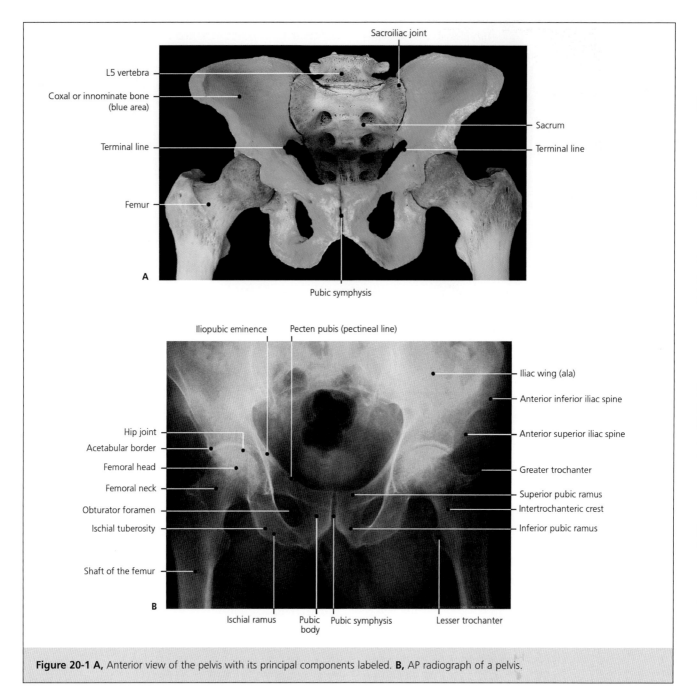

Figure 20-1 A, Anterior view of the pelvis with its principal components labeled. **B,** AP radiograph of a pelvis.

nal muscle inserts, an intermediate line for the insertion of the internal oblique muscle of the abdomen, and an external border where the external oblique muscle and the latissimus dorsi insert (Figure 20-4). The anterolateral region of the crest has a prominence, known as the iliac tubercle.

On the crest are four spines, two anterior and two posterior: the anterior superior iliac spine (insertion of the sartorius), the anterior inferior iliac spine (insertion of the rectus femoris), the posterior superior iliac spine, and the posterior inferior iliac spine.

The iliac wing is flat and has two faces, the internal and external. The external face is called the gluteal face because it serves as the origin of the gluteal muscles. In this face one can distinguish three lines—the anterior, posterior, and inferior gluteal lines. The anterior gluteal line starts close to the iliac tubercle, the inferior gluteal line starts immediately above the anterior superior iliac spine, and the posterior gluteal line starts close to the posterior superior iliac spine. The gluteus minimus originates between the anterior and inferior lines, the gluteus medius originates between the anterior and posterior lines, and the gluteus maximus originates behind the posterior gluteal line. The internal face is the iliac fossa, where the iliac muscle originates. In the posterior aspect of this face is the

Osteology

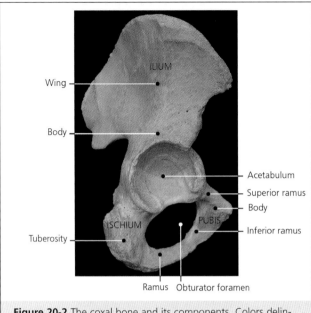

Figure 20-2 The coxal bone and its components. Colors delineate different regions.

articular zone for the sacrum, called the sacropelvic face. In it, one can see the auricular surface of the ilium, with an ear shape that articulates with a similar one in the sacrum to form the sacroiliac joint. This joint is reinforced by numerous sacroiliac ligaments that produce markings around the auricular surface, the iliac tuberosity. Finally, in the internal face of the ilium is the arcuate line, which forms part of the terminal line that separates the major and minor pelvis.

Pubis

The zone where the ilium and the pubis join is slightly raised and is called the iliopubic eminence (Figures 20-3 and 20-5). The pubic body is close to the symphysis, and from it arise two rami, one superior and one inferior, which form the margins of the obturator foramen. The superior ramus reaches the iliopubic eminence, and in its superior aspect is the pectineal crest where the pectineus muscle takes its origin. Near it, a small eminence is sometimes found, which serves as origin for the

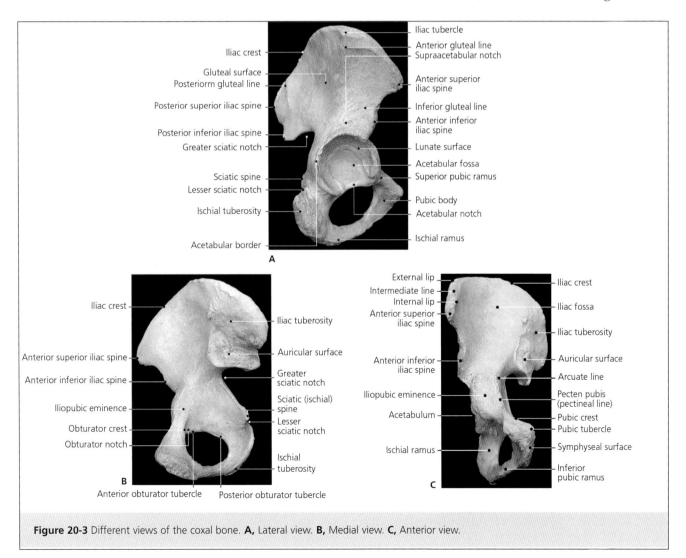

Figure 20-3 Different views of the coxal bone. **A**, Lateral view. **B**, Medial view. **C**, Anterior view.

American Academy of Orthopaedic Surgeons

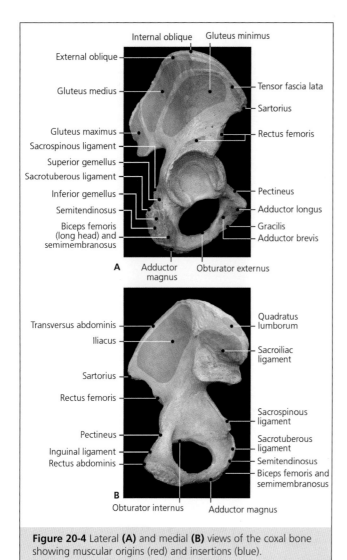

Figure 20-4 Lateral (A) and medial (B) views of the coxal bone showing muscular origins (red) and insertions (blue).

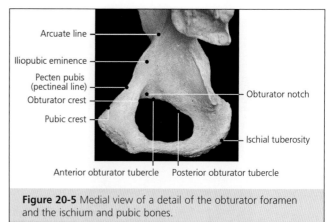

Figure 20-5 Medial view of a detail of the obturator foramen and the ischium and pubic bones.

Ischium

The ischium forms the most posterior and inferior part of the coxal bone (Figures 20-3 through 20-5). Its body is located below the acetabulum and describes an angle that forms the ischial ramus, which joins the inferior pubic ramus. In the angle of the ischium are strong impressions (or markings) of the origins of the ischiotibial and biceps femoris muscles as well as that of the adductor magnus: this is the ischial tuberosity. In its posterior region is the ischial (sciatic) spine. Ischial clearly comes from ischium, although the term *sciatic* is used as a synonym. Nevertheless, it is correct to use the adjective *ischial* to refer to the sciatic nerve.

The ischial spine is located between two notches, the greater sciatic notch above it and the lesser sciatic notch below it. These two notches become foramina delineated by the sacrotuberous and the sacrospinous ligaments, as described in chapter 21.

Femur

The femur is a robust long bone that forms the thigh skeleton (Figure 20-6). Proximally, it articulates with the acetabulum of the pelvis and forms the hip joint. Distally, it articulates with the patella and the tibia to form the knee joint.

To orient the femur and place it in anatomic position, the condyles should be distal and located in the same horizontal plane. The greater trochanter is lateral and the patellar surface is anterior.

The femoral head has a spherical shape and is covered with hyaline cartilage. In its center, facing the acetabulum, is the fovea, which contains the round ligament. The head is connected to the femoral neck and then with the diaphysis. The neck and the shaft form a neck-shaft angle of about 125° to 130° in the adult (greater in children and smaller in the elderly) (Figure 20-7).

It is important to name two important prominences: the greater trochanter, which is subcutaneous and located laterally, and the lesser trochanter, which is

psoas minor muscle. In the inferior aspect of the superior pubic ramus, the obturator crest runs from the acetabulum to the pubic tubercle; in it inserts the pubofemoral ligament. Medial to this point, near the body of the pubis, is the obturator groove, through which the obturator neurovascular bundle courses. The pubic tubercle (pubic spine) is located in the superior aspect of the pubic body and serves as the insertion of the inguinal ligament. Medially is the pubic crest, where the rectus abdominis and the pyramidal muscles insert. The pubic body and the inferior pubic ramus serve as the origin of the majority of the adductor muscles. The inferior ramus joins its homonym, the ischium. The anterior and posterior obturator tubercles also can be found in the pubis, although the latter is sometimes absent.

The two pubic bones articulate via a symphysis. Below that is the pubic arch and the subpubic angle (in men, 75°; in women, 90° to 100°).

Osteology

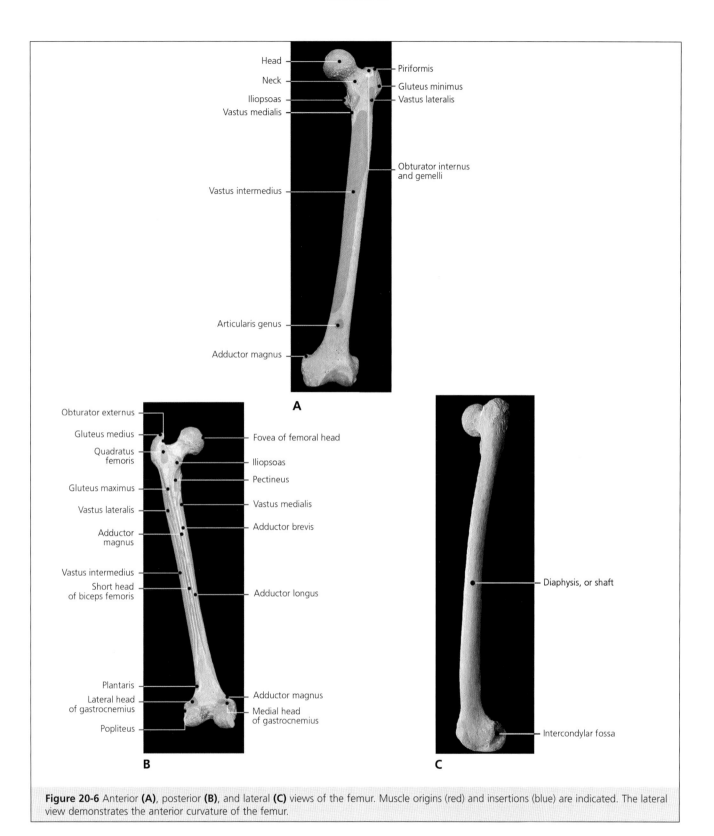

Figure 20-6 Anterior (**A**), posterior (**B**), and lateral (**C**) views of the femur. Muscle origins (red) and insertions (blue) are indicated. The lateral view demonstrates the anterior curvature of the femur.

posteromedial (Figure 20-8). The former serves as the attachment for the gluteal musculature and the latter for the iliopsoas. In the posteromedial region of the greater trochanter is the trochanteric (piriformis) fossa, where some muscles of the pelvitrochanteric group insert. Between the trochanters and posteriorly is the intertrochanteric crest and the quadrate tubercle, a blunt elevation in the crest. Anteriorly, one can find the intertrochanteric line.

The diaphysis is long, is slightly convex, and has great

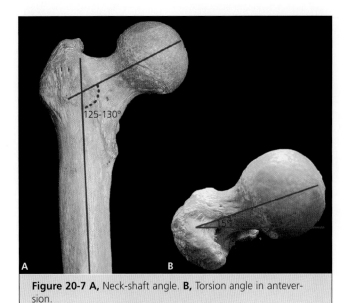

Figure 20-7 A, Neck-shaft angle. **B,** Torsion angle in anteversion.

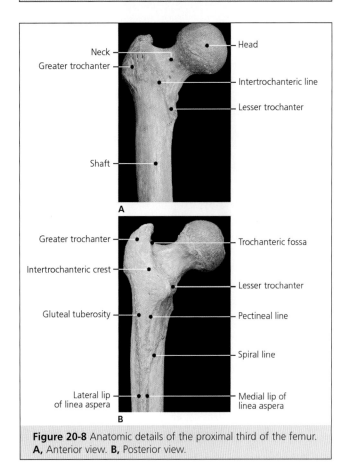

Figure 20-8 Anatomic details of the proximal third of the femur. **A,** Anterior view. **B,** Posterior view.

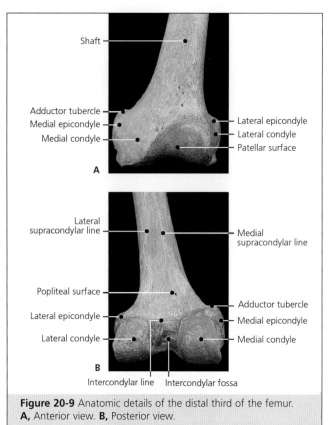

Figure 20-9 Anatomic details of the distal third of the femur. **A,** Anterior view. **B,** Posterior view.

cortices, which give it its robustness. It forms a 10° angle with a vertical line passing through the head of the femur and a torsion angle of 15° anteversion (Figure 20-7). In its anterior aspect, there are no notable details, but in the posterior aspect the linea aspera can be seen. This provides the insertion for the vastus medialis and lateralis, the adductors, and the short head of the biceps femoris. In the linea aspera, there are a medial and a lateral lip. The medial lip continues proximally as the spiral line, and above it is the pectineal line that lies between the lesser trochanter and the linea aspera and offers insertion to the pectineus muscle. Proximally, the lateral lip continues as the greater trochanter, where the gluteus maximus inserts. Sometimes there is an elevation here, known as the third trochanter.

Distally, the linea aspera opens into the medial supracondylar line, which reaches the adductor tubercle and the lateral supracondylar line (Figure 20-9). Between the two is the flat, triangular popliteal surface. The distal femoral epiphysis is characterized by the two great condyles, one medial and one lateral. Over each condyle is an osseous eminence, the medial and lateral epicondyles. In the superior aspect of the medial epicondyle is the adductor tubercle, an elevation where the longitudinal (medial) head of the adductor magnus inserts. Below the lateral epicondyle is the popliteus sulcus, formed by the popliteus muscle. Sometimes, some asperities called the supracondylar tubercles (of Grüber) can be seen above the posterior aspect of the condyles, where the heads of the gastrocnemius muscle insert.

In the posterior region, between the condyles, is the intercondylar fossa; the cruciate ligaments are located in the inferior part of this fossa. Above it is the intercondylar line.

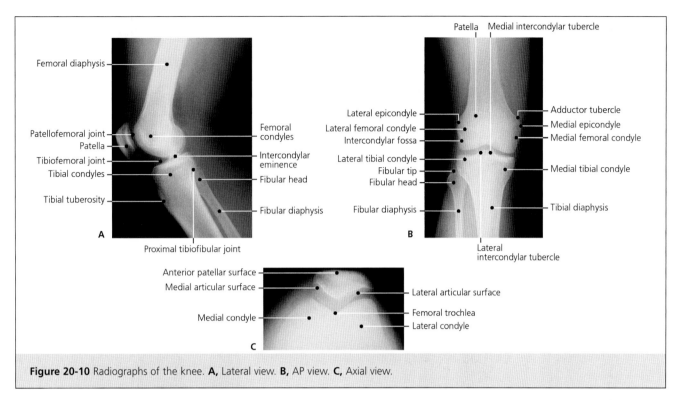

Figure 20-10 Radiographs of the knee. **A,** Lateral view. **B,** AP view. **C,** Axial view.

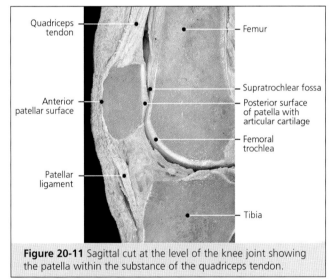

Figure 20-11 Sagittal cut at the level of the knee joint showing the patella within the substance of the quadriceps tendon.

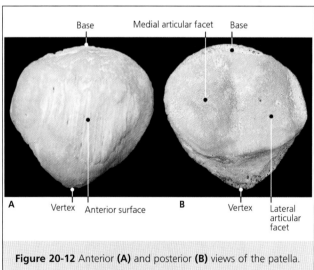

Figure 20-12 Anterior **(A)** and posterior **(B)** views of the patella.

Finally, the anterior aspect of the condyles is covered with hyaline cartilage and forms the patellar surface or femoral trochlea, which articulates with the patella when the knee is semiflexed or flexed (Figures 20-10 and 20-11). With the knee in full extension, the patella lies just above the femoral trochlea.

Patella

The patella is classified by many authors as a sesamoid bone of the quadriceps tendon (Figure 20-11). It has a shape reminiscent of a triangle (Figure 20-12), with a proximal base and a distal apex. It has a medial and a lateral border. Its dorsal (articular) surface is covered with hyaline cartilage and presents two articular facets for the femur, which are separated by a crest. The lateral facet is larger than the medial surface, which helps to orient it. The apex of its dorsal aspect is not articular but serves as the insertion of the patellar tendon.

The anterior surface is convex and is covered by the superficial fibers of the quadriceps and the patellar tendons. In the isolated bone, one can see many nutrient foramina and vertical crests.

Tibia

The tibia is a long bone that has a proximal epiphysis; a body, or diaphysis; and a distal epiphysis (Fig-

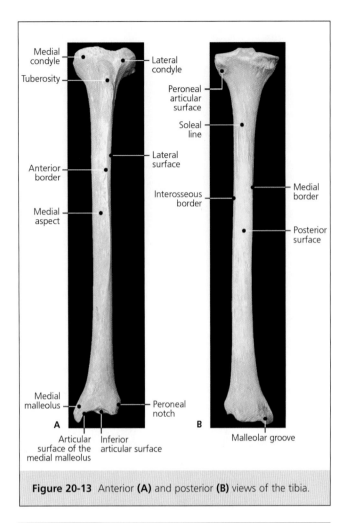

Figure 20-13 Anterior (A) and posterior (B) views of the tibia.

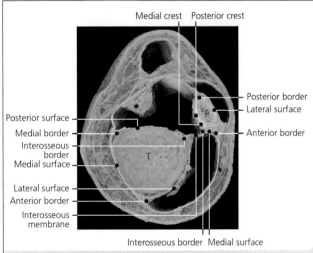

Figure 20-15 Transverse section of the leg with emptied muscular compartments. Observe the surfaces and borders of the tibia (T) and fibula (F), as well as the interosseous membrane that joins them.

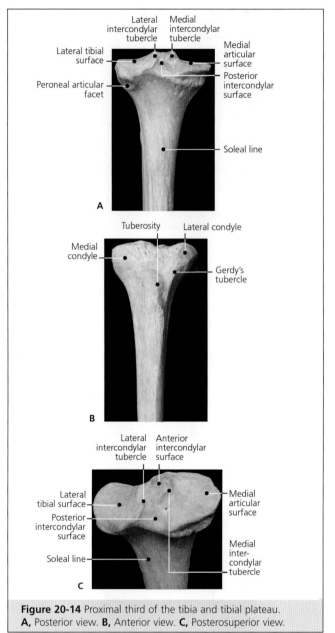

Figure 20-14 Proximal third of the tibia and tibial plateau. **A,** Posterior view. **B,** Anterior view. **C,** Posterosuperior view.

ure 20-13). With the fibula, it forms the skeleton of the leg, where it has a medial location. It articulates with the femur proximally and with the talus distally. It also articulates with the fibula in the proximal and distal tibiofibular joints.

The proximal epiphysis contains two osseous eminences, the medial and lateral condyles (Figure 20-14). The superior surface is the articular surface or tibial plateau, with two flat surfaces, the facets, covered with hyaline cartilage; these, along with the menisci, form the articular surface for the femoral condyles. Between the condyles is the intercondylar eminence, or tibial spine, which has two intercondylar tubercles—one medial and one lateral.

In front of the intercondylar eminence is an area where the anterior cruciate ligament inserts, the anterior intercondylar area. Similarly, there is a posterior intercondylar area for the insertion of the posterior cruciate ligament.

OSTEOLOGY

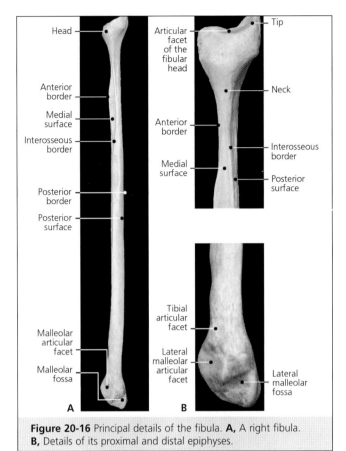

Figure 20-16 Principal details of the fibula. **A,** A right fibula. **B,** Details of its proximal and distal epiphyses.

In the anterior aspect of the tibia, below the condyles, is the tibial tuberosity, where the patellar ligament inserts. In the lateral condyle is the fibular articular surface, which with the fibula forms the proximal tibiofibular joint. Above and medial to it is Gerdy's tubercle, where the fibers of the iliotibial tract and the tibialis anterior muscle insert.

The transverse section of the shaft, or diaphysis, of the tibia is triangular (Figure 20-15); therefore, it has three borders and three surfaces: an anterior (shin), a medial, and an interosseous border; and a medial, a lateral, and a posterior surface. On the posterior surface, a curved line can be observed. This is the soleus line, located between the insertions of the soleus and popliteus muscles. Because of this, some authors call it the popliteal line. The medial surface is subcutaneous and can be palpated distally down to the medial malleolus. It has no muscular insertions.

The distal epiphysis contains the medial malleolus, a subcutaneous prominence on the medial aspect of the ankle. The tibia contains an inferior articular surface and an articular surface for the medial malleolus that articulate with the talus and forms part of the ankle joint. Laterally, the fibular notch forms part of the distal tibiofibular joint. In the posterior surface of the malleolus is the malleolar groove for the tendon

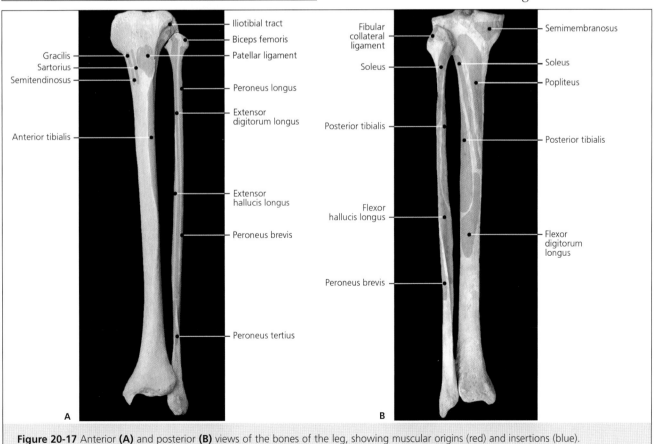

Figure 20-17 Anterior **(A)** and posterior **(B)** views of the bones of the leg, showing muscular origins (red) and insertions (blue).

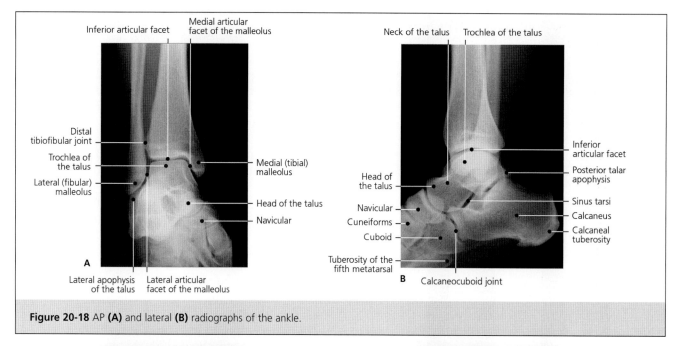

Figure 20-18 AP (A) and lateral (B) radiographs of the ankle.

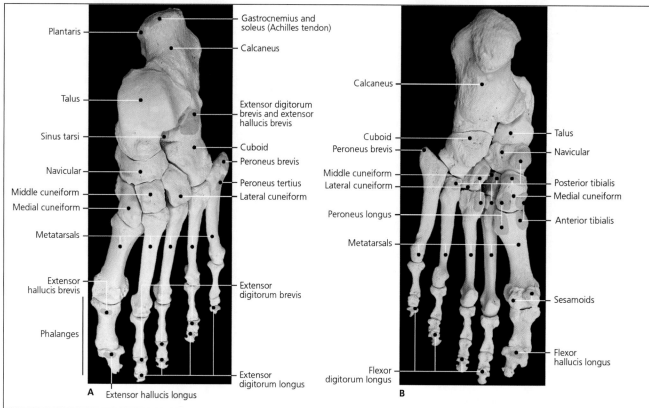

Figure 20-19 Dorsal (A) and plantar (B) views of the bones of a left foot showing the insertions (blue) of the extrinsic foot muscles. Only the origin of the extensor digitorum brevis and the extensor hallucis brevis is marked (red).

of the posterior tibial muscle. In its transverse section, the distal end of the tibia has a trapezoidal shape formed by osseous tissue and covered by a fine cortical layer. This contrasts with the thick cortices of the tibial diaphysis. Clinically, this distal epiphyseometaphyseal area is known as the tibial plafond.

Fibula

The fibula is a long bone that, along with the tibia, forms the skeleton of the leg (Figures 20-16 and 20-17). In the fibula, a proximal epiphysis, a body, and a distal epiphysis can be distinguished.

The proximal epiphysis, or head of the fibula,

OSTEOLOGY

presents a vertex (styloid apophysis) where the lateral collateral ligament of the knee and the tendon of the biceps femoris insert. It also contains a medial articular surface for articulation with the tibia to form the proximal tibiofibular joint. Its morphology and inclination are variable.

Below the head is the fibular neck, with its close relationship to the common peroneal nerve. The shaft, like that of the tibia, has a triangular shape with three surfaces and three borders (Figure 20-15). They are the medial, lateral, and posterior surfaces and the anterior, posterior, and interosseous borders. In the posterior surface is a medial crest that separates the origins of the posterior tibial and the flexor hallucis longus muscles.

The distal epiphysis is formed by the lateral malleolus, subcutaneously located in the lateral aspect of the ankle (Figure 20-18). It has an articular surface of the lateral malleolus for the talus in the ankle joint. It also has an articular surface for the tibia, with which it forms the distal tibiofibular syndesmosis. In the posterior aspect of the malleolus, the malleolar groove for the tendons of the peroneal muscles can be seen. In the posteromedial aspect, the lateral malleolar fossa forms a depression for the insertion of the posterior talofibular ligament.

By palpation, it is easy to recognize that the lateral malleolus is more distal and more posterior than the medial or tibial malleolus.

Tarsus

The tarsus is composed of the following bones: the talus; the calcaneus; the navicular; the medial, middle, and lateral cuneiforms; and the cuboid (Figure 20-19).

Because of its anatomic and clinical importance, the tarsus can be divided into the midfoot, or anterior tarsus (navicular, cuneiforms, and cuboid), and the hindfoot, or posterior tarsus (talus and calcaneus). They are separated by the mediotarsal line (of Chopart).

Talus

The talus is a short bone that, along with the tibia and the fibula, forms the ankle joint (Figures 20-18 and 20-20). It articulates with the calcaneus and the navicular as well. It is the only tarsal bone that has no muscular or tendinous insertions.

The talar head is covered with hyaline cartilage and has several articular facets. The most important are the navicular articular facet and the anterior articular facet of the calcaneus. The talar head also contains the articular surface of the plantar calcaneonavicular ligament and the calcaneonavicular portion of the bifurcate ligament.

The talar neck is narrower and follows the talar head, forming a 15° angle with the talar body. In the

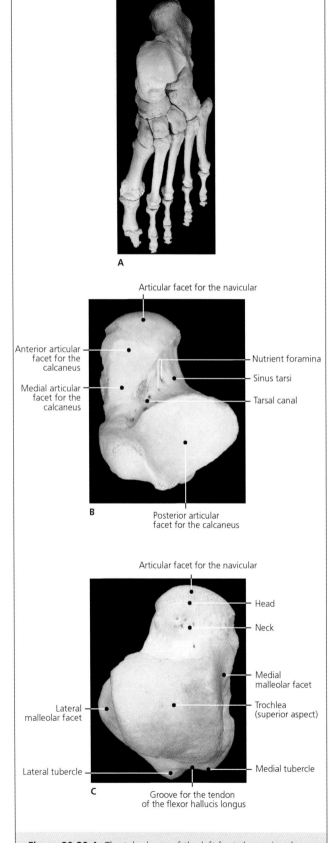

Figure 20-20 A, The talar bone of the left foot shown in relationship to the other bones of the foot. Plantar **(B)** and superior **(C)** views of the talar bone.

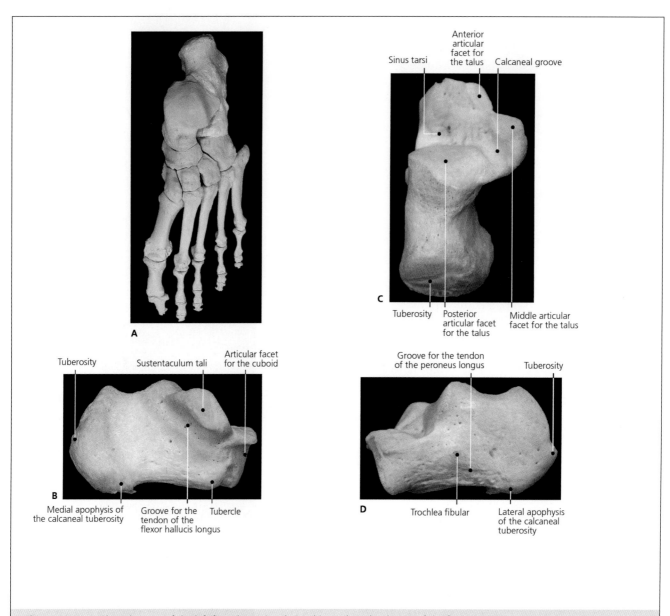

Figure 20-21 A, The calcaneus of the left foot shown in relationship to the other bones of the foot. **B,** Medial view. **C,** Superior view. **D,** Lateral view.

plantar aspect of the talar neck is the medial articular facet of the calcaneus. The most posterior part of the bone forms the body of the talus. The superior region forms the so-called talar trochlea, or talar dome, with the superior articular facet of the tibia. On either side of the trochlea are the medial and lateral malleolar surfaces of the medial and lateral malleoli, respectively. In the lateral aspect, the lateral process can be seen; dorsally, the posterior process is found. The latter presents a medial and a lateral tubercle; between the two is the groove (trough) for the tendon of the flexor hallucis longus. In the inferior aspect, covered with hyaline cartilage, is the posterior articular facet for the calcaneus and, near it, between this facet and the medial facet, is the tarsal canal.

It is easy to see that most of the talar surface is covered by articular cartilage, with the exception of the talar neck, through which the blood supply to the bone enters.

Calcaneus

The calcaneus is located under the talus and slightly medial to it. It is also a short bone that articulates with the talus and distally with the cuboid (Figure 20-21).

The posterior aspect of the calcaneus contains the calcaneal tuberosity where the calcaneal (Achilles) tendon inserts; it presents a medial and a lateral apophysis. In its plantar aspect, one can see the calcaneal tubercle for the insertion of the calcaneocuboid plantar ligament. Medially, there is an osseous prominence

Osteology

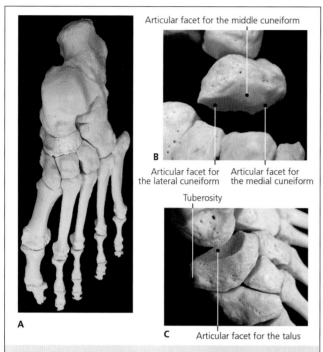

Figure 20-22 A, The navicular bone of the left foot shown in relationship to the other bones of the foot. **(B)** and **(C)** show the articular facets and tuberosity of the navicular.

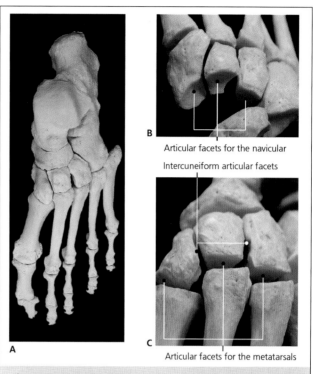

Figure 20-23 A, The three cuneiform bones of the left foot shown in relationship to the other bones of the foot. **(B)** and **(C)** show the articular facets of the cuneiforms.

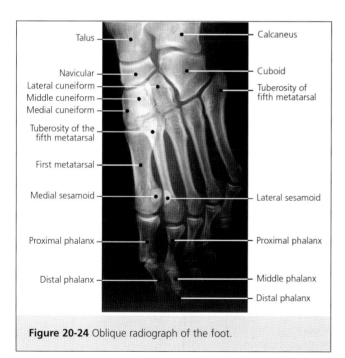

Figure 20-24 Oblique radiograph of the foot.

tarsi. Also in this lateral aspect is an osseous prominence that separates the peroneal tendons, which is known as the peroneal tubercle. Behind it is the groove for the peroneus longus, and in front of it passes the peroneus brevis. Finally, the anterior aspect contains the articular facet for the cuboid.

Navicular

The navicular (Figure 20-22) is a short bone located between the talus and the three cuneiforms. In its medial aspect is the tuberosity for the insertion of the tibialis posterior muscle. Distally, it presents flat articular surfaces separated by crests for each of the cuneiforms.

Cuneiforms

Distal to the navicular are the three cuneiforms (Figure 20-23). The medial cuneiform is characterized by a wider plantar aspect, the middle cuneiform is located between the other two, and the lateral cuneiform articulates laterally with the cuboid. Each of the three cuneiforms articulates with its respective metatarsal: the medial cuneiform with the first metatarsal, the middle with the second, and the lateral with the third. The medial and lateral cuneiforms are longer than the middle. The middle cuneiform thus serves as a keystone between the neighboring bones, giving the tarsometatarsal joint great stability (Figures 20-23 and 20-24).

resembling a platform over which the talus rests; because of this it is called sustentaculum tali. In its inferior surface is the groove for the flexor hallucis longus tendon. In its superior aspect we find the articular facets for the talus: anterior, medial, and posterior. Between the medial and posterior facets is the calcaneal canal. In the lateral aspect, the continuation of the talar and calcaneal canals where they join together is the sinus

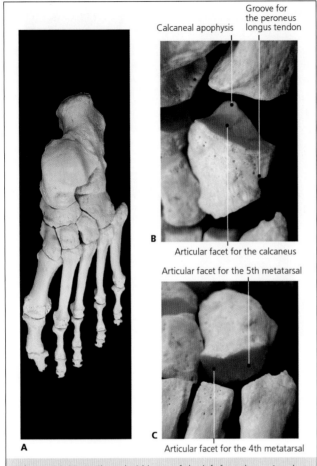

Figure 20-25 A, The cuboid bone of the left foot shown in relationship to the other bones of the foot. **(B)** and **(C)** show the articular facets of the cuboid.

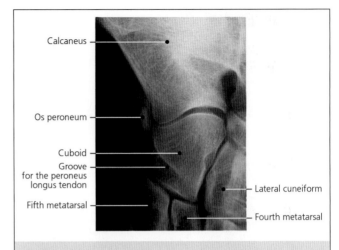

Figure 20-26 Radiograph showing the os peroneum, a very common sesamoid bone in the tendon of the peroneus longus.

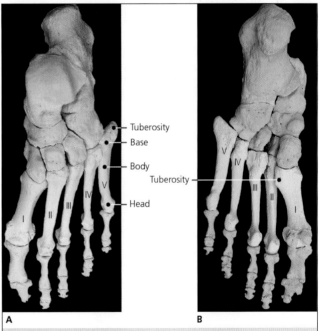

Figure 20-27 The metatarsals of the left foot shown in relationship to the other bones of the foot. **A,** Dorsal view. **B,** Plantar view.

Cuboid

The cuboid is found between the lateral cuneiform, the navicular, and the calcaneus (Figures 20-25 and 20-26). Posteriorly it articulates with the calcaneus, medially with the lateral cuneiform and the navicular, and distally with the fourth and fifth metatarsal bases. In its lateral and plantar aspects is the groove for the peroneus longus. Proximal to it is the cuboid tuberosity and the calcaneal apophysis, which extends medially to provide the articular surface for the calcaneus.

METATARSALS

The metatarsals of the foot are homologous to the metatarsals of the hand. Like them, these five long bones are numbered from medial to lateral (the first metatarsal corresponding to the hallux, or great toe) (Figure 20-27). Each metatarsal contains a base, a body, and a head.

The metatarsals articulate proximally with the tarsal bones, the first metatarsal with the first cuneiform, the second with the middle cuneiform, and the third with the lateral cuneiform. The fourth and fifth metatarsals articulate with the cuboid. Distally, the metatarsal heads articulate with the bases of their respective phalanges.

The first metatarsal is the only one to present a tuberosity (styloid apophysis) inferolaterally, and the fifth metatarsal presents one laterally for the insertion of the peroneus brevis.

BONES OF THE TOES

The skeleton of the toes is formed by the phalanges. The great toe, or hallux, contains two phalanges; the rest of the toes contain three (Figure 20-28, A).

Osteology

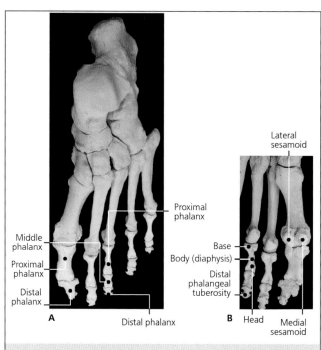

Figure 20-28 A, Dorsal view of a left foot with the phalanges shown in relationship to the other bones of the foot. **B,** The first three toes, with details of a phalanx and the sesamoid bones indicated.

There is a proximal, a middle, and a distal phalanx. The proximal articulates with the corresponding metatarsal and middle phalanx, the middle articulates with the other two phalanges, and the distal articulates proximally with the middle phalanx.

Each phalanx has a base with a proximal articular surface, a body, and a head with a trochlea. In its tip, the distal phalanx presents the distal phalanx tuberosity, with a rough surface and an arrowhead shape.

The sesamoid bones are found under the first metatarsal head and on each side of the flexor hallucis longus tendon (Figure 20-28, *B*). These have a flat lentil shape on the dorsal surface to articulate with the metatarsal.

Chapter 21

Arthrology

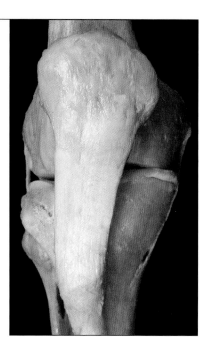

Introduction
This chapter describes the joints that join the pelvic girdle to the spinal column and those of the lower limb. Among the former are the joints of the pelvis (Figures 21-1 and 21-2); among the latter are the hip (coxofemoral), knee, and ankle joints and the joints of the foot.

Joints of the Pelvic Girdle
These joints join the two coxal bones with the spinal column, the sacrum, and the coccyx. In addition, the two pubic bones articulate to form the pubic symphysis. The ligaments and the obturator membrane also are discussed.

Iliolumbar Joint
The iliolumbar joint is formed by the junction of the transverse processes of L4-L5 with the iliac tuberosity. The strong iliolumbar ligament (Figure 21-3) consists of transverse fibers between the vertex of the transverse process and the iliac crest, oblique fibers between the inferior third of the process and the posterosuperior aspect of the iliac tuberosity, and fibers that descend from the anterior aspect of the transverse process and insert near the posterior aspect of the arcuate line of the ilium.

Sacroiliac Joint
The sacroiliac joint is the syndesmosis between the sacrum and the two iliac bones (Figures 21-1 and 21-2). It articulates the auricular surface of the sacrum with the corresponding surface of the iliac bone. Anteriorly, the articular line has the shape of an italic S. These bones join each other by their reciprocal shape, so their fibrocartilaginous cover does not regularize their surface (Figure 21-4). They are joined by a fibrous capsule that joins the borders of the auricular surfaces and several fibrous fascicles that form the anterior sacroiliac ligaments and posterior sacroiliac ligaments.

Anterior Sacroiliac Ligaments
These ligaments (Figure 21-3) extend from the anterior aspect of the base of the first two sacral foramina to the anterior and proximal aspect of the coxal bone, especially the iliac fossa. The anterior sacroiliac ligaments are thin and weak.

Posterior Sacroiliac Ligaments
These ligaments (Figure 21-5) are formed by a continuum of sacroiliac fascicles that join the sacrum and the ilium and are organized in two planes, the superficial plane and the deep plane.

The ligaments of the superficial plane are located immediately below the spinal muscles. These run from the posterior aspect of the crest and the posterior iliac spines to the intermediate and lateral crests of the sacrum. In this plane, four fascicles are superimposed vertically: (1) the first fascicle, which runs from the iliac crest to the lateral sacral crest, underlies the iliolumbar ligament and is the continuation of that ligament; (2) the vagus fascicle, or axial ligament, crosses the imaginary transverse axis. The sacrum makes its nutation-counternutation movements around this axis. This fascicle runs from the

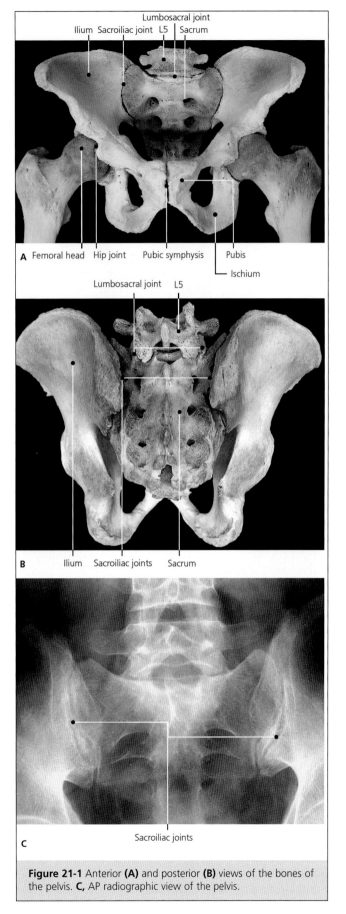

Figure 21-1 Anterior (**A**) and posterior (**B**) views of the bones of the pelvis. **C,** AP radiographic view of the pelvis.

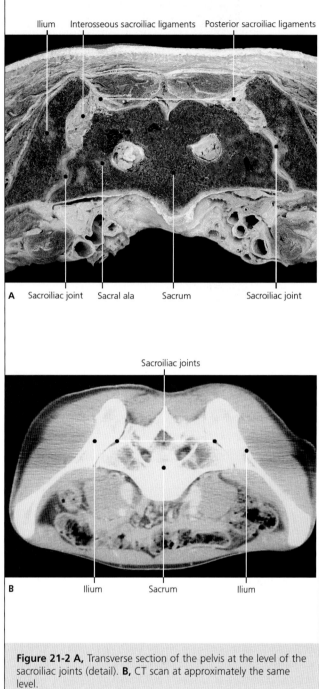

Figure 21-2 A, Transverse section of the pelvis at the level of the sacroiliac joints (detail). **B,** CT scan at approximately the same level.

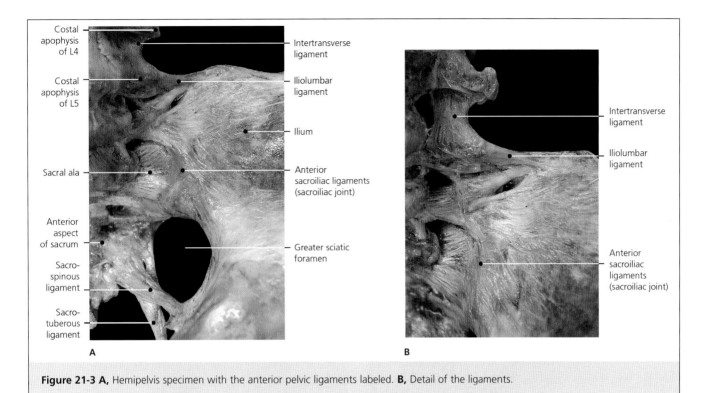

Figure 21-3 A, Hemipelvis specimen with the anterior pelvic ligaments labeled. **B,** Detail of the ligaments.

iliac tuberosity to the lateral aspect of the sacral foramen formed by S1-S2; (3) the fascicle sometimes called the Zaglas fascicle is a thick, short fascicle that runs from the posterior superior iliac spine to the tubercle lateral to the third sacral foramen; (4) the so-called sacrospinous ligament of Bichat is a cord that runs from the posterior superior iliac spine and the underlying notch to the tubercle lateral to the third sacral foramen and continues inward with the fascia covering the spinal muscles and outward with the corresponding fascicles of the sacrotuberous ligament.

The ligaments of the deep plane are the so-called interosseous sacroiliac ligaments (even though they are located outside the joint) (Figures 21-2 and 21-4). These are formed by very strong ligament fascicles that run from the sacral to the iliac tuberosity.

The sacroiliac joint has a true synovium that is formed only by the joint line.

Pubic Symphysis

The pubic symphysis is formed by the articulation of the two pubic bones (Figure 21-1) through their elliptic articular surfaces covered with hyaline cartilage.

The interpubic disk (Figure 21-6) is a fibrocartilaginous, wedge-shaped mass located between the articular surfaces of the pubic bones. It consists of a peripheral portion, which is dense, hard, and resistant, and a central portion, which is softer, with

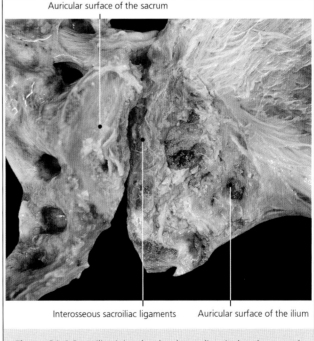

Figure 21-4 Sacroiliac joint that has been disarticulated to reveal the auricular surfaces and the interosseous ligaments.

an irregular cavity in its center. Typically, the central portion is thicker in women. It is considered a fibrocartilaginous joint.

Four peripheral ligaments—the anterior, posterior, superior, and inferior pubic ligaments—form a fibrous capsule that surrounds the joint (Figure 21-6).

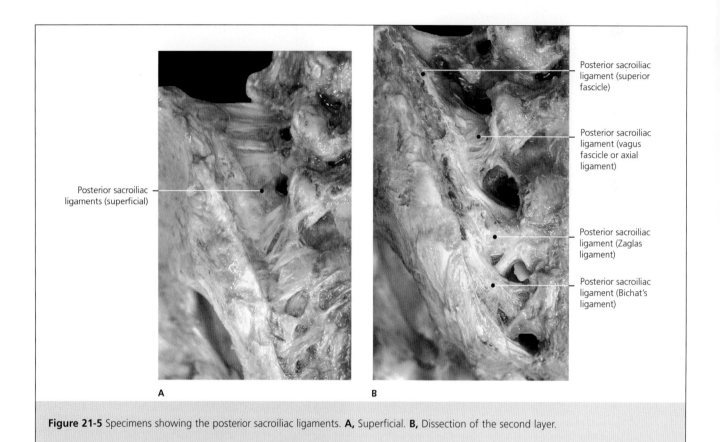

Figure 21-5 Specimens showing the posterior sacroiliac ligaments. **A,** Superficial. **B,** Dissection of the second layer.

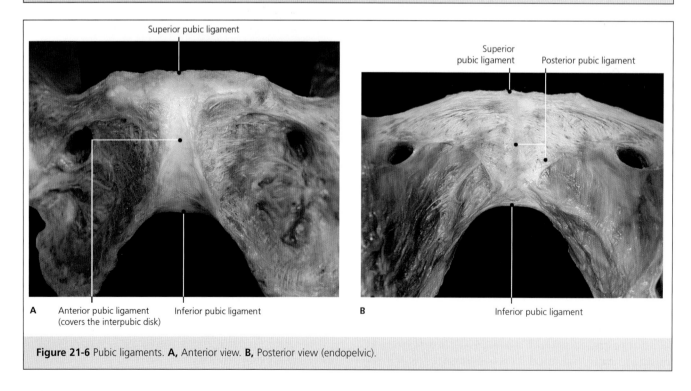

Figure 21-6 Pubic ligaments. **A,** Anterior view. **B,** Posterior view (endopelvic).

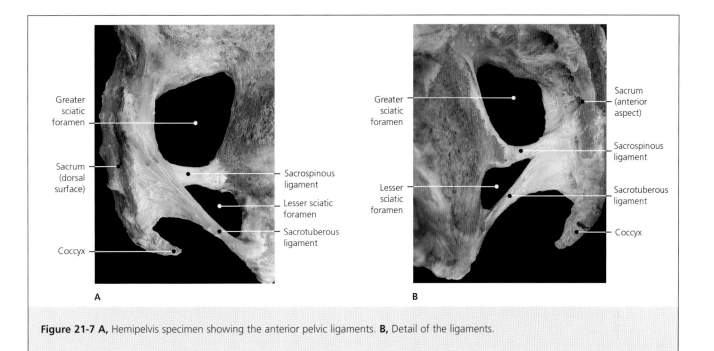

Figure 21-7 A, Hemipelvis specimen showing the anterior pelvic ligaments. **B,** Detail of the ligaments.

Anterior Pubic Ligament
This ligament is formed by multiple fibrous fascicles of very diverse strength and direction that occupy the entire anteroinferior aspect of the pubis. The superficial fascicles contain the tendinous fibers of the muscles that insert in the pubis. Deep to these are transverse fibers that connect one pubic bone to the one opposite. These fibers blend laterally with the periosteum and medially with the interpubic disk.

Posterior Pubic Ligament
Located in the intrapelvic aspect of the joint, this ligament is formed by the pelvic periosteum and is reinforced at the level of the articular line by some transverse fascicles that insert in the posterior border of the corresponding articular facet.

Superior Pubic Ligament
This yellowish, fibrous band extends horizontally over both pecten pubis, joining them.

Inferior Pubic Ligament
This ligament is formed by a very strong arcuate fibrous band that is located below the symphysis and joins the inferior pubic rami.

Other Ligaments
Sacrotuberous Ligament (Sacrosciatic Major)
This ligament (Figure 21-7) is formed by very strong and resistant fibrous fascicles. The ligament extends from the two posterior iliac spines, the most dorsal aspect of the iliac fossa, the lateral sacral crests, and the coccyx to the ischial tuberosity. Some of the fibers of this ligament are deflected and insert in the medial border of the ischial ramus, the falciform process. Most of the fibers cross each other in the shape of an X in the narrowest portion of the ligament.

Sacrospinous Ligament (Sacrosciatic Minor)
This ligament (Figure 21-7) is composed of diverse fibrous bundles mixed with a greater or lesser number of muscle fibers. It extends from the posterior aspect of the sacrum and coccyx to the sciatic spine. In its posterior aspect, the sacrospinous ligament has an intimate relationship with the sacrotuberous ligament. The superior border of the sacrospinous ligament is continued by a cellulous sheet (sacrosciatic lamina of Morestin) that diverges from the ligament and disappears below the sciatic nerve and the piriformis muscle.

The sacrotuberous and sacrospinous ligaments close the notches of the coxal bone and limit two spaces, the superior space and the inferior space.

Foramina
The superior space, called the greater sciatic foramen (Figure 21-7), is bound by the greater sciatic notch, the sacrum, the sacrospinous ligament, and the superior border of the sacrotuberous ligament. Its contents include the piriformis muscle; the superior gluteal artery, vein, and nerve; the internal pudendal artery and vein; and the pudendal, sciatic, and posterior femorocutaneous nerves.

The inferior space, called the lesser sciatic foramen (Figure 21-7), is bound by the lesser sciatic notch and the sacrospinous and sacrotuberous ligaments. It con-

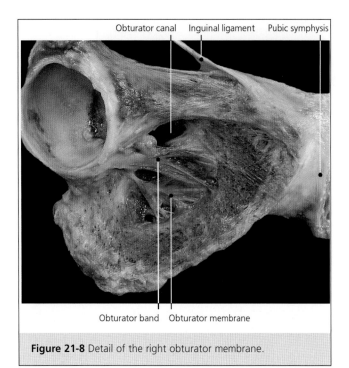

Figure 21-8 Detail of the right obturator membrane.

tains the obturator internus muscle, the internal pudendal artery and vein, and the pudendal nerve.

The obturator foramen of the coxal bone is covered with two fibrous structures, the obturator membrane and the obturator band.

Obturator Membrane

The obturator membrane (Figure 21-8) is formed by nacreous fibers that cross each other in different directions and close the majority of the obturator foramen. Its intrapelvic aspect is in contact with the obturator internus muscle, which covers it completely and takes the majority of its insertions. Its extrapelvic aspect is related to and covered by the obturator externus muscle.

The obturator band (subpubic) (Figure 21-8) is a fibrous sheet that reinforces the obturator membrane. It extends from the transverse acetabular ligament and the posterior obturator tubercle to the pubic body and the obturator membrane.

These two structures limit a channel called the obturator canal (subpubic). It corresponds to the obturator notch, which extends from the two obturator tubercles, anterior and posterior. The obturator artery, vein, and nerve course through the obturator canal.

HIP (COXOFEMORAL) JOINT

The coxofemoral joint joins the spherical head of the femur with the acetabulum of the coxal bone, forming an enarthrosis or ball-and-socket articulation that permits motion in three planes (Figures 21-1, 21-9, and 21-10).

More than half of the femoral head engages in the acetabular cavity. The cavity contains a semilunar facet covered with cartilage, which is the true articular surface. The acetabular border presents a labrum (Figure 21-9), which augments the articular surface and the congruity of both surfaces, rendering the joint very stable. At the level of the acetabular notch, the labrum completes the circle with the transverse ligament of the acetabulum (Figure 21-9), which leaves a small foramen underneath, through which courses the acetabular branch of the obturator artery for the head of the femur.

The articular capsule inserts in the region adjacent to the acetabular border and the transverse acetabular ligament. These longitudinal fibers cover the femoral head and insert principally in the intertrochanteric line. A deep portion of the fibers surround the femoral neck and are called the orbicular zone (annular ligament of Weber). These fibers can be found under the ligaments described in the next section.

Ligaments

Externally, the articular capsule of the hip presents several reinforcing ligaments that originate from the different parts of the coxal bone, namely the ilium, the ischium, and the pubis. These ligaments are called the iliofemoral, ischiofemoral, and pubofemoral ligaments.

The iliofemoral ligament reinforces the anterior aspect of the joint. It inserts in the anteroinferior iliac spine, and its fibers bifurcate in a medial and a lateral fascicle with the shape of an inverted Y (Y ligament of Bigelow). The lateral fascicle inserts in the superior portion of the intertrochanteric line to the greater trochanter. The medial fascicle goes toward the inferior portion of the intertrochanteric line near the lesser trochanter. This ligament limits hip extension to no more than 15°.

The ischiofemoral ligament reinforces the articular capsule posteriorly. Its fibers can be observed in the posterior region of the acetabulum, which forms part of the ischium. The fibers extend to the posterior portion of the femoral neck near the intertrochanteric fossa. Some fibers relate to the lateral fascicle of the iliofemoral ligament. Others blend with the orbicular zone, which is why some authors name this portion the ischiocapsular ligament. The ischiofemoral ligament limits hip flexion and adduction.

The pubofemoral ligament (Figure 21-11) reinforces the joint inferiorly. It extends from the inferior aspect of the acetabular border to the superior pubic ramus. Its oblique fibers can be observed converging and inserting in the femur above the lesser trochanter. This ligament limits hip abduction.

In dissection, when the hip joint capsule is opened and the femoral head is dislocated, the ligament of the head of the femur (round ligament) can be seen (Figure 21-9). The ligament is flat and located deep in the acetabular fossa, which contains a little pad of areolar tissue called the acetabular pulvinar (Figure 21-9). The round

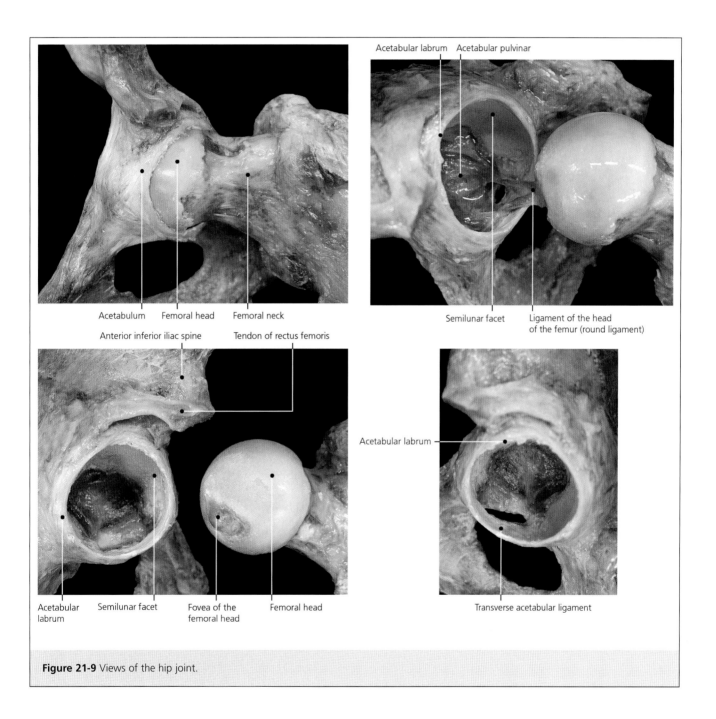

Figure 21-9 Views of the hip joint.

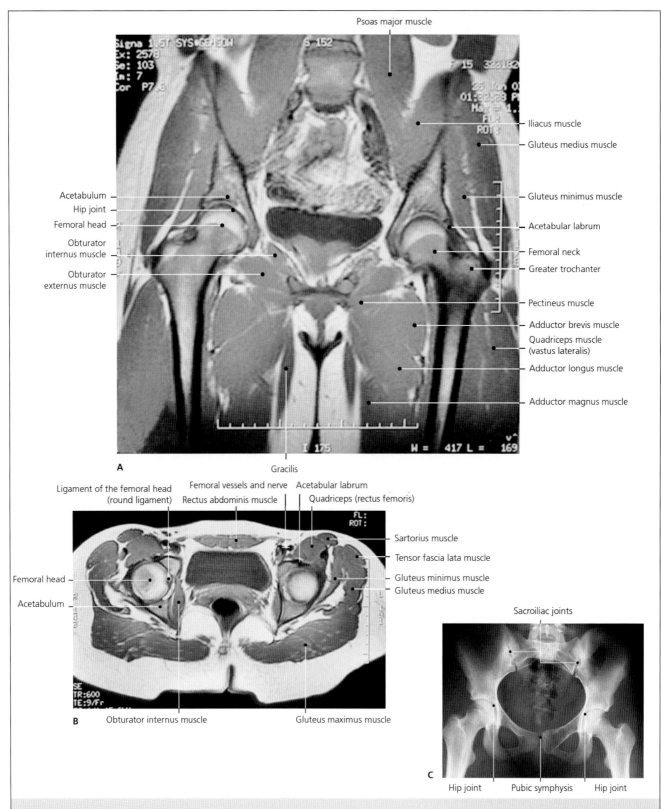

Figure 21-10 Images of the pelvis. **A,** Coronal (frontal) MRI through the pelvis and hip. **B,** Transverse MRI at the level of the hip joint. **C,** AP radiograph of the pelvic region.

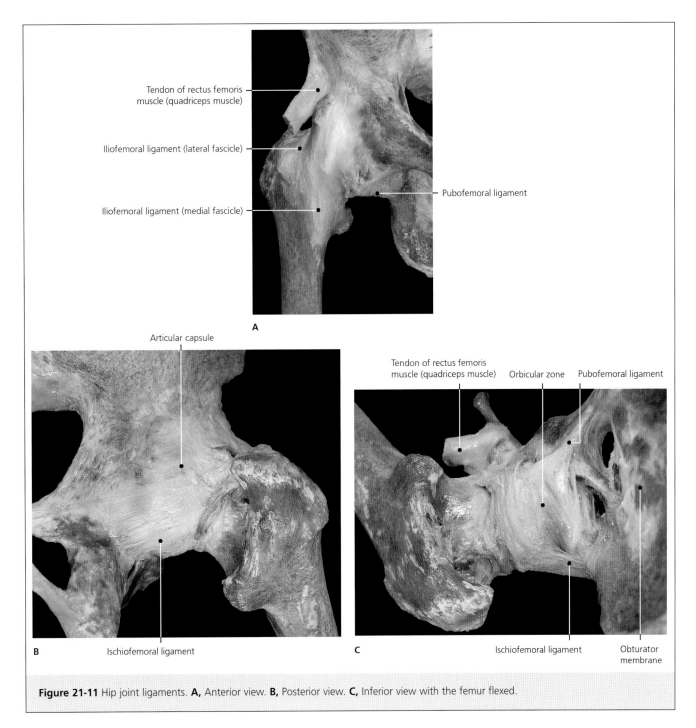

Figure 21-11 Hip joint ligaments. **A,** Anterior view. **B,** Posterior view. **C,** Inferior view with the femur flexed.

ligament extends from the acetabular fossa to the fossa in the head of the femur and is covered by a reflection of the synovial membrane. The ligament does not have a reinforcing function but instead leads the vessels mentioned earlier to the femoral head.

KNEE JOINT

The articular complex of the knee contains two joints that share the same synovial capsule, the patellofemoral joint and the tibiofemoral joint. This capsule inserts in the periphery of the femoral and tibial condyles and extends proximally, deep to the quadriceps tendon, communicating with the suprapatellar bursa (Figure 21-12). The patella and the patellar ligament are the sites of anterior insertion of the capsule, which also inserts in the menisci. The part between the menisci and the tibia is called the coronary ligament.

The knee has several adipose structures, the largest of which is the infrapatellar fat pad. Opening the capsule and lifting the patella exposes the infrapatellar synovial plica (adipose ligament), which inserts in the

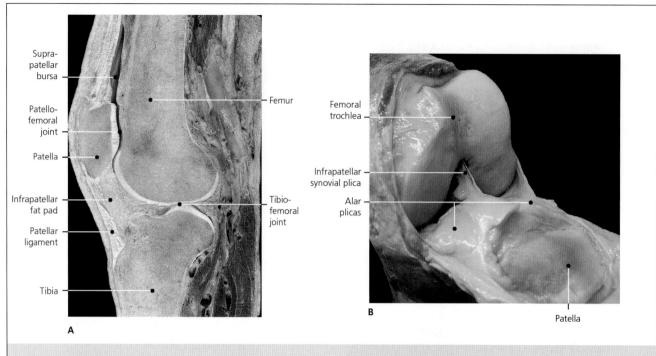

Figure 21-12 A, Sagittal cut of knee joint. **B,** The joint has been opened and the patella has been dislocated downward.

intercondylar fossa. In addition, in front of the menisci are two small fat pads, the alar plicas.

The patellofemoral joint is formed by the patella and the patellofemoral surface (femoral trochlea) (Figures 21-12 through 21-14). The patella glides in a sulcus covered by articular cartilage between the femoral condyles. The lateral articular facet of the patella is larger than the medial facet, which helps to identify sidedness. The intrinsic ligaments of the patella are called alar ligaments or transverse retinacular ligaments and consist of a series of aponeurotic fibers that join the sides of the patella with the respective condyles, medial and lateral. The other knee ligaments are discussed later.

The tibiofemoral joint is formed by the two femoral condyles and the superior articular surface of the tibia (Figures 21-13 through 21-16). The condyles are convex from anterior to posterior and from lateral to medial, which makes their congruency with the flat tibial surface difficult. However, two fibrocartilaginous articular menisci are interposed between the two articular surfaces, which insert in the superior aspect of the tibia and allow condylar gliding, increasing articular congruity (Figure 21-17). They form a condylar synovial joint (bicondylar) that is capable of flexion and extension movements as well as rotation (in flexion).

The angle between the tibial tubercle (or tuberosity) and the femur (Q angle) of 10° to 12° represents a physiologic valgus that is more pronounced in women.

The menisci insert in the tibia at their ends (horns) and peripherally in the articular capsule (Figure 21-16). The medial meniscus has a C shape and inserts anteriorly in the anterior intercondylar area and the superior aspect of the tibial tuberosity. Posteriorly, the medial meniscus inserts behind the medial intercondylar tubercle. It is less mobile than the lateral meniscus because of its connections with the tibial collateral ligament. The lateral meniscus is shaped like a more closed C (almost an O) and inserts anteriorly in front of the lateral intercondylar tubercle and posteriorly behind the same tubercle. Frequently, this meniscus presents a ligament, called the posterior meniscofemoral ligament (of Wrisberg), that originates from the posterior horn, courses behind the posterior cruciate ligament, and inserts in the internal aspect of the medial femoral condyle (Figure 21-18). The ligament also courses in front of the posterior cruciate ligament and is called the anterior meniscofemoral ligament (of Humphrey). The anterior portions of both menisci are joined by the transverse ligament of the knee (of Winslow).

In addition to the fibrous capsule, this joint is reinforced by the collateral and the cruciate ligaments. The extracapsular fibular collateral ligament looks like a fibrous cord located between the lateral femoral epicondyle and the apex of the fibular head and is intimately related with the insertion of the biceps femoris muscle (Figures 21-13 and 21-19). The tibial collateral ligament is a medial capsular thick-

Arthrology

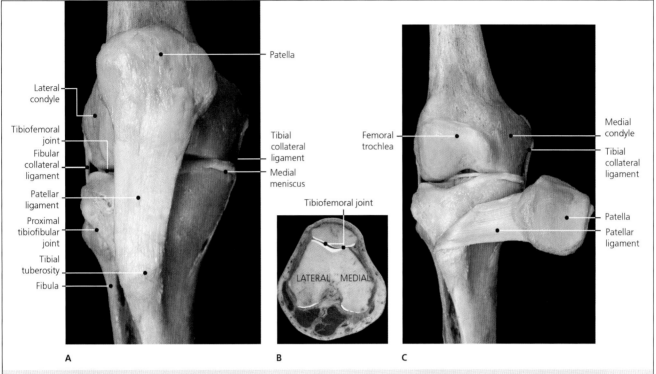

Figure 21-13 A, Anterior view of knee. **B,** Transverse section at the level of the patellofemoral joint. **C,** Knee with the patella displaced medially.

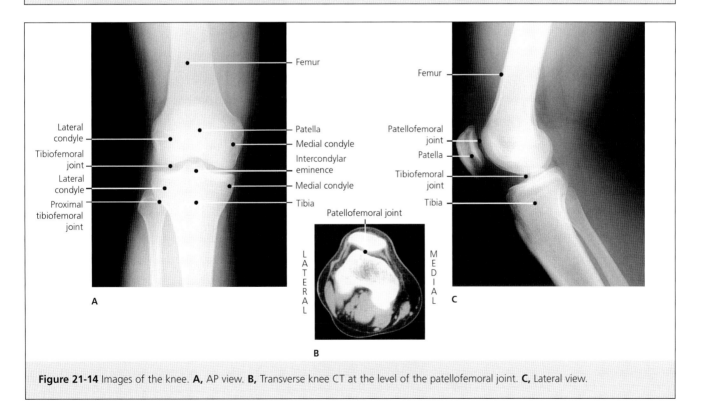

Figure 21-14 Images of the knee. **A,** AP view. **B,** Transverse knee CT at the level of the patellofemoral joint. **C,** Lateral view.

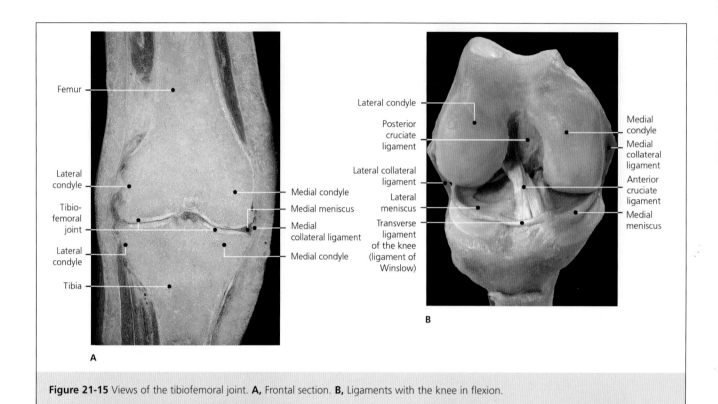

Figure 21-15 Views of the tibiofemoral joint. **A**, Frontal section. **B**, Ligaments with the knee in flexion.

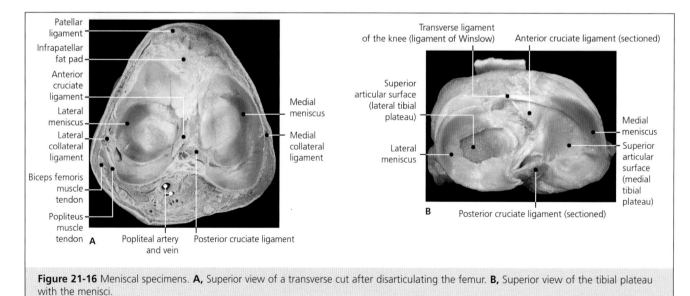

Figure 21-16 Meniscal specimens. **A**, Superior view of a transverse cut after disarticulating the femur. **B**, Superior view of the tibial plateau with the menisci.

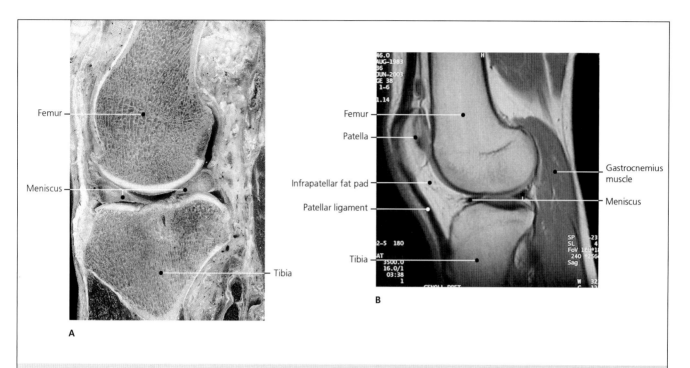

Figure 21-17 Meniscal details. **A,** Sagittal section demonstrating the characteristic triangular wedge shape between the tibia and the femur. **B,** Sagittal MRI of the knee.

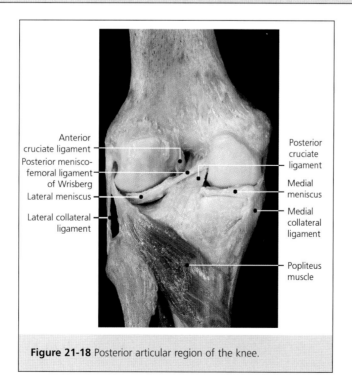

Figure 21-18 Posterior articular region of the knee.

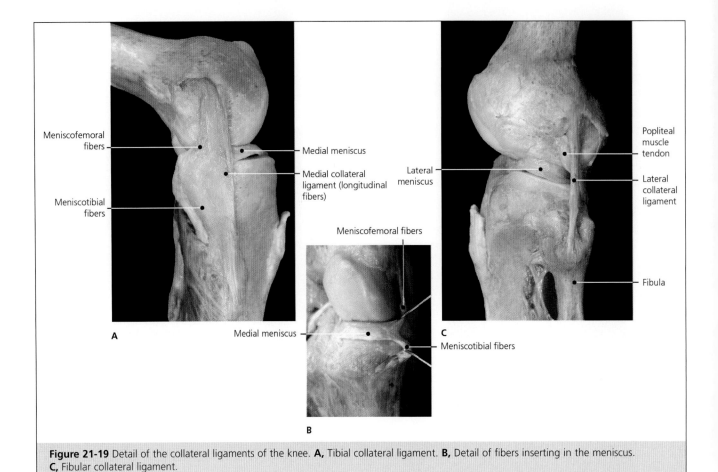

Figure 21-19 Detail of the collateral ligaments of the knee. **A,** Tibial collateral ligament. **B,** Detail of fibers inserting in the meniscus. **C,** Fibular collateral ligament.

ening that is triangular (Figures 21-13 and 21-19). It contains two types of fibers: the anterior fibers are parallel and go from the medial femoral epicondyle to the medial surface of the proximal tibial epiphysis, blending with the medial meniscus; the posterior fibers are oblique and adhere strongly to the medial meniscus through meniscofemoral and meniscotibial fibers.

The anterior and posterior cruciate ligaments, named according to where they insert on the tibia, are fibrous cords that cross with the shape of an X. They are located inside the knee joint but remain extrasynovial (Figure 21-15). Their function is to limit the anteroposterior displacement of the tibia and help direct the gliding of the knee joint during flexion and extension. The anterior cruciate ligament is located in the anterior intercondylar area between the two menisci (Figures 21-15 and 21-20). It follows an oblique course posteriorly and laterally to insert in the medial aspect of the lateral femoral condyle. Its function is to avoid anterior displacement of the tibia with respect to the femur. The posterior cruciate ligament (Figures 21-15, 21-18, and 21-20) is located in the posterior intercondylar area, and its fibers course obliquely anteriorly and medially to insert in the medial

aspect of the medial femoral condyle. Its function is to prevent posterior displacement of the tibia. When the knee joint is fully extended, the condyles effect a small turn or torsion that makes these ligaments cross each other, thereby tightening them (the so-called "screw-home" mechanism).

Anteriorly, the joint is reinforced by the quadriceps muscle through its insertions in the patella (quadriceps tendon) and tibial tuberosity through the patellar ligament (Figures 21-12 and 21-13). Moreover, aponeurotic expansions of the quadriceps course longitudinally on both sides of the patella—the medial and lateral patellar retinaculi. These retinaculi belong to a knee aponeurotic mesh with fibers of multiple trajectories; among these are the alar ligaments or transverse retinaculi.

Posteriorly, the joint possesses a strong fibrous capsule that prevents hyperextension; when the condyles are removed, robust capsular reinforcement can be seen (Figure 21-21). The articular capsule is reinforced by the popliteal ligaments. The oblique popliteal ligament is described in chapter 22, with the discussion of the semimembranosus muscle. The arcuate popliteal ligament is a capsular thickening that overlies the popliteal muscle and inserts in the fibular head.

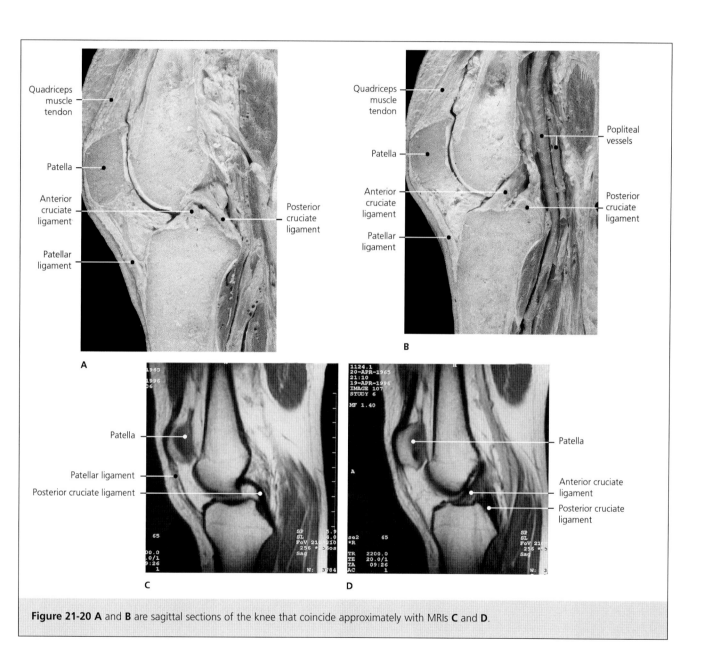

Figure 21-20 A and **B** are sagittal sections of the knee that coincide approximately with MRIs **C** and **D**.

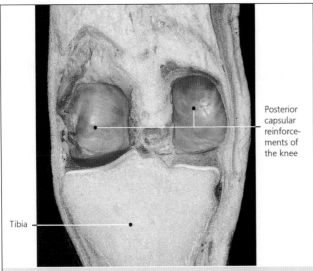

Figure 21-21 Posterior knee capsule, frontal section. The femur has been removed so that the reinforcement of the articular capsule that covers the femoral condyles can be seen.

Joints of the Leg

The skeleton of the leg is formed by the tibia and the fibula. These articulate at their two epiphyses, forming the proximal and distal tibiofibular joints. In addition, joining the two bones is the interosseous membrane that extends between the interosseous borders of the tibia and the fibula (Figure 21-22). This membrane increases the surface of origin of the leg muscles.

The proximal tibiofibular joint (Figure 21-22) is a gliding joint that joins the articular surface of the fibular head with the fibular articular facet of the tibia. It is reinforced by two ligaments: the anterior ligament of the fibular head reinforces the anterior portion of the capsule, and the posterior ligament of the fibular head reinforces the posterior portion.

The distal tibiofibular joint (Figure 21-22) is a syndesmosis (ligamentous joint), meaning that fibrous tissue joins both articular surfaces. The tibia contains a fibular notch that receives the medial and

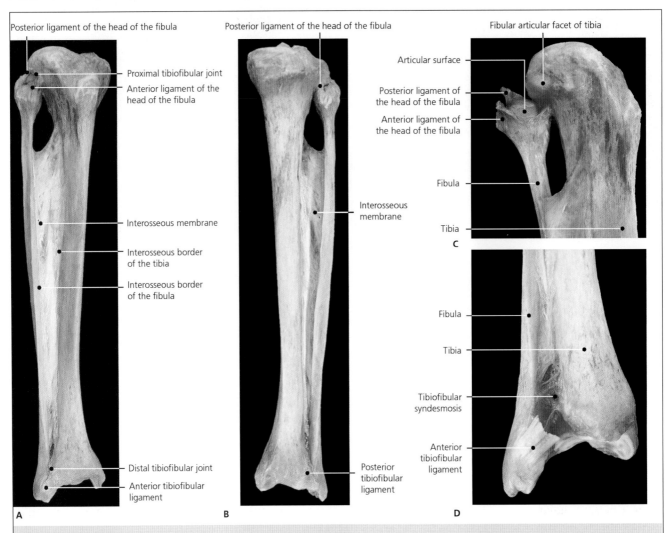

Figure 21-22 A, Anterior view of the bones of the leg with its ligaments and interosseous membrane. **B,** Posterior view. **C,** Details of the proximal ends of the bones with their ligaments. **D,** Details of the distal ends of the bones with their ligaments.

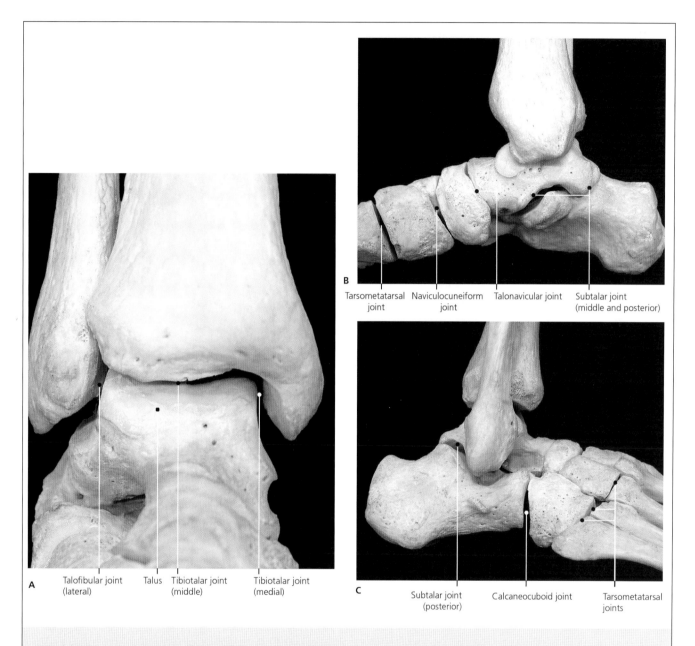

Figure 21-23 Views of the joints of the ankle and foot. **A,** Detailed anterior view of the ankle. **B,** Medial view. **C,** Lateral view.

distal portion of the fibula. There are two fibrous bundles that form the anterior tibiofibular ligament anteriorly and the posterior tibiofibular ligament posteriorly.

Ankle and Subtalar Joints

The ankle (tibiotalar) joint is formed by three bones: the tibia, the fibula, and the talus (Figures 21-23 through 21-25), which is why it is classically known as the tibiofibulotalar joint. All three bone ends are included in an articular capsule that inserts in the inferior borders of the leg bones and around the articular surfaces the talus.

The superior articular trochlea of the talus articulates with the inferior articular surface of the tibia. On both sides, the tibial and fibular malleolar articular surfaces articulate with the medial and lateral articular facets of the talus. This joint forms part of the trochlea and allows dorsiflexion and plantar flexion. The talar trochlea is wider anteriorly, which accounts for the greater joint stability in dorsiflexion than in plantar flexion because there is a tighter fit in the mortise formed by the tibia and the fibula (in dorsiflexion). Conversely, in plantar flexion, the narrower talar trochlea allows a little mobility in this joint.

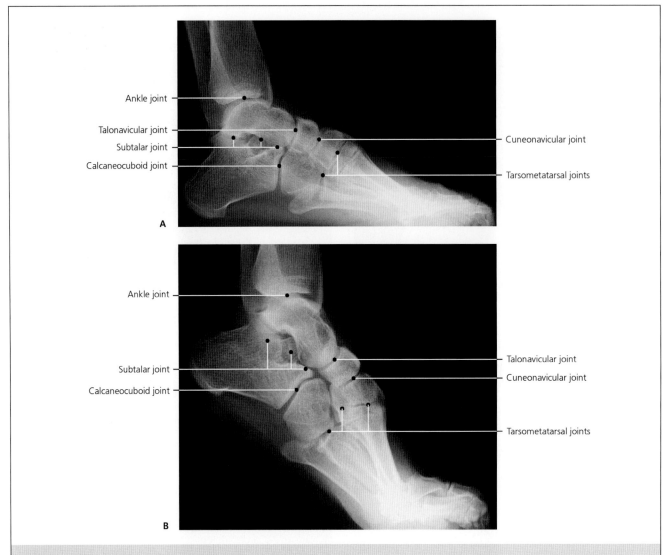

Figure 21-24 Radiographs of the ankle and foot. **A,** Weight-bearing lateral view. **B,** Weight-bearing lateral view with the patient standing on the toes.

As its name implies, the subtalar, or talocalcaneal, joint is formed by the talus and the subjacent calcaneus. They articulate at three points through the anterior, middle, and posterior articular facets. The anterior and middle articular facets are very close, and they are separated from the posterior facet by the sinus tarsi, a space formed by the sulci in the two bones. This joint permits pronation and supination of the foot and belongs in the class of pivot joints.

The ligaments of these joints are located medially, laterally, and in the interosseous space. The medial collateral ligament, also called the deltoid (Figure 21-26) because of its triangular shape, inserts in the medial malleolus. In the group of anterior fibers is its tibionavicular portion, which inserts in the superior and medial aspect of the navicular, and its anterior tibiotalar portion, containing shorter bundles that insert in the medial aspect of the talar neck. These fibers limit plantar flexion. The middle fibers correspond to the tibiocalcaneal portion that inserts in the sustentaculum tali and limit foot pronation. Finally, the posterior tibiotalar portion inserts in the medial talar apophysis and limits dorsiflexion.

The lateral collateral ligament (Figure 21-26) is formed by separate ligaments—the anterior and posterior talofibular and the calcaneofibular ligaments. The anterior talofibular ligament courses between the anterior border of the lateral malleolus and the lateral aspect of the talar neck; it limits plantar flexion and foot supination. The calcaneofibular ligament courses obliquely between the apex of the fibular malleolus and the postero-

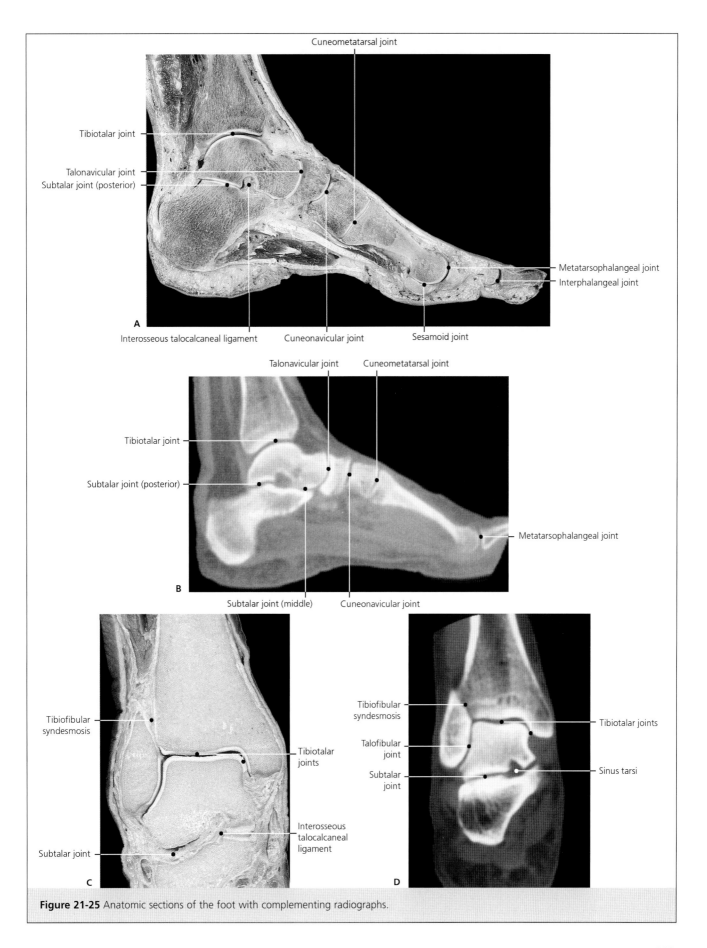

Figure 21-25 Anatomic sections of the foot with complementing radiographs.

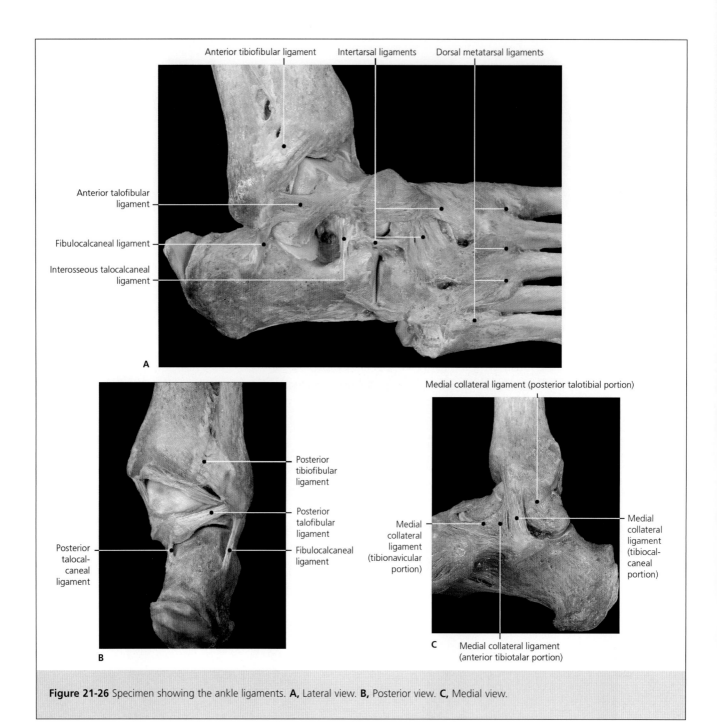

Figure 21-26 Specimen showing the ankle ligaments. **A,** Lateral view. **B,** Posterior view. **C,** Medial view.

lateral aspect of the calcaneus; the peroneal tendons glide over it. This ligament's principal function is to limit foot supination. The posterior talofibular ligament courses between the posterior aspect of the lateral malleolus and the lateral tubercle of the posterior apophysis of the talus; it limits dorsiflexion and foot supination.

In the subtalar joint are two collateral ligaments, a posterior and an interosseous ligament. The lateral talocalcaneal ligament extends from the talar trochlea to the lateral aspect of the calcaneus. The medial talocalcaneal ligament (Figure 21-27) is located under the deltoid ligament, between the medial tubercle of the posterior apophysis of the talus and the sustentaculum tali. The posterior talocalcaneal ligament (Figure 21-26) is formed by fibers that extend from the posterior apophysis of the talus to the calcaneus. Finally, the interosseous talocalcaneal ligament (Figures 21-25 through 21-27) is located between the talar and calcaneal sulci, like a fibrous wall that occupies a good part of the sinus tarsi.

Arthrology

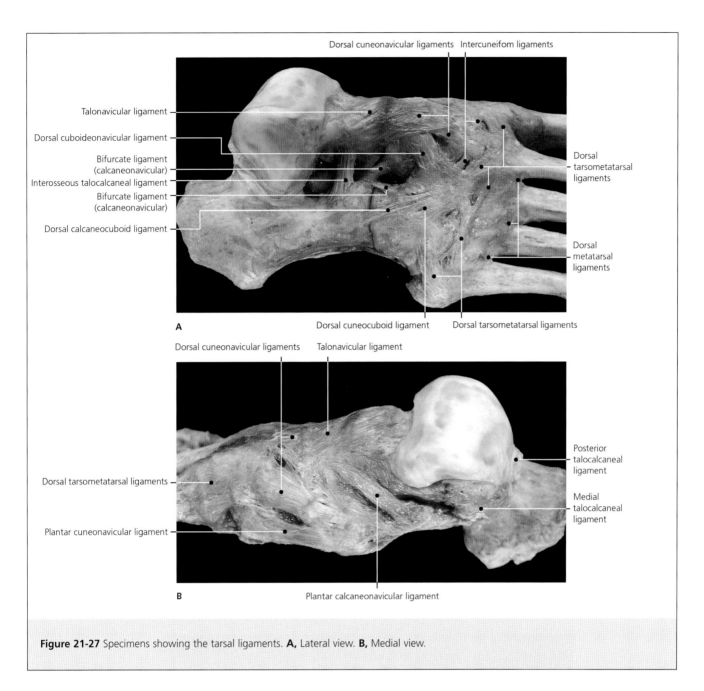

Figure 21-27 Specimens showing the tarsal ligaments. **A,** Lateral view. **B,** Medial view.

Joints of the Foot

Intertarsal Joints

There are two longitudinal columns in the foot, one medial and one lateral. The medial column is formed by the talus, navicular, and cuneiforms. The lateral column is formed by the calcaneus and the cuboid (Figures 21-28 and 21-29).

The subtalar (between the talus-calcaneus) and navicular-cuboid joints were discussed earlier. The transverse joint of the tarsus (Chopart's joint) is discussed here (Figure 21-28). This joint is divided into two joints: the talocalcaneonavicular and the calcaneocuboid.

The talocalcaneonavicular joint has an articular capsule, which is anchored in three bones. Functionally, it is a multiaxial joint (enarthrosis) (Figure 21-30). The hemispheric head of the talus articulates with the concavity formed by the navicular bone, the anterior and middle facets of the calcaneus, as well as the plantar calcaneonavicular ligament (Figures 21-27 and 21-30). This ligament is very strong, reinforces the joint inferiorly, and contributes to the formation of its bottom; it extends from the sustentaculum tali to the inferior aspect of the navicular. In addition to amplifying the articular cavity, the plantar calcaneonavicular ligament is reinforced by the tendons of the posterior tibia, the flexor digitorum longus, and the flexor hallucis longus. In this manner, the talar head has a good point of support.

The calcaneocuboid joint (Figures 21-23 and 21-28) is formed by the articular surfaces of the cuboid and the calcaneus. The articular capsule is anchored in the periphery of the articular surfaces and is reinforced by several ligaments. The long plantar ligament (Figure 21-31) extends from the calcaneus laterally toward the cuboid and the second through fifth metatarsals. A deep and short fascicle of the long plantar ligament constitutes the plantar calcaneocuboid ligament. Dorsally is found a V-shaped ligament with its vertex in the calcaneus, the bifurcate ligament. It contains two portions (Figure 21-27), the calcaneocuboid and calcaneonavicular ligaments. The first inserts in the dorsal aspect of the cuboid and the second in the lateral aspect of the navicular. The dorsal calcaneocuboid ligament is located lateral to the calcaneocuboid ligament.

Distal to the transverse tarsal joint are the cuneiform

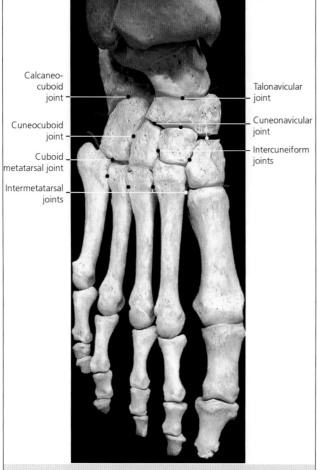

Figure 21-28 Tarsal joints. The Chopart articular line is shown in red. The Lisfranc articular line (tarsometatarsal joints) is shown in blue.

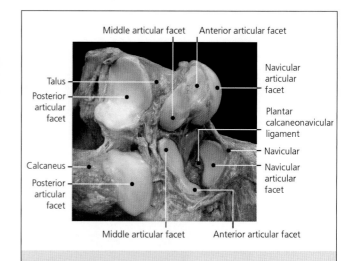

Figure 21-30 The talocalcaneonavicular joint, disarticulated to show the articular surfaces.

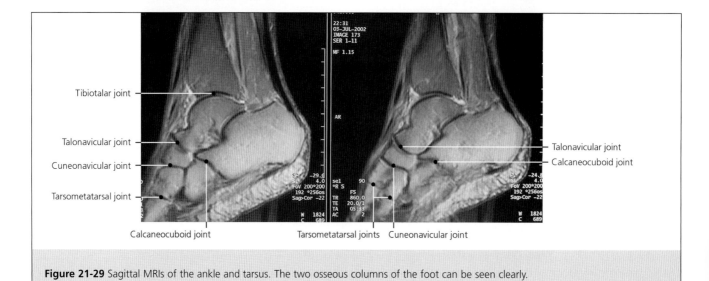

Figure 21-29 Sagittal MRIs of the ankle and tarsus. The two osseous columns of the foot can be seen clearly.

Arthrology

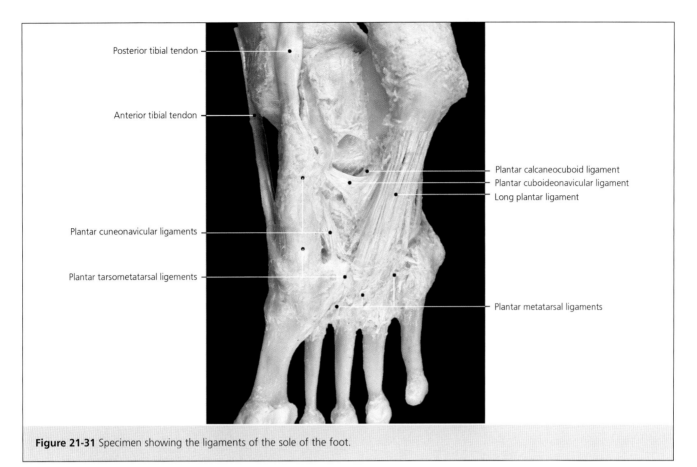

Figure 21-31 Specimen showing the ligaments of the sole of the foot.

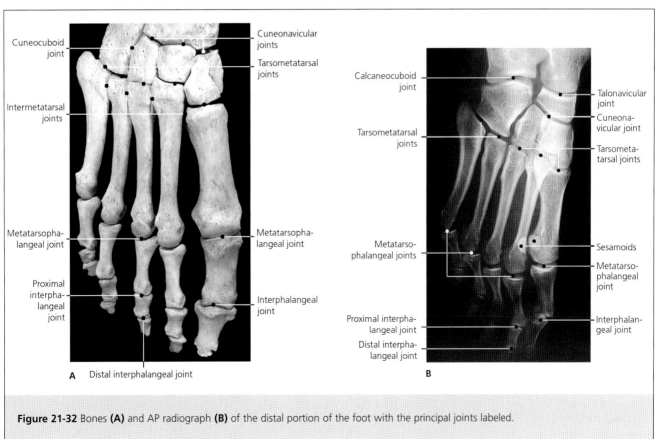

Figure 21-32 Bones **(A)** and AP radiograph **(B)** of the distal portion of the foot with the principal joints labeled.

joints (Figure 21-28). The navicular articulates by three small facets, with the three cuneiforms forming the cuneonavicular joint. The cuneiforms articulate among themselves through the intercuneiform joints. Finally, the cuboid articulates with the lateral cuneiform at the cuneocuboid joint. All of these joints are gliding joints, and their movements are limited to displacements between the articular surfaces. The three cuneiforms are joined by the plantar, dorsal, and interosseous intercuneiform ligaments (Figure 21-27). The last one is constant between the middle and the lateral cuneiforms. Between the navicular and the cuneiforms are the dorsal and plantar cuneonavicular ligaments. The ligaments that join the lateral cuneiform and the cuboid are the interosseous, dorsal, and plantar cuneocuboid ligaments. The dorsal and plantar cubonavicular ligaments join the cuboid and the navicular. Some authors also describe an interosseous cubonavicular ligament.

Tarsometatarsal Joints

The tarsometatarsal joints are formed by the three cuneiforms and the cuboid of the tarsus and the five metatarsals (Figure 21-32). Collectively, these are also called the Lisfranc joint (Figure 21-28). All are gliding joints.

The three cuneiforms articulate with their corresponding metatarsals: the medial cuneiform with the first, the middle with the second, and the lateral with the third metatarsal. The cuboid articulates with the fourth and fifth metatarsals. The tarsometatarsal joints are reinforced by the plantar and dorsal tarsometatarsal ligaments (Figures 21-27 and 21-31). In addition, between the cuneiforms and the metatarsals are the interosseous cuneometatarsal ligaments. The first ligament extends between the medial cuneiform and the second metatarsal, and the second ligament extends from the lateral cuneiform to the third metatarsal. Occasionally, there are additional ligaments.

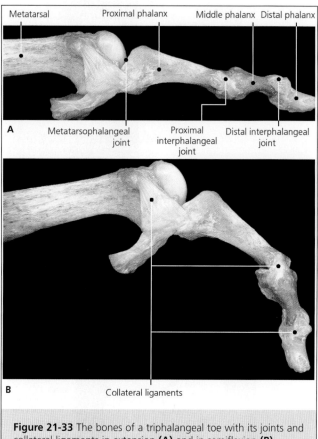

Figure 21-33 The bones of a triphalangeal toe with its joints and collateral ligaments in extension (**A**) and in semiflexion (**B**).

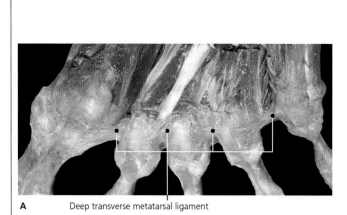

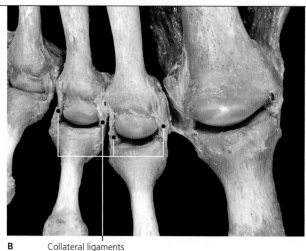

Figure 21-34 Detail of the ligaments at the level of the metatarsophalangeal joints. **A**, Plantar view. **B**, Dorsal view.

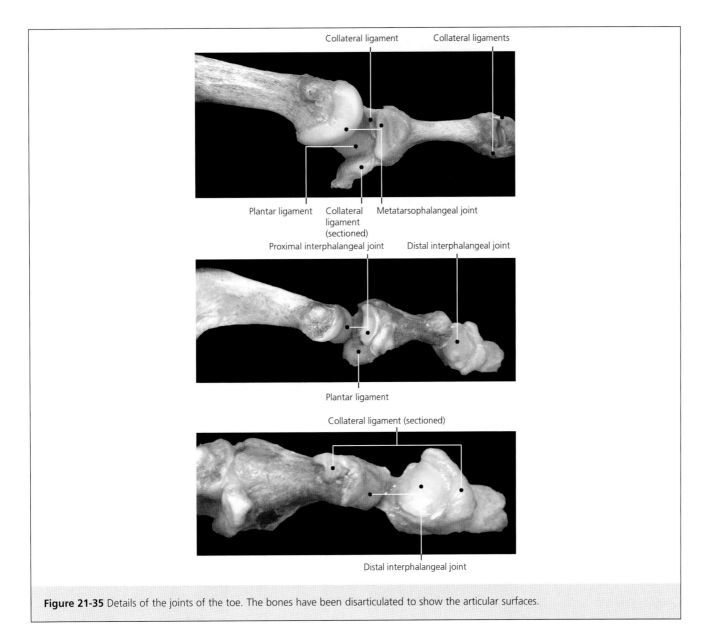

Figure 21-35 Details of the joints of the toe. The bones have been disarticulated to show the articular surfaces.

Intermetatarsal Joints

The intermetatarsal joints are gliding joints formed by the metatarsal bases (Figure 21-32). The first articulates only laterally and the fifth only medially, but the others articulate on both sides. The dorsal, plantar, and interosseous ligaments (Figures 21-27 and 21-31) hold these bones together.

The metatarsals limit the interosseous spaces, where the interosseous muscles are located.

Metatarsophalangeal Joints

The metatarsophalangeal joints are those formed by the metatarsal heads and the bases of the proximal phalanges of the toes (Figures 21-32 through 21-35). Functionally, the second through fifth metatarsophalangeal joints function like condylar joints, allowing flexion-extension and abduction-adduction movements. Because of their morphology, they could also be grouped with the ball-and-socket joints. The metatarsophalangeal joint of the great toe is a trochlear joint.

In the plantar aspect of these joints are dense, channeled plaques called plantar ligaments. These ligaments are intimately related to the deep transverse plantar ligament. There are also collateral ligaments on both sides of these joints.

Interphalangeal Joints

In toes with three phalanges, the interphalangeal joints are formed by the head of the proximal phalanx and the base of the middle phalanx (proximal interphalangeal joint) and by the head of the middle phalanx and the base of the distal phalanx (distal interphalangeal joint) (Figures 21-32, 21-33, and 21-35). Two condyles in the phalangeal heads articulate with two

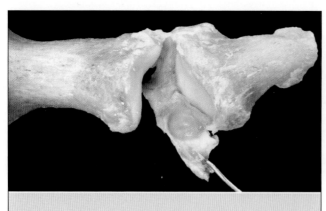

Figure 21-36 Detail of the interphalangeal joint of the great toe.

articular facets separated by a crest, allowing flexion-extension movements.

These joints also have plantar and collateral ligaments.

The great toe has only one interphalangeal joint (Figure 21-36); the head of the proximal phalanx articulates with the base of the distal phalanx. This joint is also trochlear, allowing flexion and extension movements.

Sesamoids

At the level of the head of the first metatarsal are two small sesamoid bones that articulate with the plantar facets of the condyles (Figures 20-26 and 21-37). They have articular facets to achieve small gliding movements, which is why they are gliding joints.

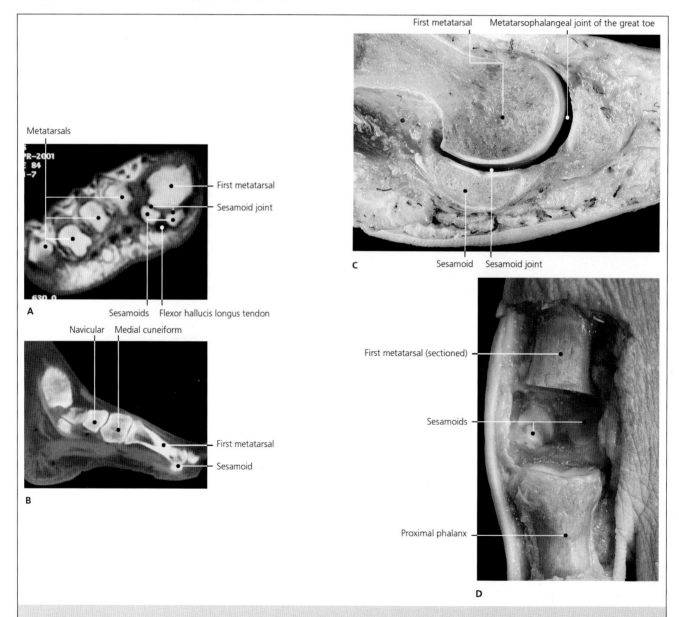

Figure 21-37 Details of the sesamoid joints. A, Coronal (frontal) MRI at the level of the sesamoids. B, Sagittal CT scan at the level of the first ray. C, Detail of a sagittal section of the sesamoid joint. D, Specimen with the first metatarsal head removed to show the sesamoids.

Chapter 22

Myology

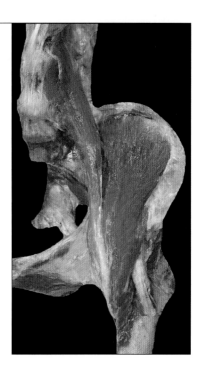

Introduction
This chapter describes the muscles of the pelvic girdle, thigh, leg, and foot. The muscles are grouped in compartments separated by intermuscular septa and fasciae.

There is a difference between the shoulder girdle and pelvic girdle muscles. Because there is no motion between the pelvis and the axial skeleton, the pelvic girdle muscles act on the femur. However, some of them act through the pelvis on the spinal column. Also, some of the muscles can act on the femur or, if the foot is firmly planted, on the hip joint. Because of this, these motions are described in relation to the hip joint.

Muscles of the Pelvic Girdle
The muscles of the pelvic girdle insert into the greater trochanter of the femur and have their origin along the spinal column or in the pelvis (Figure 22-1). There are three muscle groups: posterior, lateral, and anterior.

Posterior Muscles
The posterior muscles of the hip are listed in Table 22-1.

The largest and most superficial gluteal muscle is the gluteus maximus (Figure 22-2). It is a muscle of considerable thickness that helps form the bulk of the buttock. Its origin is from the ilium, posterior to the posterior gluteal line, the dorsal surface of the sacrum and coccyx, and sacrotuberous ligaments (Figure 22-3). Some fibers also originate from the lumbar aponeurosis of the erector spinae muscle and from the gluteal aponeurosis, which extends laterally and enters between the muscle fascicles. These fascicles travel in a caudal and lateral direction to insert into the iliotibial tract, the lateral intermuscular septum, and the gluteal tuberosity of the femur.

The actions of the gluteus maximus are primarily extension but also include some abduction and lateral rotation of the hip. In addition, as the gluteus maximus tightens the iliotibial tract, it helps with knee flexion. The muscle is innervated by the inferior gluteal nerve (L5 through S2).

Between the skin and the insertion of the gluteus maximus is the subcutaneous trochanteric bursa, which cushions the impact on this superficial bony prominence. Between the greater trochanter and this muscle lies the trochanteric bursa of the gluteus maximus. Between the insertion of the gluteus maximus in the gluteal tuberosity and the femur are the intermuscular bursae of the gluteal muscles. Finally, between the inferior portion of the gluteus maximus and the ischial tuberosity is the ischial bursa of the gluteus maximus.

Deep to the gluteus maximus lie the pelvitrochanteric muscles, which originate from the pelvis and insert onto the greater trochanter and intertrochanteric crest. They include the piriformis, the obturator internus, the superior and inferior gemelli, the quadratus femoris, the obturator externus, and the glutei medius and minimus (the last three described in the lateral group).

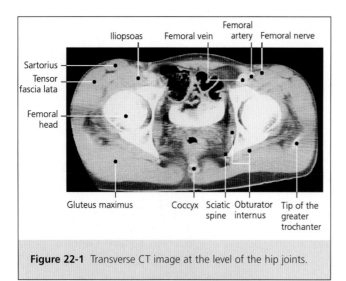

Figure 22-1 Transverse CT image at the level of the hip joints.

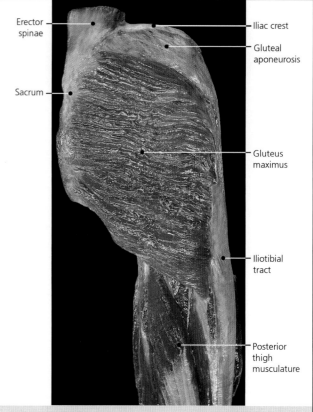

Figure 22-2 Posterior view of the muscles of the buttock and thigh region.

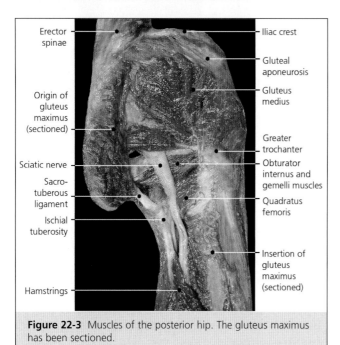

Figure 22-3 Muscles of the posterior hip. The gluteus maximus has been sectioned.

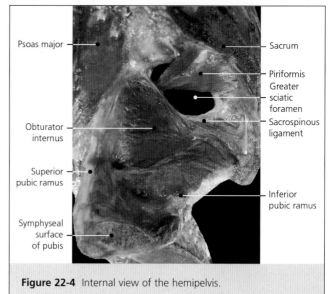

Figure 22-4 Internal view of the hemipelvis.

The piriformis originates from the anterior surface of the sacrum (S2 through S4) between the anterior sacral foramina and the sacrotuberous ligament (Figure 22-4). Its fibers converge and leave the pelvis through the greater sciatic foramen superior to the sciatic nerve (Figure 22-3). The piriformis tendon then inserts onto the greater trochanter. A bursa lies between the trochanter and its tendinous insertion.

The primary function of the piriformis is abduction and external rotation of the hip. It is innervated by a sacral plexus branch, the nerve of the piriformis muscle (S1-S2).

The obturator internus and gemelli muscles (superior and inferior) are located inferior to the piriformis (Figures 22-3 and 22-5). They are derived from a common muscle mass that shares the same insertion. For this reason, they are sometimes referred to as the triceps of the pelvis. The superior and inferior gemelli insert superior and inferior to the obturator internus. The obturator internus muscle originates from the internal aspect of the obturator membrane and adjacent regions of the pelvis (Figure 22-4). Its fibers travel toward the lesser sciatic notch, where they change

Myology

TABLE 22-1 POSTERIOR MUSCLES OF THE HIP

MUSCLE	Gluteus maximus	Piriformis	Obturator internus	Gemelli	Quadratus femoris
ORIGIN	Posterior aspect of the ileum, dorsal sacrum, gluteal aponeurosis, sacrotuberous ligament	Anterior surface of the sacrum, sacrotuberous ligament	Intrapelvic surface of the obturator membrane and surrounding pelvis	*Superior:* Ischial spine *Inferior:* Ischial tuberosity and ramus	Ischial tuberosity, lateral border
INSERTION	Gluteal tuberosity and iliotibial tract	Trochanteric fossa	Trochanteric fossa	Obturator internus tendon (trochanteric fossa)	Below intertrochanteric crest
INNERVATION	Inferior gluteal	Ventral rami of S1-S2	Nerve to the obturator internus	*Superior:* Nerve to the obturator internus *Inferior:* Nerve to the quadratus femoris	Sacral plexus nerve to the quadratus femoris
FUNCTION	Hip extension, external rotation	Hip abduction, external rotation	Hip abduction, external rotation	Hip abduction, external rotation	Hip adduction, external rotation

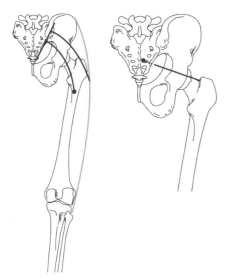
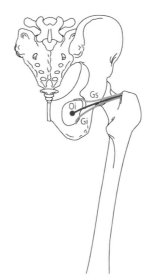
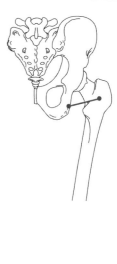

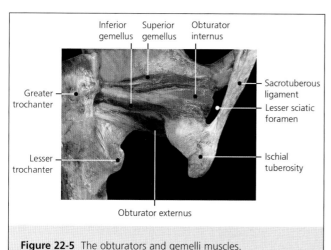

Figure 22-5 The obturators and gemelli muscles.

direction and exit the pelvis through the lesser sciatic foramen (Figure 22-5). Between the border of the sciatic notch and the muscle lies the ischial bursa of the obturator internus. The obturator internus inserts into the trochanteric fossa. The superior gemellus originates from the medial aspect of the ischial spine and crosses the lesser sciatic foramen to also insert into the trochanteric fossa (Figure 22-5). The inferior gemellus originates from the ischial tuberosity and crosses the lesser sciatic foramen, inserting onto the intertrochanteric crest. The principal function of this muscle group is the abduction and external rotation of the hip. These muscles are innervated by sacral plexus branches, the most important being the nerve to the obturator internus (L5 through S2).

The quadratus femoris muscle is the most caudal of the pelvitrochanteric muscles. As its name indicates, its shape is that of a rectangle (Figure 22-3). It origi-

TABLE 22-2 LATERAL MUSCLES OF THE HIP

MUSCLE	Gluteus medius	Gluteus minimus	Obturator externus
ORIGIN	Gluteal aspect of the ilium between the anterior and posterior gluteal lines	Gluteal aspect of the ilium between the anterior and inferior gluteal lines	External surface of the obturator membrane and surrounding pelvis
INSERTION	Greater trochanter	Greater trochanter	Trochanteric fossa
INNERVATION	Superior gluteal nerve	Superior gluteal nerve	Obturator nerve
FUNCTION	Hip abduction	Hip abduction	Hip abduction, and external rotation

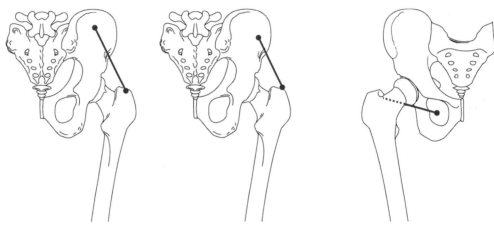

nates in the lateral portion of the ischial tuberosity, courses horizontally, and inserts onto the intertrochanteric crest. The quadratus femoris also causes external rotation of the hip, but in contrast with the other muscles in its group, it is also an adductor. It is innervated by a branch of the sacral plexus, the nerve to the quadratus femoris (L5 and S1).

Lateral Muscles

The lateral muscles of the hip are listed in Table 22-2.

The glutei medius and minimus are located in the gluteal (lateral) region of the ilium (Figure 22-6). The gluteus medius is fan shaped and is covered by the gluteal aponeurosis and the gluteus maximus (Figure 22-2). Its origin is the aponeurosis and the area between the anterior and posterior gluteal lines. Its fibers converge with its tendon inserting onto the lateral surface of the greater trochanter (Figure 22-3). The gluteus minimus, located deep to the medius, originates from between the inferior and anterior gluteal lines. Its fibers converge to insert onto the anterior surface of the greater trochanter.

The two glutei function in a manner similar to the deltoid; their principal function is abduction of the hip. However, the anterior fibers produce internal rotation, and the posterior fibers produce external rotation. They are both innervated by the superior gluteal nerve (L4 through S1).

Between their insertions are several bursae. Between the insertion of the gluteus medius and the greater trochanter is the trochanteric bursa of the gluteus medius. Another bursa of the same name is found between the insertions of the gluteus medius and the piriformis. Finally, the trochanteric bursa of the gluteus minimus is found between its insertion and the greater trochanter.

The obturator externus originates from the external surface of the obturator membrane and the adjacent regions of the pelvis. It inserts onto the trochanteric fossa (Figure 22-7). This muscle is an external rotator of the hip and is innervated by the branch of the obturator nerve (L2 through L4).

Anterior Muscles of the Hip

In the anterior region of the hip lie three muscles: the psoas major and iliacus, which comprise the iliopsoas, and the psoas minor (Table 22-3).

The iliopsoas is formed by the psoas major and the iliacus muscles (Figure 22-8). Its name is derived from the strong tendinous insertion they share onto the lesser trochanter. Both are covered by iliac fascia, thin in its superior portion and thick in its inferior portion.

The psoas major has both a deep and a superficial origin. The deep origin is from the twelfth rib and the anterolateral aspect of the bodies and rib apophyses

MYOLOGY

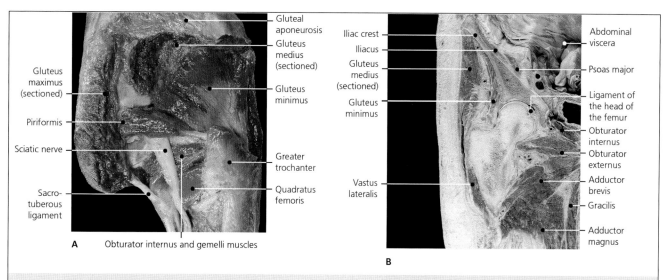

Figure 22-6 A, Specimen with the gluteus medius muscle sectioned to show the adjacent gluteus minimus. **B,** Sagittal section showing the relationship of the gluteus medius to the gluteus minimus.

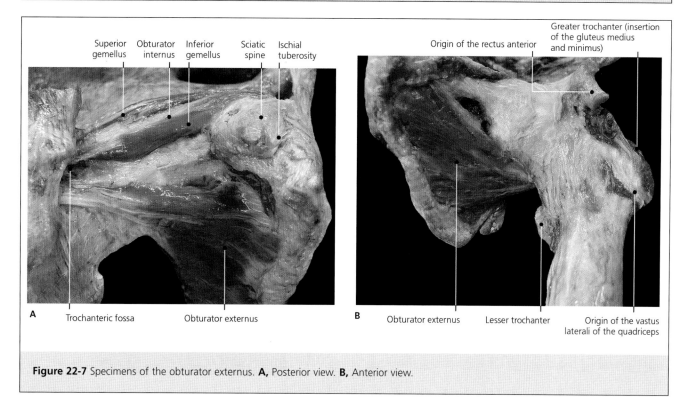

Figure 22-7 Specimens of the obturator externus. **A,** Posterior view. **B,** Anterior view.

TABLE 22-3 ANTERIOR MUSCLES OF THE HIP

MUSCLE	Iliacus	Psoas major	Psoas minor
ORIGIN	Iliac fossa	L1 through L4 bodies and rib apophyses	T12-L1 bodies
INSERTION	Lesser trochanter	Lesser trochanter	Superior pubic ramus
INNERVATION	Lumbar plexus and femoral nerve	Lumbar plexus	Lumbar plexus
FUNCTION	Hip flexion	Hip flexion	No distinct function

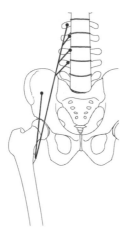
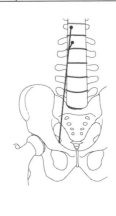

of vertebrae L1 through L4. The superficial origin occurs on the intervertebral disks and adjacent regions of the vertebrae. The psoas major allows for vessels and nerves to cross via tendinous arches. From its insertion, it forms a fusiform muscular body that crosses the muscular window, the lateral space formed between the inguinal ligament and the iliopectineal arch through which the iliopsoas and the femoral nerve pass. Distally, the psoas major merges with the iliacus and inserts onto the lesser trochanter.

The iliacus originates by a fleshy insertion from and covers the iliac fossa. The insertion coalesces with the psoas major, forming a large tendon that inserts onto the lesser trochanter.

The iliopsoas is principally a hip flexor, but it is also an external rotator. It produces lumbar lordosis and is an accessory trunk flexor when the femur is fixed. The psoas major and the iliacus are innervated by branches of the lumbar plexus. The iliacus also receives innervation from the femoral nerve.

After crossing the muscular window, the iliopsoas glides over the anterior aspect of the pelvis and the hip joint. The iliopectineal bursa interposed at this level helps to avoid friction. Another bursa between the tendon and the lesser trochanter is called the subtendinous iliac bursa.

The psoas minor is a variable muscle that originates from the vertebral bodies of T12-L1 and their corresponding intervertebral disks. It inserts into the iliacus fascia and the superior border of the superior pubic ramus (Figure 22-9).

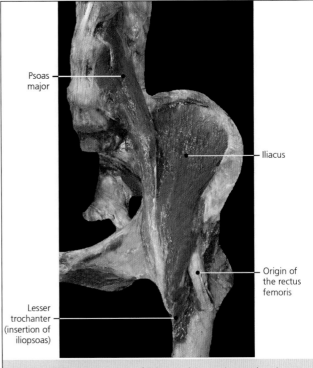

Figure 22-8 Anterior view of a hemipelvis specimen, showing the iliopsoas.

Myology

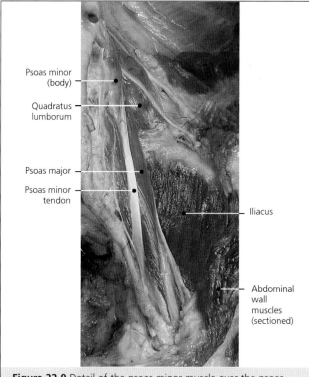

Figure 22-9 Detail of the psoas minor muscle over the psoas major. Anterior view showing the pelvis with the viscera removed.

This rudimentary muscle does not have a notable function, although it may contribute minimally to trunk flexion. It is innervated by branches of the lumbar plexus.

Muscles of the Thigh

All the thigh muscles originate from the pelvis except for the three vasti of the quadriceps and the short head of the biceps femoris.

The inguinal ligament, which separates the abdominal cavity from the thigh, is formed by a reflection of the aponeurosis of the insertion of the external oblique. Beneath the ligament lie two spaces, limited by the inguinal ligament proper, the coxal bone, and the iliopectineal arch. The lateral space is called the muscular window, briefly described earlier, through which course the iliopsoas muscle and the femoral nerve. The medial space is the vascular window; its floor is formed by the pectineus muscle, and through it course the external iliac artery and vein, which become the femoral artery and vein.

The thigh is divided into three compartments: medial, anterior, and posterior (Figures 22-10 through 22-14). All are enveloped by the superficial crural fascia (Figure 22-11) and by the deep fascia lata. In its anterosuperior portion, the crural fascia appears perforated and is known as the cribiformis fascia; through it courses the greater saphenous vein, which empties into the femoral vein. Laterally, the fascia thickens and

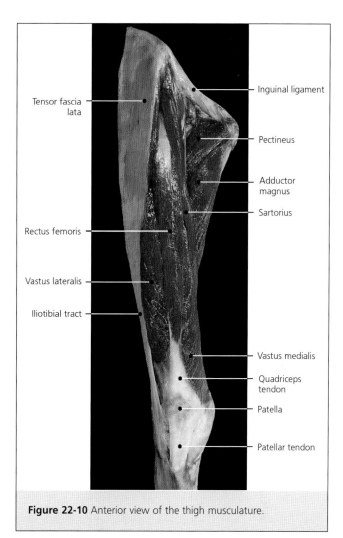

Figure 22-10 Anterior view of the thigh musculature.

forms the iliotibial tract (Figures 22-15 and 22-16), which originates from the anterolateral aspect of the iliac crest and inserts onto Gerdy's tubercle of the tibia.

The muscular compartments of the thigh are separated by intermuscular septi. The anterior and medial compartments are separated by the medial femoral intermuscular septum that courses from the fascia lata to the medial lip of the linea aspera. Likewise, the anterior and posterior compartments are separated by the lateral femoral intermuscular septum that extends from the fascia lata to the lateral lip of the linea aspera.

The muscles of the anterior and posterior compartments cross two joints. They act over both the hip and the knee. In the medial compartment, the gracilis is the only muscle that crosses these two joints.

Anterior Compartment

The tensor fascia lata, sartorius, and quadriceps comprise the anterior compartment (Table 22-4). These muscles help to flex the hip and extend the knee.

The tensor fascia lata is located in the anterolateral region of the thigh (Figure 22-15). It is a flat, thin muscle that originates from the anterior portion of the

American Academy of Orthopaedic Surgeons

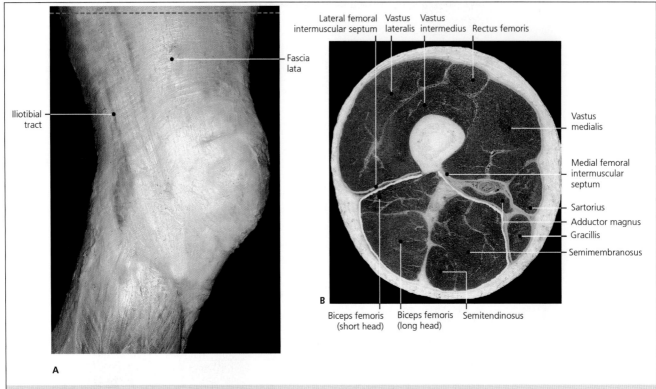

Figure 22-11 A, Thigh and knee with the covering fascias preserved. The dashed green line marks the level of the transverse section of the right thigh. **B,** Transverse section. The anterior compartment is outlined in blue, the posterior compartment is outlined in yellow, and the medial compartment is outlined in green. The medial compartment does not present all its muscles as they are proximal to the section.

iliac crest, immediately lateral to the anterior superior iliac spine. It also originates from the aponeurosis that covers the gluteus medius. Its fibers course obliquely in a dorsal caudal direction to insert into the iliotibial tract and the fascia lata (Figure 22-16). Some authors have noted that a portion of the insertion of the tendon may continue to the lateral portion of the patella and the tibia.

The tensor fascia lata is a very important lateral stabilizer of the hip. It also is an abductor and works with the gluteus medius and minimus to rotate the thigh medially. The tensor fascia lata is innervated by a branch of the superior gluteal nerve.

The sartorius is a long, thin, ribbonlike muscle spanning two joints. It originates from the anterior superior iliac spine and courses obliquely across the thigh (Figure 22-15). Distally, it inserts medially into the pes anserine, located on the proximal medial tibia. This name comes from the common insertion of three muscles, the sartorius, gracilis, and semitendinosus. Their insertion has the appearance of a goose foot, hence the name.

The sartorius was known as the tailor's muscle (from the Latin *sartor*, "tailor") because in ancient Rome tailors repaired clothes while they sat over a crossed leg (the action of the muscle).

The sartorius also forms the limit of the topographic region of the thigh called the femoral triangle (Scarpa's triangle) (Figures 22-15 through 22-17). It is a triangular fascial space in the superoanterior third of the thigh. It is bounded superiorly by the inguinal ligament, medially by the adductor longus, and laterally by the sartorius. The base of the triangle is formed by the inguinal ligament, and its apex is located where the lateral border of the sartorius crosses the medial border of the adductor longus. Its floor is formed by the pectineus and iliopsoas muscles and the roof by the cribiformis fascia. The femoral triangle contains the femoral artery and veins.

The sartorius acts on both the hip and the knee. It is a flexor, abductor, and external rotator of the hip and a flexor of the knee. It is innervated by the femoral nerve (L2 through L4).

The quadriceps femoris, as its name indicates, is formed by four heads or vasti that unite to form one common tendon (Figure 22-11).

The vastus medialis originates from the intertrochanteric line, the medial lip of the linea aspera, and the medial surface of the femoral shaft. It is the muscular portion of the quadriceps that extends the farthest distally and is in contact with the sartorius medially.

The vastus lateralis originates from the greater trochanter, the lateral lip of the linea aspera, and the lateral surface of the femoral diaphysis (Figure 22-18).

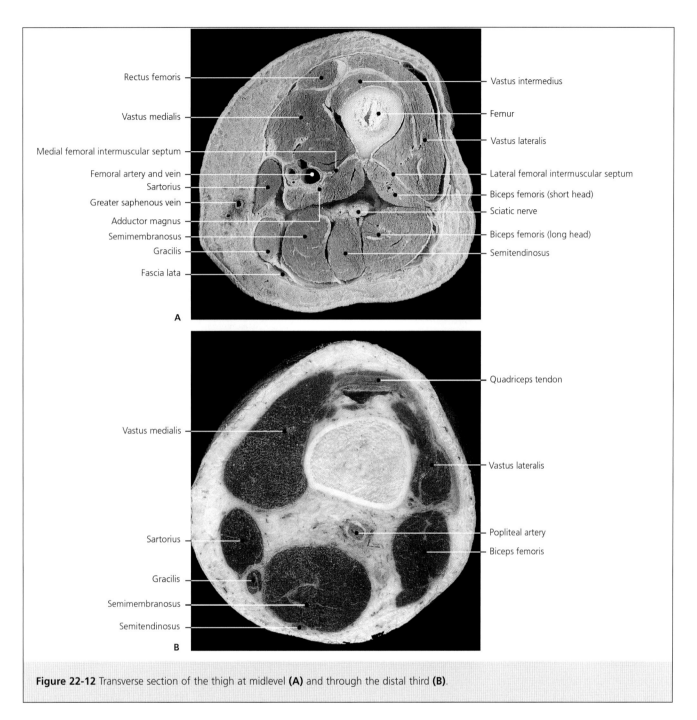

Figure 22-12 Transverse section of the thigh at midlevel **(A)** and through the distal third **(B)**.

Between the vastus medialis and lateralis is the vastus intermedius, which is located deep to the rectus femoris and originates from the anterior and lateral surfaces of the femoral diaphysis. The rectus femoris is the only portion of the quadriceps that crosses both the hip and the knee. It originates from the anteroinferior iliac spine (direct head), the supra-acetabular notch (reflected head), and the anterior capsule of the hip (Figure 22-19). Its fibers then follow the axis of the femur caudally.

The four heads of the quadriceps insert onto the superior pole of the patella through the quadriceps tendon. Some fibers of this tendon (predominantly those of the rectus femoris) cross the patella anteriorly without inserting, while others terminate onto the patella, and those of the vastus medialis and lateralis cross fibers. From the inferior pole of the patella, the patellar tendon and some direct fibers of the rectus femoris insert onto the tibial tuberosity.

From the vasti medialis and lateralis, aponeurotic expansions course along the medial and lateral aspects of the patella; these are the medial and lateral patellar retinaculi (Figure 22-20), which insert onto the respective condyles of the tibia. They are difficult to

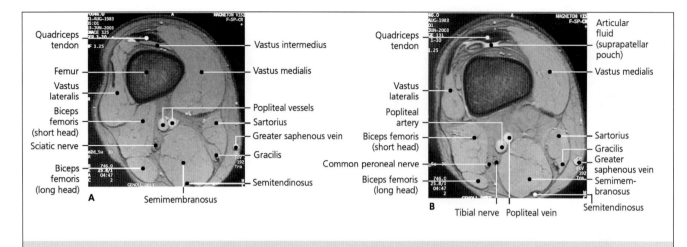

Figure 22-13 Transverse MRIs of the distal third of the thigh. **A,** More proximal cut. **B,** More distal cut.

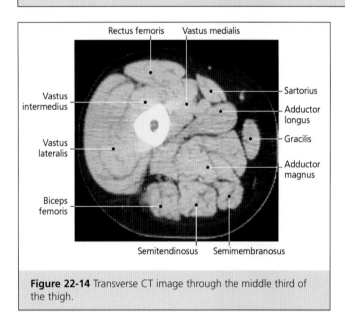

Figure 22-14 Transverse CT image through the middle third of the thigh.

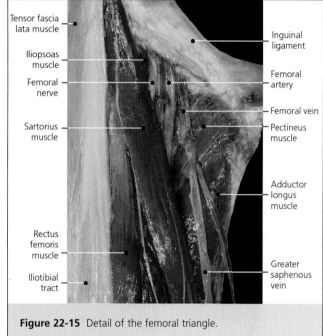

Figure 22-15 Detail of the femoral triangle.

identify in the fascial criss-crossing of this knee region.

Deep to the vastus intermedius lies a small bundle of fibers that insert in the articular capsule of the knee and thus are called the genu articularis. Its function is to tighten the capsule during extension of the knee.

The quadriceps act together as a knee extensor. In addition, as a biarticular muscle, the rectus femoris aids in hip flexion. The quadriceps muscle is innervated by the femoral nerve (L2 through L4).

Medial Compartment

The medial compartment is located in the medial region of the thigh. These muscles are also called the adductors of the hip. The muscles that form this group are the pectineus, adductor longus, adductor brevis, adductor magnus, and gracilis (Table 22-5).

The pectineus is the most proximal muscle in this compartment and forms the floor of the femoral triangle (Figures 22-15, 22-18, and 22-21). It originates from the crest and pectineal ligament of the superior pubic ramus and courses obliquely to insert onto the pectineal line of the femur just inferior to the lesser trochanter. The fascia of this muscle contributes to the formation of the iliopectineal arch and the pectineal ligament.

The principal functions of the pectineus muscle are adduction and lateral rotation of the thigh; it also assists in hip flexion. It is innervated by the femoral nerve (L2 through L4) and occasionally from the anterior branch of the obturator nerve (L2 through L4). It frequently receives innervation from the accessory obturator nerve (L3-L4).

The adductor longus originates just lateral to the pubic symphysis (Figures 22-15, 22-21, and 22-22).

MYOLOGY

TABLE 22-4 ANTERIOR MUSCLES OF THE THIGH

MUSCLE	Tensor fascia lata	Sartorius	Quadriceps
ORIGIN	Iliac crest, anterior superior iliac spine	Anterior superior iliac spine	*Rectus femoris:* Anterior inferior iliac spine, superior acetabulum, anterior hip capsule *Vastus medialis:* Medial aspect of femur, intertrochanteric line, linea aspera *Vastus intermedius:* Anterior aspect of femur *Vastus lateralis:* Lateral aspect of femur, linea aspera, greater trochanter
INSERTION	Iliotibial tract, fascia lata	Tibia (superficial pes anserine)	Base of the patella, tibial tuberosity via the patellar tendon
INNERVATION	Superior gluteal nerve	Femoral nerve	Femoral nerve
FUNCTION	Abduction, flexion, and internal rotation of the hip; assists in knee extension	Flexion, abduction, external rotation of the hip; flexion and internal rotation of the knee	Knee extension *Rectus femoris:* Hip flexion

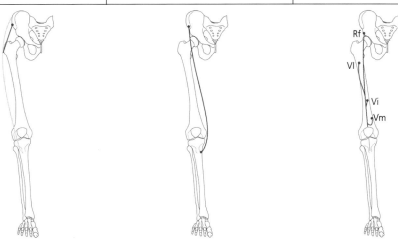

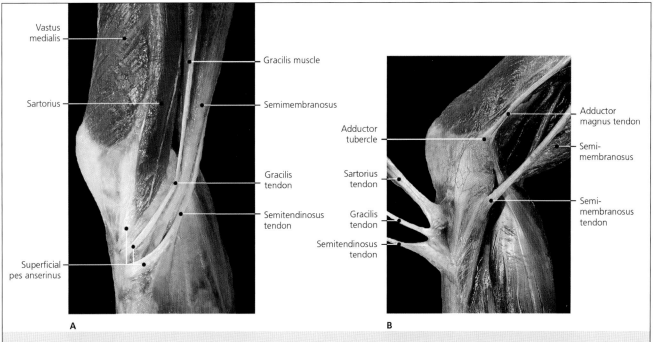

Figure 22-16 A, Medial view of the knee region showing the insertions of the superficial pes anserine. **B,** With the insertions lifted.

American Academy of Orthopaedic Surgeons

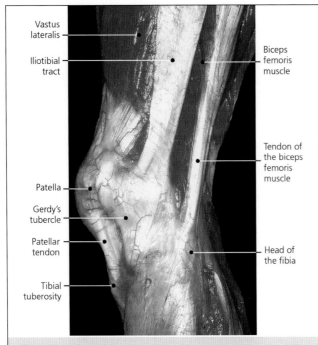

Figure 22-17 Lateral view of knee, showing the insertions of the biceps femoris, the iliotibial tract, and the vastus lateralis.

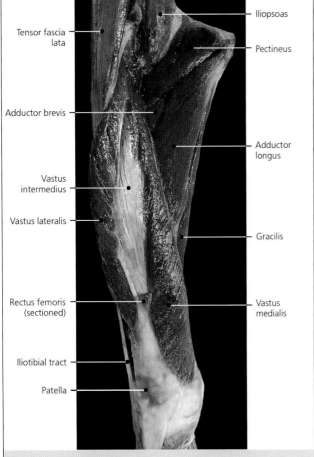

Figure 22-18 Anterior view of thigh musculature with the rectus femoris and sartorius muscles removed.

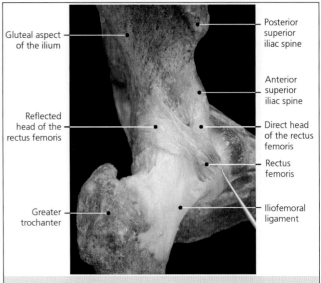

Figure 22-19 Detail of the direct and reflected origins of the rectus femoris muscle (anterolateral view).

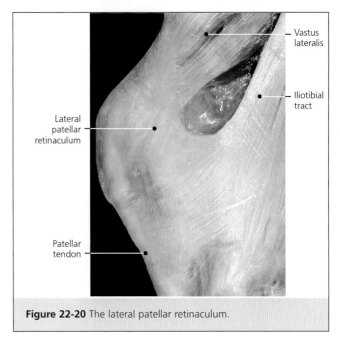

Figure 22-20 The lateral patellar retinaculum.

From this origin, this muscle's fibers diverge slightly and course distally and laterally to insert onto the middle third of the lateral border of the linea aspera.

As its name indicates, the principal function of the adductor longus is adduction of the hip, but it also contributes to external rotation and slight flexion. The adductor longus is innervated by muscular branches of the anterior branch of the obturator nerve (L2 through L4).

The adductor brevis lies deep to the pectineus and the adductor longus muscles (Figures 22-15 and 22-22). It originates from the inferior pubic ramus, and its fibers course distally and laterally. It inserts onto the proximal portion of the lateral border of

MYOLOGY

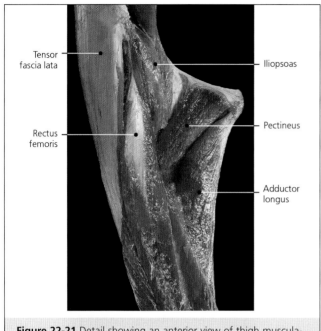

Figure 22-21 Detail showing an anterior view of thigh musculature. The sartorius has been removed.

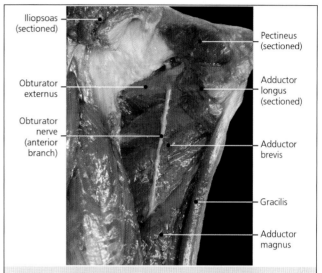

Figure 22-22 Detail of the medial compartment showing the anterior branch of the obturator nerve.

TABLE 22-5 MUSCLES OF THE MEDIAL COMPARTMENT OF THE THIGH

MUSCLE	Pectineus	Adductor longus	Adductor brevis	Adductor magnus	Gracilis
ORIGIN	Pectineal crest (superior pubic ramus)	Pubic body	Inferior pubic ramus	Ischial ramus, inferior pubic ramus	Inferior pubic ramus
INSERTION	Pectineal line	Lateral lip of linea aspera (distal)	Lateral lip of linea aspera (proximal)	Lateral lip of linea aspera and adductor tubercle	Tibia (superficial pes anserine)
INNERVATION	Femoral and obturator nerves	Obturator anterior division	Obturator anterior division	Obturator nerve (adductor) Tibial nerve (hamstring)	Obturator nerve
FUNCTION	Adductor, external rotator of thigh	Adductor, external rotator of thigh	Adductor, external rotator of thigh	Adductor, extensor, external rotator of thigh	Thigh adductor, knee flexor

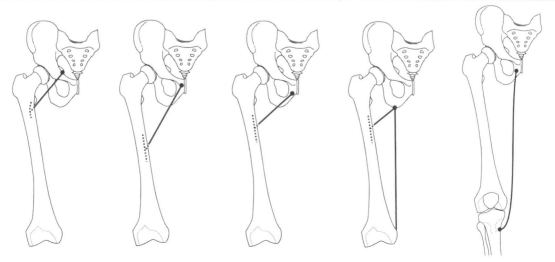

American Academy of Orthopaedic Surgeons

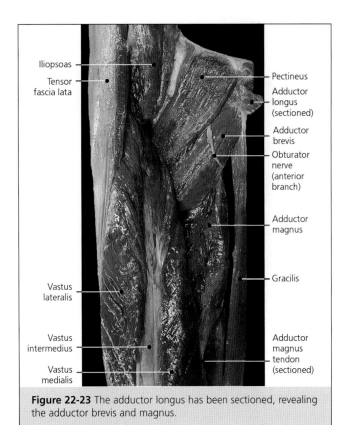

Figure 22-23 The adductor longus has been sectioned, revealing the adductor brevis and magnus.

the linea aspera proximal to the insertion of the adductor longus.

Its most important function is hip adduction, although it is also an external rotator and slight flexor. The adductor brevis is innervated by muscular branches of the anterior branch of the obturator nerve (L2 through L4).

The adductor magnus muscle is the largest muscle mass in the adductor compartment (Figure 22-23). It originates from the ischial tuberosity and the ischial ramus and forms a large muscle with two portions. The lateral portion is fan shaped, and its fibers open up laterally to insert onto the medial lip of the linea aspera. Some authors designate the adductor minimus as the proximal anterior portion of this muscle. The most medial portion has a fusiform shape and descends caudally, continuing with a tendon that inserts into the adductor tubercle of the femur (Figure 22-16). Between both insertions is the adductor hiatus.

This muscle is a potent hip adductor and external rotator (lateral portion), as well as a hip extensor (medial portion). It is innervated by muscular branches of the posterior division of the obturator nerve (L2 through L4) (lateral portion), while the medial portion is innervated by the sciatic nerve (L4).

TABLE 22-6 MUSCLES OF THE POSTERIOR COMPARTMENT OF THE THIGH

MUSCLE	Semimembranosus	Semitendinosus	Biceps femoris
ORIGIN	Ischial tuberosity	Ischial tuberosity	*Long head:* Ischial tuberosity *Short head:* Lateral lip of linea aspera
INSERTION	Posterior aspect of proximal tibia (deep pes anserinus)	Medial aspect of the tibia (superficial pes anserinus)	Fibular head
INNERVATION	Tibial division of sciatic nerve	Tibial division of sciatic nerve	*Long head:* Tibial nerve *Short head:* Peroneal nerve
FUNCTION	Knee flexion and hip extension, trunk extension	Knee flexion and hip extension, trunk extension	Knee flexion and hip extension

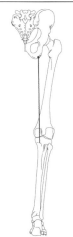  

Myology

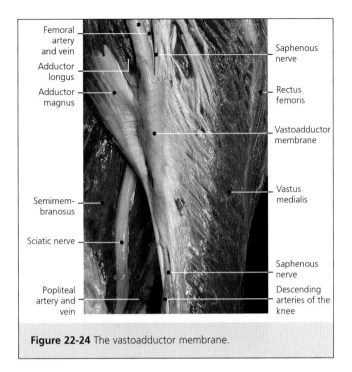

Figure 22-24 The vastoadductor membrane.

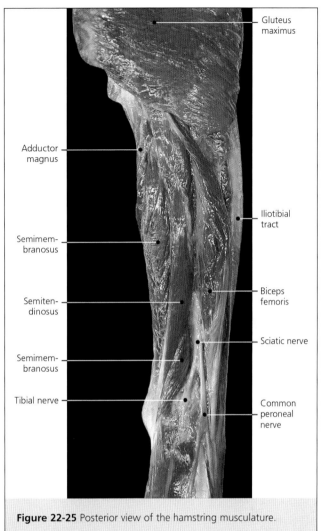

Figure 22-25 Posterior view of the hamstring musculature.

The gracilis (rectus internus) is a long, straplike muscle. It also crosses two joints. It originates from the inferior pubic ramus, below the adductor brevis. It is the most superficial of the adductors. Its fibers travel caudally in the medial aspect of the thigh (Figures 22-15 and 22-23). Distally, its tendon courses around the medial condyles of the femur and the tibia, and it inserts through the pes anserine onto the tibia medial to the tuberosity (Figure 22-16).

Crossing both the hip and the knee joints, the gracilis works as an adductor of the hip joint and a flexor and internal rotator of the knee joint. This muscle is innervated by the anterior branch of the obturator nerve (L2 through L4).

Between the vastus medialis of the quadriceps and the adductor musculature, principally the adductor magnus and the longus, is the vastoadductor membrane (Figure 22-24). This forms the roof of the adductor canal (of Hunter), through which the femoral vessels pass from an anterior position in a medial and caudal direction, through the adductor hiatus to become posterior in the popliteal fossa.

Posterior Compartment

Three muscles form the posterior musculature of the thigh: the semimembranosus, semitendinosus, and biceps femoris (Table 22-6 and Figures 22-11 and 22-25). They are also known as the hamstring muscles.

The semitendinosus is located in the posteromedial region of the thigh and is the most superficial of the hamstrings. It originates from the ischial tuberosity (Figure 22-26) as a conjoined tendon with the long head of the biceps femoris. Its fibers course obliquely in an inferomedial direction. In the distal third of the thigh, its muscular fibers are continued by a long tendon that inserts onto the tibia, medial to the tibial tuberosity through the pes anserinus.

This biarticular muscle extends the thigh and flexes the knee. It is innervated by the tibial division of the sciatic nerve (L5 through S3).

The semimembranosus is located deep to the semitendinosus and receives its name from the strong, broad, flat membrane through which it originates from the ischial tuberosity (Figures 22-26 and 22-27). In the medial third of the thigh, it forms its thick muscular mass and travels distally and medially to insert onto the posterior aspect of the tibia. This muscle inserts through a tendon with three fascicles: The direct fascicle continues in the same direction as the muscle and inserts in the posterior aspect of the medial condyle. The reflected fascicle goes toward the superior aspect of the medial plateau of the tibia, coursing under the medial collateral ligament. Finally, the recurrent fascicle courses toward the posterior

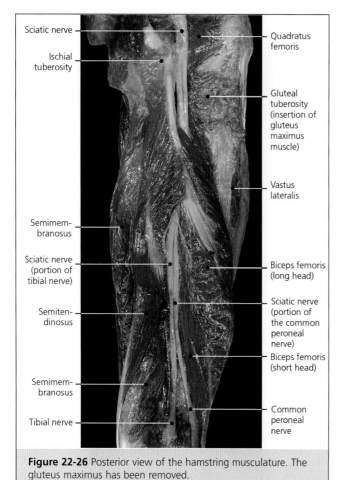

Figure 22-26 Posterior view of the hamstring musculature. The gluteus maximus has been removed.

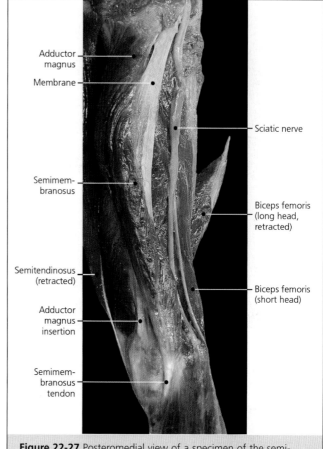

Figure 22-27 Posteromedial view of a specimen of the semimembranosus muscle.

aspect of the lateral condyle, mixing with the fibers of the capsule, thus forming the oblique popliteal ligament. Altogether, these three insertions form the so-called deep pes anserinus (Figure 22-28).

The semimembranosus muscle is a hip extensor and a knee flexor. It is innervated by the tibial division of the sciatic nerve (L5 through S3).

For diagnostic imaging, it is important to remember that the medial muscular mass in the proximal and posterior aspect of the thigh is the semitendinosus because at this level the semimembranosus is a thin membrane. Distally, however, the composition is reversed: The semitendinosus is tendinous and the semimembranosus is muscular.

The biceps femoris muscle, as its name indicates, is formed by two heads, one long and one short. The long head originates in a tendon shared with the semitendinosus from the ischial tuberosity (Figures 22-25 and 22-26). Its fibers course distally and laterally toward the fibular head. The short head originates from the distal third of the femoral shaft, the lateral lip of the linea aspera, and lateral intermuscular septum, merging with the long head (Figures 22-12, 22-26, and 22-29). The muscular fibers continue with its tendon, which has its principal insertion onto the fibular head near the insertion of the lateral collateral ligament. Aponeurotic expansions that extend to the posterior aspect of the tibia also may be present.

The two heads of the biceps are knee flexors. In addition, the long head is biarticular and acts as a hip extensor. The long head of the biceps is innervated by the tibial division of the sciatic nerve (L5 through S3), whereas the short head is innervated by the peroneal division (L4 through S2).

Muscles of the Leg

The muscles of the leg are organized into three compartments: anterior, posterior, and lateral (peroneal) (Figures 22-30 and 22-31). All are surrounded by the crural fascia that inserts into the free borders of the tibia and fibula and offers surface for muscle insertion. The crural fascia is contiguous proximally with the thigh fascia and distally with the foot fascia. The tibia and fibula, with the interosseous membrane located between them, separate the anterior and posterior compartments. The anterior and lateral compartments are separated by the anterior intermuscular septum of the leg, which extends from the anterior border of the fibula to the crural fascia. The posterior and lateral compartments are separated by the poste-

MYOLOGY

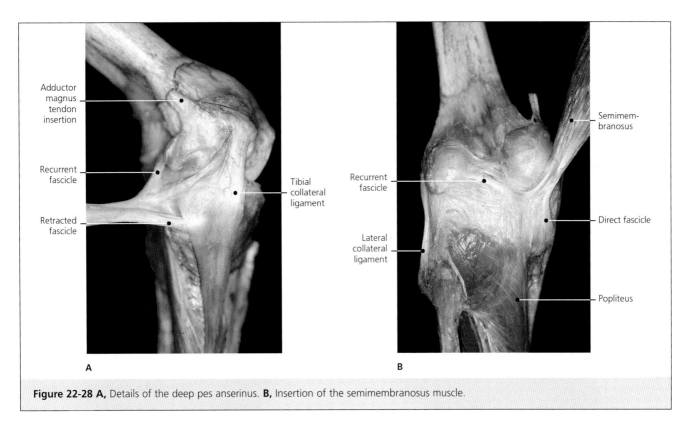

Figure 22-28 A, Details of the deep pes anserinus. **B,** Insertion of the semimembranosus muscle.

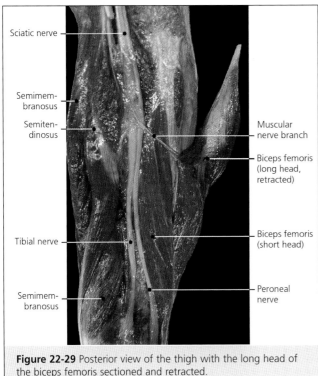

Figure 22-29 Posterior view of the thigh with the long head of the biceps femoris sectioned and retracted.

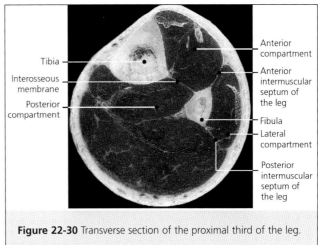

Figure 22-30 Transverse section of the proximal third of the leg.

rior intermuscular septum of the leg, which extends from the posterior border of the fibula to the crural fascia.

Most of these muscles form the extrinsic musculature of the foot.

Posterior Compartment

The posterior compartment has seven muscles: the plantaris, the gastrocnemius, the soleus, the popliteus, the flexor digitorum longus, the flexor hallucis longus, and the tibialis posterior (Table 22-7). They are primarily plantar flexors of the ankle and toes. They are all innervated by the tibial nerve and are described here from superficial to deep.

The gastrocnemius is the most superficial muscle of the posterior compartment and, with the soleus, forms the characteristic shape of the sural region, the "calf" (Figure 22-32). It originates from two heads that combine with the soleus muscle to form the triceps surae.

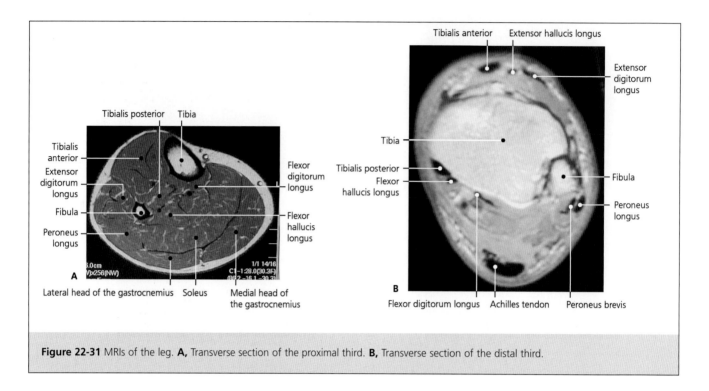

Figure 22-31 MRIs of the leg. **A,** Transverse section of the proximal third. **B,** Transverse section of the distal third.

TABLE 22-7 MUSCLES OF THE POSTERIOR COMPARTMENT OF THE LEG

MUSCLE	Gastrocnemius	Soleus	Plantaris
ORIGIN	*Medial head:* medial femoral condyle *Lateral head:* lateral femoral condyle	Posterior aspect of the fibular head, tibial condyle and body of the tibia (soleus line)	Lateral femoral condyle, oblique popliteal ligament
INSERTION	Calcaneal tuberosity through Achilles tendon	Calcaneal tuberosity through Achilles tendon	Calcaneal tuberosity
INNERVATION	Tibial	Tibial	Tibial
FUNCTION	Plantar flexion of the ankle, foot supination, knee flexion	Plantar flexion of the ankle, foot supination	Plantar flexion of the ankle, knee flexion

The lateral and medial heads originate from the posterior nonarticular aspects of the lateral and medial femoral condyles, respectively (Figure 22-27). The medial and lateral heads comprise the distal borders of the popliteal fossa. This fossa is limited proximally by the semimembranosus muscle (medially) and the biceps femoris (laterally). The contents of this fossa are the popliteal artery and vein as well as the bifurcation of the sciatic nerve into the common peroneal and the tibial nerves. The roof is formed by the popliteal fascia, which is pierced by the lesser saphenous vein, which empties into the popliteal vein. Between the gastrocnemius heads and the femoral condyles are the medial and lateral subtendinous gastrocnemius bursae. The two heads of the gastrocnemius course caudally parallel to each other, inserting onto the dorsal surface of a thick aponeurosis. This aponeurosis joins that of the soleus to form the Achilles tendon, a broad tendon with a triangular shape that inserts onto the posterior calcaneal tuberosity (Figure 22-32). Between the calcaneus and the Achilles tendon is the retrocalcaneal bursa. Posteriorly, between the tendon and the skin, is the subcutaneous calcaneal bursa.

In the deep portion of the origin of the lateral head of the gastrocnemius is a sesamoid bone called the fabella. It can be seen radiologically at the level of the lateral condyle of the femur (Figure 22-33).

The gastrocnemius acts to plantar flex the ankle and supinate the foot. Because the muscle's heads are proximal to the knee joint, it is also a knee flexor. It is innervated by muscular branches of the tibial nerve (S1-S2).

The soleus muscle is broad and flat. It is located deep to the gastrocnemius (Figure 22-34). It originates from a thick aponeurosis from the posterior aspect of the head of the fibula and posterior medial surface of the proximal tibia and its shaft (along the soleus line). The tendinous arch of the soleus is formed by its two insertions, and the tibial neurovascular bundle courses through it.

Its medial and lateral fibers insert in both sides of an aponeurotic lamina located in its posterior surface, and its intermediate fibers descend perpendicularly. This gives the muscle a piriformis morphology. The aponeurosis continues with the Achilles tendon to insert onto the calcaneal tuberosity.

TABLE 22-7 MUSCLES OF THE POSTERIOR COMPARTMENT OF THE LEG (CONT.)

Popliteus	Tibialis posterior	Flexor hallucis longus	Flexor digitorum longus
Posterior aspect of the proximal epiphysis of the tibia	Interosseous membrane and adjacent areas of the tibia and fibula	Posterior surface of the fibula	Posterior surface of the tibia
Lateral epicondyle of the femur	Navicular, cuneiforms, metatarsal bases, cuboid	Distal phalanx of the great toe	Distal phalanx of the triphalangeal toes 2-5
Tibial	Tibial	Tibial	Tibial
Flexion and internal rotation of the knee	Plantar flexion and inversion of the hindfoot	Plantar flexion of the ankle and great toe	Plantar flexion of the ankle and toes 2-5

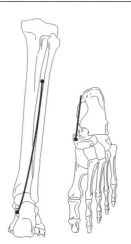
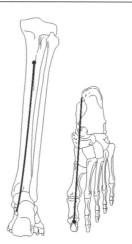
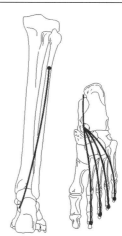

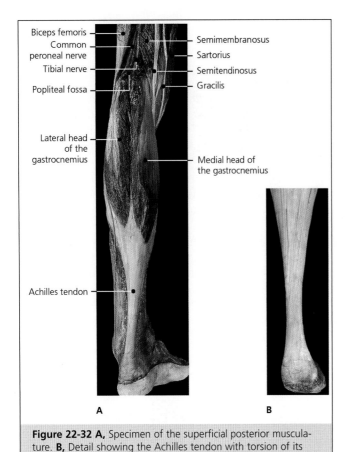

Figure 22-32 A, Specimen of the superficial posterior musculature. **B,** Detail showing the Achilles tendon with torsion of its fibers.

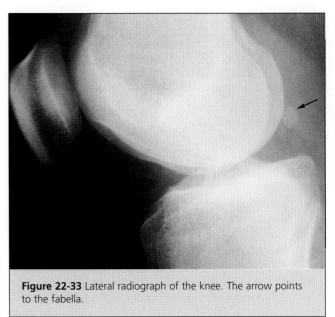

Figure 22-33 Lateral radiograph of the knee. The arrow points to the fabella.

This muscle is a plantar flexor of the ankle and foot supinator. It is innervated by muscular branches of the tibial nerve (L5-S1).

Deep and proximal to the lateral head of the gastrocnemius is the plantaris (Figure 22-34). It originates just proximal to the lateral femoral condyle. Its muscular fibers appear conical or tear shaped, with the base proximal, at its origin, and its tip distal. This muscle has a small body continued by a long, thin tendon that courses distally and medially. In the proximal third of the leg, the plantaris is located deep to the lateral head of the gastrocnemius, and at the level of the beginning of the Achilles tendon, the plantaris is located medial to it. The tendon inserts medially adjacent to the Achilles tendon in the calcaneal tuberosity.

Because of its position, the plantaris is a knee flexor and an ankle plantar flexor, but because of its size and morphology, it does not have significant function compared with its neighbors. It is innervated by muscular branches of the tibial nerve (L5-S1).

The popliteus (Figures 22-34 and 22-35) is the deepest of the proximal muscles of this compartment and forms the floor of the distal aspect of the popliteal fossa. Its proximal attachment is on the lateral epicondyle of the femur. There is a notch formed by its attachment. The tendon heads posteriorly around the lateral femoral condyle deep to the lateral collateral ligament. Between the tendon and the condyle is a subpopliteal recess. The distal muscle belly attaches to the posterior aspect of the proximal tibial epiphysis. It is covered by fascia reinforced by the semimembranosus muscle, the oblique popliteal ligament. In the transition zone between the muscular and tendinous fibers is the arcuate popliteal ligament. This muscle is an important lateral support of the knee. The muscular portion is flat and inserts onto the posterior aspect of the tibia above the soleus line, an area called the popliteal line by some authors. This is the only muscle with a distal origin and a proximal insertion.

The popliteus muscle is principally an internal rotator of the knee. It is also a knee flexor. It is innervated by muscular branches of the tibial nerve (L4 through S1).

Deep to the soleus muscle is a group of three muscles that occupy the most internal region of the posterior compartment; they originate from the interosseous membrane and the diaphysis of the tibia and fibula (Figure 22-36). The superficial and deep muscles are separated by the deep crural fascia. The deep muscles are the posterior tibialis, the flexor hallucis longus, and the flexor digitorum longus. Their homonymous muscles are in the anterior compartment.

The posterior tibialis muscle is located adjacent to the interosseous membrane (Figures 22-36 and 22-37). It is the deepest of the three and is located between the two flexors of the toes. The posterior tibialis muscle originates from the interosseous membrane and adjacent regions of the bodies of the tibia and fibula. In the distal third of the leg, it forms a thick

MYOLOGY

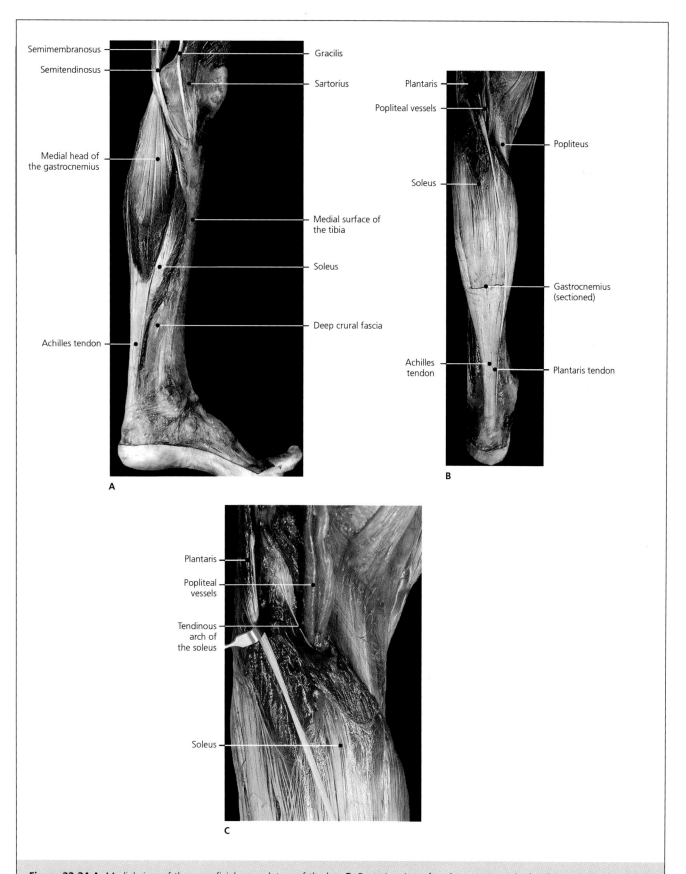

Figure 22-34 A, Medial view of the superficial musculature of the leg. **B,** Posterior view after the gastrocnemius has been removed. **C,** Detail of the arch of the soleus.

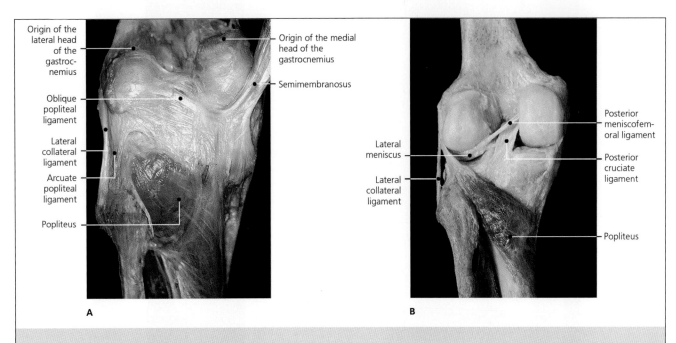

Figure 22-35 Specimens of the popliteal muscle **(A)** with the tendon of the semimembranosus and ligaments and **(B)** isolated.

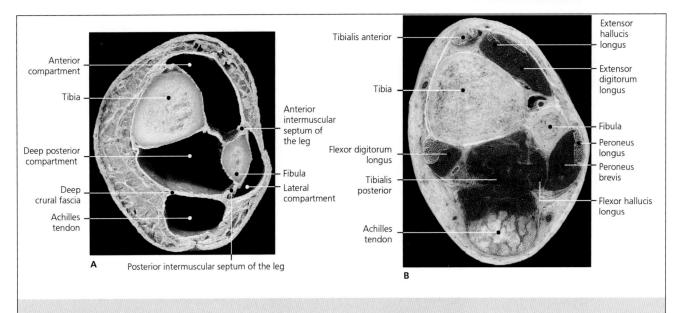

Figure 22-36 A, Compartments of the leg. **B,** Transverse section of the distal third of the leg.

MYOLOGY

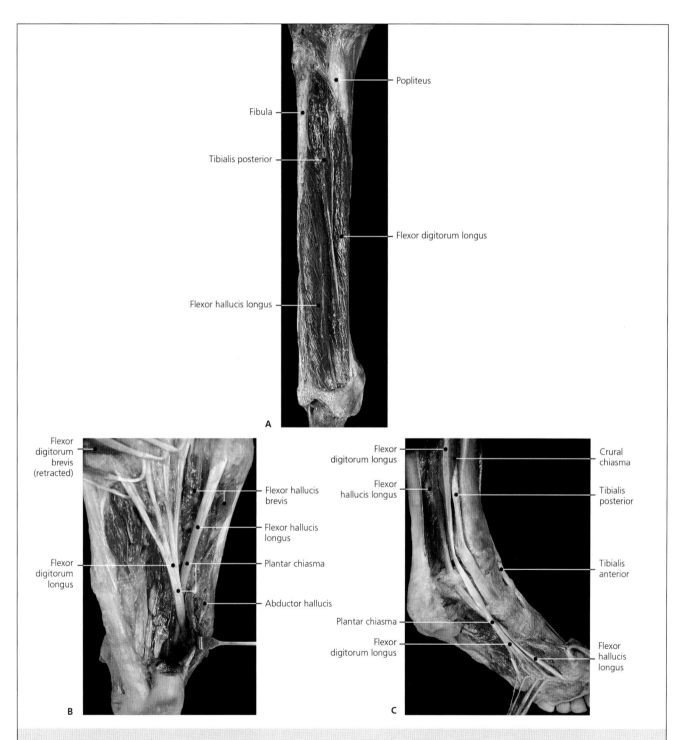

Figure 22-37 Specimens of the posterior compartment, deep layer. **A,** Muscles of this layer. **B,** Plantar view with the flexor digitorum brevis sectioned and retracted. **C,** Oblique medial view.

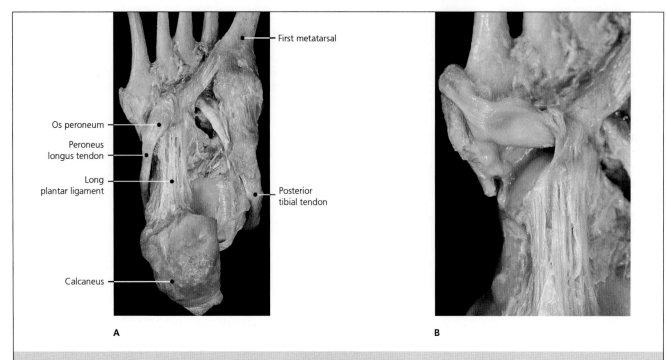

Figure 22-38 A, Plantar view of a right foot showing tendon and insertion in the sole of the peroneus longus. **B,** Enlarged view of peroneum in the same specimen.

tendon that courses distally and medially toward the medial malleolus. In the gliding zone of this tendon over the medial malleolus is the malleolar sulcus of the posterior aspect of the distal tibial epiphysis. The tendon is located superior to the sustentaculum of the talus, where it can have an insertion fascicle, and it courses toward the plantar aspect of the foot. It inserts chiefly onto the navicular, although it emits tendinous rami to the cuneiforms, cuboid, and metatarsals (Figure 22-38).

The posterior tibialis muscle is responsible for plantar flexion of the ankle and hindfoot inversion (supination, plantar flexion, and adduction). It is innervated by the tibial nerve (L5-S1).

The flexor digitorum longus is also a long muscle and is located medial to the posterior tibialis (Figures 22-36 and 22-37). The flexor digitorum longus originates from the posterior surface of the tibia, and its fibers merge into a tendon at the level of the posterior ankle. The flexor digitorum longus tendon then crosses over the posterior tibialis in the so-called crural chiasma, is located posteriorly in the retromalleolar region, and then courses medially toward the plantar aspect of the foot, where it crosses superficial to the tendon of the flexor hallucis longus. In the plantar aspect of the foot, it divides into four tendons for the lesser toes. These tendons insert onto the bases of the distal phalanges. As in the upper limb, these tendons pass through a buttonhole formed by the flexor digitorum brevis tendons.

The flexor digitorum longus is a plantar flexor of the ankle and a toe flexor for toes 2 through 5. It is innervated by the tibial nerve (L5-S1).

The flexor hallucis longus is located in the deep and lateral region of the posterior compartment of the leg (Figures 22-36 and 22-37). It originates from the posterior surface of the fibular diaphysis and adjacent interosseous membrane. Its muscular fibers then descend caudally and medially to the posteromedial region of the ankle. The flexor hallucis longus is continued by a tendon that leaves its impression in the talus and the calcaneus. Along the talus, the tendon passes through a sulcus between the medial and lateral tubercles of the posterior process of the talus. The tendon continues toward the plantar aspect of the foot, coursing through its sulcus inferior to the sustentaculum tali of the calcaneus. It courses toward the first ray, where it inserts onto the base of the distal phalanx of the great toe. En route, it crosses the flexor digitorum longus at the knot of Henry. The flexor hallucis longus tendon emits some tendinous interconnections to the flexor digitorum longus insertions of the second and third toes (Figure 22-39).

The flexor hallucis longus is a flexor of the great toe and a plantar flexor of the ankle. It is innervated by the tibial nerve (L5 through S2).

These three muscles, along with the tibial nerve and the posterior tibial vessels, course posterior to the medial malleolus through a space known as the tarsal tunnel, the roof of which is formed by the flexor retinaculum

(Figure 22-40). It consists of a fibrous expansion that extends from the medial malleolus to the calcaneus. Deep to it course the tendons protected by their sheaths. In the plantar osteofibrous channels, the flexor digitorum tendons are also protected by their respective sheaths.

Anterior Compartment

The anterior compartment is located between the lateral surface of the tibia, the interosseous membrane, and the anterior intermuscular septum of the leg (Table 22-8 and Figure 22-36). This compartment contains three muscles, which are homonymous to the deep posterior compartment muscles. These muscles are the anterior tibialis, extensor digitorum longus, and extensor hallucis longus. In addition, there is a small, more variable muscle known as the peroneus tertius. All these muscles are innervated by the deep peroneal nerve.

In the anterior aspect of the ankle are two retinaculi that restrain the anterior musculature (Figures 22-41 and 22-42). The superior retinaculum of the extensor muscles is a reinforcement of the crural fascia, which can be observed as a transverse thickening from the superior region of the medial malleolus to the inferior border of the lateral malleolus. The inferior retinaculum of the extensor muscles has a V shape. The superior band originates from the medial aspect of the medial malleolus and courses obliquely toward the apex located near the lateral surface of the calcaneus. The inferior

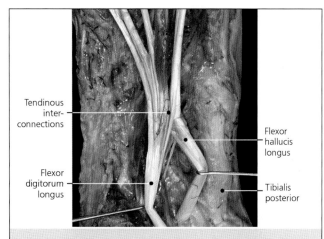

Figure 22-39 Plantar view of a right foot showing details of the tendinous interconnections between the flexor hallucis longus and the flexor digitorum longus.

TABLE 22-8 MUSCLES OF THE ANTERIOR AND LATERAL COMPARTMENTS OF THE LEG

MUSCLE	Anterior tibialis	Extensor digitorum longus	Extensor hallucis longus	Peroneus longus	Peroneus brevis
ORIGIN	Lateral aspect of the tibia and interosseous membrane	Medial aspect of the fibula and interosseous membrane	Interosseous membrane	Lateral aspect of the fibula (proximal two thirds)	Lateral aspect of the fibula (inferior two thirds)
INSERTION	Medial cuneiform and plantar aspect of the first metatarsal base	Distal phalanx of toes 2-5	Distal phalanx of the great toe	Medial cuneiform and plantar aspect of the first metatarsal	Tuberosity of the fifth metatarsal
INNERVATION	Deep peroneal	Deep peroneal	Deep peroneal	Superficial peroneal	Superficial peroneal
FUNCTION	Dorsiflexion of the ankle and inversion of the foot	Extension of toes (2-5) and ankle dorsiflexion	Extension of the great toe, ankle dorsiflexion	Plantar flexion of the ankle and eversion of the foot	Plantar flexion of the ankle and eversion of the foot

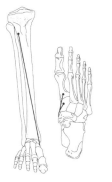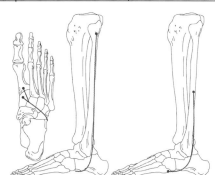

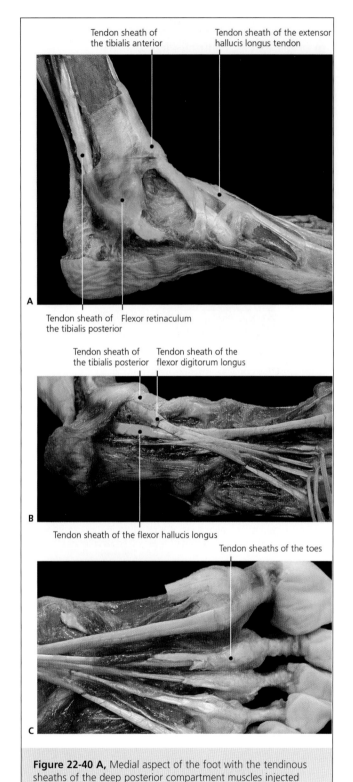

Figure 22-40 A, Medial aspect of the foot with the tendinous sheaths of the deep posterior compartment muscles injected with latex. **B,** Plantar aspect with only the sheath of one toe injected. **C,** Distal plantar aspect.

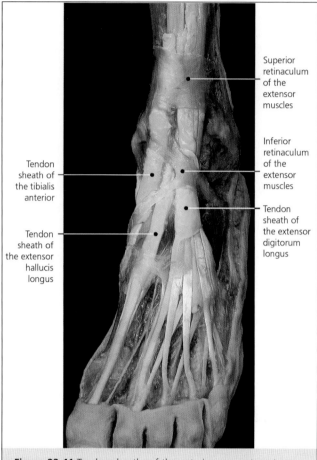

Figure 22-41 Tendon sheaths of the anterior compartment muscles injected with latex.

band crosses the foot transversely, from the apex to the medial surface of the navicular and medial cuneiform.

The tendons pass deep to the retinaculi protected by the following synovial sheaths (Figure 22-41): the tendon sheath of the anterior tibialis, the tendon sheath of the extensor hallucis longus, and the tendon sheath of the extensor digitorum longus.

The most prominent muscle of this group is the tibialis anterior (Figures 22-40 and 22-41), which originates from the lateral aspect of the tibia and the interosseous membrane. The tibialis anterior can have fibers that originate from the crural fascia as far proximal as Gerdy's tubercle. It has a large, fusiform muscle belly that continues as a tendon in the distal third of the leg, coursing distally and medially. It continues deep to the extensor retinaculi and inserts into the medial region of the foot. Between the navicular and the medial cuneiform, the tibialis anterior changes course and heads to the plantar surface of the foot. Between the tendon and the medial cuneiform is the subtendinous bursa of the tibialis anterior. It then inserts into the medial aspect of the medial cuneiform and the base of the first metatarsal.

This muscle produces dorsiflexion of the ankle and inversion of the forefoot. It is innervated by the deep peroneal nerve (L4 through S1).

The extensor digitorum longus is located laterally in the anterior compartment (Figures 22-36, 22-42, and

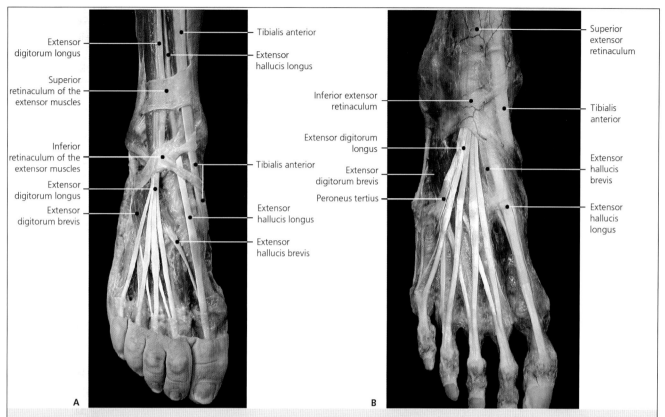

Figure 22-42 Specimens of the tendon of the tibialis anterior and other extensor tendons with their retinacula in the dorsum of the foot. Note that in **A** the tibialis anterior passes through a buttonhole between two laminae of the inferior extensor retinaculum. The peroneus tertius is absent.

22-43). Proximally, it originates from the anteroinferior region of the lateral proximal tibia and the medial surface aspect of the fibular diaphysis as well as in the neighboring areas of the interosseous membrane and the fascia that covers it. Its muscular fibers course along the fibular axis and form a long tendon proximally. This courses deep to the extensor retinaculi. Between the superior and inferior retinaculi, it divides into four tendons that diverge toward each of the lesser toes. The extensor digitorum longus inserts through an extensor apparatus similar to that of the hand, onto the dorsal aspect of the base of the distal phalanx. Unlike in the upper extremity, this extensor tendon does not have relevant intertendinous connections.

The extensor digitorum longus allows for extension of the lesser toes (second through fifth) and is an ankle dorsiflexor and foot pronator. It is innervated by the deep peroneal nerve (L4 through S1).

The peroneus tertius (anterior peroneus) can appear as a stand-alone muscle or, more commonly, as a part of the extensor digitorum longus (Figure 22-42). It is not uncommon for the peroneus tertius to be absent. It can be observed as a small tendon that separates from those that go toward the toes by coursing laterally toward the base of the fifth metatarsal. It functions as a foot dorsiflexor and evertor and is innervated by the deep peroneal nerve (L4 through S1).

The extensor hallucis longus is located deep in the anterior compartment between the extensor digitorum longus and the anterior tibialis (Figures 22-36, 22-42, and 22-43). It originates in the middle region of the leg, from the interosseous membrane and the medial surface of the fibula. Its fibers course distally, and its long tendon passes under the retinaculi between the anterior tibialis and the extensor digitorum communis. Distally, it inserts onto the distal phalynx of the great toe via the extensor apparatus.

It works as a great toe extensor and ankle dorsiflexor. It is innervated by the deep peroneal nerve (L4 through S1).

Lateral Compartment

The lateral compartment is limited by the anterior and posterior intermuscular septi of the leg and comprises two muscles that originate from the fibula, which is why they are called peroneal muscles (Table 22-8 and Figures 22-30 and 22-36). Both are innervated by the superficial peroneal nerve.

The peroneus longus originates in the proximal third of the leg (Figures 22-43 and 22-44). Its muscular

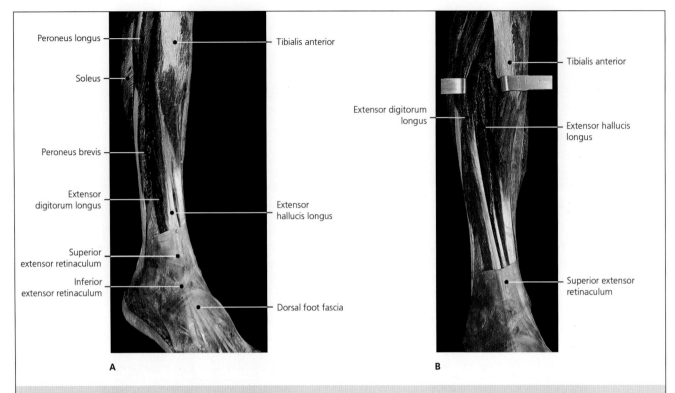

Figure 22-43 Specimens of the anterior and lateral leg musculature. In **B** the tibialis anterior and extensor digitorum longus have been separated to allow visualization of the extensor hallucis longus, which is partially hidden.

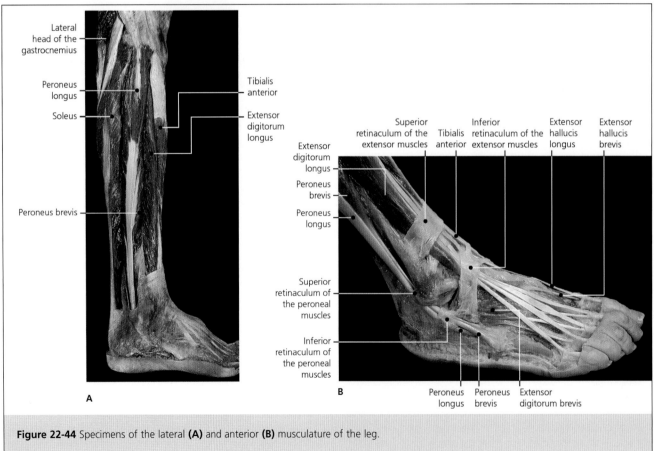

Figure 22-44 Specimens of the lateral **(A)** and anterior **(B)** musculature of the leg.

fibers originate from the lateral aspect of the fibula and the intermuscular septi of the leg. They continue with a flat tendon that is located superficial to the peroneus brevis. This tendon courses posterior to the lateral malleolus along with the peroneus brevis tendon through the tunnel formed by the superior retinaculum of the peroneal muscles. This retinaculum extends from the lateral malleolus to the lateral surface of the calcaneus. Both tendons are protected by the common tendon sheath of the peroneal tendons (Figure 22-45). After passing deep to the retinaculum, the tendon of the peroneus longus is posterior to that of the peroneus brevis. Both tendons pass under a second (inferior) retinaculum that separates them into two osteofibrous tunnels, preserving their common sheath.

The inferior retinaculum of the peroneal tendons extends from the inferior extensor retinaculum to the peroneal tubercle of the calcaneus, forming a tunnel for the peroneus brevis. From this tubercle the retinaculum heads toward the lateral and distal aspect of the calcaneus. Distal to the peroneal tubercle the peroneus longus enters a sulcus. The peroneus longus tendon keeps its sheath until it reaches the cuboid, and then it courses around the cuboid through a sulcus in this bone; it is reflected over the tuberosity of the cuboid and then heads medially. In the plantar or lateral regions, a sesamoid bone known as the os perineum is often found within the tendon's substance (Figure 22-38). In the plantar aspect of the foot, the tendon is covered by a sheath. Finally, it continues obliquely in a medial direction to insert onto the medial cuneiform and the base of the first metatarsal (Figure 22-38).

The peroneus longus is a muscle that allows ankle plantar flexion and hindfoot eversion. It is innervated by the superficial peroneal nerve (L4 through S1).

The peroneus brevis originates from the lateral aspect of the middle third of the fibular diaphysis (Figure 22-44). It is deep to the inferior fibers and tendon

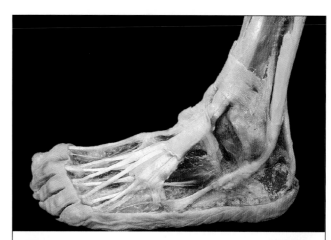
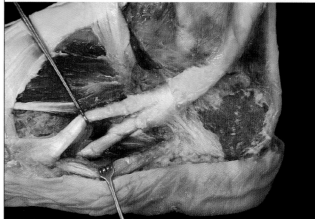

Figure 22-45 Peroneal tendinous sheaths injected with latex.

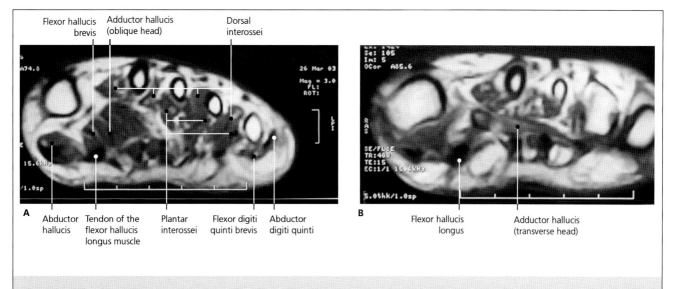

Figure 22-46 Frontal CT scans at the level of the metatarsal shafts. **A,** Proximal section. **B,** Distal section.

TABLE 22-9 DORSAL AND MEDIAL PLANTAR MUSCLES OF THE FOOT

MUSCLE	Extensor digitorum brevis and extensor hallucis brevis	Abductor hallucis	Flexor hallucis brevis	Adductor hallucis
ORIGIN	Anterolateral aspect of the calcaneus	Medial tubercle of the calcaneal tuberosity, flexor retinaculum, plantar aponeurosis	Plantar aspect of navicular, cuneiforms, and long plantar ligament	*Oblique head:* Second through fifth metatarsals, lateral cuneiform, and cuboid *Transverse head:* Region of the third through fifth metatarsophalangeal joints
INSERTION	*Extensor digitorum brevis:* Extensor apparatus of second through fourth toes *Extensor hallucis brevis:* Extensor apparatus of the great toe	Medial sesamoid and base of the proximal phalanx of the great toe	*Medial head:* medial sesamoid and proximal phalanx of the great toe *Lateral head:* Lateral sesamoid and proximal phalanx of the great toe	Lateral sesamoid and base of the proximal phalanx of the great toe
INNERVATION	Deep peroneal	Medial plantar	*Medial head:* Medial plantar *Lateral head:* Lateral plantar	Lateral plantar
FUNCTION	Toe extension	Great toe abduction	Great toe flexion	Great toe adduction

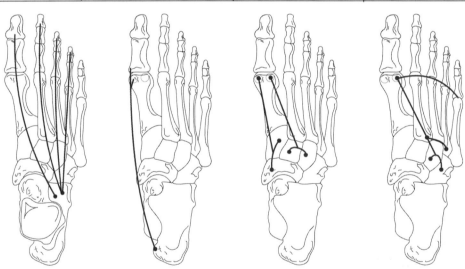

of the peroneus longus. Its muscular fibers reach the distal region of the fibula and continue as a tendon located anterior to the peroneus longus. As described earlier, the tendons share a synovial sheath and are separated by the peroneal tubercle of the calcaneus. The tendon of the peroneus brevis passes superior to the tubercle and leaves its sheath to insert into the styloid of the fifth metatarsal.

This muscle allows ankle plantar flexion and hindfoot eversion. It is innervated by the superficial peroneal nerve (L4 through S1).

Muscles of the Foot

The intrinsic (short) muscles of the foot are located in the dorsal and plantar aspects of the foot (Table 22-9 and Figures 22-46 through 22-48). The short extensors are on the dorsal surface, and the short flexors, interossei, lumbricals, and the muscles of the great and fifth toes are on the plantar surface. The intrinsic muscles of the foot help to maintain the plantar arch along with the bony structure of the foot.

Dorsal Musculature

In this region, deep to the tendons of the extensor muscles, lie the extensor digitorum brevis and the extensor hallucis brevis (Figure 22-44). They are covered by the dorsal foot fascia, which is contiguous with the crural fascia.

The extensor digitorum brevis muscle originates from the anterolateral region of the calcaneus (Figure 22-49). From its origin, it forms three small

Myology

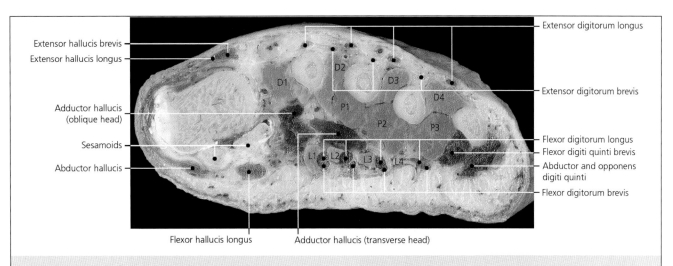

Figure 22-47 Frontal section at the level of the metatarsal shafts. The interossei of the intermediate group have been colored. D = dorsal interossei; P = plantar interossei; L = lumbricals.

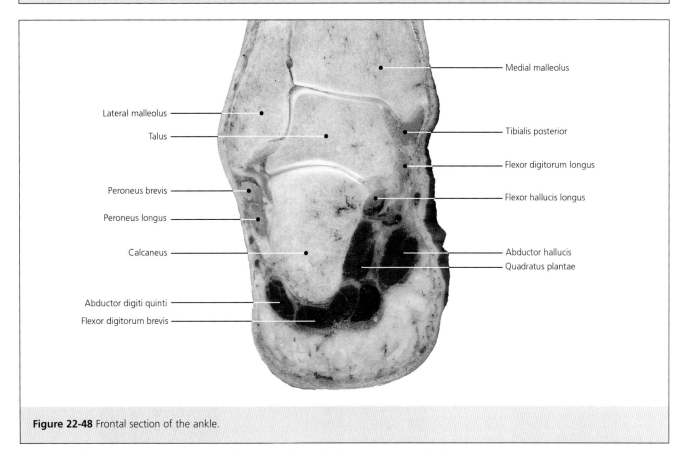

Figure 22-48 Frontal section of the ankle.

muscle bellies that continue distally as tendons to the second through fourth toes; there is no tendon for the fifth toe, although occasionally it can receive a small tendon from the peroneus brevis (Figure 22-49). These tendons insert in the toe extensor apparatus, lateral to the extensor digitorum longus.

The extensor digitorum brevis allows for the extension of the second, third, and fourth toes, in conjunction with the long extensor. It is innervated by the deep peroneal nerve (L4 through S1).

The extensor hallucis brevis muscle has small muscle bellies different from those of the extensor digitorum brevis (Figure 22-49). From its origin on the dorsolateral surface of the calcaneus, its fibers course obliquely in a medial direction. They continue distally as a tendon that inserts into the extensor apparatus of the great toe, lateral to the extensor hallucis longus.

American Academy of Orthopaedic Surgeons

The extensor hallucis brevis allows for extension of the great toe, coordinated with the extensor hallucis longus. It is innervated by the deep peroneal nerve (L4 through S1).

Plantar Musculature

The plantar musculature can be divided into three groups: medial, intermediate, and lateral.

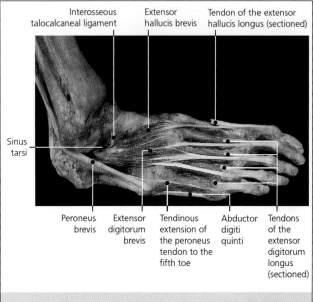

Figure 22-49 Specimen of the dorsal intrinsic musculature of the the foot after sectioning the extrinsics.

Medial Plantar Muscles

The medial plantar group includes the plantar intrinsic muscles of the great toe—the abductor hallucis and the flexor hallucis brevis. The adductor hallucis is located in the intermediate group but is discussed here for reasons explained below.

The abductor hallucis is the most medial of the plantar muscles (Figure 22-50). It originates from the plantar medial tubercle of the calcaneal and intermuscular septum. It forms a muscle belly that follows the medial border of the foot and continues as a flat tendon that inserts onto the medial sesamoid and the lateral aspect of the first proximal phalanx. It sometimes gives off small expansions to the extensor apparatus of the great toe.

As its name indicates, the abductor hallucis abducts the great toe, but it also allows a little bit of flexion. This muscle also helps maintain the plantar arch. It is innervated by the medial plantar nerve (L5-S1).

The flexor hallucis brevis is located lateral to the abductor hallucis. It is formed by two heads that originate from the plantar aspect of the cuboid, the lateral cuneiforms. The medial head is located medial to the tendon of the flexor hallucis longus (Figure 22-50) and inserts onto the medial sesamoid and the medial aspect of the first proximal phalanx. The lateral head is located lateral to the tendon of the flexor hallucis longus and inserts onto the lateral sesamoid and lateral aspect of the base of the proximal phalanx.

The flexor hallucis brevis flexes the great toe and helps maintain the plantar arch. Its medial head is

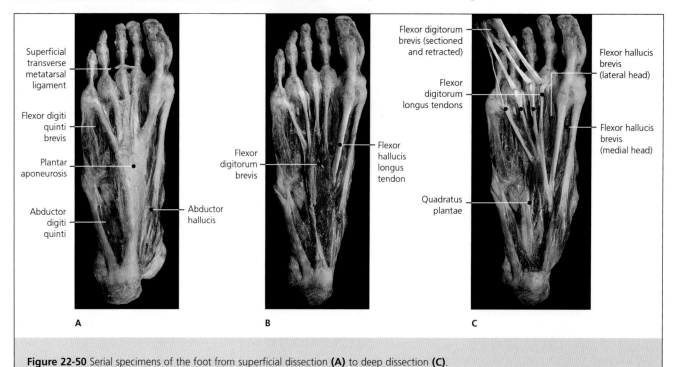

Figure 22-50 Serial specimens of the foot from superficial dissection **(A)** to deep dissection **(C)**.

TABLE 22-10 INTERMEDIATE MUSCLES OF THE FOOT

MUSCLE	Flexor digitorum brevis	Quadratus plantae	Lumbricals	Dorsal interossei	Plantar interossei
ORIGIN	Calcaneal tuberosity and plantar aponeurosis	Inferior surface of the calcaneus	Medial aspect of the tendons of the flexor digitorum longus	Metatarsals	Metatarsals
INSERTION	Middle phalanx of toes 2-5	Tendons of the flexor digitorum longus	Medial aspect of the proximal phalanx of toes 2-5	*First:* Medial aspect of second toe, proximal phalanx *Second-fourth:* Lateral aspect of second through fourth toes	Medial aspect of proximal phalanges 3 - 5
INNERVATION	Medial plantar	Lateral plantar	*First and second lumbricals:* Medial plantar *Third and fourth lumbricals:* Lateral plantar	Lateral plantar	Lateral plantar
FUNCTION	Flexion of toes 2-5	Flexion of toes 2-5 through the flexor digitorum longus	Flexion of proximal phalanges and extension of middle and distal phalanges	Separate toes (abduct digits and flex metatarsophalangeal joints)	Adduct digits 2-4 and flex metatarsophalangeal joints

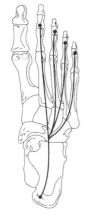  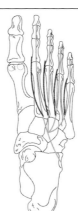  

innervated by the medial plantar nerve (L5-S1) and the lateral head by the lateral plantar nerve (L5-S1).

The adductor hallucis muscle belly is located in the intermediate plantar region, but fibers insert medially onto the lateral side of the proximal phalynx of the great toe; for this reason it is discussed with the medial group. This muscle is formed by two heads that insert onto the lateral sesamoid and the lateral aspect of the base of the first phalanx (Figure 22-51). The oblique head originates from the second through fourth metatarsals and the plantar aspect of the lateral cuneiform and the cuboid. The adductor's fibers course obliquely toward their insertion. The transverse head is shorter than the oblique head and originates at the level of the metatarsophalangeal joints of the third through fifth toes. These fibers course transversely toward their insertion into the proximal phalanx of the great toe.

As its name implies, this muscle adducts the great

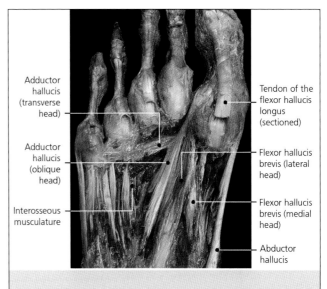

Figure 22-51 Specimen showing the deep plantar musculature.

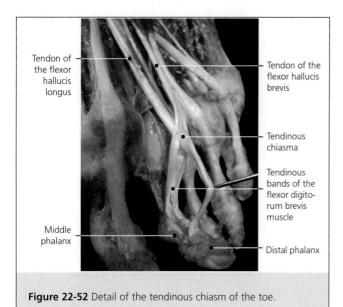

Figure 22-52 Detail of the tendinous chiasm of the toe.

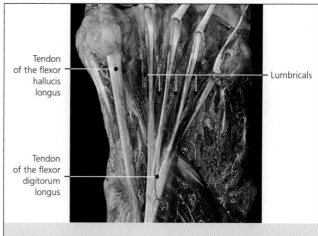

Figure 22-53 Plantar view of the flexor digitorum longus, giving origin to the lumbricals.

toe and helps maintain the plantar arch. It is innervated by the lateral plantar nerve (S1-S2).

Intermediate Plantar Muscles

The muscles of the intermediate group (Table 22-10) are located underneath the plantar fascia and are organized in different layers. They are, from superficial to deep, the flexor digitorum brevis, the quadratus plantae, the lumbricals, and the plantar and dorsal interosseous muscles.

The plantar fascia looks like a series of fibrous laminae that extend longitudinally in the medial region of the foot (Figure 22-50). It originates from the plantar medial tubercle of the calcaneal tuberosity and is distributed distally as a series of radii that reach the middle phalanges. Between the radii are transverse fascicles. The plantar fascia is an important structure that helps maintain the plantar arch.

Distal to the transverse fascicles is the superficial transverse intermetatarsal ligament (Figure 22-50), located between the metatarsal heads.

The flexor digitorum brevis originates from the inferior aspect of the calcaneal tuberosity and in the plantar fascia (Figure 22-50). It forms a broad, flat muscle belly that is located deep to the plantar aponeurosis. Distally, the flexor digitorum brevis divides into four bellies that continue as four tendons to the second through fifth toes. They correlate with the superficial and deep flexor muscles described in the forearm. The flexor digitorum brevis forms two tendons that insert in the middle phalanx of the toes. These bands form a buttonhole through which the tendon of the flexor hallucis longus courses (Figure 22-52); it is a tendinous chiasm.

This muscle flexes the second through fifth toes and helps maintain the plantar arch. It is innervated by the medial plantar nerve (L5-S1).

The flexor muscle tendons course through synovial sheaths (Figure 22-40). Deep to these sheaths are the tendinous vinculae that conduct the nutrient vessels. The fibrous sheaths, like those in the hand, have annular and cruciform pullies.

The quadratus plantae is located deep to the flexor digitorum brevis (Figure 22-50). It originates from the inferior aspect of the calcaneus and inserts in the lateral border of the flexor digitorum longus, distal to the knot of Henry. Because of this, it is also called an accessory flexor muscle.

It works in coordination with the flexors of the toes to produce toe flexion, and it also helps maintain the plantar arch. It is innervated by the lateral plantar nerve (S1-S2).

The lumbrical muscles are four elongated muscles that originate from the medial side of each of the tendons of the flexor digitorum longus (Figures 22-50 and 22-53). Their elongated fibers are continued by a tendon that inserts in the medial proximal phalanx of the toes.

Because of their location, they are flexors of the metatarsophalangeal joints and extend the proximal and distal interphalangeal joints via the extensor apparatus. These muscles are innervated by the medial plantar nerve (first and second lumbricals) (L5-S1) and lateral plantar nerve (third through fifth lumbricals) (S1-S2).

In the foot, unlike the hand, the "central" digit (corresponding to the axis of the foot) is the second toe. This toe, therefore, has a dorsal interosseous muscle on either side, whereas the third and fourth have only one laterally. This makes for four dorsal interosseous muscles. On the medial

MYOLOGY

sides of the third through fifth toes are three plantar interosseous muscles.

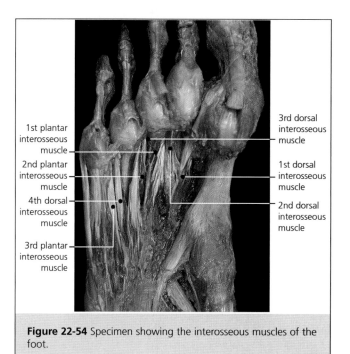

Figure 22-54 Specimen showing the interosseous muscles of the foot.

The three plantar interosseous muscles are numbered from medial to lateral. They originate from a single head in the second through fourth interosseous spaces, and they insert onto the medial side of the proximal phalanx of the third through fifth toes; ie, they insert on the same phalanx of the metatarsal from which they originate (Figure 22-54).

They produce flexion of the metatarsophalangeal joints of the third through fifth toes, and they adduct them to the second toe (adduction). They are innervated by the lateral plantar nerve (S1-S2).

The four dorsal interosseous muscles are numbered from medial to lateral. They originate from two heads (one in each metatarsal) in all the interosseous spaces (Figure 22-54). The first inserts onto the medial side of the extensor apparatus of the second toe and the proximal phalanx. The others insert onto the lateral side of the proximal phalanges of the second through fourth toes.

These muscles flex the metatarsophalangeal joints. They allow for abduction of the second through fourth digits.

Lateral Plantar Muscles

Finally, the lateral plantar muscles of the foot (Table 22-11) include the intrinsic muscles for the fifth

TABLE 22-11 LATERAL PLANTAR MUSCLES OF THE FOOT

MUSCLE	Abductor digiti quinti	Flexor digiti quinti	Opponens digiti quinti
ORIGIN	Lateral apophysis of the calcaneal tuberosity and plantar aponeurosis	Base of the fifth metatarsal and long plantar ligament	Base of the fifth metatarsal
INSERTION	Lateral aspect of the fifth toe proximal phalanx	Base of the fifth toe proximal phalanx	Body of the fifth metatarsal
INNERVATION	Lateral plantar	Lateral plantar	Lateral plantar
FUNCTION	Abduction of the fifth toe	Flexion of the metatarsophalangeal joint of the fifth toe	Opposition of the fifth toe and great toe

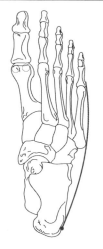  

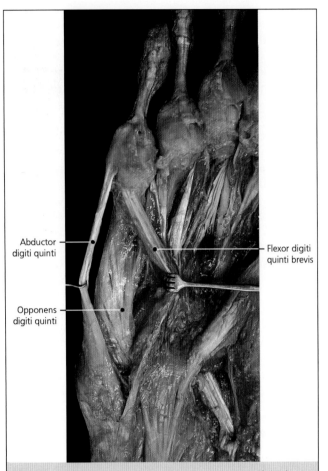

Figure 22-55 Detail of the lateral group musculature of the foot. The abductor and the flexor brevis of the fifth toe have been retracted to show the opponens.

toe, which are the abductor digiti quinti, flexor digiti quinti, and opponens digiti quinti.

The abductor digiti quinti is the most lateral muscle of the sole of the foot (Figures 22-50 and 22-55). It originates from the calcaneal tuberosity and plantar fascia. The fibers of the abductor digiti quinti follow the lateral border of the foot and are continued by a tendon that inserts onto the lateral aspect of the proximal phalanx of the fifth toe.

The muscle gets its name from its action, the abduction of the fifth toe. It is also a flexor. This muscle is innervated by the lateral plantar nerve (S1-S2).

The flexor digiti quinti is located medial to the abductor digiti quinti (Figures 22-50 and 22-55). It originates from the base of the fifth metatarsal. Its fibers course distally, reaching the base of the proximal phalanx of the fifth toe.

This muscle is a fifth toe flexor. It is innervated by the lateral plantar nerve (S1-S2).

The opponens digiti quinti is located deep to the short flexor of the fifth toe and often blends with it (Figure 22-55). It shares the same origin and inserts onto the body of the fifth metatarsal.

This muscle opposes the fifth toe to the great toe, a function preserved only by trained subjects. This muscle is innervated by the lateral plantar nerve (S1-S2).

Chapter 23

Neurology

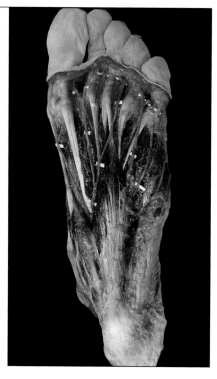

Introduction
The muscles of the pelvic girdle and the lower limb are innervated by the lumbosacral plexus (Figure 23-1). This is formed by the anterior rami of the lumbar and sacral nerve roots, in general from L1 to S3, which maintain a connection through the spinal nerve L4.

As described in chapter 17, the spinal nerves originate from an anterior ramus (the anterior rami form the plexi in the cervical and lumbosacral regions) and a posterior ramus. The latter courses dorsally and divides into medial and lateral branches. In addition, the superior cluneal nerves that innervate the lateral region of the buttock to the greater trochanter originate from the L1 through L3 nerves. Also, the medial cluneal nerves originate from the sacral and coccygeal nerves.

Lumbar Plexus
The lumbar plexus is formed by the L1 through L4 nerve roots (Figures 23-1, 23-2, and 23-3) deep to the psoas muscle. The plexus forms several nerves, most of which course anteriorly to innervate the abdominal wall and inguinal regions (Figures 23-4 and 23-5). These are as follows:
- T12-L1—forms a small trunk that rests over the anterior aspect of the quadratus lumborum and divides into two nerves, the iliohypogastric (cranial) and the ilioinguinal (caudal)
- L1-L2—genitofemoral nerve
- L2-L3—lateral femorocutaneous nerve
- L2 through L4—obturator nerve
- L3-L4—accessory obturator nerve
- L2 through L4—femoral nerve

Iliohypogastric Nerve
The iliohypogastric nerve (Figure 23-2) travels around the muscular abdominal wall between the transverse and internal oblique muscles. It gives off a cutaneous lateral branch that becomes superficial and provides sensory nerves to the lateral gluteal and the superior thigh regions. The nerve continues its course anteriorly, enters the inguinal canal, and continues toward the midline as the anterior cutaneous branch, providing sensory nerves to the hypogastric and inguinal areas.

Ilioinguinal Nerve
The ilioinguinal nerve (Figure 23-2) follows a similar course to the one just described, but caudal to the iliohypogastric. It travels around the abdominal wall following the iliac crest, enters the inguinal canal, and innervates the anterior pubic region, ending in the genital skin.

These two nerves innervate the following abdominal muscles: transverse, internal and external obliques, rectus abdominis, and pyramidal.

Genitofemoral Nerve
The genitofemoral nerve (Figures 23-2 and 23-3) crosses the muscular body of the psoas muscle on its

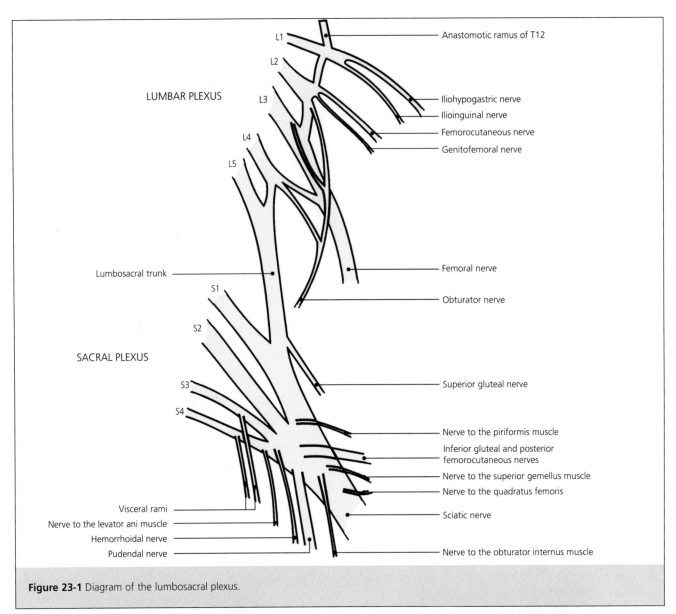

Figure 23-1 Diagram of the lumbosacral plexus.

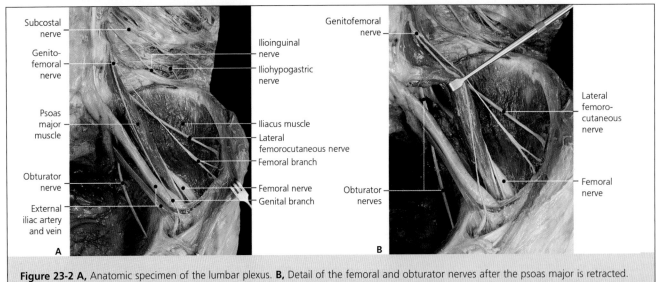

Figure 23-2 A, Anatomic specimen of the lumbar plexus. **B,** Detail of the femoral and obturator nerves after the psoas major is retracted.

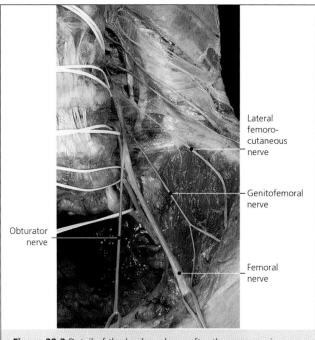

Figure 23-3 Detail of the lumbar plexus after the psoas major muscle is removed. Yellow bands surround the lumbar nerve roots.

anterior surface and is covered by the muscle fascia. This nerve divides into two branches, one genital and one femoral. The genital branch courses caudally and medially toward the inguinal canal and innervates the cremaster muscle and the skin of the genital and the superomedial regions of the thigh. The femoral branch forms part of the contents of the vascular window, coursing between the femoral artery and the iliopectineal arch. In the femoral triangle, this branch divides into multiple superficial branches that traverse the saphenous hiatus and the cribriform fascia to provide sensory nerves to the skin of this area.

Lateral Femorocutaneous Nerve

The lateral femorocutaneous nerve (Figures 23-2 and 23-3) appears in the area between the lateral border of the psoas major and the iliac crest. This nerve is located above the iliacus muscle, coursing through the pelvis internally. Finally, the lateral femorocutaneous nerve crosses the inguinal ligament near the anterior superior iliac spine and follows a caudal path. It innervates the lateral aspect of the thigh (lateral to the femoral triangle). It is not unusual for this nerve to

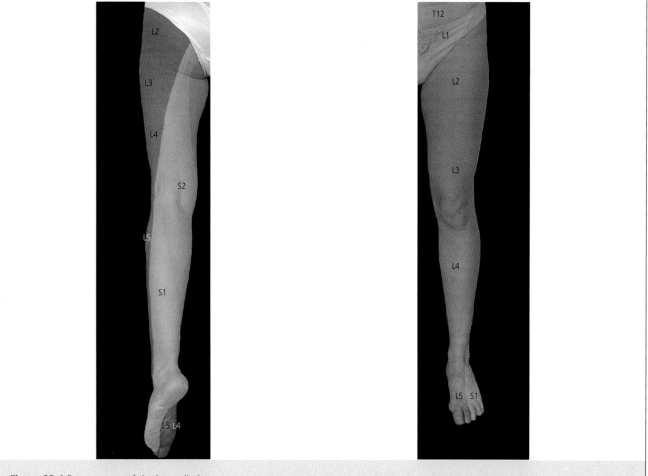

Figure 23-4 Dermatomes of the lower limb.

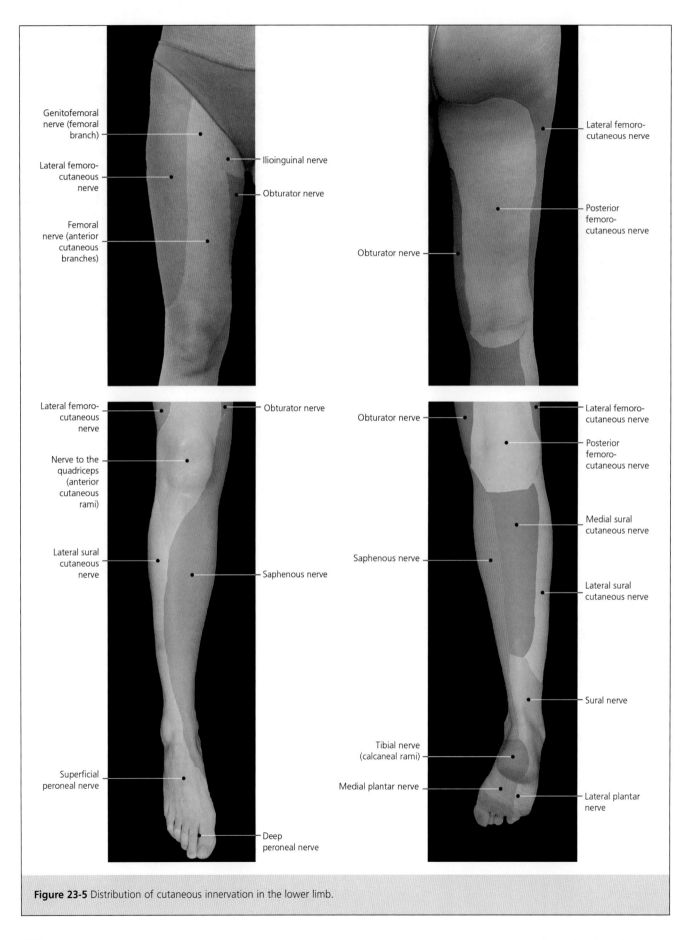

Figure 23-5 Distribution of cutaneous innervation in the lower limb.

anastomose with sensitive rami of the iliohypogastric nerve.

Obturator Nerve

The obturator nerve (Figures 23-2, 23-3, and 23-6) is located deep within the psoas, dorsal to the internal iliac artery. It courses caudally toward the obturator foramen, penetrates the obturator membrane, and innervates the obturator externus muscle. In the thigh it is located in the medial (adductor) compartment, which it innervates through muscular rami. The obturator nerve divides into anterior and posterior branches. The first is located deep to the pectineal muscle, coursing in the plane between the adductor brevis and the adductor longus. Distally, a cutaneous branch innervates the skin of the medial third of the thigh. The posterior branch, with its muscular and articular rami, is located between the adductor brevis and magnus and innervates both.

The obturator nerve innervates the following muscles: obturator externus, adductors, gracilis, and pectineus.

Accessory Obturator Nerve

The accessory obturator nerve (of Schmidt) is variable; it usually innervates the pectineus muscle and the hip joint.

Femoral Nerve

The femoral nerve originates from the L2 through L4 roots of the lumbar plexus (Figures 23-2 and 23-3). Most of this nerve is covered by the psoas muscle. At the level of the junction of the psoas with the iliacus, the nerve is located in the lateral border of the psoas and courses toward the muscular window. Accompanied by the iliopsoas muscle and lying lateral to the

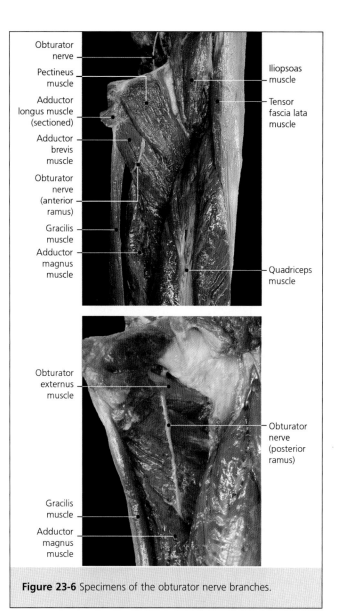

Figure 23-6 Specimens of the obturator nerve branches.

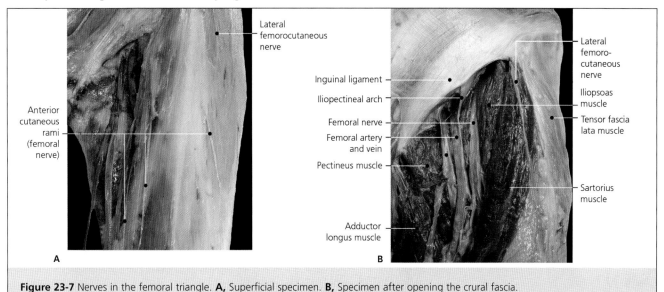

Figure 23-7 Nerves in the femoral triangle. **A,** Superficial specimen. **B,** Specimen after opening the crural fascia.

Pelvic Girdle and Lower Limb

with a descending arterial branch to the knee, and later accompanies the greater saphenous vein. The saphenous nerve has two branches. The infrapatellar branch penetrates the sartorius near the superficial pes anserine to provide sensory nerves to the infrapatellar region. Distally, in the proximal anteromedial third of the leg, the saphenous nerve penetrates the fascia of the leg and provides sensory nerves to the medial region of the leg and part of the foot through the cutaneous branches of the leg.

The femoral nerve innervates the iliopsoas, sartorius, quadriceps, and pectineus (Figure 23-9).

Sacral Plexus

The sacral plexus is formed by the anterior rami of spinal nerves L5 through S3, although there are contributions from L4 and S4 (Figure 23-10). From L4 there is an anterior ramus that joins L5 to form the lumbosacral trunk, which courses along the front of the sacral ala and joins the lumbar and the sacral plexi.

Because of the relationship with the sacrum, the sacral spinal nerves can be observed at their exit through the anterior sacral foramina and over the anterior aspect of the piriformis muscle. They distribute on the posterior regions of the pelvic girdle and lower extremity (Figures 23-4 and 23-5). These are the origins of the principal nerves:

- L4 through S1—superior gluteal nerve
- L5 through S2—inferior gluteal nerve
- S1 through S3—posterior cutaneous nerve of the thigh
- L4 through S3—sciatic nerve

The sacral plexus gives rise to muscular branches for the posterior deep pelvitrochanteric muscles. The nerve to the obturator internus muscle originates near the sciatic nerve, crosses the greater sciatic foramen below the piriformis muscle (infrapiriformis foramen), and reenters the pelvis through the lesser sciatic foramen, reaching the obturator internus and the superior gemellus (this muscle can also receive innervation through its own independent nerve). The nerve of the quadratus femoris muscle crosses the infrapiriformis foramen and reaches the quadratus femoris and inferior gemellus muscles.

The nerve of the piriformis muscle innervates that muscle directly.

Superior Gluteal Nerve

The superior gluteal nerve (Figure 23-11) crosses the greater sciatic foramen above the piriformis, along with the superior gluteal vessels. It innervates the lateral gluteal musculature, coursing between the gluteus medius and the minimus until it reaches the tensor fascia lata, which it also innervates.

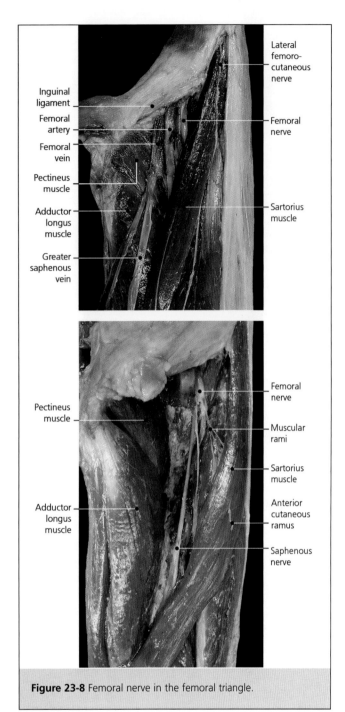

Figure 23-8 Femoral nerve in the femoral triangle.

femoral artery, the femoral nerve courses through the window to the femoral triangle.

In the femoral triangle (Figures 23-7 and 23-8), the femoral nerve divides into numerous muscular rami for the sartorius, quadriceps femoris, and pectineus. The anterior cutaneous branches penetrate the fascia lata and provide sensory nerves to the anterior aspect of the thigh.

The femoral nerve continues distally as the saphenous nerve (Figure 23-8). This is a sensory nerve, which follows the femoral artery toward Hunter's canal, penetrates the vastoadductor membrane along

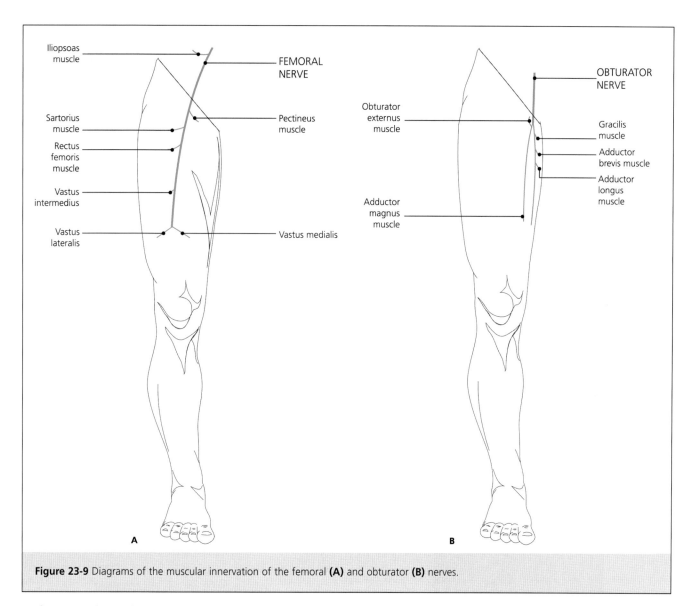

Figure 23-9 Diagrams of the muscular innervation of the femoral **(A)** and obturator **(B)** nerves.

Inferior Gluteal Nerve

The inferior gluteal nerve (Figure 23-11) crosses the greater sciatic foramen below the piriformis muscle, along with the inferior gluteal vessels. After crossing the foramen, the nerve divides into several muscular rami that innervate the gluteus maximus muscle.

Posterior Cutaneous Nerve of the Thigh

The posterior cutaneous nerve of the thigh (Figure 23-11) crosses the greater sciatic foramen along with the inferior gluteal nerve, below the piriformis. Unlike the other nerves in the sacral plexus, this is a sensory nerve that follows the axis of the limb underneath the thigh fascia, which it penetrates in the middle of the thigh to be located subcutaneously. It innervates the posterior region of the thigh to the posterior and medial popliteal area. At the level of the inferior gluteal fold, the inferior cluneal nerves branch off, penetrate the fascia, and provide sensory nerves to this area. Just proximal to these branches, the posterior cutaneous nerve of the thigh gives off perineal branches.

Sciatic Nerve

The sciatic nerve, in the posterior region of the pelvis, is very thick compared with the other nerves of the sacral plexus (Figure 23-11, B). To see this nerve, the gluteus maximus must be divided and removed. After crossing the greater sciatic foramen, this nerve lies below the piriformis muscle, resting over the obturator internus, gemelli, and quadratus femoris. The sciatic nerve descends along the posterior region of the thigh (Figure 23-12), following the thigh's axis. In the pelvic region, it is located slightly lateral to the ischial tuberosity and then under the ischiocrural muscles. Along its course through the posterior compartment of the thigh, it runs between the ischiotibial muscles and the biceps femoris until it reaches the popliteal fossa, where it bifurcates (Figure 23-13).

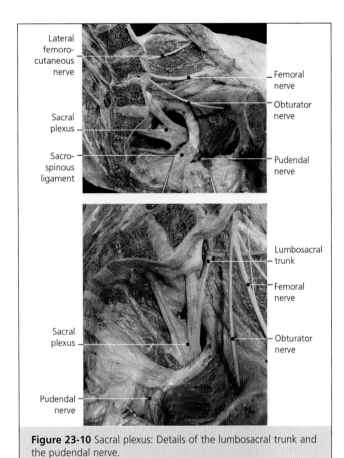

Figure 23-10 Sacral plexus: Details of the lumbosacral trunk and the pudendal nerve.

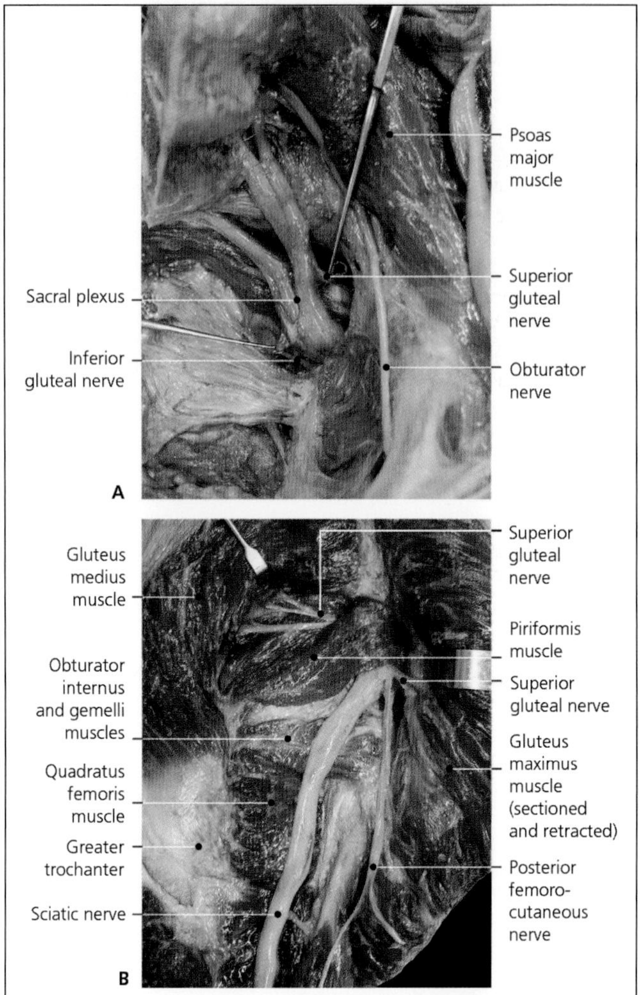

Figure 23-11 Sacral plexus, gluteal nerves. **A,** Endopelvic view. **B,** Posterior view, after the gluteus maximus has been sectioned and retracted.

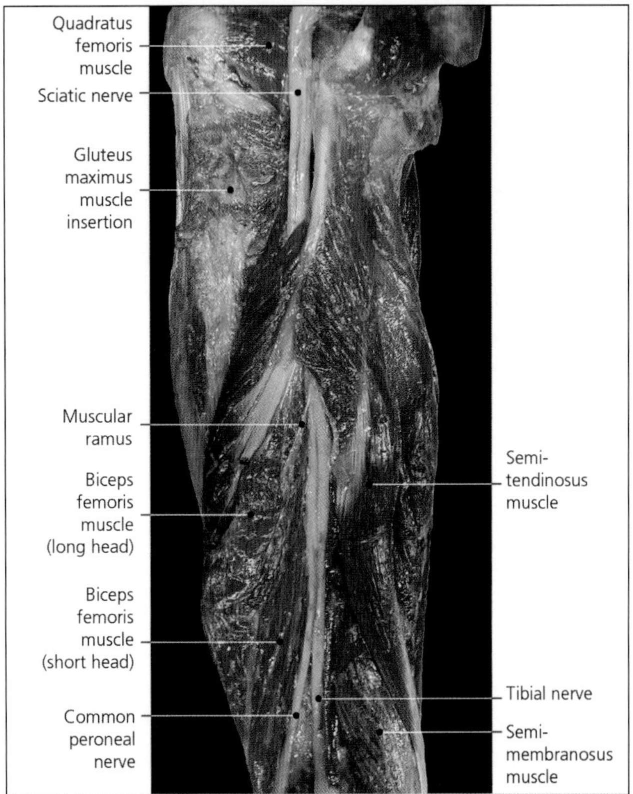

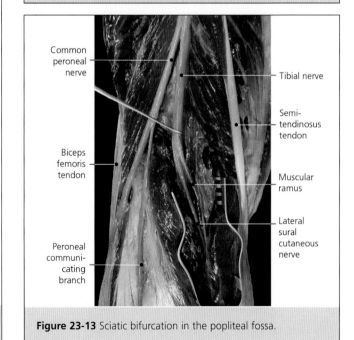

Figure 23-12 Trajectory of the sciatic nerve in the posterior thigh region.

Figure 23-13 Sciatic bifurcation in the popliteal fossa.

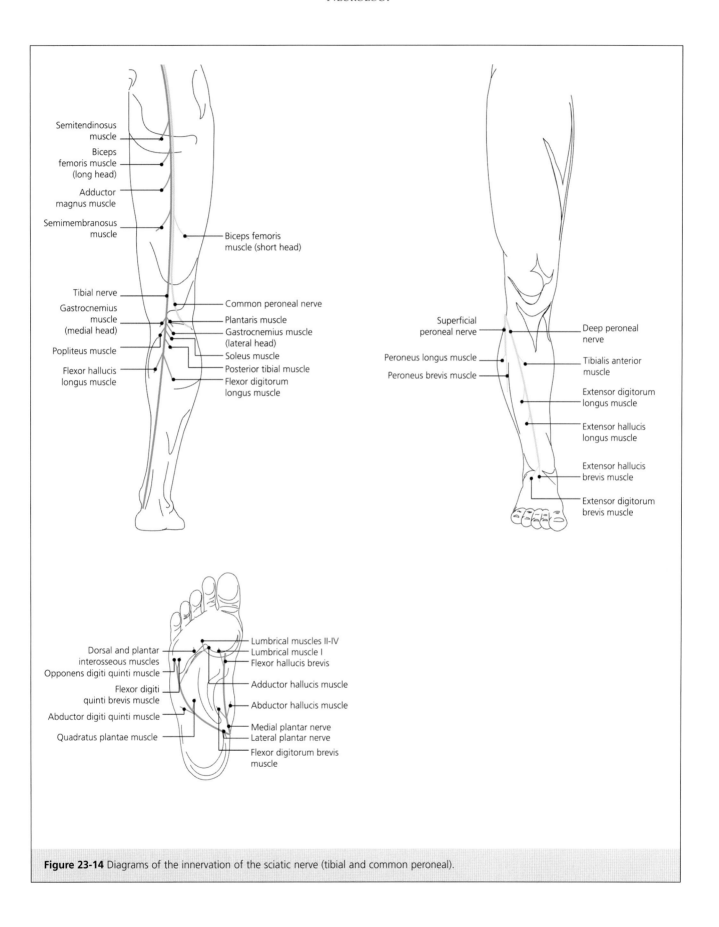

Figure 23-14 Diagrams of the innervation of the sciatic nerve (tibial and common peroneal).

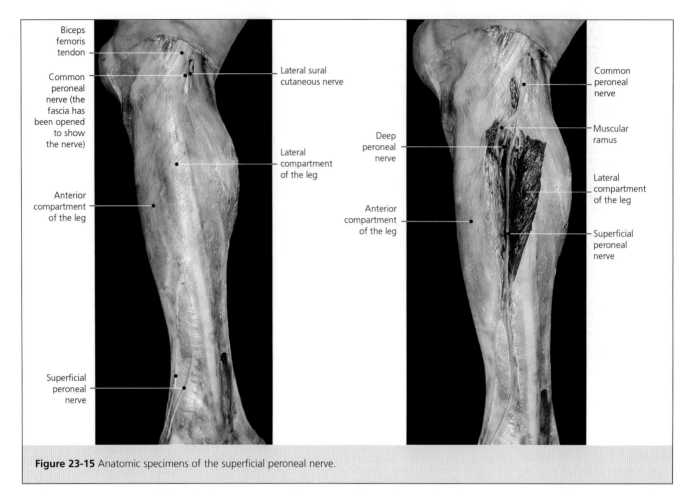

Figure 23-15 Anatomic specimens of the superficial peroneal nerve.

The sciatic nerve actually includes two nerves that can be separated proximally by blunt dissection. These two then bifurcate in the popliteal fossa into the tibial and common peroneal nerves (Figure 23-14). When they diverge, the common peroneal nerve usually traverses the fibers of the piriformis muscle. In the thigh, the tibial nerve innervates the ischiotibial muscles and the long head of the biceps, whereas the common peroneal innervates the short head of the biceps.

The common peroneal nerve (Figures 23-12, 23-13, and 23-15) is one of the components of the sciatic nerve that becomes evident in the popliteal fossa. In this region, the nerves occupy their most superficial position, and the bifurcation of the sciatic nerve can be seen.

The common peroneal nerve then courses toward the neck of the fibula. Before reaching and winding around it, the lateral sural cutaneous nerve branches off (Figure 23-16), crosses the leg fascia, and innervates the lateral leg. The peroneal communicating ramus originates from this nerve and anastomoses with the medial sural cutaneous nerve to form the sural nerve, which will be described later.

The common peroneal nerve goes around the neck of the fibula and bifurcates into the superficial and deep peroneal nerves.

The superficial peroneal nerve (Figure 23-15) is located in and innervates the lateral compartment of the leg through muscular branches to the peroneus longus and brevis. In the distal third of the leg, this nerve penetrates the fascia and is located superficially.

It courses above the extensor retinaculi and divides into two branches that provide cutaneous sensation to the dorsum of the foot and toes. The medial dorsal cutaneous nerve (Figure 23-17) courses toward the medial side of the great toe, whereas the intermediate dorsal cutaneous nerve divides into branches for the dorsum of the toes, except the lateral side of the great toe and the medial side of the second toe; it does this via the dorsal digital nerves of the toes.

The deep peroneal nerve (Figure 23-18) is located in the anterior compartment of the leg and innervates all its musculature as well as the short extensors of the toes. It courses in the leg along with the anterior tibial artery, which is why some authors call it the anterior tibial nerve. In the dorsum of the foot, it gives off muscular branches for the short extensors and continues toward the first web space, where it bifurcates into the dorsal lateral digital nerve of the great toe and the dorsal medial digital nerve of the second toe, which are both sensory nerves.

After the sciatic bifurcation, the tibial nerve (Figure 23-19) crosses the popliteal space vertically to pass under the tendinous arch of the soleus, travel-

Neurology

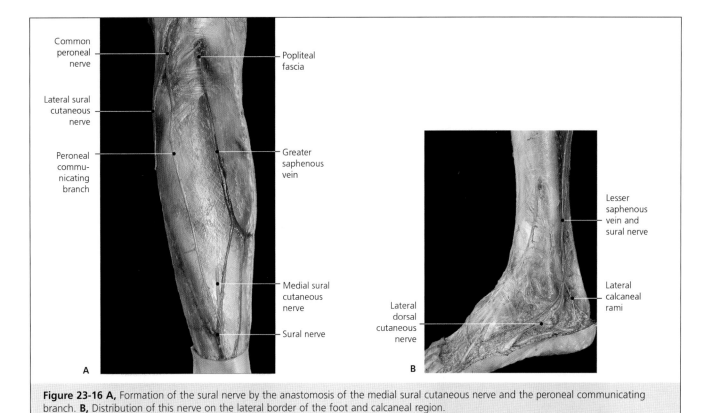

Figure 23-16 A, Formation of the sural nerve by the anastomosis of the medial sural cutaneous nerve and the peroneal communicating branch. **B,** Distribution of this nerve on the lateral border of the foot and calcaneal region.

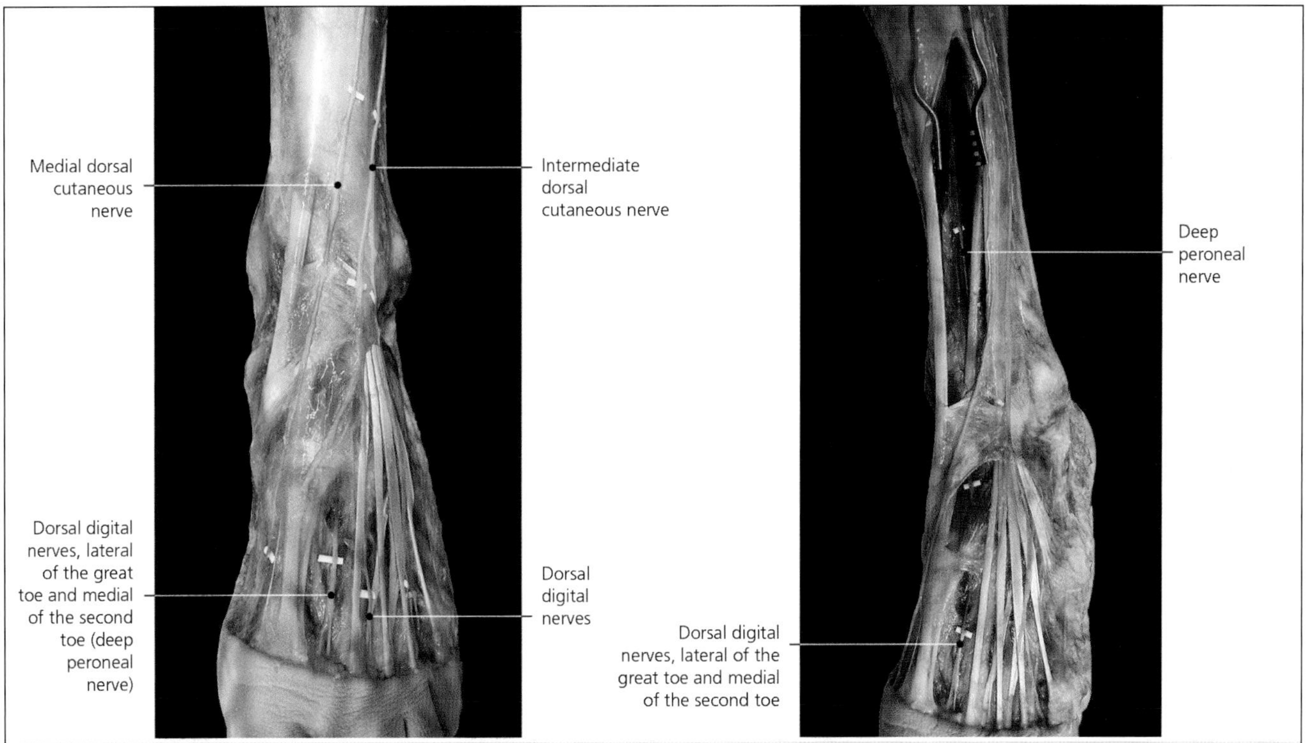

Figure 23-17 Innervation of the dorsum of the foot.

American Academy of Orthopaedic Surgeons

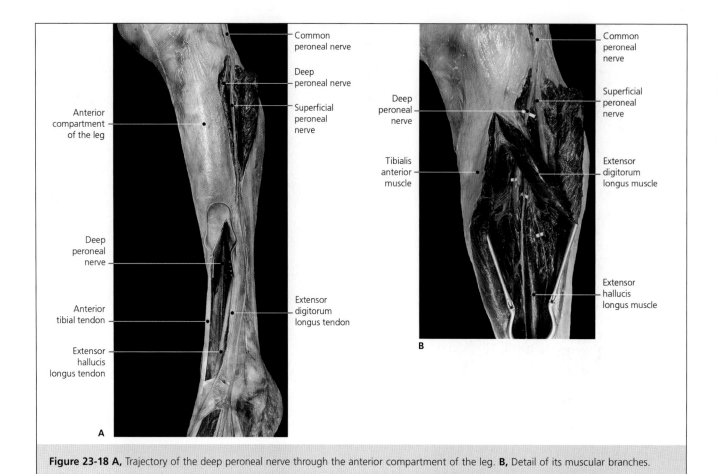

Figure 23-18 A, Trajectory of the deep peroneal nerve through the anterior compartment of the leg. **B,** Detail of its muscular branches.

ing with the posterior tibial artery in the posterior compartment of the leg. In the popliteal fossa, the tibial nerve gives rise to muscular branches for the gastrocnemius, soleus, and plantaris muscles as well as the medial sural cutaneous nerve (Figure 23-16), which is located between the two heads of the gastrocnemius. This nerve then crosses the leg fascia in its distal third and receives the anastomosis of the peroneal communicating ramus from the common peroneal nerve to form the sural nerve, which provides sensory nerves to the lateral malleolar region and the lateral border of the foot through the dorsolateral calcaneal nerve and the lateral calcaneal rami.

The tibial nerve courses through the deep posterior compartment of the leg, innervating the muscles of the region (posterior tibial, flexor digitorum longus, flexor hallucis longus) (Figure 23-20), along with the posterior tibial artery; for this reason, some authors call it the posterior tibial nerve. The nerve then branches into the interosseous nerve of the leg, which accompanies the anterior tibial artery. In the distal region of the leg, this nerve passes through the tarsal tunnel, where the medial calcaneal rami branch off (Figure 23-21). Distal to this, the interosseous nerve of the leg bifurcates into the plantar nerves.

The medial plantar nerve (Figure 23-22) is located in the sole of the foot underneath the great toe abductor, which it innervates. Distally, the medial plantar nerve divides into the common plantar digital nerves and, finally, the plantar proper digital nerves (Figure 23-23). It provides sensory nerves to the toes, first to the medial side of the fourth toe, and innervates the short muscles of the great toe, the short toe flexor, and the first lumbrical.

The lateral plantar nerve (Figure 23-22) is thinner than the medial plantar nerve. It courses obliquely between the short toe flexor and the quadratus plantaris, innervating this last one, and then coursing to the lateral region of the foot. Its superficial sensory branch courses toward the fourth and fifth toe (common plantar digital nerves). A small branch from these is located on the lateral side of the fifth toe, and the other branch is located in the space between the fourth and fifth toes. Finally, the latter branch divides into two proper digital nerves for the medial side of the fifth toe and the lateral side of the fourth. The deep branch (Figure 23-24) of the plantar nerve enters the intrinsic musculature and innervates the muscles of the fifth toe, interossei, great toe adductor, and the lateral three lumbricals.

In summary, the tibial nerve innervates all the muscles of the posterior compartment of the thigh (except the short head of the biceps), the posterior compartment of the leg, and the sole of the foot.

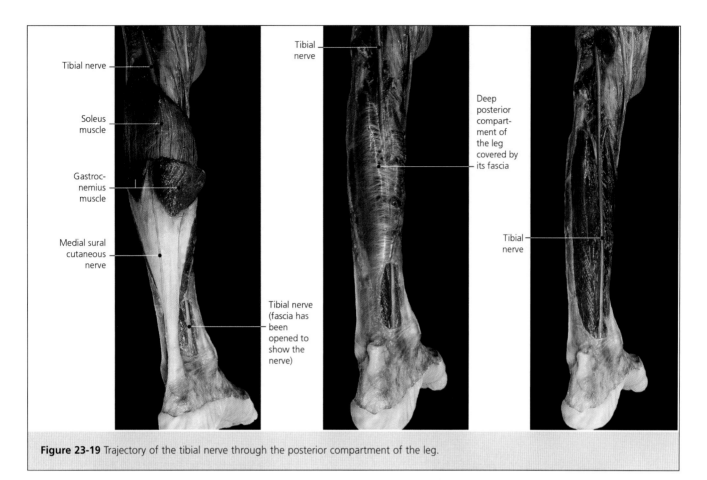

Figure 23-19 Trajectory of the tibial nerve through the posterior compartment of the leg.

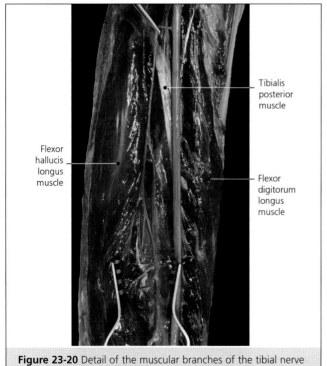

Figure 23-20 Detail of the muscular branches of the tibial nerve in the posterior compartment of the leg.

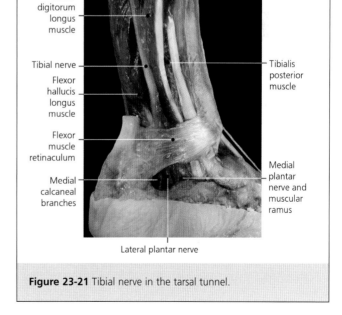

Figure 23-21 Tibial nerve in the tarsal tunnel.

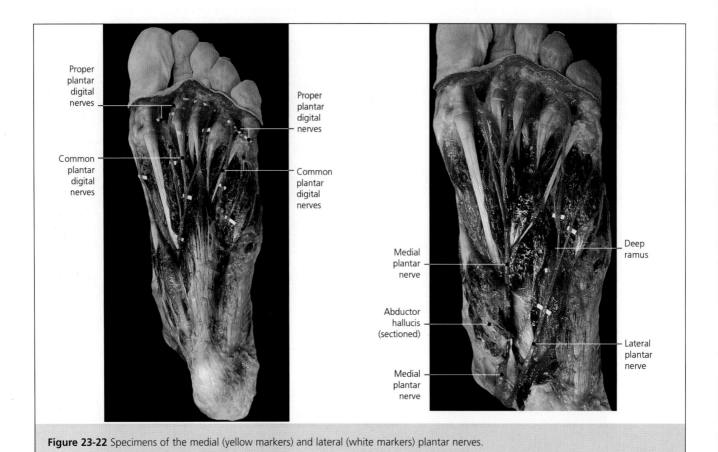

Figure 23-22 Specimens of the medial (yellow markers) and lateral (white markers) plantar nerves.

Figure 23-23 Detail of the plantar digital nerves.

Figure 23-24 Deep ramus of the lateral plantar nerve innervating the intrinsic foot musculature.

Pudendal Nerve

Finally, the pudendal nerve (S2 through S4) (Figure 23-10) exits the pelvis under the piriformis, goes around the sacrospinous ligament, and reenters the pelvis through the lesser sciatic foramen. The pudendal nerve innervates the muscles of the pelvic floor and provides sensory nerves to the perineal and genital regions. Its course follows that of the internal pudendal artery.

Coccygeal Plexus

There are only one or two coccygeal nerves that provide sensation to the skin covering the coccyx.

Chapter 24

Angiology

Arteries

The abdominal aorta bifurcates distally at the level of L4 into the two common iliac arteries (Figure 24-1), one for each lower limb, and the middle sacral artery. The common iliac artery bifurcates into the internal and external iliac arteries.

Internal Iliac Artery

The description of this artery will be limited to the collaterals and branches that supply musculoskeletal structures.

The branches of the internal iliac artery course posteriorly and distally, entering the minor pelvis, and are located on both sides of the sacrum, over the iliopsoas and piriformis muscles. One of the internal iliac artery's first bifurcations forms the iliolumbar and the superior gluteal arteries.

The iliolumbar artery divides into three branches. The iliac branch is located in the iliac fossa, supplies the iliac muscle, and finally anastomoses with the deep circumflex iliac artery. The lumbar branch ascends past the posterior region of the endopelvic aspect of the iliacus toward the psoas and quadratus lumborum muscles. The spinal branch enters the vertebral canal between the sacrum and L5.

The superior gluteal artery (Figure 24-2) crosses the greater sciatic foramen, above the piriformis, along with the superior gluteal nerve and principally supplies the lateral gluteal region. Its superficial branch courses between the gluteus maximus and medius; the deep branch courses between the gluteus medius and the minimus. The deep branch bifurcates into a superior branch, which courses through the superior portion of the gluteus minimus toward the tensor fascia lata muscle, and an inferior branch, which courses caudally toward the region of the greater trochanter.

The obturator artery (Figure 24-2) originates from the internal iliac artery at the level of the piriformis muscle and courses anteriorly toward the obturator foramen. The obturator artery penetrates the obturator membrane and distributes on the adductor musculature. It has four principal branches. The pubic branch forms a small arch and anastomoses with the obturator branch of the inferior epigastric artery. The acetabular branch enters the acetabular fossa through the acetabular notch, under the transverse ligament of the acetabulum, and courses through the round ligament of the femur to enter the femoral head. The other two branches are named according to their position relative to the adductor brevis muscle: the anterior branch courses in front of the muscle and anastomoses with the medial circumflex femoral artery; the posterior branch is posterior to the muscle.

The inferior gluteal artery (Figure 24-2) goes between the S2-S3 nerves and then courses toward the greater sciatic foramen and crosses it below the piriformis, along with the inferior gluteal nerve, to distribute deeply in the gluteus maximus muscle. This artery anastomoses with several of the regional arteries, namely the superior gluteal, the obturator, and the femoral circumflex arteries. The artery of the sciatic nerve originates from the inferior gluteal artery.

External Iliac Artery

The external iliac artery supplies the lower extremity. Before becoming the femoral artery, two important collaterals branch off: the inferior epigastric and the deep circumflex iliac arteries.

The inferior epigastric artery (Figure 24-3) originates medial to the deep inguinal foramen and courses cranially under the rectus abdominis to anastomose with the internal thoracic artery, which was described in chapter 12. Its important branch is the pubic branch, which courses toward the pubic bone and sends off a collateral, the obturator branch, which anastomoses with the obturator artery.

The deep circumflex iliac artery (Figures 24-3 and 24-4) has an arching shape and distributes from front to back along the iliac crest. This artery sends off an ascending branch that courses between the transverse and the internal oblique muscles of the abdomen and anastomoses with the iliolumbar artery.

Femoral Artery

The external iliac artery becomes the femoral artery when it passes under the inguinal ligament. This artery is located in the femoral triangle (of Scarpa) (Figures 24-4 and 24-5) between the femoral nerve (lateral) and the femoral vein (medial). A series of collateral arteries branch off immediately and distribute in the inguinal and genital regions. This chapter focuses on the inguinal branches.

The superficial epigastric artery courses cranially over the anterior abdominal wall toward the umbilicus. The superficial circumflex iliac artery (Figure 24-4) courses

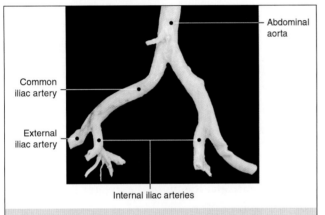

Figure 24-1 Isolated abdominal aorta with the bifurcation into the common iliac arteries.

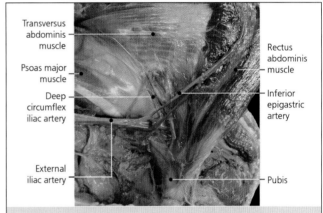

Figure 24-3 Internal view of the branches of the external iliac artery.

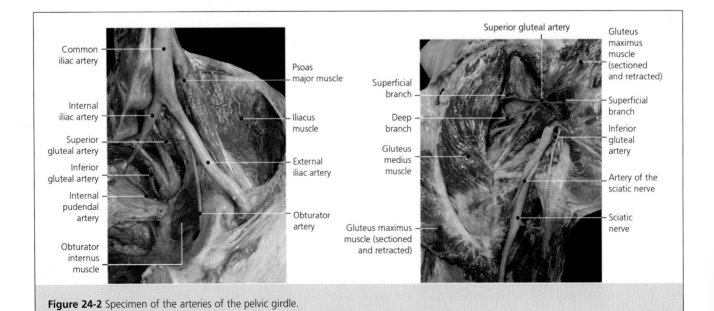

Figure 24-2 Specimen of the arteries of the pelvic girdle.

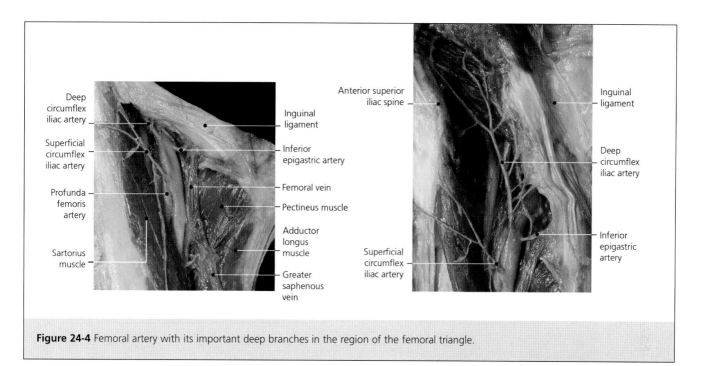

Figure 24-4 Femoral artery with its important deep branches in the region of the femoral triangle.

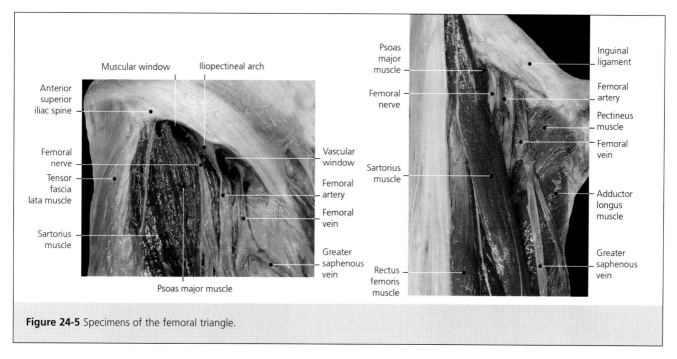

Figure 24-5 Specimens of the femoral triangle.

along the inguinal ligament toward the anterosuperior iliac spine.

The femoral artery follows a distal course toward the tip of the femoral triangle, deep to the sartorius muscle. In this region, the artery enters the adductor canal (of Hunter), between the vastus lateralis and adductor longus muscles, and is covered by the vastoadductor membrane. The femoral artery crosses the adductor hiatus and enters the popliteal fossa, where it changes into the popliteal artery (Figures 24-6 and 24-7). In the adductor canal, it sends off the descending branch of the knee (anastomotic magna), which crosses the vastoadductor membrane and is located in the medial region of the knee. Two branches originate from this artery: the saphenous branch, which follows the saphenous nerve and vein; and the articular branches, which descend toward the knee and are intimately related to the vastus medialis.

In the deep region of the femoral triangle, the deep femoral (or profunda femoris) artery branches from the femoral artery (Figures 24-4 and 24-8). The deep femoral artery originates from the lateral portion of

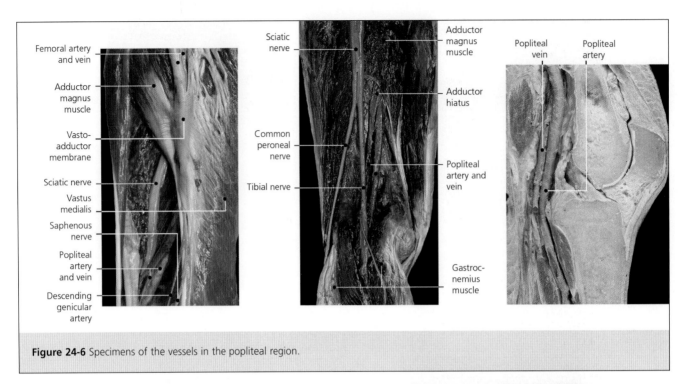

Figure 24-6 Specimens of the vessels in the popliteal region.

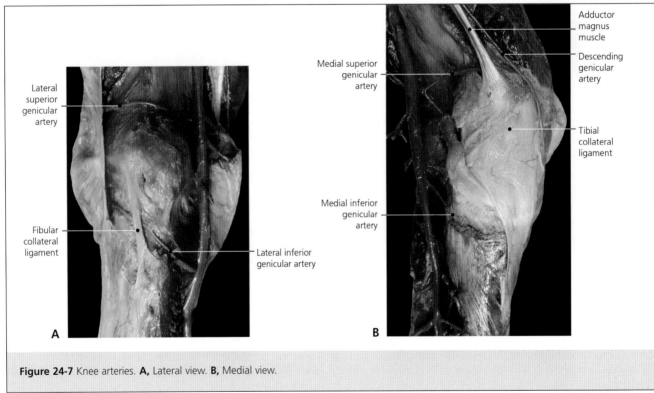

Figure 24-7 Knee arteries. **A,** Lateral view. **B,** Medial view.

the femoral artery and is located posterior to it along the thigh. It gives the medial and lateral femoral circumflex arteries, which course around the proximal epiphysis of the femur and originate from the deep femoral artery.

The medial femoral circumflex artery (Figure 24-8) has an arching course from medial to posterolateral, crossing the space between the iliopsoas and the pectineus muscles. Along its course, it sends off five branches: the superficial branch courses over the anterior aspect of the pectineus and adductor longus; the ascending branch follows a proximal and medial course toward the obturator externus and the origin of the adductor magnus and brevis, finally anastomos-

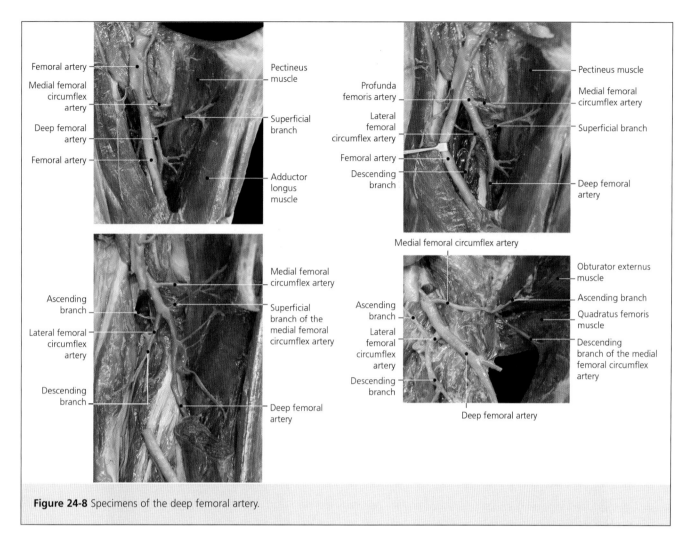

Figure 24-8 Specimens of the deep femoral artery.

ing with the obturator artery; the descending branch courses between the quadratus femoris and adductor magnus until it reaches the ischiocrural musculature; the acetabular branch courses toward the acetabulum, entering through the acetabular notch toward the ligamentum teres; and the deep branch follows a dorsal course toward the lesser trochanter, anastomosing with the gluteal arteries.

The lateral femoral circumflex artery (Figure 24-8) follows a lateral course under the rectus femoris muscle. It sends off three branches that are distributed over the lateral region of the thigh. The ascending branch is located under the sartorius, rectus femoris, and tensor fascia lata muscles, coursing around the femoral neck and anastomosing with the terminal portion of the medial femoral circumflex artery. The transverse branch enters the vastus lateralis muscle and courses around the femur at the level of the lesser trochanter, anastomosing with terminal branches of the medial femoral circumflex artery. Finally, the descending branch courses along the lateral region of the thigh toward the knee.

The deep femoral artery ends with the perforating arteries, which turn posterior when they cross the small tendinous hiatus formed by the insertion of the adductor magnus. They supply the posterior thigh musculature and the femur, through the nutrient arteries of the femur.

Popliteal Artery

After passing through the adductor hiatus, the femoral artery becomes the popliteal artery (Figures 24-6, 24-9, and 24-10). The popliteal artery is located in the region of the popliteal fossa, deep and medial to the popliteal vein.

This artery sends off a series of collateral branches that form the articular arterial network of the knee anteriorly. There are two superior branches, two inferior branches, and a middle branch, in addition to the sural artery.

The lateral superior genicular artery (Figures 24-7 and 24-9) is located deep in the popliteal fossa and courses around the lateral condyle of the femur in an anterior direction. The medial superior genicular artery (Figures 24-7 and 24-9) courses around the medial femoral condyle anteriorly to end in the arterial articular network of the knee.

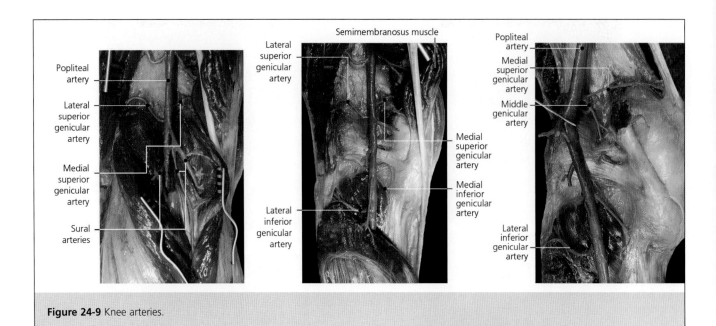

Figure 24-9 Knee arteries.

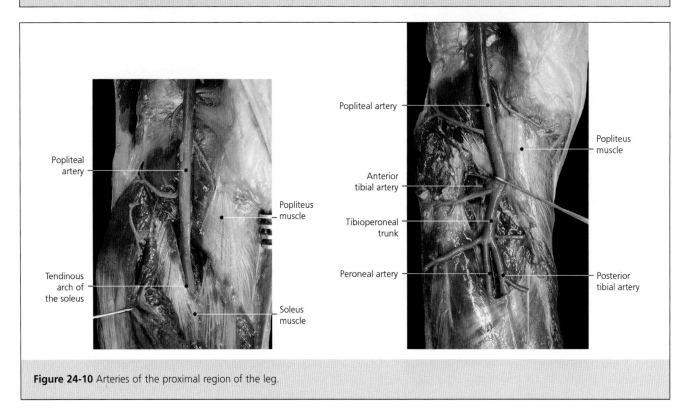

Figure 24-10 Arteries of the proximal region of the leg.

The lateral inferior genicular artery (Figures 24-7 and 24-9) courses around the lateral condyle of the tibia deep to the lateral head of the gastrocnemius and the lateral collateral ligament. The medial inferior genicular artery (Figures 24-7 and 24-9) courses around the medial condyle of the tibia under the medial head of the gastrocnemius and the medial collateral ligament to join the arterial articular network of the knee as well.

The middle genicular artery (Figure 24-9) originates between the superior and inferior geniculars below the sural arteries. It is located in the midline and becomes deep, reaching the cruciate ligaments.

There are two sural arteries (Figure 24-9), one on each side, which are distributed on the triceps surae.

The popliteal artery passes under the tendinous arch of the soleus and bifurcates into the anterior tibial artery and the tibioperoneal trunk (Figure 24-10). The tibioperoneal trunk in turn bifurcates into the posterior tibial and peroneal arteries. These three arteries distribute in the leg compartments, the posterior tib-

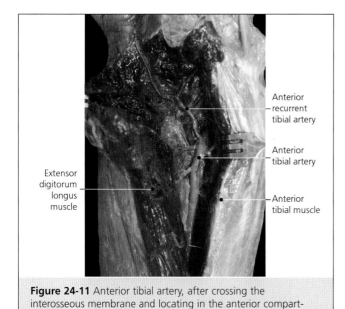

Figure 24-11 Anterior tibial artery, after crossing the interosseous membrane and locating in the anterior compartment.

ial and peroneal artery in the posterior compartment, and the anterior tibial artery in the anterior compartment.

Anterior Tibial Artery

The anterior tibial artery (Figures 24-10 through 24-13) originates at the level of the popliteal muscle in the posterior region of the knee, passes below it, and crosses the interosseous membrane to enter the anterior compartment of the leg. From it arises the anterior recurrent tibial artery (Figure 24-11), which ascends to anastomose with the arterial articular network of the knee. The anterior tibial artery descends in the anterior compartment between the anterior tibial and the extensor digitorum longus muscles. Distally, it is located medially between the anterior tibial and the extensor hallucis longus muscles. At the level of the malleoli, it sends off two branches, one for each malleolus (Figure 24-12), which are named the anterior medial and lateral malleolar arteries and form part of the medial and lateral malleolar networks, respectively.

After passing under the extensor retinaculi, the anterior tibial artery is called the dorsal artery of the foot (pedal artery) (Figure 24-13). It is located between the tendons of the extensor digitorum longus and the extensor hallucis longus, where its pulse can be felt. This artery bifurcates (Figures 24-12 and 24-13) into the medial tarsal artery, distributed over the medial border of the foot, and the lateral tarsal artery, which courses from the talar region toward the lateral border of the foot under the short toe extensor muscles. The medial and lateral tarsal arteries communicate with each other through the arcuate artery (Figure 24-12), forming a dorsal arterial arch. From this arch originate the dorsal metatarsal arteries (Figures 24-12 through

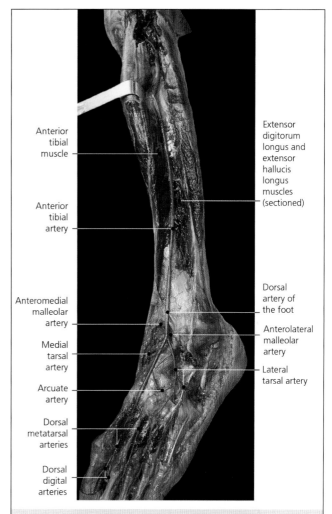

Figure 24-12 Trajectory of the anterior tibial artery. The short extensor muscles of the toes have been removed. The arcuate artery in this case is very small and proximal.

24-14), which are located in the interosseous spaces between the toes. At the level of the metatarsophalangeal joints, these arteries each bifurcate into two dorsal digital arteries that course toward both sides of the toes in each interosseous space. The lateral border of the fifth toe is vascularized by an artery that originates in the same arcuate artery, whereas the medial border of the great toe is vascularized by plantar arteries.

The deep plantar artery (Figure 24-14) is a perforating artery that connects the dorsal metatarsal arteries with the deep plantar arch through the first interosseous space.

Peroneal Artery

The peroneal artery is located behind the fibula (Figure 24-15) and courses distally toward the calcaneus. At the level of the fibular diaphysis, the nutrient artery of the fibula branches from the peroneal artery. Distally, a communicating branch from the peroneal artery con-

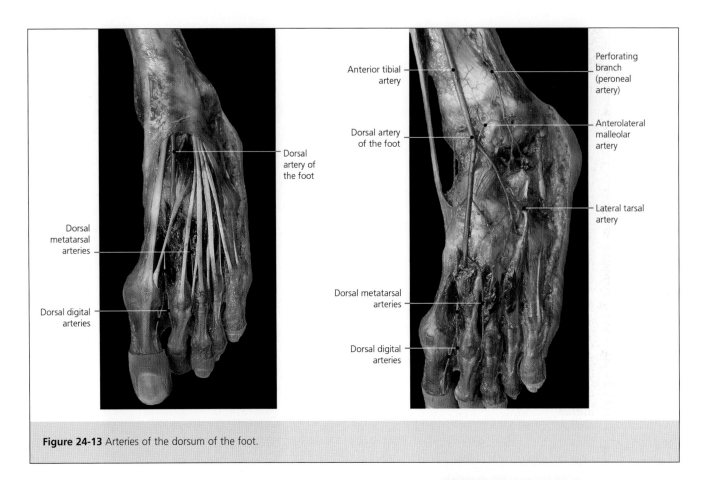

Figure 24-13 Arteries of the dorsum of the foot.

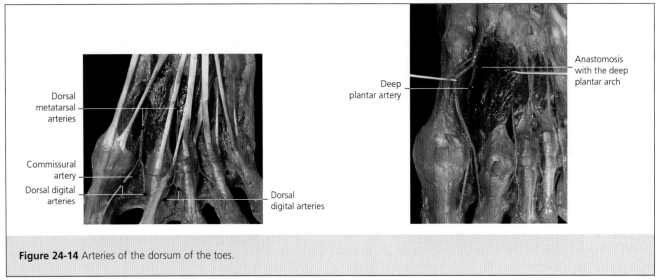

Figure 24-14 Arteries of the dorsum of the toes.

nects with the posterior tibial artery. Immediately distal to this branch, the peroneal artery produces the perforating branch (Figures 24-13 and 24-15), crosses the interosseous membrane, and distributes over the lateral malleolus and the dorsum of the foot. Near the lateral malleolus, it sends off the lateral malleolar branches, which anastomose with those of the anterior tibial artery. Finally, the calcaneal branches help form the calcaneal network in the posterior aspect of this bone.

Posterior Tibial Artery

The posterior tibial artery is distributed in the posterior compartment of the leg (Figure 24-15), between the superficial and the deep groups. Shortly after its origin, the nutrient artery of the tibia branches off. Along its course, the posterior tibial artery becomes more medial as it goes toward the tarsal tunnel. At the level of the medial malleolus, the medial malleolar branches help form the medial malleolar network.

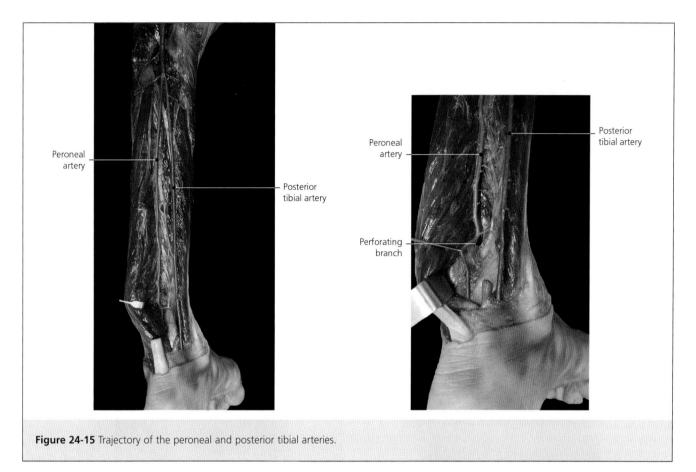

Figure 24-15 Trajectory of the peroneal and posterior tibial arteries.

The tarsal tunnel (Figure 24-16) is formed by the distal tibial epiphysis, the talus, and the calcaneus, which are joined by the medial collateral ligament. This canal is covered by the flexor retinaculum. The tendinous floor of this region is formed by the tendons of the flexor digitorum longus and the posterior tibial muscle with their tendon sheaths. The neurovascular contents are the posterior tibial artery, vein, and nerve that distribute in the sole of the foot.

In the tarsal tunnel, the posterior tibial artery bifurcates into the medial and lateral plantar arteries. The medial plantar artery (Figures 24-16 and 24-17) bifurcates, in turn, into a deep branch that courses between the abductor and the short flexor of the great toe toward the first interosseous space; the superficial branch follows the medial border of the foot to the great toe. The lateral plantar artery (Figures 24-16 through 24-18) courses obliquely in the plane between the short flexor of the toes and the quadratus plantaris toward the lateral region of the foot. Both arteries anastomose with each other at the midlevel of the metatarsals, forming the deep plantar arch. From this arch originate the plantar metatarsal arteries, which course in the interosseous spaces of the foot and are continued by the common plantar digital arteries (Figure 24-19). In addition, a pair of perforating branches course toward the dorsum of the foot.

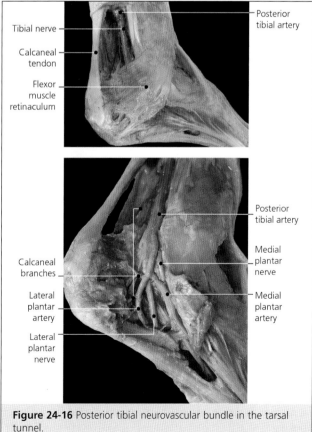

Figure 24-16 Posterior tibial neurovascular bundle in the tarsal tunnel.

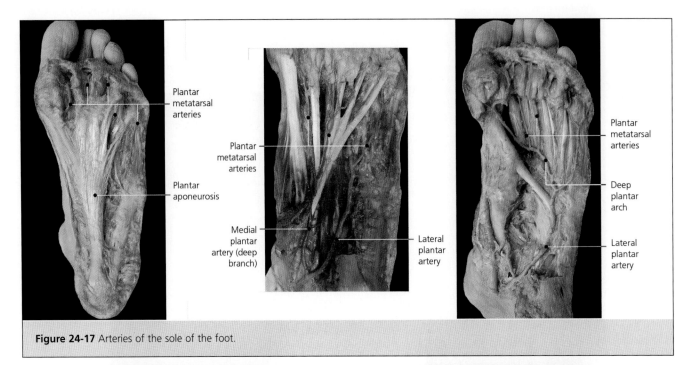

Figure 24-17 Arteries of the sole of the foot.

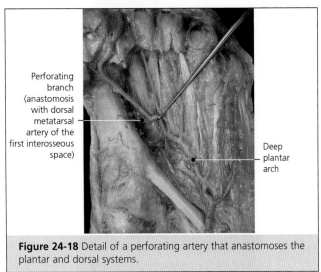

Figure 24-18 Detail of a perforating artery that anastomoses the plantar and dorsal systems.

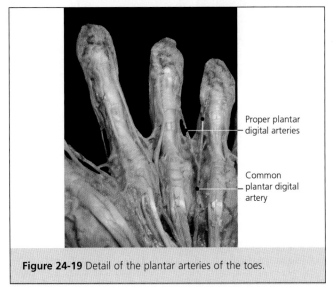

Figure 24-19 Detail of the plantar arteries of the toes.

Near the metatarsophalangeal joints, the common digital arteries bifurcate into the proper digital arteries, which extend on each side of the toes that form each interosseous space. The medial artery of the great toe and the lateral of the fifth toe originate directly from the deep plantar arch.

A superficial anastomosis between the plantar arteries is variably present. It is called the superficial plantar arch.

Veins and Lymphatics

All the veins of the pelvic girdle and the lower limb flow into the common iliac veins, which join to form the inferior vena cava.

The veins are classified into superficial and deep (Figure 24-20). The latter are usually concomitant veins, which travel with and follow the arteries described in this chapter.

Deep and Superficial Veins of the Foot

The deep and superficial veins of the foot are described from distal to proximal, following the direction of their flow. The superficial veins of the foot are located in the dorsum, whereas the deep veins are located plantarly. They have numerous communications, however.

The venous blood of the toe region is collected by the digital plantar veins and the digital dorsal veins of the foot. These flow into the plantar and dorsal metatarsal veins of the foot (Figure 24-21), respectively. In turn, the metatarsal veins flow into the plantar venous arch and the dorsal arch of the foot. In the

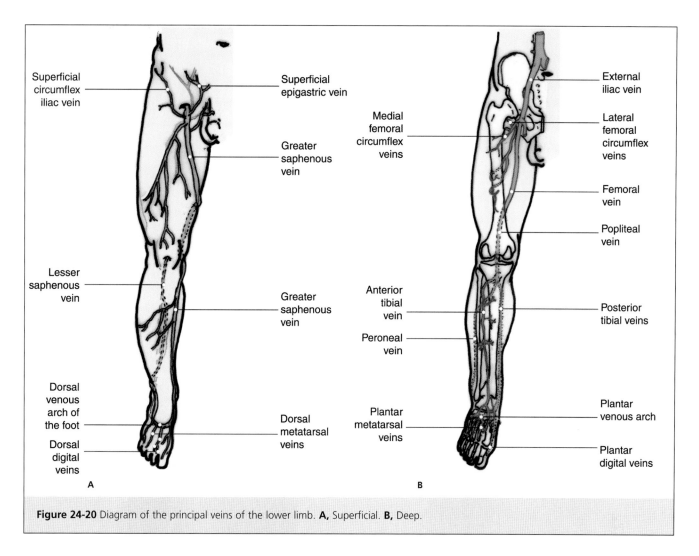

Figure 24-20 Diagram of the principal veins of the lower limb. **A,** Superficial. **B,** Deep.

toes are a series of communications between the plantar and dorsal systems through the intercapitular veins at the level of the interosseous spaces and the medial marginal vein (on the medial side of the great toe) and the lateral marginal vein (on the lateral side of the fifth toe). Moreover, a series of perforating veins communicate the subcutaneous with the subfascial veins of the leg, joining the superficial and deep systems.

Deep Veins of the Leg and Thigh

The anterior tibial veins (Figure 24-20) accompany the anterior tibial artery in the anterior compartment of the leg. They cross the interosseous membrane to become posterior. They collect the blood of the dorsum of the foot and flow into the posterior tibial vein to jointly form the popliteal vein.

The posterior tibial veins (Figure 24-20) collect the venous blood from the sole of the foot, including that of the plantar superficial venous network. The posterior tibial veins accompany the posterior tibial artery through the posterior compartment of the leg and join the anterior tibial vein to form the popliteal vein.

The peroneal veins (Figure 24-20) originate at the level of the lateral calcaneal region and ascend, following the peroneal artery until they join the posterior tibial vein in the proximal third of the leg.

Located superficial to the artery in the popliteal fossa, the popliteal vein forms from the confluence of the anterior and posterior tibial veins and the peroneal veins (Figures 24-20 and 24-21). It receives the genicular veins (of the knee), the sural and the lesser saphenous veins, which will be described later with the superficial veins. The popliteal vein continues as the femoral vein when it passes into the adductor canal though the adductor hiatus.

The femoral vein (Figure 24-20) follows the femoral artery from the adductor hiatus to the femoral triangle, where it is located medial to the artery. At the level of the femoral triangle, it receives the deep femoral vein and the greater saphenous vein. Proximally, near the inguinal ligament, it receives the external pudendal, superficial circumflex iliac, and superficial epigastric veins (Figure 24-22), with other venous branches named for the concomitant arteries of the same name.

The deep femoral vein joins the femoral vein and

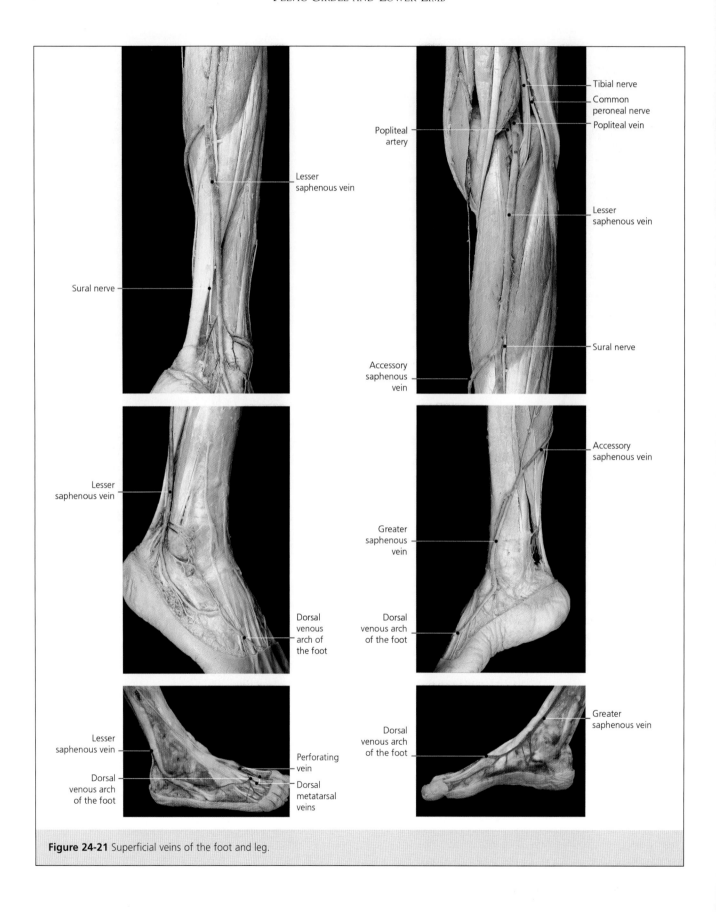

Figure 24-21 Superficial veins of the foot and leg.

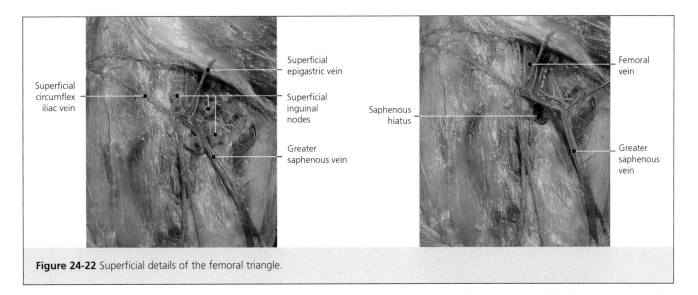

Figure 24-22 Superficial details of the femoral triangle.

receives blood from the medial and lateral femoral circumflex veins as well as from the perforating veins from the adductor magnus.

Superficial Veins of the Leg and the Thigh

The superficial veins of the lower limb are called saphenous veins. The lesser saphenous vein (Figures 24-20 and 24-21) originates in the lateral region of the foot, collects blood from the dorsal venous network of the foot and venous dorsal arch, and ascends behind the lateral malleolus in a medial direction toward the popliteal fossa. The lesser saphenous vein crosses the popliteal fascia in the inferior angle of the popliteal fossa and joins the popliteal vein.

The greater saphenous vein (Figures 24-5, 24-20, and 24-23) originates in the medial region of the foot, collects blood from the dorsal venous network of the foot and venous dorsal arch, and ascends in front of the medial malleolus and the medial side of the leg and thigh. In the proximal third of the thigh, it crosses the cribiformis fascia (roof of the femoral triangle) through the saphenous hiatus (fascia lata foramen) (Figure 24-22) to join the femoral vein near the inguinal ligament.

In the inguinal region are two spaces under the inguinal ligament separated by the iliopectineal arch (Figure 24-5), which is a thickening of the iliac fascia that extends between the iliopectineal eminence and the inguinal ligament. The muscular window is lateral, and through it course the iliopsoas muscle and the femoral nerve. The vascular window is medial and is crossed by the femoral artery and vein as well as the characteristic lymph node of Cloquet. These neurovascular structures represent the contents of the femoral triangle.

Veins of the Pelvic Girdle

The femoral vein continues as the external iliac vein after crossing the vascular window. This receives blood from

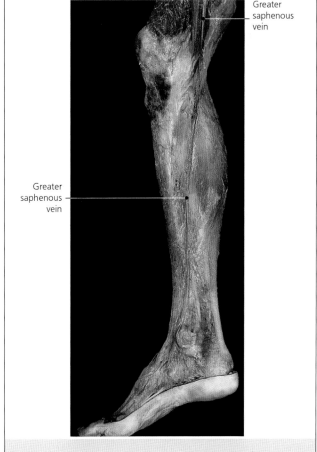

Figure 24-23 Greater saphenous vein in the leg and foot.

the inferior epigastric and the deep circumflex veins, homonyms of the arteries that have been described.

The internal iliac veins receive blood from the internal pudendal, superior and inferior gluteal, and obturator veins. They join the external iliac veins to form the common iliac veins, which join to form the inferior vena cava. The inferior vena cava also receives the flow from the middle sacral vein. The iliolumbar vein flows into the common iliac vein, although it can also flow into the internal iliac vein.

In their lumen, both the superficial and deep veins have characteristic venous valves that translate into small external dilatations.

LYMPHATICS

The principal lymph nodes of the lower extremity are organized into superficial and deep in the inguinal and popliteal regions.

Within the inguinal region are the superomedial and superolateral lymph nodes (Figure 24-22), which are located under the inguinal ligament, and the inferior inguinal nodes, which are located along the exit of the greater saphenous vein when it crosses the cribiformis fascia. The deep inguinal nodes are located subfascially, medial to the femoral vein (node of Cloquet).

The superficial popliteal nodes are located in the popliteal fossa, at the level of the junction of the lesser saphenous and the popliteal vein. The deep popliteal nodes are located deep to the popliteal artery.

Index

*Page numbers with *f* indicate figures; page numbers with *t* indicate tables.

A

Abdomen
  arteries of, 288–289, 289f
  muscles of, 270t, 271, 271f
    anterolateral, 271–273, 271f, 275
    posterior, 275, 275f
  veins of, 289–290, 290f, 291f
Abdominal aorta, 283–284, 391, 392f
Abduction, 21f
Abductor digiti quinti
  of the foot, 372f, 376, 375t, 376f
  of the hand, 159, 160f, 162t
Abductor hallucis, 370t, 372, 372f, 373
Abductor indicis, 163
Abductor pollicis brevis, 156–157, 157ft, 161t, 185
Abductor pollicis longus, 107, 148f, 148t, 149f, 151–152, 151t, 152f
Accessory collateral ligament, 112
Accessory fascicle of Gantzer, 141, 142f
Accessory obturator nerve, 381
Accessory process, 224, 224f
Acetabular fossa, 299, 391
Acetabular notch, 299
Acetabular pulvinar, 320, 321f
Acetabulum, 299
Achilles tendon, 359
Acromial angle, 54
Acromial beak, 54
Acromioclavicular joint, 81, 83f, 117

Acromioclavicular ligaments, 81, 83f
Acromion, 49, 54
Adductor brevis, 352, 353tf, 354, 381, 391
Adductor canal (of Hunter), 355, 393
Adductor hallucis, 370t
Adductor hiatus, 354, 393
Adductor longus, 295, 350, 350f, 352, 353tf, 354, 381
Adductor magnus, 304, 351f, 353t, 354, 354f, 355f, 395, 395f
Adductor pollicis, 157f, 158, 158f, 159f, 160f, 161t, 190
Adductor pollicis brevis, 156–157, 157f, 158, 158f, 159f, 160f
Adductor tubercle, 304
Afferent innervation, 29
Afferent nerve fibers, 31
Amphiarthrosis, 18, 217, 249, 256
Amphiarthrotic joint, 18
Anatomic position, 3, 4f
Anatomic snuffbox, 206
Anconeus muscle, 145, 146f, 146t, 180
Angular muscle of scapula, 120
Ankle. *See also* specific structures
  bones of, 308-311, 308f, 309f, 310f
  joints of, 331-335, 331f, 332f, 333f, 334f, 335f
  ligaments of, 332-335, 334f, 335f
Annular epiphysis, 217
Annular ligament of radius, 88, 90f
Annular ligament of Weber, 320

Annulus fibrosis, 244
Antebrachial fascia, 133
Antecubital fossa, 45–46, 47f
Antepulsion, 22, 23f
Anterior arcuate ligament, 97
Anterior circumflex humeral artery, 201, 202f
Anterior cruciate ligament, 326f, 328, 329f
Anterior inferior iliac spine, 300
Anterior longitudinal ligament, 240f, 244, 246f
Anterior meniscofemoral ligament (of Humphrey), 324
Anterior pubic ligament, 319
Anterior sacroiliac ligaments, 315, 317f
Anterior scalene muscles, 265t
Anterior superior iliac spine, 300
Anterior talofibular ligament, 332
Anterior tibial artery, 396f, 397, 397f, 398f
Anterior tibialis, 365, 365t
Anteversion, 20
Aorta, 283–286, 284f, 285f, 286f
  abdominal, 283–284, 391, 392f
  ascending, 283, 284f
  descending, 283
  thoracic, 283, 284f
Aortic arch, 283, 284f, 287
Aortic hiatus, 276, 283
Aortic valve, 283
Aponeurosis, 29
Appendicular skeleton, 7
Arcade of Frohse, 192
Arcade of Osborne, 135, 187
Arcuate line (of Douglas), 272, 301

Index

Arcuate popliteal ligament, 328, 360
Arm. *See also* specific structures
   arteries of, 201-206, 202f, 203f, 204f, 205f, 206f
   bones of, 54-67, 56f, 57f, 58f, 59f, 60f, 61f, 62f, 63f, 64f, 65f, 66f, 67f
   joints of, 93-96, 95f, 96f
   ligaments of, 101-102, 101f
   muscles of, 129-156, 130t, 131f, 132f, 133f, 134f, 135f, 136t, 137f, 139f, 145t, 146f, 147f, 148t, 151t, 152f, 154t, 155f, 156f
   nerves of, 173-194, 178f, 179f, 180f, 181f, 182f, 183f, 184f, 185f, 186f, 187f, 188f, 189f, 190f, 191f, 192f, 193f, 194f
   topography of, 45-46, 47f
Arteries, 37, 38–39, 38f, 39f. *See also* specific arteries
   adventitia, or external layer, 38
   axillary, 39
   brachial, 39
   conducting, 38
   elastic, 38
   femoral, 39
   internal thoracic, 288, 289f
   intima, or internal layer, 38
   media, or middle layer, 38
Arterioles, 39
Arteriovenous anastomoses, 41
Articular capsule, 79, 81, 83, 83f, 328
Articular cartilage, 9, 9f
Articular congruity, 324, 327f
Articular disks, 18, 18f, 217
Articular fibrocartilage, 18, 18f, 49, 51f
Articular meniscus, 18, 18f
Articular motion, range of, 23
Articular nerves, 19
Articular processes, 217
Articulation, 15

Ascending cervical artery, 287
Atlantoaxial joint, 240f, 241
   anterior atlantoaxial ligament, 241–242
   lateral atlantoaxial ligaments, 241
   posterior atlantoaxial ligament, 242
   synovium, 242
Atlanto-occipital joint, 239, 240f
   capsular ligaments, 239, 240f
   anterior atlanto-occipital membrane, 239–240, 240f
   posterior atlanto-occipital membrane, 240, 240f
Atlas, 219, 220f, 239
Atrium, right, 40
Auricular region, 213
Autonomic nervous system, 35
Axial plane, 4–5
Axial skeleton, 7, 217
Axillary artery, 39, 195, 198–199
   branches lateral to pectoralis minor
     anterior circumflex humeral artery, 201, 202f
     posterior circumflex humeral artery, 201, 202f
   branches medial to pectoralis minor
     superior thoracic artery, 199
     thoracoacromial artery, 199, 199f, 200f
   branches posterior to pectoralis minor
     lateral thoracic artery, 199, 200f
     subscapular artery, 199–201, 200f
Axillary fossa, 45
Axillary nerve, 179f, 180, 181f
Axillary region, 45, 46f
Axillary vein, 209
Axis (vertebra), 219–220, 220f, 239

Axis of body, 5f
   anteroposterior, 4
   horizontal, 4
   longitudinal, 4
   main, 4
   sagittal, 4, 5, 21, 21f, 22f
   transverse, 4, 20–21
   vertical, 4
Axons, 31
Azygos system, 40
Azygos vein, 289, 290

B

Back. *See* Dorsal region (the back)
Bardinet's ligament, 92
Basilic vein, 209
Biaxial joints, 19, 20f, 22
Biceps brachii, 26f, 129, 130, 130t, 131f, 132, 132f, 178, 179f, 201
Biceps brachii aponeurosis, 130
Biceps femoris, 324, 325f, 328f, 354t, 355f, 356, 356f, 357f, 359, 383
Bicipital groove, 57
Bicipital tuberosity, 62, 130, 132f
Bicipitoradialis bursa, 130
Bifurcate ligament, 335f, 336
Blood vessels. *See also* Arteries; Veins
Bone collar, 13
Bones. *See also* names of specific bones
   cancellous, 10f, 12, 12f
   classification of, 10–11
   compact, 12
   cortical, 10f, 12, 12f, 217
   flat, 10, 10f, 15
   functions of, 7
   layers of coverage, 9–10
   marrow of, 9
   matrix of, 11–12
   structure of, 7, 8f
   vascularity of, 12, 13f
Brachial artery, 39, 201–204, 202f

Brachialis, 64, 129, 130t, 131f, 132, 132f, 178, 179f
Brachial plexus, 172f
  anterior branches
    lateral pectoral nerve, 172, 175f, 176f
    medial pectoral nerve, 171–172, 175f, 176f
    nerve for subclavius muscles, 171, 175f
  collateral branches, 33, 171, 172f, 173f, 175f, 264, 279
  medial cord of, 171
  posterior branches
    dorsal scapular nerve, 173, 177f
    long thoracic nerve, 173, 175f, 176f, 178f
    subscapular nerves, 173, 178f
    suprascapular nerve, 172, 175f, 176f, 178
    thoracodorsal nerve, 173, 178f
  terminal nerves (arm region)
    axillary nerve, 179f, 180, 181f
    medial brachial cutaneous nerve, 178, 179f
    median nerve, 173, 178, 178f, 179f
    musculocutaneous nerve, 173, 178, 178f, 179f
    radial nerve, 178, 179f, 180
    ulnar nerve, 178, 178f, 179f
  terminal nerves (forearm region)
    lateral antebrachial cutaneous nerve, 180–181, 182f
    medial antebrachial cutaneous nerve, 179f, 181–182, 182f, 183f
    median nerve, 176f, 182–185, 182f, 183f, 184f, 185f, 186f, 187f
    radial nerve, 192, 192f, 193f, 194, 194f
    ulnar nerve, 187–190, 189f, 190f, 191f
Brachial veins, 209, 209f
Brachiocephalic trunk, 38, 39f, 195, 283
Brachiocephalic vein, 289, 290
Brachioradialis muscle, 45, 62, 154, 154t, 155f
Brodie's ligament, 84, 84f
Buccal region, 213, 214f
Bursa
  bicipitoradialis, 130
  calcaneal, 359
  iliopectineal, 346
  interosseous, of elbow, 130
  ischial, 341
  retrocalcaneal, 359
  subacromial, 125
  subdeltoid, 124
  subtendinous iliac, 346
  suprapatellar, 323, 324f
  trochanteric, 341, 344
Buttock, 295

C

Calcaneal apophysis, 312
Calcaneal canal, 311
Calcaneal tuberosity, 310
Calcaneocuboid joint, 331f, 335, 336, 336f
Calcaneocuboid plantar ligament, 310
Calcaneofibular ligament, 332
Calcaneus, 310–311, 310f, 399, 399f
Capitate, 70, 72f
Capitellotrochlear sulcus, 59
Capitellum, 59
Cardiac muscle, 25
Cardiovascular system, 37
Carotid arteries, 38, 39f, 219
Carotid triangle, 213
Carotid tubercle, 219
Carpal bones, 97f
Carpal joints, 102-107
  intercarpal joints, 102, 102f
    distal row, 103
    proximal row, 102
  midcarpal joint, 97f, 102f, 103, 105, 106f, 107, 107f, 108f
  midcarpal ligaments, 103, 103f, 104f, 105f
Carpal radial collateral ligament, 101f, 102
Carpal tunnel, 63, 67, 68, 105, 107, 184–185, 185f
Carpal ulnar collateral ligament, 101f, 102
Carpometacarpal joints, 107–108, 109f, 110f, 111, 112–113
Carpus, 46, 65-71
Cartilage, articular, 9, 9f
  elastic, 9
  fibrocartilage, 9
  hyaline, 9
Cauda equina, 33
Caudal vertebra, 217
Cavus foot, 296
Central nervous system (CNS), 31, 32f
Cephalic region. See Head.
Cephalic vein, 199, 209
Cephalic vertebra, 217
Cervical nerves, 31
Cervical plexus, 119, 279–280, 279f, 280f
Cervical spine, 239
Cervical vertebrae, 219–220, 239
Cervicothoracic rhombus, 117
Chassaignac tubercle, 219
Checkrein ligament, 114f, 116
Chest
  arteries of, 288–289, 289f
  muscles of, 268-270
    dorsal, 262f, 268
    intercostal, 268–270, 269f
Chopart's joint, 335, 336f
Circumflex scapular artery, 200f, 201, 201f
Clavicle, 45, 46f, 49, 50f, 51f, 79, 229
  acromial end of, 117, 118f

Clavicular notches, 79, 229
Clavipectoral fascia, 49, 121, 171, 199
Cleidomastoid fascicle, 119
Cleidooccipital fascicle, 119
Cleland's ligament, 168–169, 168f
Coccygeal fossa or fovea, 250
Coccygeal glomus, 286
Coccygeal nerves, 31
Coccygeal plexus, 390
Coccygeal vertebrae, 226
Coccyx, 225f, 226, 226f, 249, 299, 315, 341
Collateral ligaments, of metacarpophalangeal joints, 112
Columbar fascia, 273
Common carotid artery, 287, 288f
Common iliac arteries, 286
Common peroneal nerve, 384f, 386, 386f, 387f
Condylar process, 278
Conjoined tendon, 128
Conoid ligament, 81
Conoid tubercle, 49
Cooper's ligament, 92
Coracoacromial ligament, 79, 80f, 81f
Coracobiceps, 128, 128f
Coracobrachialis, 45, 49, 57, 127t, 128, 128f, 178, 201
Coracoclavicular ligaments, 49, 81, 83f
Coracoglenoid ligament, 85
Coracohumeral ligament, 85, 86f
Coracoid notch, 49, 51
Coracoid process, 49, 81, 81f, 130, 199
Coronary arteries, 283
Coronoid apophysis, 59
Coronoid humeral fossa, 64
Coronoid process, 64
Costal notch, 79, 229
Costal process, 224, 224f
Costocervical trunk, 198

Costochondral joints, 15, 17f, 254, 255f
Costochondral junctions, 270, 270f
Costochondral synchondrosis, 122
Costoclavicular ligament, 49, 79, 82f
Costolaminar ligament, 253
Costotransverse joint, 251–253, 253f, 254f
Costotransverse ligament, 252, 253f
Costovertebral joint, 250–251, 250f, 252f, 253f
Costoxyphoid ligaments, 255, 255f
Coxal (innominate) bone, 299, 301f, 347
  ilium, 299–301, 301f
  ischium, 301f, 302, 302f
  obturator foramen of, 320
  pubis, 301–302, 301f, 302f
Coxofemoral joint, 299, 316f, 320, 321f, 322f
Cranial nerves, 31
Cranioscapular muscles
  sternocleidomastoid, 121t
  trapezius, 121t
Craniothoracoscapular muscles, 117-122
  levator scapulae, 119–120, 119f, 120f
  omohyoid, 122
  pectoralis minor, 121–122, 121f
  rhomboid, 119, 120f
  serratus anterior, 120, 120f
  sternocleidomastoid, 118–119, 118f, 119f
  subclavius, 122
  trapezius, 117–118, 118f
Cribriform fascia, 379
Cribriform fossa, 225
Cruciform ligament, 240, 241, 242f, 243
Crural chiasma, 364

Crural fascia, 347
Crural region, 296, 297f
Cubital tunnel, 92
Cuboid, 310, 310f, 312, 312f, 338, 364
  tuberosity, 312
Cuneiforms, 311, 311f, 336, 336f, 338, 364
Cuneonavicular joint, 338

D

Deep cervical artery, 287
Deep circumflex iliac artery, 392, 392f, 393f
Deep femoral artery, 395, 395f
Deep femoral vein, 401, 403
Deep palmar arch, 207
Deep palmar venous arch, 209
Deep peroneal nerve, 372, 386, 388f
Deep pes anserinus, 356, 357f
Deep plantar artery, 397, 398f
Deep popliteal nodes, 404
Deep transverse, 111f, 113–114
Deep veins, 41
  of foot, 400–401, 402f
  of leg, 401, 401f, 403
  of thigh, 401, 401f, 403
Deltoid muscles, 54, 57, 124, 124f, 125f, 126t, 180, 332, 334f
Deltoid region, 45
Deltoid tuberosity, 57
Deltopectoral interval, 199
Deltopectoral triangle, 45, 124
Dens (odontoid process), 220
Dentate joint, 15, 16f
Denucé's ligament, 88, 91f
Dermatome, 31
Descending aorta, 283
Deviation, ulnar, 21
Diaphragm, 271, 275–277, 275f, 276f, 277f
Diaphysis, 7, 303–304
Diarthrosis, 15, 245
Diarthrotic joints, 18
Digastric buttonhole, 266

Digastric fossa, 265
Digastric muscles, 27, 27f, 265, 267tf
Diploë, 10, 10f
Directions, anatomic, 3-4
Distal biceps, 45–46
Distal carpal row, 103
Distal epiphysis, 62, 62f, 63f, 65, 65f, 307–308, 308f, 309
Distal femoral epiphysis, 304
Distal humeral epiphysis, 57, 57f, 58f, 59f
Distal phalanx, 77, 77f, 313
Distal phalanx tuberosity, 313
Distal radioulnar joint, 95–97, 96f
Distal tibial epiphysis, 399, 399f
Distal tibiofibular joint, 330–331, 330f
Distal tibiofibular syndesmosis, 309
Dorsal calcaneocuboid ligament, 336
Dorsal carpometacarpal ligaments, 108, 110, 111f
Dorsal cutaneotendinous fibers, 169, 169f
Dorsal intercarpal ligament, 103, 103f
Dorsal intermetacarpal ligament, 107
Dorsal interossei muscles, 163–164, 165f, 166f, 166t, 373t
Dorsal interosseous muscles, 375, 375f
Dorsal radial ligament, 107
Dorsal radiocarpal ligament, 99
Dorsal radioscaphoid ligament, 99
Dorsal ramus, 33, 34f, 35
Dorsal region (the back), 3, 4, 215, 215f
    infrascapular, 215
    lumbar, 215
    Petit's triangles, 215
    sacral, 215
    scapular, 215
Dorsal root, 31

Dorsal scapular nerve, 119, 120, 173, 177f
Dorsal spinous process, 217
Dorsal ulnocarpal ligament, 99, 102
Dorsiflexion, 20, 21
Double condylar joint, 103

E

Elastic arteries, 38, 39
Elastic cartilage, 9
Elbow. *See also* specific structures
    arteries of, 201-205, 202f, 203f, 204f, 205f
    bones of, 65-67, 67f
    joints of, 86-94, 87f, 88f, 89f, 90f, 91f, 92f, 93f, 94f
    ligaments of, 88-93, 91f, 92f, 93f
    muscles of, 129-139, 130t, 131f, 132f, 133f, 134f, 135f, 136t, 137f, 139f, 146t
    topography of, 45-46
Eminences, 11
Enarthroses, 19
Endochondral ossification, 13, 13f
Endomysium, 25, 26f
Endoneurium, 31, 32f
Endosteum, 10
Entheses, 10
Epicondylar muscles, 148
Epimysium, 25, 26f
Epineurium, 31, 32f
Epiphyseal line, 9, 9f, 13
Epiphyseal plate, 13
Epiphyseal vessels, 12, 13f
Epiphysis, 7, 13
Epitrochlea, 59, 133
Epitrochlear intermuscular septum, 135
Epitrochlear muscles, 133
Erb's point, 280
Erector spinae, 257, 257f
Esophageal hiatus, 276
Extensor carpi radialis brevis, 147f, 148t, 154t, 155–156, 155f, 156f

Extensor carpi radialis longus, 147f, 148f, 148t, 154–155, 154t, 155f
Extensor carpi ulnaris, 65, 74, 146t, 147–148, 147f, 148f, 148t, 149f
Extensor digiti quinti, 146–147, 146t, 147f, 148f, 148t, 149f, 150f
Extensor digitorum brevis, 370–371, 370t, 372f
Extensor digitorum communis, 107, 145–146, 146t, 147f, 148ft
Extensor digitorum longus, 362f, 365, 365t, 366–367, 367f
Extensor hallucis brevis, 370t, 371–372, 372f
Extensor hallucis longus, 362f, 365, 365t, 367
Extensor indicis proprius, 107, 148t, 151t, 152f, 153, 153f
Extensor mechanism, 164–165, 166f, 167, 167f, 168f
Extensor pollicis brevis, 107, 148t, 151t, 152, 153f
Extensor pollicis longus, 107, 148f, 148t, 149f, 151t, 152, 152f, 153f
Extensor retinaculum (dorsal carpal ligament), 107, 108f
Extensor tendons, 46
External carotid artery, 287, 288, 288f
External iliac artery, 347, 392, 392f, 393f
External iliac vein, 347
External jugular veins, 289
External oblique, 270t, 272, 272f
External occipital tuberosity, 117
Extracapsular fibular collateral ligament, 324, 328
Extracapsular ligaments, 15, 17f

F

Fabella, 359
Facial artery, 213

INDEX

Facial region, 213
False ribs, 227
False vertebrae, 224
Fascia lata, 347, 348f
Femoral artery, 39, 379, 392–395, 393f
Femoral nerve, 378f, 379f, 381–382, 382f
Femoral triangle, 295, 348, 350f, 379, 382, 382f, 383f, 392, 393, 393f, 401
Femoral vein, 401, 401f
Femur, 10, 302–305, 303f, 304f, 305f
Fibrocartilage, 9, 15
Fibrous septum, 125
Fibula, 307f, 308–309, 331
Fingers, ligaments of skin of, 162f, 167–169, 168f, 169f
Flat foot, 296
Flexor carpi radialis, 134, 135f, 136t, 137f, 138f
Flexor carpi ulnaris, 135, 136f, 136t, 139f, 187
Flexor digiti quinti, 372f, 375t, 376, 376f
Flexor digiti quinti brevis, 162t
Flexor digitorum brevis, 364, 372f, 373t, 374
Flexor digitorum communis, 145t
Flexor digitorum longus, 359t, 363f, 364, 388, 389f
Flexor digitorum profundus, 77, 138, 140–141, 141f, 145t, 183, 187, 188f
Flexor hallucis brevis, 370t, 372–373
Flexor hallucis longus, 310, 313, 313f, 359t, 362f, 363f, 364, 372, 374, 374f, 388, 389f
Flexor pollicis brevis, 157, 157f, 158f, 159f, 161t, 184
Flexor pollicis longus, 68, 95, 141–142, 142f, 145t, 183
Flexor retinaculum, 105, 107f, 108f, 158, 159, 188, 364–365

Foot. *See also* specific structures
 arteries of, 397–400, 398f, 399f, 400f
 bones of, 309-313, 308f, 309f, 310f, 311f, 312f, 313f
 joints of, 335-340, 335f, 336f, 337f, 338f, 339f, 340f
  intermetatarsal, 337f, 339
  interphalangeal, 337f, 338f, 339–340, 339f
  intertarsal, 335–336
  metatarsophalangeal, 338f, 339, 339f
  tarsometatarsal, 337f, 338
 muscles of, 369-376, 369f, 370t, 371f, 372f, 373tf, 374f, 375tf, 376f
  dorsal, 368f, 370–372, 372f
  plantar, 372
   intermediate muscles, 372f, 373t, 374–375, 374f
   lateral muscles, 375–376, 375t
   medial muscles, 372–374, 372f, 373f
 topography of, 296, 297f
 veins of, 400–401, 402f
Foramen magnum, 263
Foramina, 319–320, 319f
Forearm, 46, 47f
 muscles of, 132
  anterior compartment, 133, 135f
   first layer, 133–135, 136t
    flexor carpi radialis, 134, 135f, 137f, 138f
    flexor carpi ulnaris, 135, 136f, 139f
    palmaris longus, 135f, 139f, 1345–135
    pronator teres, 133–134, 137f
   second layer—flexor digitorum superficialis, 135, 138, 139f, 140f
   third layer, 138
    flexor digitorum profundus, 138, 140–141, 141f
    flexor pollicis longus, 141–142, 142f
   fourth layer—pronator quadratus, 144, 144f, 145t
  posterior compartment, 144
   deep layer, 137f, 148–151, 151ft
    abductor pollicis longus, 148f, 149f, 151–152, 152f
    extensor indicis proprius, 152f, 153, 153f
    extensor pollicis brevis, 152, 153f
    extensor pollicis longus, 148f, 149f, 152, 152f, 153f
    supinator, 137f, 149, 151f, 152f
   superficial layer, 144–148, 146t
    anconeus, 145, 146f
    extensor carpi ulnaris, 147–148, 147f, 148f, 149f
    extensor digiti quinti, 146–147, 147f, 148f, 149f, 150f
    extensor digitorum communis, 145–146, 147f, 148ft
  posterior compartment (radial aspect), 154, 154t
   brachioradialis, 154, 155f
   extensor carpi radialis brevis, 147f, 155–156, 155f, 156f
   extensor carpi radialis longus, 147f, 148f, 154–155, 155f
François-Franck, sympathetic nerve of, 219

G

Gantzer's muscle, 95, 141–142, 142f, 183
Gastrocnemius, 296, 357, 385t, 359, 360f

Gemelli muscles, 342, 342f, 343f, 343t
Geniohyoid, 266, 267t, 268f
Genitofemoral nerve, 377, 378f, 379f, 380
Genu articularis, 350
Gerdy's tubercle, 307, 366
Glenohumeral joint, 54, 81, 83–85, 84f, 85f
Glenohumeral joint capsule, 180, 180f
Glenoid
  cavity, 53
  labrum, 83, 84f
  tubercle, 53
Gluteus maximus, 295, 300, 304, 341, 342f, 343t, 344, 345f, 391
Gluteus medius, 295, 300, 341, 344, 344t, 345f, 391
Gluteus minimus, 295, 300, 341, 344t
Gracilis, 350f, 353t, 355, 355f
Grayson's ligament, 169, 169f
Greater auricular nerve, 280
Greater occipital nerve (nerve of Arnold), 279, 279f
Greater saphenous vein, 347, 393f, 401f, 403, 403f
Greater sciatic foramen, 319, 319f, 383, 391–392, 392f
Greater sciatic notch, 302
Greater sigmoid cavity, 63, 64
Greater supraclavicular fossa, 214
Greater trochanter, 302–303, 303, 304, 341, 342f, 344, 377
Greater tuberosity, 57, 125
Great toe, 296
Growth plate, 13
Guyon's canal, 105, 188, 190

H

Hallux, 296, 312, 312f
Hamate, 70-71, 72f, 103, 135, 159
Hamstrings, 296

Hand. *See also* specific structures
  bones of, 65-78, 67f, 68f, 69f, 70f, 71f, 72f, 73f, 74f, 75f, 76f, 77f, 78f
  joints of, 107-116, 107f, 108f, 109f, 110f, 111f, 112f, 113f, 114f, 115f, 116f
  ligaments of, 107-116, 107f, 108f, 109f, 110f, 111f, 112f, 113f, 114f, 115f, 116f, 167-169, 168f, 169f
  muscles of, 156-168, 157f, 158f, 159f, 160tf, 162tf, 163f, 164f, 165f, 166tf, 167tf, 167f, 168f
  topography of, 47-48
Head
  arteries of, 286–288, 286f, 287f, 288f
  veins of, 289, 290f
Heart, 37, 38f, 40
Hematopoiesis, 7, 9
Hemiazygos vein, 276, 289–290
Henle, nerves of, 187
Hiatus
  aortic, 276, 283
  of Bochdaleck, 276
  esophageal, 276
  of Larrey, 276
Hilton's law, 19
Hip. *See also* specific structures
  bones of, 299–302, 300f, 301f, 302f
  joints of, 320–323, 321f, 322f, 323f
  ligaments of, 320, 323f
  muscles of, 341–347, 342f, 343tf, 344t, 345f, 346tf, 347f
Humeral condyle, 59
Humeral head, 54, 56f, 83
Humeral trochlea, 59
Humerohumeral fibers, 91, 92f
Humeroradial meniscus, 93, 94f
Humerus, 8f, 10, 54, 56f, 57, 57f, 58f, 59, 59f, 60f, 61, 91, 92f
Hunter's canal, 382
Hyaline cartilage, 9, 243, 243f

Hyoid bone, 229–230, 229f, 266
Hypoglossal nerve, 213, 279
Hypothenar eminence, 47, 156, 158, 161, 162t, 190, 190f
  muscles of, 162t

I

Iliac artery, 286
Iliac crest, 299, 300
Iliac fossa, 300
Iliac tubercle, 300
Iliac tuberosity, 315
Iliac wing, 299, 300
Iliacus muscles, 344, 346, 346f, 346t
Iliocostalis, 258t, 259, 260f
Iliocostalis cervicalis, 259
Iliocostalis lumborum, 259
Iliocostalis thoracis, 259
Iliofemoral ligament, 320
Iliohypogastric nerve, 377, 378f
Ilioinguinal nerve, 377, 378f
Iliolumbar artery, 391
Iliolumbar joint, 315, 317f
Iliolumbar ligament, 315, 317f
Iliolumbar vein, 404
Iliopectineal arch, 347, 379, 393f, 403
Iliopectineal bursa, 346
Iliopsoas, 344, 346
Iliopubic eminence, 301, 301f, 302f
Iliotibial tract, 341, 347, 350f, 351f
Ilium, 299–301, 301f
Inferior epigastric artery, 288–289, 392, 392f
Inferior epigastric vein, 289
Inferior gluteal artery, 391–392, 392f
Inferior gluteal nerve, 383, 384f
Inferior lateral ligament of Arnold, 241
Inferior lumbar arteries, 286
Inferior oblique muscle, 263, 263f
Inferior phrenic artery, 284

Inferior pubic ligament, 319
Inferior thyroid artery, 195, 287
Inferior transverse scapular ligament, 79, 80f
Inferior vena cava, 40
Infraclavicular fossa, 45
Infraglenoid tubercle, 53
Infrahyoid muscles, 122, 266, 267t
Infrapalpebral grooves, 213
Infrapatellar fat pad, 323
Infrapatellar synovial plica, 323–324
Infrapiriformis foramen, 382
Infraspinatus fossa, 54, 125, 201
Infraspinatus muscle, 54, 57, 124f, 125, 126f, 126t, 172, 178
Inguinal ligament, 295, 347
Innominate bone, 299
Intercarpal joints, 102, 102f
Interchondral joints, 255–256, 255f
Interclavicular ligament, 79, 82f
Intercoccygeal joints, 249–250
Intercondylar eminence, 306
Intercondylar line, 304
Intercostal arteries, 283, 285f, 287, 288, 289f
Intercostal muscles, 268–270, 269f
Intercostal nerves, 35, 279, 280, 281f, 282, 282f
Intercostal neurovascular bundle, 270
Intercostal spaces, 283, 285f
Intercostobrachial nerves, 178, 282
Intercuneiform joints, 338
Interdigital arches, 161
Intergluteal cleft, 295
Intermetacarpal joints, 110–111, 110f, 111f
Intermetatarsal joints, 337f, 339
Intermuscular septi, 347
Internal carotid artery, 287
Internal iliac artery, 286, 286f, 391–392
Internal jugular vein, 289, 290f
Internal lip of iliac crest, 273
Internal oblique, 270t, 272–273, 274f
Internal thoracic artery, 195, 286, 288
Internal thoracic veins, 289
Interosseous artery, 205–206, 205f, 206f
Interosseous bursa of elbow, 130
Interosseous carpometacarpal ligament, 108
Interosseous intermetacarpal spaces, 160
Interosseous ligaments, 15
Interosseous membrane, 62, 93, 183, 184f
Interosseous metacarpal spaces, 111
Interosseous muscles, 163–164, 165f, 166f
Interosseous sacroiliac ligaments, 316f, 317, 317f
Interosseous scapholunate ligament, 98
Interosseous space, 90f, 93, 95
Interosseous talocalcaneal ligament, 333f, 334, 334f, 335f
Interosseous tendon, 167
Interphalangeal artery, 208
Interphalangeal joints
  of the foot, 337f, 338f, 339–340, 339f
  of the hand, 113f, 114f, 115–116, 115f, 116f, 157
Interpubic disk, 17f, 317, 318f
Interscalene space, 228
Interscalene triangle, 171, 264
Interspinalis muscles, 261t, 262
  interspinalis capitis, 262
  interspinalis cervicis, 262
  interspinalis thoracis, 262
Interspinous ligaments, 246, 249f
Intertarsal joints, 335–336, 336f, 337f, 338
Intertendinous fibers, 160
Intertransverse ligaments, 247, 249, 249f, 250f, 259t
Intertrochanteric crest, 303, 341, 343
Intertubercular groove, 57
Intervertebral disk, 217, 244, 245f, 246f
Intervertebral foramen, 33, 34f, 219
Intervertebral symphysis, 243–244, 245f
Intra-articular sternocostal ligaments, 254
Intrafossal ridge, 63
Ischiofemoral ligament, 320
Ischium, 301f, 302, 302f

J

Joints
  amphiarthrotic, 18
  biaxial, 19, 20f
  cartilaginous, 15, 17f
  classification of
    according to function, 18
    according to structure, 15, 16f, 17f, 18
  costochondral, 17f
  cylindrical, 19, 19f
  defined, 15
  fibrous, 15, 16f
  ginglymus, 19, 19f
    carpometacarpal, 107–108, 109f, 110f
    intermetacarpal, 110–111, 110f, 111f
    interphalangeal, 113f, 114f, 115–116, 115f, 116f
    metacarpophalangeal, 111–115, 112f, 113f, 114f
  saddle, 19, 20f
  sternochondral, 17f
  synarthrotic, 18
  synovial, 15, 17f
  triaxial, 19, 20f
  trochlear, 19, 19f
  trochoid, 19, 19f
Jugular notch, 229

Jugular process, 263
Jugular vein, 289, 290f
Junctura tendinum, 145

K

Knee. *See also* Patella
   arteries of, 393, 394, 394f
   bones of, 305, 305f
   joints of, 323-330, 324f, 325f, 326f, 327f, 328f, 329f, 330f
   ligaments of, 323-330, 326f, 328f
Knot of Henry, 364, 374

L

Labrum
   acetabular, 320, 321f
   glenoid, 18, 18f, 83, 84f
Lacertus fibrosus, 130
Lacunar ligament (of Gimbernat), 272
Lamina, 217
   superior, 217
Lateral antebrachial cutaneous nerve, 180–181, 182f
Lateral bicipital groove, 45, 180, 182f, 192
Lateral collateral ligament
   of the elbow, 92, 93f
   of the ankle, 332, 334f
Lateral condyle, 307
Lateral costotransverse ligament, 252
Lateral epicondyle of the humerus, 59
Lateral femoral circumflex artery, 395, 395f
Lateral femorocutaneous nerve, 378f, 379, 379f, 381
Lateral inferior genicular artery, 394f, 396, 396f
Lateral intermuscular septum, 341
Lateral intertransverse muscles, 261
Lateral mammary rami, 282
Lateral patellar retinaculi, 349, 352f

Lateral pectoral nerve, 172, 175f, 176f
Lateral plantar artery, 399, 399f, 400f
Lateral plantar nerve, 374, 376, 388, 390f
Lateral sacral arteries, 286
Lateral superior genicular artery, 394f, 395, 396f
Lateral supracondylar line, 304
Lateral supracondylar ridge, 59
Lateral talocalcaneal ligament, 334
Lateral thoracic artery, 199, 200f
Latissimus dorsi, 45, 53, 57, 122–123, 123t, 124f, 198, 300, 302f
Left ventricle, 283
Leg, 296
   joints of, 330–331, 330f
   muscles of, 356–357, 357f, 358f
     anterior compartment, 362f, 365–367, 365t, 366f, 367f
     lateral compartment, 365t, 367, 369–370
     posterior compartment, 357, 358t, 360, 360f, 362f, 363f, 364–365, 364f, 365f
   veins of
     deep, 401, 401f, 403
     superficial, 403
Lesser occipital nerve, 280
Lesser saphenous vein, 401f, 402f, 403
Lesser sciatic foramen, 319, 343, 343f, 382
Lesser sciatic notch, 342–343
Lesser sigmoid notch, 64, 88
Lesser trochanter, 302–303, 346
Lesser tuberosity, 57
Levatores costarum, 262f, 268
Levator scapulae, 53, 119–120, 119f, 120, 120f, 122t, 173, 197
Ligament of Henle, 272
Ligament of Struthers, 59
Ligaments, 15-17. *See also* names of specific ligaments

Ligamentum flavum, 245–246, 249f
Ligamentum nuchae, 244f, 247
Ligamentum teres, 299
Linea alba, 272
Linea aspera, 304, 352, 354
Linguofacial trunk, 287
Lisfranc joint, 336f, 338
Lister's tubercle, 63
Longissimus, 259–260, 258t, 260f
   capitis, 260
   cervicis, 260
   thoracis, 260
Long plantar ligament, 336, 337f
Long radiolunate ligament, 97
Long thoracic nerve, 120, 173, 175f, 176f, 178f
Longus capitis, 239, 263–264, 264, 264tf
Longus colli, 264–265, 266f
Lumbar aponeurosis, 341, 342f, 343t
Lumbar arteries, 284
Lumbar lordosis, 225, 346
Lumbar nerves, 31
Lumbar plexus, 377, 378f, 379f
   accessory obturator nerve, 381
   femoral nerve, 378f, 379f, 381–382, 382f
   genitofemoral nerve, 377, 378f, 379f, 380
   iliohypogastric nerve, 377, 378f
   ilioinguinal nerve, 377, 378f
   lateral femorocutaneous nerve, 378f, 379, 379f, 381
   obturator nerve, 378f, 379f, 381, 382f
Lumbar triangle, 273
Lumbar vertebrae, 224, 224f, 275
Lumbocostal ligament, 273
Lumbocostal triangle, 276
Lumbosacral joint, 249, 250f
Lumbosacral plexus, 33, 279, 377, 378f

Index

Lumbosacral spine, arthrology of
costotransverse joint, 251–253, 253f, 254f
costovertebral joint, 250–251, 250f, 252f, 253f
intercoccygeal joints, 249–250
lumbosacral joint, 249, 250f
sacrococcygeal joint, 249, 252f
Lumbrical muscles, 161, 163, 163f, 164f, 166t, 190, 372f, 373t, 374, 374f
Lunate, 65, 68, 69f, 97, 102, 103, 105
Lymphatic system, 209
Lymph nodes, 404

M

Major supraclavicular fossa, 45
Malleoli, 296
Mammary region, 45, 46f
Mandible, 266
Mandibular nerve, 265, 278
Manubriosternal symphysis, 250f, 256, 256f
Manubrium, 229
Martin-Gruber, anastomosis of, 183–184
Masseter muscle, 277–278, 277f, 278f
Mastication, muscles of, 277–278, 277f, 278f
Mastoid region, 213
Maxillary artery, 288
Medial accessory bundle, 253
Medial antebrachial cutaneous nerve, 179f, 181–182, 182f, 183f
Medial atlantoaxial joint (atlanto-dens articulation), 240f, 242–243, 242f, 243f
Medial brachial cutaneous nerve, 178, 179f
Medial collateral ligament, of the ankle, 332, 334f
Medial compartment, 261–262, 261t
Medial cord, 178, 178f

Medial dorsal cutaneous nerve, 386, 387f
Medial epicondyle, of the humerus, 57, 59, 91, 133, 135
Medial femoral circumflex artery, 394–395, 395, 395f
Medial inferior genicular artery, 394f, 396, 396f
Medial intercondylar tubercle, 324
Medial malleolus, 307, 398
Medial meniscus, 324
Medial patellar retinaculi, 349, 352f
Medial pectoral nerve, 171–172, 172, 175f, 176f
Medial plantar artery, 399, 399f, 400f
Medial plantar nerve, 388, 390f
Medial superior genicular artery, 394f, 395, 396f
Medial supracondylar ridge, 59
Medial sural cutaneous nerve, 387f, 388
Medial talocalcaneal ligament, 334, 335f
Medial tarsal artery, 397
Median nerve, 138, 157, 163, 173, 176f, 178, 178f, 179f, 182–185, 182f, 183f, 184f, 185f, 186f, 187f
Medullary canal, 9, 10f
Meniscosternal cavity, 79
Mesotenon, 29
Metacarpals, 47–48, 71–75, 73f, 145, 148f
  fifth, 75, 75f
  first, 73–74, 73f
  fourth, 74, 75f, 108, 110, 163, 164
  second, 74, 74f, 108, 110, 158
  third, 74, 74f, 108, 110, 158, 163
Metacarpophalangeal capsule, 163
Metacarpophalangeal joints, 111–115, 112f, 113f, 114f, 157, 158, 159

Metacarposesamoid ligaments, 113
Metaphyseal vessels, 12, 13f
Metaphysis, 7, 8f, 9
Metatarsal arteries, 397, 397f, 398f
Metatarsals, 8f, 296, 312, 312f, 339, 364
Metatarsophalangeal joints, 338f, 339, 339f, 375, 397
Midcarpal joint, 97f, 102f, 103, 105, 106f, 107, 107f, 108f
  ligaments, 103, 103f, 104f, 105f
Middle genicular artery, 396, 396f
Middle glenohumeral ligament, 85
Middle palmar region, 156, 159, 166t
  muscles of, 159, 166t
Middle sacral artery, 286
Middle scalene muscles, 265t
Minor supraclavicular fossa, 45, 46f
Mixed spinal nerve, 31
Multifidi, 261t, 262, 263f
Muscle bellies and tendons, 25, 27, 27f
Muscles. *See also* names of specific muscles
  anatomic classification systems, 25, 27
  antagonist, 28, 28f
  arrangement of fibers, 27–28
  cardiac, 25
  digastric, 27, 27f
  functions of, 28–29, 28f, 29f
  fusiform, 27, 27f
  innervation, 28–29
  multipennate, 27f, 28
  muscular compartments of, 25, 26f
  omohyoid, 27, 27f
  pennate, 27–28, 27f, 28
  perforated, 128
  polygastric, 27, 27f

semipennate, 27, 27f
skeletal, 25
smooth, 25
vascularity of, 28–29, 28f
Musculocutaneous nerve, 128, 173, 178, 178f, 179f
Musculophrenic epigastric artery, 288
Mylohyoid muscle, 213, 266, 267tf, 268f
Myocardial circulation, 283

N

Nasal region, 213
Nasolabial grooves, 213
Natatory ligament, 162f, 167–168
Navicular-cuboid joints, 335
Navicular joint, 311, 338, 364
Neck
  arteries of, 286–288, 286f, 287f, 288f
  muscles of, 264
    infrahyoid, 266, 267t
    lateral, 264, 265f, 266f
      anterior musculature, 264–265, 266f
    suprahyoid, 265–266, 267t
  veins of, 289, 290f
Nerve fibers
  afferent, 31
  efferent, 31
Nerves, 19. See also specific
  intercostal, 35
  peripheral, 31
  spinal (rachidial), 31, 32f, 33, 33f, 34f, 35, 35f
Nervous system, 31
  central, 31
  peripheral, 31
Neural arch, 217
Neuromuscular function, 28
Nuchal ligaments, 117
Nucleus pulposus, 244
Nutrient artery, 12
Nutrient foramen, 9, 9f

O

Oblique cord, 90f, 95
Oblique muscles, 263, 264t
Oblique popliteal ligament, 360
Oblique retinacular ligament (of Landsmeer), 167, 168f
Obturator artery, 299, 391, 392f
Obturator band, 320, 320f
Obturator canal, 320
Obturator crest, 302
Obturator externus muscle, 341, 344, 344t, 345f
Obturator foramen, 299, 320, 381, 391
Obturator internus, 341, 342–343, 342f, 343, 343f, 343t, 382
Obturator membrane, 299, 320, 320f
Obturator muscles, 299
Obturator nerve, 35, 378f, 379f, 381, 382f
Occipital artery, 287
Occipital condyles, 219
Occipital region, 213
Occipitoaxial joint, 240–241
  occipito-odontoid ligaments, 241, 242f
  tectorial membrane, 240–241, 241f
Occipito-occipital bundle, 241
Occipito-odontoid ligaments, 241, 242f
Odontoid process, 241f, 243, 243f
Olecranon, 63, 91, 92f
Olecranon-epitrochlear tunnel, 92
Olecranon fossa, 59, 61, 63
Omoclavicular triangle, 45, 214
Omohyoid, 122
Omohyoideus muscle, 51
Omohyoid muscle, 27, 27f, 45, 197, 266, 267tf, 268
Omotracheal (muscular) triangle, 213
Omotricipital triangle, 201
Opponens digiti quinti, 159, 162t, 376, 376f

Opponens pollicis, 157–158, 158f, 161t
Orbital region, 213
Ossification, 7, 9, 13
  endochondral, 13, 13f
  intramembranous, 13
Osteoblasts, 12
Osteoclasts, 12
Osteocytes, 12
Osteoprogenitor cells, 13

P

Palmar aponeurosis, 157f, 158, 159–161, 161f, 162f, 163f
Palmar carpal branch, 206
Palmar carpal ligament, 105
Palmar carpometacarpal ligaments, 108
Palmar digital nerves, 185
Palmar fascia, 47–48
Palmar interossei, 164, 165f, 166f, 166t
Palmaris brevis, 158–159, 160f, 188, 190f
Palmaris longus, 134–135, 135f, 136t, 139f
Palmar ligament, 112
Palmar radiocarpal ligament, 97
Palmar radioulnar ligament, 99
Palmar ulnar ligament, 107
Palmar ulnocarpal ligament, 99
Paratenon, 29
Parathyroid gland, 288
Patella, 9f, 305, 305f
Patellar tendon, 305
Patellofemoral joint, 323, 324, 324f
Pectineal ligament (of Cooper), 272
Pectineal line, 304
Pectineus, 347, 350, 350f, 352f, 353tf
Pectoralis major, 45, 49, 57, 121, 123, 123t, 124f
Pectoralis minor, 45, 49, 121–122, 121f, 122t, 171
Pectoralis nerves

lateral, 123
medial, 123
Pedal artery, 397, 398f
Pelvic diaphragm, 271
Pelvic girdle, 295, 299
  joints of, 315
    foramina, 319–320, 319f
    iliolumbar, 315, 317f
    obturator membrane, 320, 320f
    pubic symphysis, 316f, 317, 318f
      anterior ligament, 319
      inferior ligament, 319
      posterior ligament, 319
      sacrospinous ligament, 319, 319f
      sacrotuberous ligament, 319, 319f
      superior ligament, 319
    sacroiliac, 315, 316f
      anterior ligaments, 315, 317f
      posterior ligaments, 315, 317, 318f
  muscles of, 341, 342f
    anterior, of hip, 344, 344t, 346–347
    lateral, 344, 344t, 345f
    posterior, 341–344, 342f, 343t
  veins of, 403–404
Pelvis, 299, 300f
Pelvitrochanteric muscles, 341
Perforated muscle, 128
Pericardiophrenic artery, 288
Pericardium, 277, 280, 280f, 288
Perimysium, 25, 26f
Perineurium, 31, 32f
Periosteal branches, 12
Periosteum, 9
Peripheral nervous system, 31
Peroneal artery, 397–398, 398f, 399f
Peroneal nerves, 296
Peroneal tubercle, 311

Peroneal veins, 401, 401f
Peroneus brevis, 311, 365t, 368f, 369–370
Peroneus longus, 311, 365t, 367, 368f, 369
Peroneus tertius, 365, 367, 367f
Petit's triangle, 215, 273
Phalanges, 10, 48, 75, 76f, 77, 77f, 78f
Phrenic center, 276
Phrenic nerves, 277, 280, 280f, 287
Physiologic scoliosis, 217
Piriformis muscle, 341, 342, 343t, 382, 383, 391, 392f
Pisiform, 67, 70f, 71, 102, 135
Pisocapitate ligament, 105
Pisohamate ligament, 105, 135, 159
Pisometacarpal ligament, 135
Planes, of body, 4, 5f
Plantar arch, 296, 372, 374
Plantar calcaneonavicular ligament, 335–336, 335f, 336f
Plantar fascia, 372f, 374
Plantar flexion, 20
Plantar interossei, 373t
Plantaris, 358t, 360, 361f, 362f
Plantar ligaments, 339
Platysma, 265, 266f
Pollicis brevis, 156–157, 157t
Polygastric muscles, 27, 27f
Popliteal artery, 296, 393, 394f, 395–397, 396f
Popliteal fascia, 359
Popliteal fossa, 295–296, 296, 393
Popliteal line, 307
Popliteus, 359t, 360, 361f, 362f
Popliteus sulcus, 304
Posterior arch, 217
Posterior auricular artery, 287
Posterior cervical triangle, 171, 173f, 214
Posterior circumflex humeral artery, 201, 202f
Posterior compartment, 144
  muscles of, 154t

Posterior cruciate ligament, 323f, 327f, 328
Posterior cubital region, 45
Posterior cutaneous nerve of thigh, 383, 384f
Posterior inferior iliac spine, 300
Posterior longitudinal ligament, 244–245, 246f
Posterior meniscofemoral ligament (of Wrisberg), 324, 327f
Posterior pubic ligament, 319
Posterior sacroiliac ligaments, 315, 317, 318f
Posterior scalene muscles, 265t
Posterior superior iliac spine, 300
Posterior talocalcaneal ligament, 334, 334f
Posterior talofibular ligament, 334
Posterior tibial artery, 398–400, 399f
Posterior tibialis muscle, 360, 362f, 363f, 364
Posterior tibial veins, 401, 401f
Precapillary sphincter, 39, 39f
Prescalenic subclavian artery, 195, 196f
  internal thoracic, 195
  thyrocervical trunk, 195
    ascending cervical, 195, 197
    inferior thyroid, 195
    suprascapular, 198, 199f
    transverse cervical, 197–198, 198f
  vertebral, 195
Presternal region, 45
Princeps pollicis, 207
Pronation, 23, 88
Pronator quadratus, 63, 144, 144f, 145t, 183
Pronator teres, 133–134, 136t, 137f, 182, 183f
Pronator tuberosity, 62, 134
Proximal carpal row, 102, 105
Proximal epiphysis, 54, 57, 61, 61f, 62f, 72, 306, 308–309
Proximal radioulnar joint, 87–88, 89f, 90f, 91f

Proximal tibiofibular joint, 307, 330, 330f
Proximal transverse digital artery, 208
Psoas major, 344–345, 346, 346t
Psoas minor, 346, 346t, 348
Psoatic arch, 276
Pterygoid fossa, 278
Pterygoid muscle, 278, 278f
Pterygoid process, 278
Pubic arch, 299, 302
Pubic symphysis, 315, 316f, 317, 318f
Pubic tubercle, 302
Pubis, 301–302, 301f, 302f
Pubofemoral ligament, 320, 321f, 323f
Pudendal nerve, 384f, 390
Pulmonary artery, 37
Pulmonary circulatory system, 37, 38f
Pulmonary valve, 40, 40f
Pulvinar, 299
Pyramidalis muscle, 271f, 272

Q

Q angle, 324
Quadrate ligament, 88, 91f
Quadrate tubercle, 303
Quadratus femoris, 341, 343–344, 343ft, 382
Quadratus lumborum, 270t, 274f, 275
Quadratus plantae, 373tf, 374
Quadriceps, 26f, 305, 323, 349, 351t
Quadriceps femoris, 348, 348f
Quadrilateral space (of Velpeau), 180

R

Radial artery, 204–208, 204f, 205f
Radial deviation, 156
Radial fossa, 59
Radial groove, 57
Radial nerve, 57, 132, 145, 146, 149, 151, 152, 153, 154, 155, 178, 179f, 180, 192, 192f, 193f, 194, 194f
  motor branches of, 194, 194f
Radial notch, 64
Radial sesamoid, 77
Radial styloid, 181
Radial tuberosity, 62, 130, 132f
Radiocarpal joint, 97, 97f, 98f
  ligaments, 97, 99f, 100f, 101f, 105
    carpal radial collateral, 101f, 102
    carpal ulnar collateral, 101f, 102
    dorsal radiocarpal, 99
    dorsal radioscaphoid, 99
    dorsal ulnocarpal, 99, 102
    long radiolunate, 97
    radioscaphocapitate, 97
    radioscapholunate (of Testut and Kuenz), 97–98
    short radiolunate, 98–99
    ulnocapitate, 99, 100f
    ulnolunate, 99
    ulnotriquetral, 99
Radiohumeral joint, 87, 88f, 89f
Radioscapholunate ligament (of Testut and Kuenz), 97–98
Radioscaphocapitate ligament, 97
Radioulnar syndesmosis, 93, 95, 95f
Radius, 10, 61–63, 61f, 62f, 63f
  lateral rotation of, 151
Ramus perforans, 208
Rectus abdominis, 270t, 271–272, 271f, 289
Rectus capitis, 239, 279
Rectus capitis anterior, 264t
Rectus capitis lateralis, 264t
Rectus capitis posterior major, 262–263, 263f, 264t
Rectus capitis posterior minor, 263, 263f, 264t
Rectus femoris, 349
Reticular bone, 10f, 12
Retinacular system, 167

Retinaculum, 29
Retrocalcaneal bursa, 359
Retropulsion, 22
Rhomboid major, 54, 119
Rhomboid minor, 54, 119
Rhomboid muscles, 51, 119, 120f, 122t, 173, 198
Ribs, 11, 227–228, 227f, 228f
  cartilage, 121f, 122
  false, 227
Riche and Cannieu, anastomosis of, 185
Right atrium, 40
Right ventricle, 40
Rotator cuff, 128–129, 129f
  muscles of, 129, 130t, 261t, 262, 263f
    biceps brachii, 130, 130t, 131f, 132, 132f
    brachialis, 130t, 131f, 132, 132f
    triceps brachii, 130t, 132, 132f, 133f, 134f

S

Sacciform recess, 96
Sacral foramina, 342
Sacral hiatus, 226
Sacral kyphosis, 225
Sacral nerves, 31
Sacral plexus, 379f, 380f, 382, 383, 384f
  inferior gluteal nerve, 383, 384f
  posterior cutaneous nerve of thigh, 383, 384f
  pudendal nerve, 384f, 390
  sciatic nerve, 383, 384f, 385f, 386, 386f, 387f, 388f, 389f, 390f
  superior gluteal nerve, 382, 384f
Sacral tuberosity, 225
Sacrococcygeal joint, 249, 252f
Sacroiliac joint, 299, 301, 315, 316f
Sacropelvic face, 301

Sacrosciatic lamina of Morestin, 319, 319f
Sacrospinous ligament, 302, 317, 319, 319f
Sacrotuberous ligament, 299, 302, 319, 319f, 341, 342, 342f
Sacrum, 224–226, 224f, 225f, 299, 315, 341
Saddle joints, 19, 20f, 79
Saphenous hiatus, 379, 403, 403f
Saphenous nerve, 382, 382f, 393
Saphenous veins, 403
Sartorius muscle, 295, 348, 350f, 351t
Scalene muscles, 264, 265f, 266f
Scalene tubercle, 219
Scalenus, 197
Scaphocapitate ligament, 103
Scaphoid, 65, 67–68, 68, 69f, 97, 103
Scapholunate articulation, 97–98
Scapholunate joint, 65, 68, 69f, 71, 97, 102, 103, 105
Scaphotrapeziotrapezoid ligament, 103
Scapula, 10, 49, 51, 51f, 52f, 53–54, 53f, 54f, 55f, 117, 119, 198
Scapular acromion, 49
Scapular neck, 53
Scapular nerve, dorsal, 119, 120
Scapular notch, 49, 51, 79, 172
Scapular pillar, 53
Scapular region, 215
Scapulohumeral muscles, 124, 126–127t, 126t
   coracobrachialis, 127t, 128, 128f
   deltoid, 124f, 125f, 126t
   infraspinatus, 124f, 125, 126f, 126t
   subscapularis, 127f, 127t, 128, 128f
   supraspinatus, 119f, 120f, 124–125, 126f, 126t
   teres major, 126, 126f, 127f, 127t, 128f
   teres minor, 124f, 125–126, 126f, 127t
Scapulothoracic joint, 85–86, 86f, 120
Scarpa's triangle, 295
Schindylesis, 15, 16f
Sciatic nerve, 342, 342f, 356, 383, 384f, 385f, 386, 386f, 387f, 388f, 389f, 390f
Sclerotome, 31
Scoliosis, physiologic, 217
Screw-home mechanism, 328
Semilunar line (of Spiegel), 273, 274f
Semimembranosus, 328, 355–356, 356f, 359, 360
Semispinalis, 261f, 262, 262f
   capitis, 262, 279, 287
   cervicis, 262
   thoracis, 262
Semitendinosus, 355–356f, 356
Serratus anterior, 45, 51, 53, 120, 120f, 122t
Serratus lateralis, 120
Serratus posterior interior, 260
Serratus posterior superior, 260, 260f
Sesamoid bones, 11, 11f, 77, 305, 313, 334f, 340, 340f
Sharpey's fibers, 10, 244
Short radiolunate ligament, 98–99
Shoulder. See also specific structures
   bones of, 49–55, 51f, 52f, 53f, 54f, 55f
   joints of, 79–86, 80f, 81f, 82f, 83f, 84f, 85f, 86f
   ligaments of, 79–86, 80f, 81f, 82f, 83f, 84f, 85f, 86f
   muscles of, 117–129, 118f, 119f, 120f, 121tf, 122t, 123t, 124f, 125f, 126tf, 127tf, 128f, 129f
   nerves of, 171–181, 172f, 173f, 174f, 175f, 176f, 177f, 178f, 179f, 180f, 181f
Skeleton
   appendicular, 7
   axial, 7, 217
Skull, 230, 230f–238f
Soleus line, 307
Soleus muscle, 359–360, 361f
Spinal cord, 31, 32f, 33
Spinalis, 260f, 262, 262f
   capitis, 262
   cervicis, 262
   thoracis, 262
Spinal nerves, 31, 32f, 33, 33f, 34f, 35, 35f
Spinal trigone, 54
Spine. See also specific structures
   bones of, 217–226, 218f, 219f, 220f, 221f, 222f, 223f, 224f, 225f, 226f
   joints of, 239–248, 240f, 241f, 242f, 243f, 244f, 245f, 246f, 247f, 248f
   ligaments of, 245–247, 249, 249f, 250f, 251f
   muscles of, 257–264, 257f, 258t, 259t, 260f, 261t, 262f, 263f, 264t
   nerves of, 279–282, 280f, 281f, 282f
Spinoglenoid notch, 54, 79, 172
Spiral groove, 57, 179f, 180
Splenius capitis, 260f, 261
Splenius cervicis, 197, 260–271, 260f
Squamous suture, 15
Sternal manubrium, 49
Sternal synchondroses, 256
Sternochondral joints, 17f
Sternochondral synchondrosis, 254, 255f
Sternoclavicular joint, 79, 81f, 82f, 195
Sternoclavicular ligaments, 79, 82f

Sternocleidomastoid, 45, 49, 118–119, 118f, 119f, 121t, 280
Sternocostal angle, 288
Sternocostal joints, 250f, 254–255, 255f
Sternocostal triangle, 276
Sternocostoclavicular complex, 79
Sternohyoid, 49, 266, 267t, 268f
Sternomastoid fascicles, 118
Sterno-occipital fascicles, 118
Sternothyroid muscle, 266, 267t, 268f
Sternum, 79, 228–229, 228f
Stylohyoid, 266, 267t, 268f
Styloid apophysis, 62, 309, 312
Styloid process, 99
Subacromial bursa, 125
Subclavian arteries, 38, 39f, 195, 196f, 228, 264, 286, 286f, 287, 289
  prescalenic, 195, 196f
Subclavian vein, 208–209, 228, 264, 289
Subclavius muscle, 49, 122, 122t
Subclavius nerve, 122, 171, 175f
Subcostal artery, 284
Subcostal muscles, 270
Subcostal nerves, 282
Subcostal vein, 289–290
Submandibular salivary gland, 213
Submandibular triangle, 213
Submental triangle, 213
Suboccipital musculature, 262, 264t
  anterior muscles, 263–264, 265f
  lateral muscles, 263, 264f
  posterior group, 262–263, 263f
Suboccipital nerve, 263
Subpopliteal recess, 360
Subpubic angle, 302
Subscapular artery, 199–201, 200f
Subscapular fossa, 51f, 53

Subscapularis muscle, 45, 53, 127f, 127t, 128, 128f, 179f, 180
Subscapular nerves, 128, 173, 178f
Subtalar joints, 331–332, 331f, 332f, 333f, 334, 334f, 335, 335f
Subtendinous bursa, 84–85, 125, 126, 128, 155–156
Subtendinous iliac bursa, 346
Superficial circumflex iliac artery, 392–393, 393f
Superficial crural fascia, 347, 348f
Superficial epigastric artery, 392–393, 393f
Superficial palmar arch, 207
Superficial peroneal nerve, 367, 386, 386f
Superficial plantar arch, 400
Superficial popliteal nodes, 404
Superficial temporal artery, 288, 288f
Superficial transverse intermetacarpal ligaments, 160
Superficial transverse metacarpal ligament, 162f, 168
Superficial veins, 40, 41f
  of foot, 400–401, 402f
  of leg, 403
  of thigh, 403
Superficial venous palmar arch, 209
Superior epigastric artery, 288
Superior gemellus, 341, 343, 382
Superior gluteal artery, 391, 392f
Superior gluteal nerve, 344, 382, 384f
Superior oblique muscle, 263, 263f, 279
Superior pubic ligament, 319
Superior pubic ramus, 346, 347f
Superior retinaculum of extensor muscles, 365
Superior thoracic artery, 199
Superior thyroid artery, 287
Superior transverse ligaments, 51
Superior transverse scapular ligament, 79, 80f

Superior trunk, 171, 174f
Superior ulnar collateral artery, 202–203, 202f, 204f
Superior vena cava, 289
Superolateral lymph nodes, 403f, 404
Superomedial lymph nodes, 403f, 404
Supination, 22–23, 23f, 88
Supinator longus, 154. *See also* Brachioradialis
Supinator muscle, 64, 137f, 148–151, 151t
Supra-acetabular notch, 299
Supraclavicular nerve, 280
Supracondylar tubercles (of Grüber), 304
Supraglenoid tubercle, 53
Suprahyoid muscles, 265–266, 267t, 268f
Suprapalpebral grooves, 213
Suprapatellar bursa, 323, 324f
Suprascapular artery, 51, 79, 198, 199f
Suprascapular nerve, 51, 79, 125, 172, 175f, 176f, 178
Suprascapular vein, 51, 79
Supraspinatus muscle, 54, 57, 119f, 120f, 124–125, 126f, 126t, 172, 176f
Supraspinatus fossa, 53–54
Supraspinous ligaments, 246–247
Suprastyloid crest, 62
Supratendinous transverse sheath, 165, 168f
Sural region, 296
Suture (type of joint)
  dentate, 15, 16f
  flat, 15, 16f
  schindylesis, 15, 16f
  squamous, 15, 16f
Sustentaculum tali, 311
Symphyses, 15
Symphysis pubis, 17f, 299
Synarthrotic joint, 18
Synchondroses, 15

INDEX

Syndesmosis, 15, 17f
Synovial bursae, 29, 29f
Synovial fluid, 29, 83
Synovial joints, 15, 17f
  movements at, 19–23
  types of, 18–19, 19f
Synovial sheaths, 29, 29f, 366f, 374
Synovium, 239, 242

T

Tailor's muscle, 348
Talar dome, 310
Talar trochlea, 310, 331
Talocalcaneal joint, 332
Talocalcaneonavicular joint, 335, 336f
Talocrural region, 296
Talus, 308f, 309–310, 309f, 331, 399, 399f
  calcaneus, 310–311, 310f
  cuboid, 312, 312f
  cuneiforms, 311, 311f
  navicular, 311
  talus, 308f, 309–310, 309f
Tarsal tunnel, 364–365, 366f, 399, 399f
Tarsometatarsal joints, 311, 311f, 335f, 336f, 337f, 338
Tarsus, 308f, 309
  calcaneus, 310–311, 310f
  cuboid, 312, 312f
  cuneiforms, 311, 311f
  navicular, 311
  talus, 308f, 309–310, 309f
Tectorial membrane, 240–241, 241f
Temporalis, 277, 277f, 288
Temporomandibular joint, 278
Tendinous chiasma, 140, 141f, 142f
Tensor fascia lata, 347–348, 350f, 351t
Teres major, 45, 51, 54, 57, 126, 126f, 127f, 127t, 128f
Teres minor, 51, 54, 57, 124f, 125–126, 126f, 127t

Terminal line, 299
Terminal nerves (arm region)
  axillary nerve, 179f, 180, 181f
  medial brachial cutaneous nerve, 178, 179f
  median, 173, 178, 178f, 179f
  musculocutaneous, 173, 178, 178f, 179f
  radial nerve, 178, 179f, 180
  ulnar, 178, 178f, 179f
Terminal nerves (forearm region)
  lateral antebrachial cutaneous nerve, 180–181, 182f
  medial antebrachial cutaneous nerve, 179f, 181–182, 182f, 183f
  median nerve, 176f, 182–185, 182f, 183f, 184f, 185f, 186f, 187f
  radial nerve, 192, 192f, 193f, 194, 194f
  ulnar nerve, 187–190, 189f, 190f, 191f
Terminal vessels, 39, 39f
Thenar eminence, 47, 156, 161, 161t
  adductor pollicis, 157f, 158, 158f, 159f, 160f
  flexor pollicis brevis, 157, 157f, 158f, 159f
  muscles of, 161t
  opponens pollicis, 157–158, 158f
  pollicis brevis, 156–157, 157t
Thenar region, 47
Thermogenesis, 25
Thigh. *See also* specific structures
  arteries of, 392–395, 393f, 394f, 395f
  bones of, 302–305, 303f, 304f, 305f
  muscles of, 347–357, 347f, 348f, 349f, 350f, 351tf, 352f, 353tf, 354tf, 355f, 356f, 357f
  nerves of, 377–386, 378f, 379f, 380f, 381f, 382f, 383f, 384f, 385f

  topography of, 295, 296f
Third trochanter, 304
Thoracic aorta, 283, 284f
Thoracic nerves, 31
Thoracic skeleton, 227, 227f
  hyoid bone, 229–230, 229f
  ribs, 227–228, 227f, 228f
  skull, 230, 230f–238f
  sternum, 228–229, 228f
Thoracic vertebrae, 22f, 222–223, 223f
Thoracoacromial artery, 199, 199f, 200f
Thoracodorsal nerve, 123t, 173, 178f, 200–201
Thoracohumeral muscles, 122, 123t
  latissimus dorsi, 122–123, 123t, 124f
  pectoralis major, 123, 123t, 124f
Thoracoscapular muscles, 122t
  levator scapulae, 122t
  pectoralis minor, 122t
  rhomboids, 122t
  serratus anterior, 122t
  subclavius, 122t
Thorax, arthrology of, 250
Thyrocervical trunk, 195, 287
  ascending cervical, 195, 197
  inferior thyroid, 195
  suprascapular, 198, 199f
  transverse cervical, 197–198, 198f
Thyrocervicoscapular trunk, 195
Thyrohyoid muscle, 266, 267t, 268f
Tibia, 3, 10, 305–308, 306f, 331
Tibialis anterior, 366, 366f
Tibialis posterior, 359t
Tibial nerves, 296, 386, 388, 389f
Tibial plafond, 308
Tibiofemoral joint, 323, 324, 325f, 326f
Tibiofibulotalar joint, 331

Tibioperoneal trunk, 396, 396f
Tibiotalar joint, 331
Toes, bones of, 312–313, 313f
Trabecular (cancellous) bone, 217
Transverse anterior carpal artery, 206
Transverse carpal ligament, 105
Transverse cervical artery, 197–198, 198f, 287
Transverse cervical nerve, 280
Transverse humeral ligament, 57, 84, 84f
Transverse processes, 219
Transverse retinacular ligament, 167, 168f
Transversospinal muscles, 262
Transversus abdominis, 270t, 273, 274f, 275
Trapeziometacarpal joint, 107, 152
Trapezium, 69, 71f
Trapezius, 54, 103, 117–118, 118f, 121t, 197, 279
Trapezoid, 70, 71f
Trapezoid ligament, 81
Trasversus abdominis muscle, 270
Triangular lamina, 167, 168f
Triangular sternal muscle, 270
Triaxial joints, 19, 20f, 22
Triceps brachii, 26f, 130t, 132, 132f, 133f, 134f, 180
Triceps surae, 356f, 357, 359, 360f
Trigeminal nerve, 265
Triquetrocapitate ligament, 103
Triquetrohamate ligament, 105
Triquetrum (lunatotriquetral joint), 65, 68, 69f, 71, 97, 99, 102, 103
Trochanteric bursa, 341, 344
Trochanteric (piriformis) fossa, 303, 343, 343f, 344, 345f
Trochiter, 57
Trochlea, 59
Trochlear notch, 63

True ribs, 227
True vertebrae, 224
Trunk
　muscles of, 257
　　dorsal region, 257, 257f, 259
　　lateral compartment, 259–261, 259t, 260f
　　medial compartment, 261–262, 261t
　veins of, 289–290, 290f, 291f
Tubercles, 11
Tuberosities, 11

U

Ulna, 63–65, 64f, 65f, 151
Ulnar artery, 134, 204–208, 204f, 205f
Ulnar collateral ligament, 71, 91, 93f
Ulnar deviation, 21
Ulnar head, 65, 65f
Ulnar nerve, 92, 140, 157, 159, 164, 178, 178f, 179f, 185, 186f, 187–190, 189f, 190f, 191f
　deep branch of, 190
　dorsal branch of, 187–188, 189f
　motor branches of, 190, 191f
Ulnar notch, 63
Ulnar sesamoid, 77
Ulnar styloid, 65
Ulnocapitate ligament, 99, 100f
Ulnohumeral joint, 63–64, 87, 87f, 88f, 89f
Ulnolunate ligament, 99
Ulnotriquetral ligament, 99
Umbilicus, 272
Uncovertebral joints, 245, 247f
　of Luschka, 219, 219f

V

Vagus nerve, 279
Vasa nervorum, 35, 37
Vasa vasorum, 37
Vastoadductor membrane, 393
Vastus lateralis, 348–349, 351f, 352f
Vastus medialis, 348

Veins, 39–41, 40f, 41f. See also specific veins
　capillary, 40
　deep, 41
　of pelvic girdle, 403–404
　superficial, 40, 41f
Velpeau, topographic spaces of, 132
Vena caval foramen, 276
Venae comitantes, 41
Venous-lymphatic anastomoses, 41
Veno-venous anastomoses, 41
Ventral ramus, 35
Ventral root, 31
Ventricle, right, 40
Vertebrae, 217, 218f, 219
　cervical, 219–220, 239
　false, 224
　lumbar, 224, 224f
　true, 224
Vertebral artery, 195, 219, 286, 287, 287f
Vertebral bodies, 283
Vertebral foramen, 217–218
Vertebra prominens, 219
Vincula longa, 144
Vincula tendina, 144
Viscerotome, 31
Volar plates, 112, 114f, 116

W

Weitbrecht's foramen, 85
Weitbrecht's ligament, 90f, 95
White matter, 31
Wolff's law, 12
Wrist. *See also* Carpus; names of specific structures
　bones of, 65–72, 67f, 68f, 69f, 70f, 71f, 72f
　joints of, 95–108, 96f, 97f, 98f, 99f, 100f, 101f, 102f, 103f, 104f, 105f, 106f, 107f, 108f
　ligaments of, 97–108, 98f, 99f, 100f, 101f, 102f, 103f, 104f, 105f, 106f, 107f, 108f
　topography of, 46

X
Xiphisternal symphysis, 229
Xiphoid foramen, 229
Xiphoid process, 229

Y
Y ligament of Bigelow, 320

Z
Zaglas fascicle, 317
Zygapophyseal joints, 217, 219, 245, 246f, 247f, 248f
Zygomatic arch, 277
Zygomatic region, 213